Surrogates

CHAPMAN & HALL/CRC
Texts in Statistical Science Series
Joseph K. Blitzstein, *Harvard University, USA*
Julian J. Faraway, *University of Bath, UK*
Martin Tanner, *Northwestern University, USA*
Jim Zidek, *University of British Columbia, Canada*

Recently Published Titles

Statistics in Engineering
With Examples in MATLAB and R, Second Edition
Andrew Metcalfe, David A. Green, Tony Greenfield, Mahayaudin Mansor, Andrew Smith, and Jonathan Tuke

Introduction to Probability, Second Edition
Joseph K. Blitzstein and Jessica Hwang

Theory of Spatial Statistics
A Concise Introduction
M.N.M van Lieshout

Bayesian Statistical Methods
Brian J. Reich and Sujit K. Ghosh

Sampling
Design and Analysis, Second Edition
Sharon L. Lohr

The Analysis of Time Series
An Introduction with R, Seventh Edition
Chris Chatfield and Haipeng Xing

Time Series
A Data Analysis Approach Using R
Robert H. Shumway and David S. Stoffer

Practical Multivariate Analysis, Sixth Edition
Abdelmonem Afifi, Susanne May, Robin A. Donatello, and Virginia A. Clark

Time Series: A First Course with Bootstrap Starter
Tucker S. McElroy and Dimitris N. Politis

Probability and Bayesian Modeling
Jim Albert and Jingchen Hu

Surrogates: Gaussian Process Modeling, Design, and Optimization for the Applied Sciences
Robert B. Gramacy

For more information about this series, please visit: https://www.crcpress.com/Chapman--Hall-CRC-Texts-in-Statistical-Science/book-series/CHTEXSTASCI

Surrogates

Gaussian Process Modeling, Design, and Optimization for the Applied Sciences

Robert B. Gramacy

Virginia Tech

CRC Press
Taylor & Francis Group
Boca Raton London New York

CRC Press is an imprint of the
Taylor & Francis Group, an **informa** business

A CHAPMAN & HALL BOOK

Cover image: Lake Wanaka, New Zealand in winter.

CRC Press
Taylor & Francis Group
6000 Broken Sound Parkway NW, Suite 300
Boca Raton, FL 33487-2742

First issued in paperback 2021

© 2020 by Taylor & Francis Group, LLC
CRC Press is an imprint of Taylor & Francis Group, an Informa business

No claim to original U.S. Government works

ISBN 13: 978-0-367-41542-6 (pbk)
ISBN 13: 978-1-03-224255-2 (hbk)

Visit the Taylor & Francis Web site at
http://www.taylorandfrancis.com

and the CRC Press Web site at
http://www.crcpress.com

Publisher's Note
The publisher has gone to great lengths to ensure the quality of this reprint but points out that some imperfections in the original copies may be apparent.

Best fam ever

Leah, Natalia, Kaspar

Contents

Preface

Computer simulation experiments are essential to modern scientific discovery. Barriers to computing have come way down. Meanwhile all the low-hanging fruit has been picked from the mathematical tree of cute closed-form solutions serving as crude approximations to reality. Occam's Razor[1] is nice philosophy, but the real world isn't always simple. On that Wikipedia page, Isaac Newton is quoted as saying the following about how simple the world must be.

We are to admit no more causes of natural things than such as are both true and sufficient to explain their appearances. Therefore, to the same natural effects we must, as far as possible, assign the same causes.

If Newton believes in parsimony, he's sure chosen a complicated way of saying so. Given otherwise equivalent competing explanations for puzzling phenomena, I agree simpler is better. But we live in a world exhibiting fine balance, disequilibrium and chaotic behavior all at once. Inherent complexity rules the day. Solving interesting problems at high fidelity requires intricate numerics. Fortunately, in our modern age, lots of highly modular public libraries are available. It's never been easier to patch together a simulation to entertain "what ifs?", discover emergent behavior in novel circumstances, challenge hypotheses with data, and stress-test scenarios; that is, assuming you can code and tolerate a bit of iteration or Monte Carlo. Computer simulations aren't just for physics and chemistry anymore. Biology, epidemiology, ecology, economics, business, finance, engineering, sociology even politics are experiencing a renaissance of mathematical exploration through simulation.

Trouble is, while vastly greater than just a few decades ago, computing capacity isn't infinite. Simulation experiments must be carefully planned to make the most of a finite resource, input configurations chosen to span representative scenarios, and appropriate meta-models fit in order to effectively utilize simulations towards the advance of science. That's where surrogates come in; as meta models of computer simulations used to solve mathematical systems that are too intricate to be worked by hand. Gaussian process (GP) regression has percolated up to the canonical position in this arena. It sounds hard, but it's actually quite straightforward and supremely flexible at the same time. One of the main purposes of this text is to expose the beauty and potential held by GPs in a variety of contexts. Our emphasis will be on GP surrogates for computer simulation experiments, but we'll draw upon and exemplify many successes with similar tools deployed in geostatistics and machine learning.

Emphasis is on methods, recipes and reproducibility. The latter two make this book unique by

---

[1]https://en.wikipedia.org/wiki/Occams_razor

many measures, but especially in the subjects of surrogate modeling and GPs. Methodology wise, this monograph is somewhat less broad than Santner et al. (2018), a standard text on design and analysis of computer experiments. But it offers more depth on core subjects, particularly on computing and implementation in R (R Core Team, 2019), and is more modern in its connection to methodology from machine learning, industrial statistics and geostatistics. Every subject, with the exception of a few references to ancillary material, is paired with illustration in worked code. There's not a single table or figure (which is not a drawing) in the book that's not supported by code on the page. Everything is fully reproducible. What you see is what you get. This wouldn't be possible without modern extensions to R such as RStudio[2]. Specifically, this book is authored in Rmarkdown via **bookdown** (Xie, 2016, 2018a) on CRAN, combining **knitr** (Xie, 2015, 2018b) and **rmarkdown** (Allaire et al., 2018) packages.

One downside to Rmarkdown is that sometimes, e.g., when illustrations are based on randomly generated data, it's hard to precisely narrate the outcome of a calculation. I hope that readers will appreciate the invitation this implies. You're encouraged to cut-and-paste the example into your own session and see what I mean when I say something like "It's hard to comment precisely about outcomes in this Rmarkdown build." In a small handful of places, random number generator (RNG) seeds are fixed to "freeze" the experiments and enhance specificity, even though that's technically cheating.[3] Uncertainty quantification (UQ) is a major theme in this book. A disappointingly vague narrative represents a perfect opportunity to catch a glimpse at how important, and difficult, it can be to appropriately quantify salient uncertainties.

Use of R, rather than say Python or MATLAB®, signals that this book is statistical in nature, as opposed to computer science/machine learning, engineering or applied math. That's true, to a point. I'm a professor of statistics, but I was trained as a mathematician, computer scientist, and engineer first. I moved to stats later primarily as a means of latching onto interesting applications from other areas of applied science. This book is the product of that journey. It's mathematical language is statistical because surrogate modeling involves random variables, estimators, uncertainty, conditioning and inference. But that's where it ends. This book has (almost) none of the things practitioners hate about statistics: *p*-values, sampling distributions, asymptotics, consistency, and so on. The writing is statistical in form, but the subjects are not about statistics. They're about prediction and synthesis of model and data under uncertainty, about visualization and analysis of information, about design and decision making, about computing, and about implementation. Crucially, it's about all of those things in the context of experimentation through simulation. The target audience is PhD students and post-doctoral scientists in the natural and engineering sciences, in which I include statistics and computer science. The social sciences are increasingly mathematical and computational and I think this book will appeal to folks there as well.

There's nothing special about R here, except that I know R and CRAN packages for surrogate modeling best. Many good tools exist in Python and MATLAB, and pointers are provided. R is lingua franca in the statistical surrogate modeling world, with MATLAB on somewhat of a decline and Python picking up pace. Any coded examples in the book which don't leverage highly customized CRAN libraries would be trivial to port to any high-level language. Illustrations emphasize algorithmic execution first, using basic subroutines, and library-based

---

[2]https://www.rstudio.com/

[3]Seeds are not provided, in part because RNG sequences can vary across R versions. Conditional expressions involving floating point calculations can change across architectures and lead to different results in stochastic experimentation even with identical pseudorandom numbers. It's impossible to fully remove randomness from the experience of engaging with the book material, which inevitably thwarts precise verbiage at times.

automation second. An effort is made to strip the essence of numerical calculations into digestible component parts. I view code readability as at least as important as efficiency. I don't make use of Tideyverse[4], just ordinary R.[5] Anyone with experience coding, not only R experts, should have no trouble following the examples. Reproducibility, and careful engineering of clean and well-documented code are important to me, and I intend this book as a showcase, benchmark and template for young coders.

The progression of subjects is as follows. Chapters 1–2 offer a gentle introduction comprised of historical perspective followed by an overview of four challenging real data/simulator applications. Links to data and simulation code are provided on the book web page: http://bobby.gramacy.com/surrogates[6]. These motivating examples are revisited periodically in the remainder of the text, but mostly in later chapters. Chapter 3 covers classical response surface methodology (RSM), primarily developed before computer simulation modeling and GP surrogates became mainstream. Most of the exposition here is a fly-by of Chapters 5–6 from Myers et al. (2016) with refreshed examples in R. I'm grateful to Christine Anderson–Cook for help on some of the details. Chapter 4 begins a transition to modern surrogate modeling by introducing appropriate experiment designs. Chapter 5 is on GP regression, starting simple and building up slowly, extolling virtues but not ignoring downsides, and offering several competing perspectives on almost magical properties. Material here served as the basis of a webinar[7] I gave for the American Statistical Association's (ASA) Section on Physical and Engineering Sciences (SPES) in 2017. Chapter 6 revisits design aspects in Chapter 4 from a GP perspective, motivating sequential design as modus operandi and setting the stage for Bayesian optimization (BO) in Chapter 7. Data acquisition as a decision problem, for the purpose of learning and optimization under uncertainty, is one of the great success stories of GP surrogates, combining accurate predictions with autonomous action. Chapter 8 covers calibration and input sensitivity analysis, two important applications of GP surrogates leveraging their ability to synthesize sources of information and to sensibly quantify uncertainty. Smooshing these somewhat disparate themes into a single chapter may seem awkward. My intention is to feature them as two examples of things people do with GP surrogates where UQ is key. Other texts like Santner et al. (2018) present these in two separate chapters. Chapter 9 addresses many of the drawbacks alluded to in Chapter 5, tackling computational bottlenecks limiting training data sizes, scaling up modeling fidelity, hybridizing and dividing-and-conquering with trees, and approximating with highly-parallelizable local GP surrogates. Chapter 10 discusses recent upgrades to address surrogate modeling and design for highly stochastic, low signal-to-noise, simulations in the face of heteroskedasticity (input-dependent noise). Appendix A discusses linear algebra libraries that are all but essential when working with larger problems; Appendix B introduces a game that helps reinforce many of the ideas expounded upon in this text.

While intended for instruction at the PhD level, I hope you'll find this book to be a useful reference as well. Excepting Chapters 1–2 which target perspective, overview and motivation, the technical progression within and between chapters is highly linear. Methodological development and examples within a chapter build upon one another. Later chapters assume familiarity of concepts introduced earlier, with appropriate context and pointers provided.

---

[4]https://www.tidyverse.org/

[5]Tidyverse is a very important part of the R ecosystem, and its introduction has helped keep the R community on the cutting edge of analytics and data science. Its target audience is data wranglers. Mine is methodological developers and practitioners of applied science. I feel strongly that building from a simple base is essential to effective communication and portability of code.

[6]http://bobby.gramacy.com/surrogates

[7]https://www.youtube.com/watch?v=XxqVPzb_sGM&feature=youtu.be

Each chapter's R examples execute in a novel, standalone R session.[8] Chapter 3 on RSM is relatively self-contained, and not essential for subsequent methodological development, except perhaps as a straw man. Instructors wishing to cut material in order to streamline content should consider Chapter 3 first, possibly encouraging students to skim these sections. Relying on simple linear models, basic calculus and linear algebra, material here is the most intuitive, least mathematically and computationally challenging. Nevertheless, RSM works astonishingly well and is used widely in industry. (These techniques are highly effective on the game in Appendix B.) Chapters 4–8 are the "meat", with Chapters 9–10 demarcating the surrogate modeling frontier. Homework exercises are provided at the end of each chapter. These have been vetted in the classroom over two semesters. Many are deliberately open-ended, framed as research vignettes where students are invited to fill in the gaps. For assignments, I try to strike a balance between mathematical and computational problems (e.g., do #1, #3 and two others of your choosing ...) in a way that allows students to play to their strengths while avoiding crutches. Fully reproducible solutions in Rmarkdown are available from me upon request.

There are many subjects that are not in this book, but very well could be. GP surrogates are king here, but they're by no means the only game in town. Polynomial chaos[9] and deep neural networks[10] are popular alternatives, but they're not covered in this text. My opinion is that both fall short from a UQ perspective, although they offer many other attractive features, especially in big data contexts. Even limiting to GPs, the presentation is at times deliberately narrow. Chosen methods and examples are unashamedly biased toward what I know well, to problems and methods I've worked on, and to R packages in wide use and available on CRAN. Many of those are my own contribution. If it looks like shameless self-promotion, it probably is. I like my work and want to share it with you. Although I've tried to provide pointers to related material when relevant, this book makes no attempt to serve as a systematic review of anything. Books like Santner et al. (2018) are much better in this regard. I hope that readers of my book will appreciate that its value lies in the recipes and intuition it provides, combining math and code in a (hopefully) seamless way, and as a demonstration that reproducibility in science is well within reach.

Before we get started, there are plenty of folks to thank. Let's start with family. Where would I be without Mama and those sweet kiddos? Thank you Leah, Natalia and Kaspar for letting me be proud of you and for helping me be proud of myself. This book is the outcome of confidence's virtuous cycle more than any other single thing. Thanks to my parents for encouraging me in school and for asking "who's paying for that?" every time they called to say hi (to the kids) only to find I'm out of town. Thanks to the Universities of California (Santa Cruz), Cambridge, Chicago and Virginia Tech, for supporting my research and for nurturing my career, and thanks to the US National Aeronautics and Space Administration (NASA), UK Engineering and Physical Sciences Research Council (EPSRC), US National Science Foundation (NSF) and the US Department of Energy (DOE) for funding over the years. Kudos to the Virginia Tech Department of Statistics for inviting me to teach a graduate course on the subject of my choosing, and thereby planting the seed for this book in my mind. Many thanks to students in my Fall 2016 and Spring 2019 classes on Response Surface Methods and Surrogate Modeling, for being my guinea pigs and for helping me refine presentation and fix typos along the way; shout outs to Sierra Merkes, Valeria Quevedo and Ryan Christianson in particular. I appreciated invitations to give short courses to the Statistics Department at Brigham Young University in 2017, a summer program at

---

[8]An tacit library(knitr) begins each chapter for pretty table printing, as with kable.
[9]https://en.wikipedia.org/wiki/Polynomial_chaos
[10]https://en.wikipedia.org/wiki/Deep_learning

Lawrence Livermore National Laboratory in 2017, the 2017 Fall Technical Conference and a 2018 DataWorks meeting. Huge thanks to Max Morris (IA State) and Brian Williams (LANL) for going above and beyond with their reviews for CRC.

Robert B. Gramacy
Blacksburg, VA

1

Historical Perspective

A *surrogate* is a substitute for the real thing. In statistics, draws from predictive equations derived from a fitted model can act as a surrogate for the data-generating mechanism. If the fit is good – model flexible yet well-regularized, data rich enough and fitting scheme reliable – then such a surrogate can be quite valuable. Gathering data is expensive, and sometimes getting exactly the data you want is impossible or unethical. A surrogate could represent a much cheaper way to explore relationships, and entertain "what ifs?". How do surrogates differ from ordinary statistical modeling? One superficial difference may be that surrogates favor faithful yet pragmatic reproduction of dynamics over other things statistical models are used for: interpretation, establishing causality, or identification. As you might imagine, that characterization oversimplifies.

The terminology came out of physics, applied math and engineering literatures, where the use of mathematical models leveraging numerical solvers has been commonplace for some time. As such models became more complex, requiring more resources to simulate/solve numerically, practitioners increasingly relied on meta-models built off of limited simulation campaigns. Often they recruited help from statisticians, or at least used setups resembling ones from stats. Data collected via expensive computer evaluations tuned flexible functional forms that could be used in lieu of further simulation. Sometimes the goal was to save money or computational resources; sometimes to cope with an inability to perform future runs (expired licenses, off-line or over-impacted supercomputers). Trained meta-models became known as surrogates or *emulators*, with those terms often used interchangeably. (A surrogate is designed to emulate the numerics coded in the solver.) The enterprise of design, running and fitting such meta-models became known as a *computer experiment*.

So a computer experiment is like an ordinary statistical experiment, except the data are generated by computer codes rather than physical or field observations, or surveys. Surrogate modeling is statistical modeling of computer experiments. Computer simulations are generally cheaper than physical observation, so the former could be entertained as an alternative or precursor to the latter. Although computer simulation can be just as expensive as field experimentation, computer modeling is regarded as easier because the experimental apparatus is better understood, and more aspects may be controlled. For example many numerical solvers are deterministic, whereas field observations are noisy or have measurement error. For a long time noise was the main occupant in the gulf between modeling and design considerations for surrogates, on the one hand, and more general statistical methodology on the other. But hold that thought for a moment.

Increasingly that gulf is narrowing, not so much because the nature of experimentation is changing (it is), but thanks to advances in machine learning. The canonical surrogate model, a fitted *Gaussian process (GP)* regression, which was borrowed for computer experiments from the geostatistics' *kriging*[1] literature of the 1960s, enjoys wide applicability in contexts where prediction is king. Machine learners exposed GPs as powerful predictors for all sorts

---

[1]https://en.wikipedia.org/wiki/Kriging

of tasks[2], from regression to classification, active learning/sequential design, reinforcement learning and optimization, latent variable modeling, and so on. They also developed powerful libraries, lowering the bar to application by non-expert practitioners, especially in the information technology world. Facebook uses surrogates to tailor its web portal and apps to optimize engagement; Uber uses surrogates trained to traffic simulations to route pooled ride-shares in real-time, reducing travel and wait time.

Round about the same time, computer simulation as a means of scientific inquiry began to blossom. Mathematical biologists, economists and others had reached the limit of equilibrium-based mathematical modeling with cute closed-form solutions. They embraced simulation as a means of filling in the gap, just as physicists and engineers had decades earlier. Yet their simulations were subtly different. Instead of deterministic solvers based on finite elements[3], Navier–Stokes[4] or Euler methods[5], they were building stochastic simulations[6], and agent-based models[7], to explore predator-prey (Lotka–Voltera[8]) dynamics, spread of disease, management of inventory or patients in health insurance markets. Suddenly, and thanks to an explosion in computing capacity, software tools, and better primary school training in STEM[9] subjects (all decades in the making), simulation was enjoying a renaissance. We're just beginning to figure out how best to model these experiments, but one thing is for sure: the distinction between surrogate and statistical model is all but gone.

If there's (real) *field data*, say on a historical epidemic, further experimentation may be almost entirely limited to the mathematical and computer modeling side. You can't seed a real community with Ebola and watch what happens. Epidemic simulations, and surrogates built from a limited number of expensive runs where virtual agents interact and transmit infection, can be calibrated to a limited amount of physical data. Doing that right and getting something useful out of it depends crucially on surrogate methodology and design. Classical statistical methods offer little guidance. The notion of population is weak at best, and causation is taken as given. Mechanisms engineered into the simulation directly "cause" the outputs we observe as inputs change. Many classically statistical considerations take a back seat to having trustworthy and flexible prediction. That means not just capturing the essence of the simulated dynamics under study, but being hands-off in fitting while at the same time enabling rich diagnostics to help criticize that fit; understanding its sensitivity to inputs and other configurations; providing the ability to optimize and refine both automatically and with expert intervention. And it has to do all that while remaining computationally tractable. What good is a surrogate if it's more work than the original simulation? Thrifty meta-modeling is essential.

This book is about those topics. It's a statistics text in form but not really in substance. The target audience is both more modern and more diverse. Rarely is emphasis on properties of a statistic. We won't test many hypotheses, but we'll make decisions – lots of them. We'll design experiments, but not in the classical sense. It'll be more about active learning and optimization. We'll work with likelihoods, but mostly as a means of fine-tuning. There will be no asymptopia. Pragmatism and limited resources are primary considerations. We'll talk about big simulation but not big data. Uncertainty quantification (UQ) will play a huge role. We'll visualize confidence and predictive intervals, but rarely will those point-wise summaries

---

[2] http://research.cs.aalto.fi/pml/software/gpstuff/
[3] https://en.wikipedia.org/wiki/Finite_element_method
[4] https://en.wikipedia.org/wiki/NavierStokes_equations
[5] https://en.wikipedia.org/wiki/Euler_method
[6] https://en.wikipedia.org/wiki/Stochastic_simulation
[7] https://en.wikipedia.org/wiki/Agent-based_model
[8] https://en.wikipedia.org/wiki/Lotka-Volterra_equations
[9] https://en.wikipedia.org/wiki/Science,_technology,_engineering,_and_mathematics

be the main quantity of interest. Instead, the goal is to funnel a corpus of uncertainty, to the extent that's computationally tractable, through to a decision-making framework. Synthesizing multiple data sources, and combining multiple models/surrogates to enhance fidelity, will be a recurring theme – once we're comfortable with the basics, of course.

Emphasis is on GPs, but we'll start a little old school and finish by breaking outside of the GP box, with tangents and related methodology along the way. This chapter and the next are designed to set the stage for the rest of the book. Below is historical context from two perspectives. One perspective is so-called response surface methods (RSMs), a poster child from industrial statistics' heyday, well before information technology became a dominant industry. Here surrogates are crude, but the literature is rich and methods are tried and tested in practice, especially in manufacturing. Careful experimental design, paired with a well understood model and humble expectations, can add a lot of value to scientific inquiry, process refinement, optimization, and more. These ideas are fleshed out in more detail in Chapter 3. Another perspective comes from engineering. Perhaps more modern, but also less familiar to most readers, our limited presentation of this viewpoint here is designed to whet the reader's appetite for the rest of the book (i.e., except Chapter 3). Chapter 2 outlines four real-world problems that would be hard to address within a classical RSM framework. Modern GP surrogates are essential, although in some cases the typical setup oversimplifies. Fully worked solutions will require substantial buildup and will have to wait until the last few chapters of the book.

1.1 Response surface methodology

RSMs are a big deal to the Virginia Tech Statistics Department, my home. Papers and books by Ray Myers, his students and colleagues, form the bedrock of best statistical practice in design and modeling in industrial, engineering and physical sciences. Related fields of design of experiments, quality, reliability and productivity are also huge here (Geoff Vining, Bill Woodall, JP Morgan). All three authors of my favorite book on the subject, *Response Surface Methodology* (Myers et al., 2016), are Hokies[10]. Much of the development here and in Chapter 3 follows this highly accessible text, sporting fresh narrative and augmented with reproducible R examples. Box and Draper (2007)'s *Response Surfaces, Mixtures, and Ridge Analyses* is perhaps more widely known, in part because the authors are household names (in stats households). Methods described in these texts are in wide use in application areas ranging from materials science, manufacturing, applied chemistry, and climate science, to name just a few.

1.1.1 What is it?

Response surface methodology (RSM) is a collection of statistical and mathematical tools useful for developing, improving, and optimizing processes. Applications historically come from industry and manufacturing, focused on design, development, and formulation of new products and the improvement of existing products but also from (national) laboratory research, and with obvious military application. The overarching theme is a study of how

---

[10]https://en.wikipedia.org/wiki/HokieBird

input variables controlling a product or process potentially influence a *response* measuring performance or quality characteristics.

Consider the relationship between the response variable yield (y) in a chemical process and two process variables: reaction time (ξ_1) and reaction temperature (ξ_2). R code below synthesizes this setting for the benefit of illustration.

```
yield <- function(xi1, xi2)
 {
   xi1 <- 3*xi1 - 15
   xi2 <- xi2/50 - 13
   xi1 <- cos(0.5)*xi1 - sin(0.5)*xi2
   xi2 <- sin(0.5)*xi1 + cos(0.5)*xi2
   y <- exp(-xi1^2/80 - 0.5*(xi2 + 0.03*xi1^2 - 40*0.03)^2)
   return(100*y)
 }
```

Seasoned readers will recognize the form above as a variation on the so-called "banana function". Figure 1.1 shows this yield response plotted in perspective as a surface above the time/temperature plane.

```
xi1 <- seq(1, 8, length=100)
xi2 <- seq(100, 1000, length=100)
g <- expand.grid(xi1, xi2)
y <- yield(g[,1], g[,2])
persp(xi1, xi2, matrix(y, ncol=length(xi2)), theta=45, phi=45,
   lwd=0.5, xlab="xi1 : time", ylab="xi2 : temperature",
   zlab="yield", expand=0.4)
```

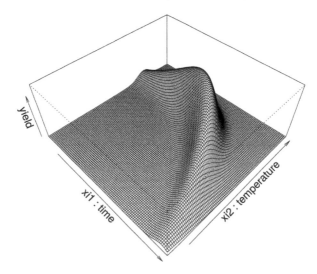

FIGURE 1.1: Banana yield example as a function of time and temperature.

Although perhaps not as pretty, it's easier to see what's going on in an image–contour plot. Figure 1.2 utilizes heat colors where white is hotter (higher) and red is cooler (lower).

```
cols <- heat.colors(128)
image(xi1, xi2, matrix(y, ncol=length(xi2)), col=cols,
  xlab="xi1 : time", ylab="xi2 : temperature")
contour(xi1, xi2, matrix(y, ncol=length(xi2)), nlevels=4, add=TRUE)
```

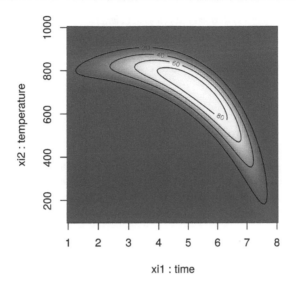

FIGURE 1.2: Alternative heat map view of banana yield.

By inspection, yield is optimized near $(\xi_1, \xi_2) = (5 \text{ hr}, 750°\text{C})$. Unfortunately in practice, the true response surface is unknown. When `yield` evaluation is not as simple as a toy banana function, but a process requiring care to monitor, reconfigure and run, it's far too expensive to observe over a dense grid. Moreover, measuring `yield` may be a noisy/inexact process.

That's where stats comes in. RSMs consist of experimental strategies for exploring the space of the process (i.e., independent/input) variables (above ξ_1 and ξ_2); for empirical statistical modeling targeted toward development of an appropriate approximating relationship between the response (`yield`) and process variables local to a study region of interest; and optimization methods for sequential refinement in search of the levels or values of process variables that produce desirable responses (e.g., that maximize yield or explain variation).

Suppose the true response surface is driven by an unknown physical mechanism, and that our observations are corrupted by noise. In that setting it can be helpful to fit an empirical model to output collected under different process configurations. Consider a response Y that depends on controllable input variables $\xi_1, \xi_2, \ldots, \xi_m$. Write

$$Y = f(\xi_1, \xi_2, \ldots, \xi_m) + \varepsilon$$
$$\mathbb{E}\{Y\} = \eta = f(\xi_1, \xi_2, \ldots, \xi_m)$$

where ε is treated as zero mean idiosyncratic noise possibly representing inherent variation, or the effect of other systems or variables not under our purview at this time. A simplifying assumption that $\varepsilon \sim \mathcal{N}(0, \sigma^2)$ is typical. We seek estimates for f and σ^2 from noisy observations Y at inputs ξ.

Inputs $\xi_1, \xi_2, \ldots, \xi_m$ above are called *natural variables* because they're expressed in their natural units of measurement, such as degrees Celsius (°C), pounds per square inch (psi), etc.

We usually transform these to *coded variables* $x_1, x_2, \ldots, x_m$ to mitigate hassles and confusion that can arise when working with a multitude of scales of measurement. Transformations offering dimensionless inputs $x_1, \ldots, x_m$ in the unit cube, or scaled to have a mean of zero and standard deviation of one, are common choices. In that space the empirical model becomes

$$\eta = f(x_1, x_2, \ldots, x_m).$$

Working with coded inputs x will be implicit throughout this text except when extra code is introduced to explicitly map from natural coordinates. (Rarely will ξ_j notation feature beyond this point.)

1.1.2   Low-order polynomials

Learning about f is lots easier if we make some simplifying approximations. Appealing to Taylor's theorem[11], a *low-order polynomial* in a small, localized region of the input (x) space is one way forward. Classical RSM focuses on disciplined application of local analysis and sequential refinement of "locality" through conservative extrapolation. It's an inherently hands-on process.

A *first-order model*, or sometimes called a *main effects model* , makes sense in parts of the input space where it's believed that there's little curvature in f.

$$\eta = \beta_0 + \beta_1 x_1 + \beta_2 x_2$$
$$\text{for example} \quad = 50 + 8x_1 + 3x_2$$

In practice, such a surface would be obtained by fitting a model to the outcome of a designed experiment. Hold that thought until Chapter 3; for now the goal is a high-level overview.

To help visualize, code below encapsulates that main effects model ...

```
first.order <- function(x1, x2)
  {
    50 + 8*x1 + 3*x2
  }
```

... and then evaluates it on a grid in a double-unit square centered at the origin. These coded units are chosen arbitrarily, although one can imagine deploying this approximating function nearby $x^{(0)} = (0, 0)$,

```
x1 <- x2 <- seq(-1, 1, length=100)
g <- expand.grid(x1, x2)
eta1 <- matrix(first.order(g[,1], g[,2]), ncol=length(x2))
```

Figure 1.3 shows the surface in perspective (left) and image–contour (right) plots, again with heat colors: white is hotter/higher, red is cooler/lower.

---

[11]https://en.wikipedia.org/wiki/Taylor's_theorem

```
par(mfrow=c(1,2))
persp(x1, x2, eta1, theta=30, phi=30, zlab="eta", expand=0.75, lwd=0.25)
image(x1, x2, eta1, col=heat.colors(128))
contour(x1, x2, matrix(eta1, ncol=length(x2)), add=TRUE)
```

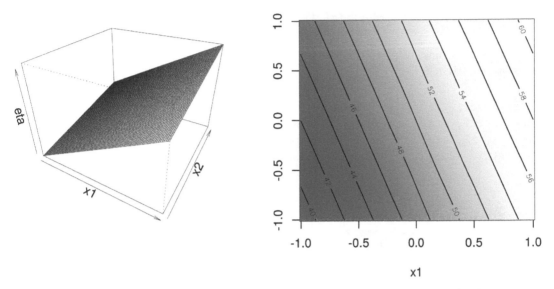

FIGURE 1.3: Example of a first-order response surface via perspective (left) and heat map (right).

Clearly the development here serves as a warm up. I presume that you, the keen reader, know that a first-order model in 2d traces out a plane in $y \times (x_1, x_2)$ space. My aim with these passages is twofold. One is to introduce classical RSMs. But another, perhaps more important goal, is to introduce the style of Rmarkdown presentation adopted throughout the text. There's a certain flow that (at least in my opinion) such presentation demands. This book's style is quite distinct compared to textbooks of just ten years ago, even ones supported by code. Although my presentation here closely follows early chapters in Myers et al. (2016), the R implementation and visualization here are novel, and are engineered for compatibility with the diverse topics which follow in later chapters.

Ok, enough digression. Back to RSMs. A simple first-order model would only be appropriate for the most trivial of response surfaces, even when applied in a highly localized part of the input space. Adding curvature is key to most applications. A *first-order model with interactions* induces a limited degree of curvature via different rates of change of y as x_1 is varied for fixed x_2, and vice versa.

$$\eta = \beta_0 + \beta_1 x_1 + \beta_2 x_2 + \beta_{12} x_1 x_2$$
$$\text{for example} \quad = 50 + 8x_1 + 3x_2 - 4x_1 x_2$$

To help visualize, R code below facilitates evaluations for pairs (x_1, x_2) ...

```
first.order.i <- function(x1, x2)
  {
```

```
  50 + 8*x1 + 3*x2 - 4*x1*x2
}
```

... so that responses may be observed over a mesh in the same double-unit square.

```
eta1i <- matrix(first.order.i(g[,1], g[,2]), ncol=length(x2))
```

Figure 1.4 shows those responses on the z-axis in a perspective plot (left), and as heat colors and contours (right).

```
par(mfrow=c(1,2))
persp(x1, x2, eta1i, theta=30, phi=30, zlab="eta", expand=0.75, lwd=0.25)
image(x1, x2, eta1i, col=heat.colors(128))
contour(x1, x2, matrix(eta1i, ncol=length(x2)), add=TRUE)
```

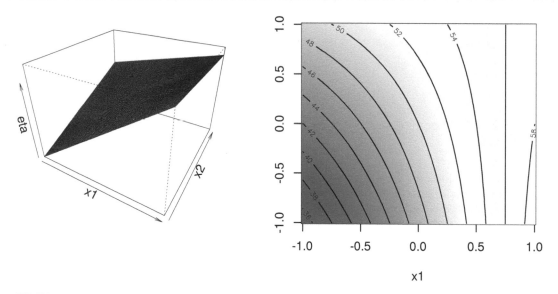

FIGURE 1.4: Example of a first-order response surface with interaction(s).

Observe that the mean response η is increasing marginally in both x_1 and x_2, or conditional on a fixed value of the other until x_1 is 0.75 or so. Rate of increase slows as both coordinates grow simultaneously since the coefficient in front of the interaction term x_1x_2 is negative. Compared to the first-order model (without interactions), such a surface is far more useful locally. Least squares regressions – wait until §3.1.1 for details – often flag up significant interactions when fit to data collected on a design far from local optima.

A *second-order model* may be appropriate near local optima where f would have substantial curvature.

$$\eta = \beta_0 + \beta_1 x_1 + \beta_2 x_2 + \beta_{11} x_1^2 + \beta_{22} x_2^2 + \beta_{12} x_1 x_2$$
$$\text{for example} \quad = 50 + 8x_1 + 3x_2 - 7x_1^2 - 3x_2^2 - 4x_1 x_2$$

The code below implements this function ...

```
simple.max <- function(x1, x2)
  {
    50 + 8*x1 + 3*x2 - 7*x1^2 - 3*x2^2 - 4*x1*x2
  }
```

... which is then evaluated on our grid for visualization.

```
eta2sm <- matrix(simple.max(g[,1], g[,2]), ncol=length(x2))
```

Panels in Figure 1.5 show that this surface has a maximum near about $(0.6, 0.2)$.

```
par(mfrow=c(1,2))
persp(x1, x2, eta2sm, theta=30, phi=30, zlab="eta", expand=0.75, lwd=0.25)
image(x1, x2, eta2sm, col=heat.colors(128))
contour(x1, x2, eta2sm, add=TRUE)
```

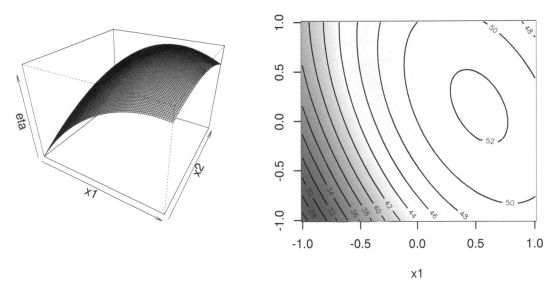

FIGURE 1.5: Second-order simple maximum surface.

Not all second-order models would have a single stationary point, which in RSM jargon is called a *simple maximum*. (In this "yield maximizing" setting we're presuming that our response surface is concave down from a global viewpoint, even though local dynamics may be more nuanced). Coefficients in front of input terms, their interactions and quadratics, must be carefully selected to get a simple maximum. Exact criteria depend upon the eigenvalues of a certain matrix built from those coefficients, which we'll talk more about in §3.2.1. Box and Draper (2007) provide a beautiful diagram categorizing all of the kinds of second-order surfaces one can encounter in an RSM analysis, where finding local maxima is the goal. An identical copy, with permission, appears in Myers et al. (2016) too. Rather than duplicate those here, a third time, we shall continue with the theme of R-ification of that presentation.

An example set of coefficients describing what's called a *stationary ridge* is provided by the R code below.

```
stat.ridge <- function(x1, x2)
 {
   80 + 4*x1 + 8*x2 - 3*x1^2 - 12*x2^2 - 12*x1*x2
 }
```

Let's evaluate that on our grid ...

```
eta2sr <- matrix(stat.ridge(g[,1], g[,2]), ncol=length(x2))
```

... and then view the surface with our usual pair of panels in Figure 1.6.

```
par(mfrow=c(1,2))
persp(x1, x2, eta2sr, theta=30, phi=30, zlab="eta", expand=0.75, lwd=0.25)
image(x1, x2, eta2sr, col=heat.colors(128))
contour(x1, x2, eta2sr, add=TRUE)
```

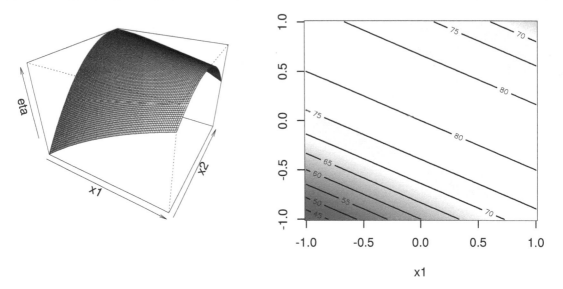

FIGURE 1.6: Example of a stationary ridge.

Observe how there's a ridge – a whole line – of stationary points corresponding to maxima: a 1d submanifold of the 2d surface, if I may be permitted to use vocabulary I don't fully understand. Assuming the approximation can be trusted, this situation means that the practitioner has some flexibility when it comes to optimizing, and can choose the precise setting of (x_1, x_2) either arbitrarily or (more commonly) by consulting some tertiary criteria.

An example of a *rising ridge* is implemented by the R code below ...

```
rise.ridge <- function(x1, x2)
 {
   80 - 4*x1 + 12*x2 - 3*x1^2 - 12*x2^2 - 12*x1*x2
 }
```

... and evaluated on our grid as follows.

```
eta2rr <- matrix(rise.ridge(g[,1], g[,2]), ncol=length(x2))
```

Notice in Figure 1.7 how there's a continuum of (local) stationary points along any line going through the 2d space, excepting one that lies directly on the ridge.

```
par(mfrow=c(1,2))
persp(x1, x2, eta2rr, theta=30, phi=30, zlab="eta", expand=0.75, lwd=0.25)
image(x1, x2, eta2rr, col=heat.colors(128))
contour(x1, x2, eta2rr, add=TRUE)
```

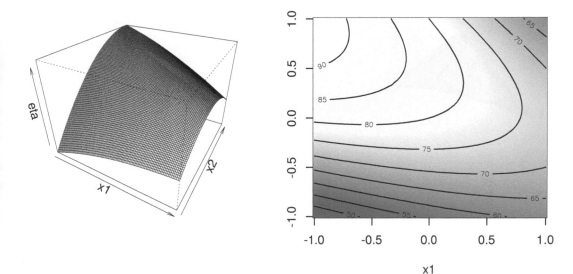

FIGURE 1.7: Example of a rising ridge.

In this case, the stationary point is remote to the study region. Although there's comfort in learning that an estimated response will increase as you move along the axis of symmetry toward its stationary point, this situation indicates either a poor fit by the approximating second-order function, or that the study region is not yet precisely in the vicinity of a local optima – often both. The inversion of a rising ridge is a *falling ridge*, similarly indicating one is far from local optima, except that the response decreases as you move toward the stationary point. Finding a falling ridge system can be a back-to-the-drawing-board affair.

Finally, we can get what's called a *saddle* or *minimax* system. In R ...

```
saddle <- function(x1, x2)
 {
   80 + 4*x1 + 8*x2 - 2*x1 - 12*x2 - 12*x1*x2
 }
```

... and ...

```
eta2s <- matrix(saddle(g[,1], g[,2]), ncol=length(x2))
```

Panels in Figure 1.8 show that the (single) stationary point is either a local maxima or minima, depending on your perspective.

```
par(mfrow=c(1,2))
persp(x1, x2, eta2s, theta=30, phi=30, zlab="eta", expand=0.75, lwd=0.25)
image(x1, x2, eta2s, col=heat.colors(128))
contour(x1, x2, eta2s, add=TRUE)
```

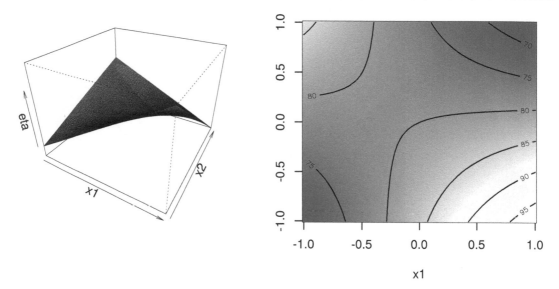

FIGURE 1.8: Example of a saddle or minimax system.

Likely further data collection, and/or outside expertise, is needed before determining a course of action in this situation. Finding a simple maximum, or stationary ridge, represents ideals in the spectrum of second-order approximating functions. But getting there can be a bit of a slog. Using models fitted from data means uncertainty due to noise, and therefore uncertainty in the type of fitted second-order model you're dealing with. A *ridge analysis* [see §3.2.2] attempts to offer a principled approach to navigating uncertainties when one is seeking local maxima. The two-dimensional setting exemplified above is convenient for visualization, but rare in practice. Complications compound when studying the effect of more than two process variables.

1.1.3   General models, inference and sequential design

The general first-order model on m process variables $x_1, \ldots, x_m$ is

$$\eta = \beta_0 + \beta_1 x_1 + \cdots + \beta_m x_m,$$

and the general second-order model thus

$$\eta = \beta_0 + \sum_{j=1}^{m} \beta_j x_j + \sum_{j=1}^{m} \beta_{jj} x_j^2 + \sum_{j=2}^{m} \sum_{k=1}^{j} \beta_{kj} x_k x_j.$$

Inference from data is carried out by *ordinary least squares (OLS)*. I won't review OLS, or indeed any of the theory for linear modeling, although properties will be introduced as needed. For an excellent review including R examples, see Sheather (2009). However we'll do plenty of fully implemented examples in Chapter 3. Throughout I shall be rather cavalier in my haphazard interchangeable use of OLS and maximum likelihood estimators (MLEs) in the typical Gaussian linear modeling setup, as the two are basically equivalent.

Besides serving to illustrate RSM methods in action, we shall see how important it is to organize the data collection phase of a response surface study carefully. A *design* is a choice of x's where we plan to observe y's, for the purpose of approximating f. Analyses and designs need to be carefully matched. When using a first-order model, some designs are preferred over others. When using a second-order model to capture curvature, a different sort of design is appropriate. Design choices often contain features enabling modeling assumptions to be challenged, e.g., to check if initial impressions are supported by the data ultimately collected.

Although a substantial portion of this text will be devoted to design for GP surrogates, especially sequential design, design for classical RSMs will be rather more limited. There are designs which help with *screening*, to determine which variables matter so that subsequent experiments may be smaller and/or more focused. Then there are designs tailored to the form of model (first- or second-order, say) in the screened variables. And then there are more designs still. In our empirical examples we shall leverage off-the-shelf choices introduced at length in other texts (Myers et al., 2016; Box and Draper, 2007) with little discussion.

Usually RSM-based experimentation begins with a first-order model, possibly with interactions, under the presumption that the current process is operating far from optimal conditions. Based on data collected under designs appropriate for those models, the so-called method of *steepest ascent* deploys standard error and gradient calculations on fitted surfaces in order to a) determine where the data lie relative to optima (near or far, say); and b) to inch the "system" closer to such regimes. Eventually, if all goes well after several such carefully iterated refinements, second-order models are entertained on appropriate designs in order to zero-in on ideal operating conditions. Again this involves a careful analysis of the fitted surface – a ridge analysis, see §3.2.2 – with further refinement using gradients of, and standard errors associated with, the fitted surfaces, and so on. Once the practitioner is satisfied with the full arc of design(s), fit(s), and decision(s), a small experiment called a *confirmation test* may be performed to check if the predicted optimal settings are indeed realizable in practice.

That all seems sensible, and pretty straightforward as quantitative statistics-based analysis goes. Yet it can get complicated, especially when input dimensions are moderate in size. Design considerations are particularly nuanced, since the goal is to obtain reliable estimates of main effects, interaction and curvature while minimizing sampling effort/expense. Textbooks devote hundreds of pages to that subject (e.g., Morris, 2010; Wu and Hamada, 2011), which is in part why we're not covering it. (Also, I'm by no means an expert.) Despite intuitive appeal, several downsides to this setup become apparent upon reflection, or after one's first attempt to put the idea into practice. The compartmental nature of sequential decision making is inefficient. It's not obvious how to re-use or update analysis from earlier phases, or couple with data from other sources/related experiments. In addition to being local in experiment-time, or equivalently limited in memory (to use math programming jargon), it's local in experiment-space. Balance between exploration (maybe we're barking up the wrong tree) and exploitation (let's make things a little better) is modest at best. Interjection of expert knowledge is limited to hunches about relevant variables (i.e., the screening phase), where to initialize search, how to design the experiments. Yet at the same time classical RSMs rely heavily on constant scrutiny throughout stages of modeling and design and on the instincts of seasoned practitioners. Parallel analyses, conducted according to the same best intentions, rarely lead to the same designs, model fits and so on. Sometimes that means they lead to different conclusions, which can be cause for concern.

In spite of those criticisms, however, there was historically little impetus to revise the status quo. Classical RSM was comfortable in its skin, consistently led to improvements or compelling evidence that none can reasonably be expected. But then in the late 20th century came an explosive expansion in computational capability, and with it a means of addressing many of those downsides.

These days people are getting more out of smaller "field experiments" or "tests" and more out of their statistical models, designs and optimizations by coupling with *mathematical models* of the system(s) under study. And by mathematical model I don't mean $F = ma$, although that's not a bad place to start pedagogically. Gone are the days where simple equations are regarded as sufficient to describe real-world systems. Physicists figured that out fifty years ago; industrial engineers followed suit. Biologists, social scientists, climate scientists and weather forecasters, have jumped on the bandwagon rather more recently. Systems of equations are required, solved over meshes (e.g., finite elements), or you might have stochastically interacting agents acting out predator–prey dynamics in habitat, an epidemic spreading through a population in a social network, citizens making choices about health care and insurance. Goals for those simulation experiments are as diverse as their underlying dynamics. Simple optimization is common. Or, one may wish to discover how a regulatory framework, or worst-case scenario manifests as emergent behavior from complicated interactions. Even economists and financial mathematicians, famously favoring equilibrium solutions (e.g., efficient markets[12]), are starting to notice. An excellent popular science book called *Forecast* by Buchanan (2013) argues that this revolution is long overdue.

Solving systems of equations, or exploring the behavior of interacting agents, requires numerical analysis[13] and that means computing. Statistics can be involved at various stages: choosing the mathematical model, solving by stochastic simulation (Monte Carlo), designing the computer experiment, smoothing over idiosyncrasies or noise, finding optimal conditions, or calibrating mathematical/computer models to data from field experiments. Classical RSMs are not well-suited to any of those tasks, primarily because they lack the fidelity required to model these data. Their intended application is too local. They're also too hands-on. Once computers are involved, a natural inclination is to automate – to remove humans from the loop and set the computer running on the analysis in order to maximize computing throughput, or minimize idle time. New response surface methodology is needed.

1.2 Computer experiments

Mathematical models implemented in computer codes are now commonplace as a means of avoiding expensive field data collection. Codes can be computationally intensive, solving sys-

---

[12]https://en.wikipedia.org/wiki/Efficient-market_hypothesis
[13]https://en.wikipedia.org/wiki/Numerical_analysis

tems of differential equations, finite element analysis, Monte Carlo quadrature/approximation, individual/agent based models (I/ABM), and more. Highly nonlinear response surfaces, high signal-to-noise ratios (often deterministic evaluations) and global scope demands a new approach to design and modeling compared to a classical RSM setting. As computing power has grown, so too has simulation fidelity, adding depth in terms of both accuracy and faithfulness to the best understanding of the physical, biological, or social dynamics in play. Computing has fueled ambition for breadth as well. Expansion of configuration spaces and increasing input dimension yearn for ever-bigger designs. Advances in high performance computing (HPC) have facilitated distribution of solvers to an unprecedented degree, allowing thousands of runs where only tens could be done before. That's helpful with big input spaces, but shifts the burden to big models and big training data which bring their own computational challenges.

Research questions include how to design computer experiments that spend on computation judiciously, and how to meta-model computer codes to save on simulation effort. Like with classical RSM, those two go hand in hand. The choice of surrogate model for the computer codes, if done right, can have a substantial effect on the optimal design of the experiment. Depending on your goal, whether descriptive or response-maximizing in nature, different model–design pairs may be preferred. Combining computer simulation, design, and modeling with field data from similar, real-world experiments leads to a new class of computer model *calibration* or *tuning* problems. There the goal is to learn how to tweak the computer model to best match physical dynamics observed (with noise) in the real world, and to build an understanding of any systematic biases between model and reality. And as ever with computers, the goal is to automate to the extent possible so that HPC can be deployed with minimal human intervention.

In light of the above, many regard computer experiments as distinct from RSM. I prefer to think of them as a modern extension. Although there's clearly a need to break out of a local linear/quadratic modeling framework, and associated designs, many similar themes are in play. For some, the two literatures are converging, with self-proclaimed RSM researchers increasingly deploying GP models and other techniques from computer experiments. On the other hand, researchers accustomed to interpolating deterministic computer simulations are beginning to embrace stochastic simulation, thus leveraging designs resembling those for classical RSM. Replication for example, which would never feature in a deterministic setting, is a tried and true means of separating signal from noise. Traditional RSM is intended for situations in which a substantial proportion of variability in the data is just noise and the number of data values that can be acquired can sometimes be severely limited. Consequently, RSM is intended for a somewhat different class of problems, and is indeed well-suited for their purposes.

There are two very good texts on computer experiments and surrogate modeling. *The Design and Analysis of Computer Experiments*, by Santner et al. (2018) is the canonical reference in the statistics literature. *Engineering Design via Surrogate Modeling* by Forrester et al. (2008) is perhaps more popular in engineering. Both are geared toward design. Santner et al. is more technical. My emphasis is a bit more on modeling and implementation, especially in contemporary big- and stochastic-simulation contexts. Whereas my style is more statistical, like Santner et al. the presentation is more implementation-oriented like Forrester, et al. (They provide extensive MATLAB® code; I use R.)

TABLE 1.1: Wing weight parameters.

Symbol	Parameter	Baseline	Minimum	Maximum
S_w	Wing area (ft^2)	174	150	200
W_{fw}	Weight of fuel in wing (lb)	252	220	300
A	Aspect ratios	7.52	6	10
Λ	Quarter-chord sweep (deg)	0	-10	10
q	Dynamic pressure at cruise (lb/ft^2)	34	16	45
λ	Taper ratio	0.672	0.5	1
R_{tc}	Aerofoil thickness to chord ratio	0.12	0.08	0.18
N_z	Ultimate load factor	3.8	2.5	6
W_{dg}	Final design gross weight (lb)	2000	1700	2500

1.2.1 Aircraft wing weight example

To motivate expanding the RSM toolkit towards better meta-modeling and design for a computer-implemented mathematical model, let's borrow an example from Forrester et al. Besides appropriating the example setting, not much about the narrative below resembles any other that I'm aware of. Although also presented as a warm-up, Forrester et al. utilize this example with a different pedagogical goal in mind.

The following equation has been used to help understand the weight of an unpainted light aircraft wing as a function of nine design and operational parameters.

$$W = 0.0365 S_w^{0.758} W_{fw}^{0.0035} \left(\frac{A}{\cos^2 \Lambda} \right)^{0.6} q^{0.006} \lambda^{0.04} \left(\frac{100 R_{tc}}{\cos \Lambda} \right)^{-0.3} (N_z W_{dg})^{0.49} \qquad (1.1)$$

Table 1.1 details each of the parameters and provides reasonable ranges on their natural scale. A baseline setting, coming from a Cessna C172 Skyhawk aircraft, is also provided.

I won't go into any detail here about what each parameter measures, although for some (wing area and fuel weight) the effect on overall weight is obvious. It's worth remarking that Eq. (1.1) is not really a computer simulation, although we'll use it as one for the purposes of this illustration. Utilizing a true form, but treating it as unknown, is a helpful tool for synthesizing realistic settings in order to test methodology. That functional form was derived by "calibrating" known physical relationships to curves obtained from existing aircraft data (Raymer, 2012). So in a sense it's itself a surrogate for actual measurements of the weight of aircrafts. It was built via a mechanism not unlike one we'll expound upon in more depth in a segment on computer model calibration in §8.1, albeit in a less parametric setting. Although we won't presume to know that functional form in any of our analysis below, observe that the response is highly nonlinear in its inputs. Even when modeling the logarithm, which turns powers into slope coefficients and products into sums, the response would still be nonlinear owing to the trigonometric terms.

Considering the nonlinearity and high input dimension, simple linear and quadratic response surface approximations will likely be insufficient. Of course, that depends upon the application of interest. The most straightforward might simply be to understand input–output relationships. Given the global purview implied by that context, a fancier model is all but essential. For now, let's concentrate on that setting to fix ideas. Another application might be optimization. There might be interest in minimizing weight, but probably not without some constraints. (We'll need wings with non-zero area if the airplane is going to fly.) Let's hold that thought until Chapter 7, where I'll argue that global perspective, and thus flexible modeling, is essential in (constrained) optimization settings.

The R code below serves as a genuine computer implementation "solving" a mathematical model. It takes arguments coded in the unit cube. Defaults are used to encode baseline settings from Table 1.1, also mapped to coded units.

```r
wingwt <- function(Sw=0.48, Wfw=0.28, A=0.38, L=0.5, q=0.62, l=0.344,
  Rtc=0.4, Nz=0.37, Wdg=0.38)
{
  ## put coded inputs back on natural scale
  Sw <- Sw*(200 - 150) + 150
  Wfw <- Wfw*(300 - 220) + 220
  A <- A*(10 - 6) + 6
  L <- (L*(10 - (-10)) - 10) * pi/180
  q <- q*(45 - 16) + 16
  l <- l*(1 - 0.5) + 0.5
  Rtc <- Rtc*(0.18 - 0.08) + 0.08
  Nz <- Nz*(6 - 2.5) + 2.5
  Wdg <- Wdg*(2500 - 1700) + 1700

  ## calculation on natural scale
  W <- 0.036*Sw^0.758 * Wfw^0.0035 * (A/cos(L)^2)^0.6 * q^0.006
  W <- W * l^0.04 * (100*Rtc/cos(L))^(-0.3) * (Nz*Wdg)^(0.49)
  return(W)
}
```

Compute time required by the `wingwt` "solver" is trivial, and approximation error is minuscule – essentially machine precision. Later we'll imagine a more time consuming evaluation by mentally adding a `Sys.sleep(3600)` command to synthesize a one-hour execution time, say. For now, our presentation will utilize cheap simulations in order to perform a sensitivity analysis (see §8.2), exploring which variables matter and which work together to determine levels of the response.

Plotting in 2d is lots easier than 9d, so the code below makes a grid in the unit square to facilitate sliced visuals. (This is basically the same grid as we used earlier, except in $[0, 1]^2$ rather than $[-1, 1]^2$. The coding used to transform inputs from natural units is largely a matter of taste, so long as it's easy to undo for reporting back on original scales.)

```r
x <- seq(0, 1, length=100)
g <- expand.grid(x, x)
```

Now we can use the grid to, say, vary N_z and A, with other inputs fixed at their baseline values.

```r
W.A.Nz <- wingwt(A=g[,1], Nz=g[,2])
```

To help interpret outputs from experiments such as this one – to level the playing field when comparing outputs from other pairs of inputs – code below sets up a color palette that can be re-used from one experiment to the next.

```
cs <- heat.colors(128)
bs <- seq(min(W.A.Nz), max(W.A.Nz), length=129)
```

Figure 1.9 shows the weight response as a function of N_z and A with an image–contour plot. Slight curvature in the contours indicates an interaction between these two variables. Actually, this output range (180–320 approximately) nearly covers the entire span of outputs observed from settings of inputs in the full, 9d input space.

```
image(x, x, matrix(W.A.Nz, ncol=length(x)), col=cs, breaks=bs,
  xlab="A", ylab="Nz")
contour(x, x, matrix(W.A.Nz, ncol=length(x)), add=TRUE)
```

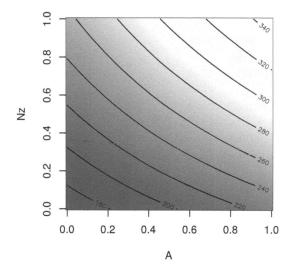

FIGURE 1.9: Wing weight over an interesting 2d slice.

Apparently an aircraft wing is heavier when aspect ratios A are high, and designed to cope with large g-forces (large N_z), with a compounding effect. Perhaps this is because fighter jets cannot have efficient (light) glider-like wings. How about the same experiment for two other inputs, e.g., taper ratio λ and fuel weight W_{fw}?

```
W.l.Wfw <- wingwt(l=g[,1], Wfw=g[,2])
```

Figure 1.10 shows the resulting image–contour plot, utilizing the same color palette as in Figure 1.9 in order to emphasize a stark contrast.

```
image(x, x, matrix(W.l.Wfw,ncol=length(x)), col=cs, breaks=bs,
  xlab="l", ylab="Wfw")
contour(x, x, matrix(W.l.Wfw,ncol=length(x)), add=TRUE)
```

Apparently, neither input has much effect on wing weight, with λ having a marginally greater

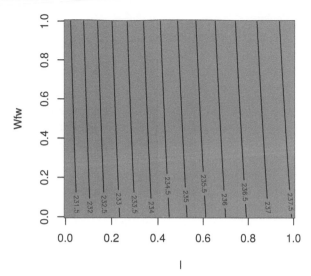

FIGURE 1.10: Wing weight over an ineffectual 2d slice.

effect, covering less than 4% of the span of weights observed in the $A \times N_z$ plane. There's no interaction evident in $\lambda \times W_{\text{fw}}$.

Well that's all fine and good. We've learned about two pairs of inputs, out of 36 total pairs. For each pair we evaluated `wingwt` 10,000 times. Doing the same for all pairs would require 360K evaluations, not a reasonable number with a real computer simulation that takes any non-trivial amount of time to evaluate. Even at just 1s per evaluation, presuming speedy but not instantaneous numerical simulation in a slightly more realistic setting, we're talking > 100 hours. Many solvers take minutes/hours/days to execute a single run. Even with great patience, or distributed evaluation in an HPC setting, we'd only really know about pairs. How about main effects or three-way interactions? A different strategy is needed.

1.2.2 Surrogate modeling and design

Many of the most effective strategies involve (meta-) modeling the computer model. The setting is as follows. Computer model $f(x) : \mathbb{R}^p \to \mathbb{R}$ is expensive to evaluate. For concreteness, take f to be `wingwt` from §1.2.1. To economize on expensive runs, avoiding a grid in each pair of coordinates say, the typical setup instead entails choosing a small design $X_n = \{x_1, \ldots, x_n\}$ of locations in the full m-dimensional space, where $m = 9$ for `wingwt`. Runs at those locations complete a set of n example evaluation pairs (x_i, y_i), where $y_i \sim f(x_i)$ for $i = 1, \ldots, n$. If f is deterministic, as it is with `wingwt`, then we may instead write $y_i = f(x_i)$. Collect the n data pairs as $D_n = (X_n, Y_n)$, where X_n is an $n \times m$ matrix and Y_n is an n-vector. Use these data to train a statistical (regression) model, producing an *emulator* $\hat{f}_n \equiv \hat{f} \mid D_n$ whose predictive equations may be used as a *surrogate* $\hat{f}_n(x')$ for $f(x')$ at novel x' locations in the m-dimensional input space.

Often the terms surrogate and emulator are used interchangeably, as synonyms. Some refer to the fit $\hat{f}_n$ as emulator and the evaluation of its predictive equations $\hat{f}_n(x')$ as surrogate, as I have done above. Perhaps the reasoning behind that is that the suffix "or" on emulator makes it like *estimator* in the statistical jargon. An estimator holds the potential to provide estimates when trained on a sample from a population. We may study properties of an estimator/emulator through its sampling distribution, inherited from the data-generating mechanism imparted on the population by the model, or through its Bayesian posterior distribution. Only after providing an x' to the predictive equations, and extracting some particular summary statistic like the mean, do we actually have a surrogate $\hat{f}(x')$ for $f(x')$, serving as a substitute for a real simulation.

If that doesn't make sense to you, then don't worry about it. Emulator and surrogate are the same thing. I find myself increasingly trying to avoid emulator in verbal communication because it confuses folks who work with another sort of computer emulator[14], virtualizing a hardware architecture in software. In that context the emulator is actually more cumbersome, requiring more flops-per-instruction than the real thing. You could say that's exactly the opposite of a key property of effective surrogate modeling, on which I'll have more to say shortly. But old habits die hard. I will not make any substantive distinction in this book except as verbiage supporting the narrative – my chatty writing style – prefers.

The important thing is that a good surrogate does about what f would do, and quickly. At risk of slight redundancy given the discussion above, good meta-models for computer simulations (dropping n subscripts)

a. provide a predictive distribution $\hat{f}(x')$ whose mean can be used as a surrogate for $f(x')$ at new x' locations and whose variance provides uncertainty estimates – intervals for $f(x')$ – that have good coverage properties;
b. may interpolate when computer model f is deterministic;
c. can be used in any way f could have been used, qualified with appropriate *uncertainty quantification* (i.e., bullet #a mapped to the intended use, say to optimize);
d. and finally, fitting $\hat{f}$ and making predictions $\hat{f}(x')$ should be much faster than working directly with $f(x')$.

I'll say bullet #d a third (maybe fourth?) time because it's so important and often overlooked when discussing use of meta-models for computer simulation experiments. There's no point in all this extra modeling and analysis if the surrogate is slower than the real thing. Much of the discussion in the literature focuses on #a–c because those points are more interesting mathematically. Computer experiments of old were small, so speed in training/evaluation were less serious concerns. These days, as experiments get big, computational thriftiness is beginning to get more (and much deserved) attention.

As ever in statistical modeling, choosing a design X_n is crucial to good performance, especially when data-gathering expense limits n. It might be tempting to base designs X_n on a grid. But that won't work in our 9-dimensional `wingwt` exercise, at least not at any reasonable resolution. Even having a modest ten grid elements per dimension would balloon into $n = 10^9 \equiv$ 1-billion runs of the computer code! *Space-filling* designs were created to mimic the spread of grids, while sacrificing regularity in order to dramatically reduce size. Chapter 4 covers several options that could work well in this setting. As a bit of foreshadowing, and to have something concrete to work with so we can finish the `wingwt` example, consider a so-called *Latin Hypercube sample* (LHS, see §4.1). An LHS is a random design which is better than totally uniform (say via direct application of `runif` in R) because it limits potential for "clumps" of runs in the input space. Maximal spread cannot be guaranteed with LHSs, but they're much easier to compute than *maximin* designs (§4.2) which maximize the minimum distance between design elements x_i.

Let's use a library routine to generate a $n = 1000$-sized LHS for our wing weight example in 9d.

---

[14]https://en.wikipedia.org/wiki/Emulator

```
library(lhs)
n <- 1000
X <- data.frame(randomLHS(n, 9))
names(X) <- names(formals(wingwt))
```

Figure 1.11 offers a projection of that design down into two dimensions, with red horizontal bars overlaid as a visual aid.

```
plot(X[,1:2], pch=19, cex=0.5)
abline(h=c(0.6, 0.8), col=2, lwd=2)
```

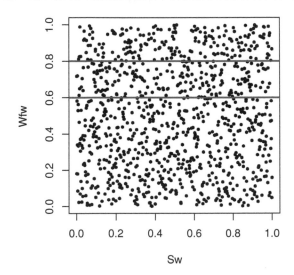

FIGURE 1.11: Projection of 9d LHS down to two dimensions with red bars demarking 20 percent of the spatial area.

Those horizontal lines, partitioning away one-fifth of the two-dimensional projection, help illustrate a nice property of LHS designs. Namely, an LHS guarantees marginals that are space-filling in a certain sense.

```
inbox <- X[,1] > 0.6 &  X[,1] < 0.8
sum(inbox)/nrow(X)
```

```
## [1] 0.2
```

Twenty percent of the design resides in a (contiguous) region occupying 20% of the volume of the study region. A grid could offer the same guarantee, but without such diversity in settings along the margin. Projections of a grid down into lower dimensions creates replicates. Something similar happens under maximin designs, although to a lesser degree. Such properties, and other points of comparison and contrast are left to Chapter 4. For now, let's take this design as given, as a decent choice, and feed it through our simulator.

```
Y <- wingwt(X[,1], X[,2], X[,3], X[,4], X[,5], X[,6], X[,7], X[,8], X[,9])
```

Ok now, what to do with that? We have 1000 evaluations Y_n at design X_n in 9d. How can that data be used to learn about input–output dynamics? We need to fit a model. The simplest option is a linear regression. Although that approach wouldn't exactly be the recommended course of action in a classical RSM setting, because the emphasis isn't local and we didn't utilize an RSM design, it's easy to see how that enterprise (even done right) would be fraught with challenges, to put it mildly. Least squares doesn't leverage the low/zero noise in these deterministic simulations, so this is really curve fitting[15] as opposed to genuine regression. The relationship is obviously nonlinear and involves many interactions. A full second-order model could require up to 54 coefficients. Looking at Eq. (1.1) it might be wise to model $\log Y_n$, but that's cheating!

Just to see what might come up, let's fit a parsimonious first-order model with interactions, cheating with the log response, using backward step-wise elimination with BIC[16]. Other alternatives include AIC[17] with k=2 below, or F-testing[18].

```
fit.lm <- lm(log(Y) ~ .^2, data=data.frame(Y,X))
fit.lmstep <- step(fit.lm, scope=formula(fit.lm), direction="backward",
  k=log(length(Y)), trace=0)
```

Although it's difficult to precisely anticipate which variables would be selected in this Rmarkdown build, because the design X_n is random, typically about ten input coordinates are selected, mostly main effects and sometimes interactions.

```
coef(fit.lmstep)
```

```
## (Intercept)          Sw           A           q           l
##    5.082505    0.218905    0.303856   -0.008359    0.030108
##         Rtc          Nz         Wdg        q:Nz
##   -0.237338    0.407365    0.188313    0.022064
```

Although not guaranteed, again due to randomness in the design, it's quite unlikely that the A:Nz interaction is chosen. Check to see if it appears in the list above. Yet we know that that interaction is present, as it's one of the pairs we studied in §1.2.1. Since this approach is easy to dismiss on many grounds it's not worth delving too deeply into potential remedies. Rather, let's use it as an excuse to recommend something new.

Gaussian process (GP) models have percolated up the hierarchy for nonlinear nonparametric regression in many fields, especially when modeling real-valued input–output relationships believed to be smooth. Machine learning, spatial/geo-statistics, and computer experiments are prime examples. Actually GPs can be characterized as linear models, in a certain sense, so they're not altogether new to the regression arsenal. However GPs privilege modeling through a covariance structure, rather than through the mean, which allows for more fine control over signal-versus-noise and for nonlinearities to manifest in a relative (i.e., velocity or differences) rather than absolute (position) sense. The details are left to Chapter 5. For now let's borrow a library I like.

---

[15]https://en.wikipedia.org/wiki/Curve_fitting
[16]https://en.wikipedia.org/wiki/Bayesian_information_criterion
[17]https://en.wikipedia.org/wiki/Akaike_information_criterion
[18]https://en.wikipedia.org/wiki/F_test

```
library(laGP)
```

Fitting, for a certain class of GP models, can be performed with the code below.

```
fit.gp <- newGPsep(X, Y, 2, 1e-6, dK=TRUE)
mle <- mleGPsep(fit.gp)
```

A disadvantage to GPs is that inspecting estimated coefficients isn't directly helpful for understanding. This is a well-known drawback of nonparametric methods. Still, much can be gleaned from the predictive distribution, which is what we would use as a surrogate for new evaluations under the fitted model. Code below sets up a predictive matrix XX comprised of baseline settings for seven of the inputs, combined with our 100×100 grid in 2d to span combinations of the remaining two inputs: A and Nz.

```
baseline <- matrix(rep(as.numeric(formals(wingwt)), nrow(g)),
  ncol=9, byrow=TRUE)
XX <- data.frame(baseline)
names(XX) <- names(X)
XX$A <- g[,1]
XX$Nz <- g[,2]
```

That testing design can be fed into predictive equations derived for our fitted GP.

```
p <- predGPsep(fit.gp, XX, lite=TRUE)
```

Figure 1.12 shows the resulting surface, which is visually identical to the one in Figure 1.9, based on 10K direct evaluations of wingwt.

```
image(x, x, matrix(p$mean, ncol=length(x)), col=cs, breaks=bs,
  xlab="A", ylab="Nz")
contour(x, x, matrix(p$mean, ncol=length(x)), add=TRUE)
```

What's the point of this near-duplicate plot? Well, I think its pretty amazing that 1K evaluations in 9d, paired with a flexible surrogate, can do the work of 10K in 2d! We've not only reduced the number of evaluations required for a pairwise input analysis, but we have a framework in place that can provide similar surfaces for all other pairs without further wingwt evaluation.

What else can we do? We can use the surrogate, via **predGPsep** in this case, to do whatever wingwt could do! How about a sensitivity analysis via main effects (§8.2)? As one example, the code below reinitializes XX to **baseline** settings and then loops over each input coordinate replacing its configuration with the elements of a one-dimensional grid. Predictions are then made for all XX, with means and quantiles saved. The result is nine sets of three curves which can be plotted on a common axis, namely of coded units in $[0, 1]$.

```
meq1 <- meq2 <- me <- matrix(NA, nrow=length(x), ncol=ncol(X))
for(i in 1:ncol(me)) {
```

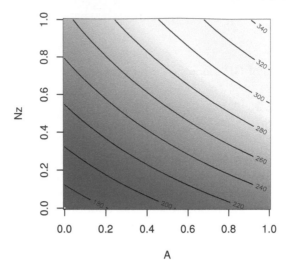

FIGURE 1.12: Surrogate for wing weight over an interesting 2d slice; compare to Figure 1.9.

```
  XX <- data.frame(baseline)[1:length(x),]
  XX[,i] <- x
  p <- predGPsep(fit.gp, XX, lite=TRUE)
  me[,i] <- p$mean
  meq1[,i] <- qt(0.05, p$df)*sqrt(p$s2) + p$mean
  meq2[,i] <- qt(0.95, p$df)*sqrt(p$s2) + p$mean
}
```

Figure 1.13 shows these nine sets of curves. On the scale of the responses, quantiles summarizing uncertainty are barely visible around their means. So it really looks like nine thick lines. Uncertainty in these surrogate evaluations is low despite being trained on a rather sparse design of just $n = 1000$ points in 9d. This is, in part, because the surrogate is interpolating deterministically observed data values.

```
matplot(x, me, type="l", lwd=2, lty=1, col=1:9, xlab="coded input")
matlines(x, meq1, type="l", lwd=2, lty=2, col=1:9)
matlines(x, meq2, type="l", lwd=2, lty=2, col=1:9)
legend("topleft", names(X)[1:5], lty=1, col=1:5, horiz=TRUE,
  bty="n", cex=0.5)
legend("bottomright", names(X)[6:9], lty=1, col=6:9, horiz=TRUE,
  bty="n", cex=0.5)
```

Some would call that a main effects plot. Others might quibble that main effects must integrate over the other (in this case eight) coordinates, rather than fixing them at baseline values. See §8.2.2 for details. Both are correct – what you prefer to look at is always a matter of perspective. Regardless, many would agree that much can be gleaned from plots such as the one in Figure 1.13. For example, we see that W_{fw}, Λ, q, and λ barely matter, at least in terms of departure from baseline. Only R_{tc} has a negative effect. Nonlinearities are slight

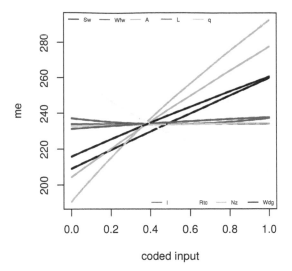

FIGURE 1.13: Main effects for the nine wing weight inputs.

in terms of main effects, as they were with pairwise interactions. It's hard to visualize in higher dimension, but there are lots more tools to help here.

1.3 A road map

We'll learn in detail about GPs and what they can and can't do. They've revolutionized machine learning, spatial statistics, and computer simulation experiments. But they're no panacea. They can be slow, which you may have noticed if you tried some of the examples above in your own R session. That's because they can involve big-matrix calculations. Even though GPs are super flexible, particularly by comparison to ordinary linear models, sometimes they're too rigid. They can over-smooth. Whether or not that's a big deal depends on your application domain. As kriging in geostatistics (Matheron, 1963), where input spaces are typically two-dimensional (longitude and latitude), these limitations are more easily hurdled. Low-dimensional input spaces emit several alternative formulations which are more convenient to work with both mathematically and computationally. §5.4.1 covers *process convolutions* which form the basis of many attractive alternatives in this setting.

When the computer experiments crowd caught on, and realized GPs were just as good in higher dimension, computer simulation efforts were relatively small by modern trends. Surrogate GP calculations were trivial – cubic $n \times n$ matrix decompositions with $n = 30$-odd runs is no big deal (Morris et al., 1993). Nobody worried about nonstationarity. With such a small amount of data there was already limited scope to learn a single, unified dynamic, let alone dynamics evolving in the input space. In the late 1990s, early 2000s, the machine learning community latched on but their data were really big, so they had to get creative. To circumvent big-matrix calculations they induced sparsity in the GP covariance structure, and leveraged sparse-matrix linear algebra libraries. Or they created designs X_n with special structure that allowed big-matrix calculations to be avoided all together. Several schemes revolving around partition-based regression via trees, Voronoi tessellations and nearest neighbors have taken hold, offering a divide-and-conquer approach which addresses computational and flexibility limitations in one fell swoop. This has paved the way for even thriftier approximations that retain many of the attractive features of GPs (non-linear, appropriate out-of-sample coverage) while sacrificing some of their beauty (e.g., smooth surfaces). When computer experiments started getting big too, around the late noughties (2000s), industrial and engineering scientists began to appropriate the best machine learning ideas for surrogate modeling. And that takes us to where we are today.

The goal for the remaining chapters of this book is to traverse that arc of methodology. We'll start with classical RSM, and then transition to GPs and their contemporary application in several important contexts including optimization, sensitivity analysis and calibration, and finally finish up with high-powered variations designed to cope with simulation data and modeling fidelity at scale. Along the way we shall revisit themes that serve as important pillars supporting best practice in scientific discovery from data. Interplay between mathematical models, numerical approximation, simulation, computer experiments, and field data, will be core throughout. We'll allude to the sometimes nebulous concept of uncertainty quantification[19], awkwardly positioned at the intersection of probabilistic and dynamical modeling. You'd think that stats would have a monopoly on quantifying uncertainty, but not all sources of uncertainty are statistical in nature (i.e., not all derive from sampling properties of observed data). Experimental design will play a key role in our developments, but not in the classical RSM sense. We'll focus on appropriate designs for GP surrogate modeling, emphasizing sequential refinement and augmentation of data toward particular learning goals. This is what the machine learning community calls active learning[20] or reinforcement learning[21].

To motivate, Chapter 2 overviews four challenging real-data/real-simulation examples benefiting from modern surrogate modeling methodology. These include rocket boosters, groundwater remediation, radiative shock hydrodynamics (nebulae formation) and satellite drag in low Earth orbit. Each is a little vignette illustrated through working examples, linked to data provided as supplementary material. At first the exposition is purely exploratory. The goal is to revisit these later, once appropriate methodology is in place.

Better than the real thing

If done right, a surrogate can be even better than the real thing: smoothing over noisy or chaotic behavior, furnishing a notion of derivative far more reliable than other numerical approximations for optimization, and more. Useful surrogates are not just the stuff of a dystopian future in science fiction[22]. This monograph is a perfect example. That's not to say I believe it's the best book on these subjects, though I hope you'll like it. Rather, the book is a surrogate for me delivering this material to you, in person and in the flesh. I get tired and cranky after a while. This book is in most ways, or as a teaching tool, better than actual me; i.e., better than the real thing. Plus it's hard to loan me out to a keen graduate student.

What are the properties of a good surrogate? I gave you my list above in §1.2.2, but I hope you'll gain enough experience with this book to come up with your own criteria. Should the surrogate do exactly what the computer simulation would do, only faster and with error-bars? Should a casual observer be able to tell the difference between the surrogate and

---

[19]https://en.wikipedia.org/wiki/Uncertainty_quantification
[20]https://en.wikipedia.org/wiki/Active_learning_(machine_learning)
[21]https://en.wikipedia.org/wiki/Reinforcement_learning
[22]https://en.wikipedia.org/wiki/Surrogates

the real thing? Are surrogates one-size-fits all, or is it sensible to build surrogates differently for different use cases? Politicians often use surrogates – community or business leaders, other politicians – as a means of appealing to disparate interest groups, especially at election time. By emphasizing certain aspects while downplaying or smoothing over others, a diverse set of statistical surrogate models can sometimes serve the same purpose: pandering to bias while greasing the wheels of (scientific) progress. Your reaction to such tactics, on ethical grounds, may at first be one of visceral disgust. Mine was too when I was starting out. But I've learned that the best statistical models, and the best surrogates, give scientists the power to focus, infer and explore simultaneously. Bias is all but unavoidable. Rarely is one approach sufficient for all purposes.

1.4 Homework exercises

These exercises are designed to foreshadow themes from later chapters in light of the overview provided here.

#1: Regression

The file wire.csv[23] contains data relating the pull strength (**pstren**) of a wire bond (which we'll treat as a response) to six characteristics which we shall treat as design variables: die height (**dieh**), post height (**posth**), loop height (**looph**), wire length (**wlen**), bond width on the die (**diew**), and bond width on the post (**postw**). (Derived from exercise 2.3 in Myers et al. (2016) using data from Table E2.1.)

  a. Write code that converts natural variables in the file to coded variables on the unit cube. Also, normalize responses to have a mean of zero and a range of 1.
  b. Use model selection techniques to select a parsimonious linear model for the coded data including, potentially, second-order and interaction effects.
  c. Use the fitted model to make a prediction for pull strength, when the explanatory variables take on the values c(6, 20, 30, 90, 2, 2), in the order above, with a full accounting of uncertainty. Make sure the predictive quantities are on the original scale of the data.

#2: Surrogates for sensitivity

Consider the so-called piston simulation function[24] which was at one time a popular benchmark problem in the computer experiments literature. (That link, to Surjanovic and Bingham (2013)'s Virtual Library of Simulation Experiments (VLSE)[25], provides references and further detail. VLSE is a great resource for test functions and computer simulation experiment data; there's a page for the wing weight example[26] as well.) Response $C(x)$ is the cycle time, in seconds, of the circular motion of a piston within a gas-filled cylinder, the configuration of which is described by seven-dimensional input vector $x = (M, S, V_0, k, P_0, T_a, T_0)$.

---

[23]http://bobby.gramacy.com/surrogates/wire.csv
[24]http://www.sfu.ca/~ssurjano/piston.html
[25]http://www.sfu.ca/~ssurjano/index.html
[26]http://www.sfu.ca/~ssurjano/wingweight.html

TABLE 1.2: Piston parameters.

Symbol	Parameter	Baseline	Minimum	Maximum
M	Piston weight (kg)	45	30	60
S	Piston surface area (m^2)	0.0125	0.005	0.020
V_0	Initial gas volume (m^2)	0.006	0.002	0.010
k	Spring coefficient (N/M)	3000	1000	5000
P_0	Atmospheric pressure (N/m^2)	100000	90000	110000
T_a	Ambient temperature (K)	293	290	296
T_0	Filling gas temperature (K)	350	340	3600

$$C(x) = 2\pi\sqrt{\frac{M}{k + S^2 \frac{P_0 V_0}{T_0} \frac{T_a}{V^2}}}, \quad \text{where} \quad V = \frac{S}{2k}\left(\sqrt{A^2 + 4k\frac{P_0 V_0}{T_0}T_a} - A\right)$$

$$\text{and} \quad A = P_0 S + 19.62M - \frac{kV_0}{S}$$

Table 1.2 describes the input coordinates of x, their ranges, and provides a baseline value derived from the middle of the specified range(s).

Explore $C(x)$ with techniques similar to those used on the wing weight example (§1.2.1). Start with a space-filling (LHS) design in 7d and fit a GP surrogate to the responses. Use predictive equations to explore main effects and interactions between pairs of inputs. In your solution, rather than showing all $\binom{7}{2} = 21$ pairs, select one "interesting" and another "uninteresting" one and focus your presentation on those two. How do your surrogate predictions for those pairs compare to an exhaustive 2d grid-based evaluation and visualization of $C(x)$?

#3: Optimization

Consider two-dimensional functions f and c, defined over $[0,1]^2$; f is a re-scaled version of the so-called Goldstein–Price[27] function, and is defined in terms of auxiliary functions a and b.

$$f(x) = \frac{\log\left[(1 + a(x))(30 + b(x))\right] - 8.69}{2.43} \quad \text{with}$$

$$a(x) = (4x_1 + 4x_2 - 3)^2$$
$$\times \left[75 - 56(x_1 + x_2) + 3(4x_1 - 2)^2 + 6(4x_1 - 2)(4x_2 - 2) + 3(4x_2 - 2)^2\right]$$
$$b(x) = (8x_1 - 12x_2 + 2)^2$$
$$\times \left[-14 - 128x_1 + 12(4x_1 - 2)^2 + 192x_2 - 36(4x_1 - 2)(4x_2 - 2) + 27(4x_2 - 2)^2\right]$$

Separately, let a "constraint" function c be defined as

$$c(x) = \frac{3}{2} - x_1 - 2x_2 - \frac{1}{2}\sin(2\pi(x_1^2 - 2x_2)).$$

a. Evaluate f on a grid and make an image and/or image contour plot of the surface.

---

[27]http://www.sfu.ca/~ssurjano/goldpr.html

b. Use a library routine (e.g., `optim` in R) to find the global minimum. When optimizing, pretend you don't know the form of the function; i.e., treat it as a "blackbox". Initialize your search randomly and comment on the behavior over repeated random restarts. About how many evaluations does it take to find the local optimum in each initialization repeat; about how many to reliably find the global one across repeats?

c. Now, re-create your plot from #a with contours only (no image), and then add color to the plot indicating the region(s) where $c(x) > 0$ and $c(x) \leq 0$, respectively. *To keep it simple, choose white for the latter, say.*

d. Use a library routine (e.g., `nloptr` in R) to solve the following constrained optimization problem: min $f(x)$ such that $c(x) \leq 0$ and $x \in [0, 1]^2$. Initialize your search randomly and comment on the behavior over repeated random restarts. About how many evaluations does it take to find the local valid optimum in each initialization repeat; about how many to reliably find the global one across repeats?

2

Four Motivating Datasets

This chapter aims to whet the appetite for modern surrogate modeling technology by introducing four challenging real-data settings. Each comes with a brief description of the data and application and a cursory exploratory analysis. Domains span aeronautics, groundwater remediation, satellite orbit and positioning, and cosmology. A small taste of "methods-in-action" is offered, focused on one or more of the typical goals in each setting. Together, these data and domains exhibit features spanning many of the hottest topics in surrogate modeling.

For example, settings may involve limited or no field data on complicated physical processes, which in turn must be evaluated with computationally expensive simulation. Simulations might require evaluation on supercomputers to produce data on a scale adequate for conventional analysis. Goals might range from understanding, to optimization, sensitivity analysis and/or uncertainty quantification (UQ). In each case, a pointer to the data files or archive is provided, or a description of how to build libraries (and run them) to create new data from live simulations.

These data sets, or their underlying processes, will be revisited throughout subsequent chapters/homework exercises to motivate and illustrate methods therein. With most of our early examples being toy in nature, carefully crafted to offer a controlled look at particular aspects of methodology – and at times combining several ideas interlocked in a complex weave – these motivating datasets can simultaneously serve as anchors to the real world, and delicious treats offered up as rewards at the end of a long slog of arduous development and implementation.

2.1 Rocket booster dynamics

Before the space shuttle program was terminated, the National Aeronautics and Space Administration (NASA) proposed a re-usable rocket booster that could be recovered after depositing its payload into orbit. One project was called the Langley Glide-Back Booster (LGBB). The idea was to have the LGBB glide back to Earth and be cheaply refurbished, rather than simply plummeting into the ocean and becoming scrap metal. Before building prototypes, at enormous expense to taxpayers, NASA designed a computer simulation to explore the dynamics of the booster in a variety of synthesized environments and design configurations. Simulations entailed solving computational fluid dynamics (CFD) systems of differential equations, and the primary study regime focused on dynamics upon re-entry into the atmosphere.

More on the simulations, models, and other details can be found in several publications (Rogers et al., 2003; Pamadi et al., 2004b,a). The brief description here is a caricature by comparison, but the data and properties revealed lack no degree of realism or intricacy. The simulator had three inputs which describe configuration of the booster at re-entry: speed (`mach`), angle of attack (`alpha`) and slide-slip angle (`beta`). The simulator utilized an Euler solver via Cart3D[1] and, for each input setting, provided six aeronautically relevant outputs describing forces exerted on the rocket in that configuration: lift, drag, pitching moment, side-force, yaw and roll. Circa late 1990s, when LGBB simulations were first being performed, and refined, each input configuration took between five and 20 hours of wall-clock time to evaluate. Larger experimental designs (i.e., comprising many triples of inputs) required substantial effort to distribute over nodes of the Columbia supercomputer[2].

To build intuition visually, imagine the solver "flying" a virtual rocket booster through a mesh, and forces accumulating at the boundary between booster and mesh. Figure 2.1 shows a virtual rocket booster flying through a mesh customized for subsonic (`mach < 1`) cases.

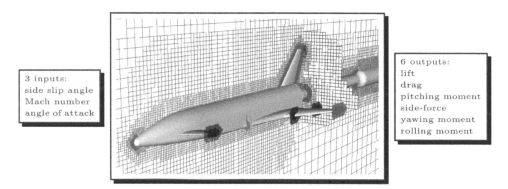

FIGURE 2.1: Drawing of the LGBB computational fluid dynamics computer model simulation. Adapted from Rogers et al. (2003); used with permission from the authors.

2.1.1   Data

There are two historical versions of the LGBB data, and one "surrogate" version, recording collections of input–output pairs gathered on various input designs and under a cascade of improvements to meshes used by the underlying Cart3D solver. The first, oldest version of the data was derived from a less reliable code implementing the solver. That code was evaluated on hand-designed input configuration grids built-up in batches, each offering a refinement on certain locales of interest in the input space. Researchers at NASA determined, on the basis of visualizations and regressions performed along the way, that denser sampling was required in order to adequately characterize input–output relationships in particular regions. For example, several batches emphasized the region nearby the sound barrier, transitioning between subsonic and supersonic regimes (at `mach=1`).

```
lgbb1 <- read.table("lgbb/lgbb_original.txt", header=TRUE)
names(lgbb1)
```

```
## [1] "mach"  "alpha" "beta"  "lift"  "drag"  "pitch" "side"  "yaw"
## [9] "roll"
```

---

[1]https://www.nas.nasa.gov/publications/software/docs/cart3d/
[2]https://en.wikipedia.org/wiki/Columbia_(supercomputer)

```
nrow(lgbb1)
```

```
## [1] 3167
```

Observe that inputs reside in the first three columns, with six outputs in subsequent columns. All together these data record results from 3167 simulations. Figure 2.2 provides a 2d visual of the first `lift` response after projecting over the third input `beta`. Simple linear interpolation via the `akima` library on CRAN (Akima et al., 2016) provides a degree of smoothing onto a regular grid for `image` plots. Lighter/whiter colors are higher values. Dots indicate the location of inputs.

```
library(akima)
g <- interp(lgbb1$mach, lgbb1$alpha, lgbb1$lift, dupl="mean")
image(g, col=heat.colors(128), xlab="mach", ylab="alpha")
points(lgbb1$mach, lgbb1$alpha, cex=0.25, pch=18)
```

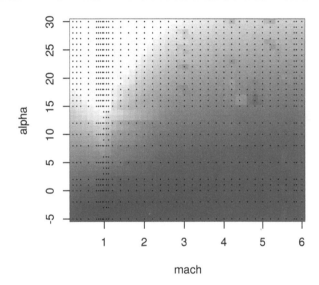

FIGURE 2.2: Heat plot of the lift response projecting over side-slip angle with design indicated by dots.

Projecting over `beta` tarnishes the utility of this visualization; however, note that this coordinate has relatively few unique values.

```
apply(lgbb1[,1:3], 2, function(x) { length(unique(x)) })
```

```
##   mach alpha  beta
##     37    33     6
```

Focused sampling at the sound barrier (`mach=1`) for large angles of attack (`alpha`), where the response is interesting, is readily evident in the figure. But working with grids has its drawbacks. Despite a relatively large number (3167) of rows in these data, observations are only obtained for a few dozen unique `mach` and `alpha` values, which may challenge extrapolation or visualization along lower-dimensional slices.

Numerical stability is also a concern. To illustrate both grid and numerical drawbacks, consider a subset of these data where `alpha == 1`. Two variations on that slice are shown in Figure 2.3, one based on `beta == 0` in solid-black and another where `beta != 0` in dashed-red.

```
a1b0 <- which(lgbb1$alpha == 1 & lgbb1$beta == 0)
a1bn0 <- which(lgbb1$alpha == 1 & lgbb1$beta != 0)
a1b0 <- a1b0[order(lgbb1$mach[a1b0])]
a1bn0 <- a1bn0[order(lgbb1$mach[a1bn0])]
plot(lgbb1$mach[a1b0], lgbb1$lift[a1b0], type="l", xlab="mach",
  ylab="lift", ylim=range(lgbb1$lift[c(a1b0, a1bn0)]), lwd=2)
lines(lgbb1$mach[a1bn0], lgbb1$lift[a1bn0], col=2, lty=2, lwd=2)
text(4, 0.3, paste("length(a1b0) =", length(a1b0)))
text(4, 0.25, paste("length(a1bn0) =", length(a1bn0)))
legend("topright", c("beta = 0", "beta != 0"), col=1:2, lty=1:2, lwd=2)
```

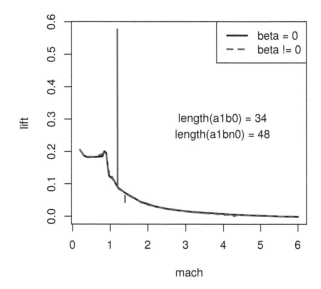

FIGURE 2.3: Lift slice with angle of attack fixed at 1 and side-slip angle fixed at zero (solid-black) and one (dashed-red). Counts of grid locations provided as overlayed text.

Clearly there are some issues with `lift` outputs when `beta != 0`. Also note the relatively low resolution, with each "curve" being traced out by just a handful of values – fewer than fifty in both cases. Consequently the input–output relationship looks blocky in the subsonic region (`mach < 1`). A second iteration on the experiment attempted to address all three issues simultaneously: a) an adaptive design without gridding; b) better numerics (improvements to Cart3D); and c) paired with an ability to back-out a high resolution surface, smoothing out the gaps, based on relatively few total simulations.

2.1.2 Sequential design and nonstationary surrogate modeling

The second version of the data summarizes results from that second experiment. Improved simulations were paired with model-based sequential design (§6.2) under a treed Gaussian process (TGP, §9.2.2) in order to obtain a more adaptive, automatic input "grid". These data may be read in as follows.

```
lgbb2 <- read.table("lgbb/lgbb_as.txt", header=TRUE)
```

A glimpse into the adaptive design of that experiment, which is again projected over the beta axis, is provided in Figure 2.4.

```
plot(lgbb2$mach, lgbb2$alpha, xlab="mach", ylab="alpha", pch=18, cex=0.5)
```

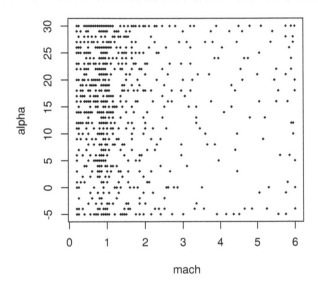

FIGURE 2.4: Adaptive LGBB design projected over side-slip angle.

In total there were 780 unique simulations (i.e., 780 dots in the figure), or less than 25% as many as the previous experiment. For reasons to do with the NASA simulation system, there was a grid underlying candidates for selection in this adaptive design, but one much finer than that used in the first experiment, particularly in the mach coordinate.

```
apply(lgbb2[,2:4], 2, function(x) { length(unique(x)) })
```

```
##   mach alpha  beta
##    110    36     9
```

Since the design was much smaller, slices like the one shown in Figure 2.5, mimicking Figure 2.3 but this time projecting over all beta-values in the design, appear blocky in raw form.

```
a2 <- which(lgbb2$alpha == 1)
a2 <- a2[order(lgbb2$mach[a2])]
plot(lgbb2$mach[a2], lgbb2$lift[a2], type="l", xlab="mach",
  ylab="lift", lwd=2)
text(4, 0.15, paste("length(a2) =", length(a2)))
```

The adaptive grid has a lower degree of axis alignment. Although deliberate, a downside is

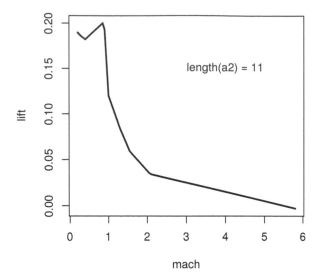

FIGURE 2.5: Lift versus angle of attack projected over side-slip angle in the adaptive design; compare to Figure 2.3.

that it's hard to tell if numerical instabilities are repaired in the update to Cart3D, or if instead resolution in the 1d slice is simply too low to reveal any issues. For now we'll have to suspend disbelief somewhat, at least when it comes to details of how the figures below were constructed from the 780 simulations. Suffice it to say that the TGP model can provide requisite extrapolations to reveal a smooth surface, in 3d or along any lower-dimensional slice desired. For example, Figure 2.6 shows a slice over `mach` and `alpha` when `beta=1`. The data behind this visual comes from an `.RData` file containing evaluations of the TGP surrogate, trained to 780 `lift` evaluations, on a dense predictive grid in the input space. The contents of this file represent the third, "surrogate data" source mentioned above.

```
load("lgbb/lgbb_fill.RData")
lgbb.b1 <- lgbb.fill[lgbb.fill$beta == 1, ]
g <- interp(lgbb.b1$mach, lgbb.b1$alpha, lgbb.b1$lift)
image(g, col=heat.colors(128), xlab="mach [beta=1]", ylab="alpha [beta=1]")
```

Notice how this view reveals a nice smooth surface with simple dynamics in high-speed regions, and a more complex relationship near the sound barrier – in particular for low speeds (`mach`) and high angles of attack (`alpha`). A suite of 1d slices shows a similar picture. Figure 2.7 utilizes the sequence of unique `alpha` values in the design, showing a prediction from TGP where each is paired with the full range of `mach` values.

```
plot(lgbb.b1$mach, lgbb.b1$lift, type="n", xlab="mach",
  ylab="lift [beta=1]")
for(ub in unique(lgbb.b1$alpha)) {
  a <- which(lgbb.b1$alpha == ub)
  a <- a[order(lgbb.b1$mach[a])]
  lines(lgbb.b1$mach[a], lgbb.b1$lift[a], type="l", lwd=2)
}
```

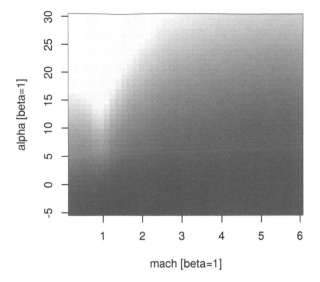

FIGURE 2.6: Heat plot slicing the lift response through side-slip angle one, illustrating an adaptive-grid surrogate version of Figure 2.2.

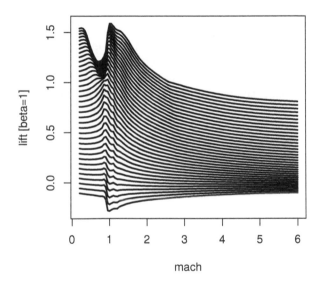

FIGURE 2.7: Multi-slice angle of attack analog of Figure 2.6.

This view conveys nuance along the sound barrier more clearly than the previous image plot did. Apparently it was worthwhile sampling more heavily in that region of the input space, relatively speaking, compared to say `mach > 2` for any angle of attack (`alpha`). At the time, building designs that automatically detected the interesting part of the input space was revolutionary. Actually, the real advance is in modeling. We need an apparatus that's simultaneously flexible enough to learn relevant dynamics in the data, but thrifty enough to accommodate calculations required for inference in reasonable time and space (i.e., computer memory). Once an appropriate model is in place, the problem of design becomes one of backing out relevant uncertainty measures, depending on the goal of the experiment(s). In this case, where understanding dynamics is key, design is a simple matter of putting more runs where uncertainty is high.

Now this discussion treated only the `lift` output – there were five others. A homework exercise (§2.6) invites the curious reader to create a similar suite of visuals for the other responses. Later we shall use these data to motivate methodology for nonstationary spatial modeling and sequential design, and as a benchmark in comparative exercises. Examples are scattered throughout the latter chapters, with the most complete treatment being in §9.2.2.

2.2 Radiative shock hydrodynamics

Radiative shocks arise from astrophysical phenomena, e.g., supernovae and other high temperature systems. These are shocks where radiation from shocked matter dominates energy transport and results in a complex evolutionary structure. See, e.g., McClarren et al. (2011), and Drake et al. (2011). The University of Michigan's Center for Radiative Shock Hydrodynamics (CRASH) is tasked with modeling a particular high-energy laser radiative shock system. They developed a mathematical model and computer implementation that simulates a field apparatus, located at the Omega laser facility[3] at the University of Rochester. That apparatus was used to conduct a limited real experiment involving twenty runs.

The basic setup of the experiment(s) is as follows. A high-energy laser irradiates a beryllium disk located at the front end of a xenon (Xe) filled tube, launching a high-speed shock wave into the tube. The left panel in Figure 2.8 shows a schematic of the apparatus. The shock is said to be a radiative shock if the energy flux emitted by the hot shocked material is equal to, or larger than, the flux of kinetic energy into the shock. Each physical observation is a radiograph image, shown in the right panel of the figure, and a quantity of interest is shock location: distance traveled after a predetermined time.

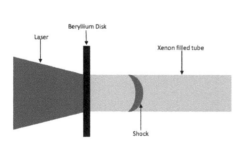
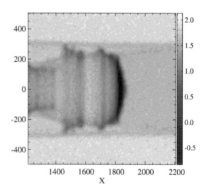

FIGURE 2.8: Schematic of the CRASH experimental apparatus (left); radiograph image of a shock as it moves through the Xe filled tube (right). Adapted from Goh et al. (2013) and used with permission from the American Statistical Association.

---

[3]http://www.lle.rochester.edu/omega_facility/

TABLE 2.1: CRASH experiment parameters.

Design Parameter	CE1	CE2	Field Design
Be thick (microns)	[18, 22]	21	21
Xe fill press (atm)	[1.1, 1.2032]	[0.852, 1.46]	[1.032, 1.311]
Time (nano-secs)	[5, 27]	[5.5, 27]	6-values in [13, 28]
Tube diam (microns)	575	[575, 1150]	{575, 1150}
Taper len (microns)	500	[460, 540]	500
Nozzle len (microns)	500	[400, 600]	500
Aspect ratio (microns)	1	[1, 2]	1
Laser energy (J)	[3600, 3900]		[3750, 3889.6]
Effective laser energy (J)		[2156.4, 4060]	

2.2.1 Data

The experiments involve nine design variables, listed in Table 2.1 along with ranges or values used in the field experiment in the final column. The first three variables specify thickness of the beryllium disk, xenon fill pressure in the tube and observation time for the radiograph image, respectively. The next four variables are related to tube geometry and the shape of the apparatus at its front end. Most of the physical experiments were performed on circular shock tubes with a small diameter (in the area of 575 microns), and the remaining experiments were conducted on circular tubes with a diameter of 1150 microns or with different nozzle configurations. Aspect ratio describes the shape of the tube: circular or oval. In the field experiment the aspect ratios are all 1, indicating a circular tube. However there's interest in extrapolating to oval-shaped tubes with an aspect ratio of 2. Finally, laser energy is specified in Joules. Effective laser energy, and its relationship to ordinary laser energy, requires some back story.

R code below reads in data from the field experiment. Thickness of the beryllium disk was not recorded in the data file, so this value is manually added in.

```
crash <- read.csv("crash/CRASHExpt_clean.csv")
crash$BeThickness <- 21
names(crash)
```

```
## [1] "LaserEnergy"    "GasPressure"   "AspectRatio"   "NozzleLength"
## [5] "TaperLength"    "TubeDiameter"  "Time"          "ShockLocation"
## [9] "BeThickness"
```

The field experiment is rather small, despite interest in exploring a rather large number (9) of inputs.

```
nrow(crash)
```

```
## [1] 20
```

Two computer experiment simulation campaigns were performed (CE1 and CE2 in Table 2.1) on supercomputers at Lawrence Livermore and Los Alamos National Laboratories. The second and third columns of the table reveal differing input ranges in the two computer experiments. CE1 explores small, circular tubes; CE2 investigates a similar input region, but

TABLE 2.2: CRASH calibration parameters.

Calibration Parameter	CE1	CE2	Field Design
Electron flux limiter	[0.04, 0.10]	0.06	
Energy scale factor	[0.40, 1.10]	[0.60, 1.00]	

also varies tube diameter and nozzle geometry. Both input plans were derived from Latin hypercube samples (LHSs, see §4.1). Thickness of the beryllium (Be) disk could be held constant in CE2 thanks to improvements in manufacturing in the time between simulation campaigns.

The computer simulator required two further inputs which could not be controlled in the field, i.e., two *calibration parameters*: electron flux limiter and laser energy scale factor, whose ranges are described in Table 2.2. It's quite typical for computer models to contain "knobs" which allow practitioners to vary aspects of the dynamics which are unknown, or can't be controlled in the field. In this particular case, electron flux limiter is an unknown constant involved in predicting the amount of heat transferred between cells of a space–time mesh used by the code. It was held constant in CE2 because in CE1 the outputs were found to be relatively insensitive to this input. Laser energy scale factor accounts for discrepancies between amounts of energy transferred to the shock in the simulations and experiments, respectively.

In the physical system the laser energy for a shock is recorded by a technician. Things are a little more complicated for the simulations. Before running CE1, CRASH researchers speculated that simulated shocks would be driven too far down the tube for any specified laser energy. So effective laser energy – the laser energy actually entered into the code – was constructed from two input variables, laser energy and a scale factor. For CE1 these two inputs were varied over ranges specified in the second column of Table 2.2. CE2 used effective laser energy directly. R code below reads in the data from these two computer experiments. Electron flux limiter was miscoded in the data file, being off by a factor of ten. A correction is applied below.

```
ce1 <- read.csv("crash/RS12_SLwithUnnormalizedInputs.csv")
ce2 <- read.csv("crash/RS13Minor_SLwithUnnormalizedInputs.csv")
ce2$ElectronFluxLimiter <- 0.06
```

Using both CE1 and CE2 data sources requires reconciling the designs of the two experiments. One way forward entails expanding CE2's design by gridding values of laser energy scale factor and pairing them with values of laser energy deduced from effective laser energy values contained in the original design. The gridding scheme implemented in the code below constrains scale factors to be less than one but no smaller than value(s) which, when multiplied by effective laser energy (in reciprocal), imply a laser energy of 5000 Joules.

```
sfmin <- ce2$EffectiveLaserEnergy/5000
sflen <- 10
ce2.sf <- matrix(NA, nrow=sflen*nrow(ce2), ncol=ncol(ce2) + 2)
for(i in 1:sflen) {
  sfi <- sfmin + (1 - sfmin)*(i/sflen)
  ce2.sf[(i - 1)*nrow(ce2) + (1:nrow(ce2)),] <-
```

```
    cbind(as.matrix(ce2), sfi, ce2$EffectiveLaserEnergy/sfi)
}
ce2.sf <- as.data.frame(ce2.sf)
names(ce2.sf) <- c(names(ce2), "EnergyScaleFactor", "LaserEnergy")
```

Figure 2.9 provides a visualization of that expansion.

```
plot(ce2.sf$LaserEnergy, ce2.sf$EnergyScaleFactor, ylim=c(0.4, 1.1),
   xlab="Laser Energy", ylab="Energy Scale Factor")
points(ce1$LaserEnergy, ce1$EnergyScaleFactor, col=2, pch=19)
legend("bottomleft", c("CE2", "CE1"), col=1:2, pch=c(21,19))
```

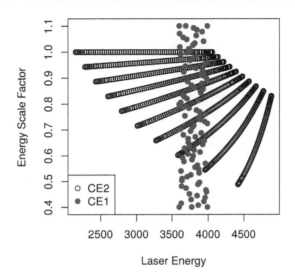

FIGURE 2.9: Expansion of inputs to resolve laser energy with its scale factor.

Subsequent combination with CE1 led to 26,458 input–output combinations.

2.2.2 Computer model calibration

What are typical goals for data/experiments of this kind? One challenge is to identify a modeling apparatus that can cope with data sizes like those described above, while maintaining the richness of fidelity required to describe and learn underlying dynamics. This is a serious challenge because many canonical methods for nonlinear modeling don't cope well with big data (i.e., more than a few thousand runs) when input dimensions are modest-to-large in size (e.g., bigger than 2d). Supposing that first hurdle is surmountable, some context-specific goals include a) learning settings of the (in this case) two-dimensional calibration parameter; and b) simultaneously determining the nature of bias in computer model runs, relative to field data observations, under those setting(s). Some specific questions might be: Are field data informative about that setting? Is down-scaling of laser energy necessary in CE1?

One may ultimately wish to furnish practitioners with a high-quality predictor for field data measurements in novel input conditions. We may wish to utilize the calibrated and

TABLE 2.3: Linear regression summary for field data.

	Estimate	Std. Error	t value	Pr(>|t|)
(Intercept)	2.507e+03	6.275e+03	0.3995	0.6951
LaserEnergy	-3.968e-01	1.491e+00	-0.2661	0.7938
GasPressure	-1.971e+02	8.477e+02	-0.2325	0.8193
TubeDiameter	-3.420e-02	4.068e-01	-0.0842	0.9340
Time	1.040e+11	1.567e+10	6.6406	0.0000

bias-corrected surrogate to extrapolate forecasts to oval-shaped disks, heavily leaning on the computer model simulations under those regimes and with full UQ.

To demonstrate potential, but also expose challenges inherent in such an enterprise, let's consider simple linear modeling of field and computer simulation data. The field dataset is very small, especially relative to its input dimension. Moreover, only four of the explanatory variables (i.e., besides the response `ShockLocation`) have more than one unique value.

```
u <- apply(crash, 2, function(x) { length(unique(x)) })
u
```

```
##    LaserEnergy   GasPressure   AspectRatio  NozzleLength   TaperLength
##             13            11             1             1             1
##   TubeDiameter          Time  ShockLocation   BeThickness
##              2             6            20             1
```

A linear model indicates that only `Time` has a substantial main effect. See Table 2.3.

```
fit <- lm(ShockLocation ~ ., data=crash[, u > 1])
kable(summary(fit)$coefficients,
  caption='Linear regression summary for field data.')
```

This is perhaps not surprising: the longer you wait the farther the shock will progress down the tube. In fact, `Time` mops up nearly all of the variability in these data with $R^2 = 0.97$, which is nicely illustrated by the visualization of data and fit provided in Figure 2.10.

```
fit.time <- lm(ShockLocation ~ Time, data=crash)
plot(crash$Time, crash$ShockLocation, xlab="time", ylab="location")
abline(fit.time)
```

It would appear that there isn't much scope for further information coming from data on the field experiment alone. Now let's turn to data from computer simulation. To keep the exposition simple, consider just CE1 which varied all but four parameters. A homework exercise (see §2.6) targets data combined from both computer experiments.

```
ce1 <- ce1[,-1] ## first col is FileNumber
u.ce1 <- apply(ce1, 2, function(x) { length(unique(x)) })
u.ce1
```

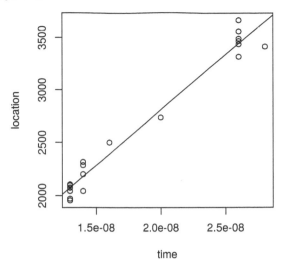

FIGURE 2.10: Time dominates predictors in a linear model fit to field data alone.

TABLE 2.4: Summary of linear regression fit to CE1.

	Estimate	Std. Error	t value	Pr(>|t|)
(Intercept)	-4.601e+02	1.321e+02	-3.483	0.0005
BeThickness	-7.595e+01	2.104e+00	-36.101	0.0000
LaserEnergy	3.153e-01	2.150e-02	14.672	0.0000
GasPressure	-3.829e+02	8.129e+01	-4.710	0.0000
Time	1.344e+11	4.998e+08	268.844	0.0000
ElectronFluxLimiter	4.126e+02	1.400e+02	2.947	0.0032
EnergyScaleFactor	1.776e+03	1.214e+01	146.249	0.0000

```
##          BeThickness        LaserEnergy          GasPressure
##                   96                 96                   96
##          AspectRatio       NozzleLength          TaperLength
##                    1                  1                    1
##         TubeDiameter               Time  ElectronFluxLimiter
##                    1                 24                   96
##    EnergyScaleFactor      ShockLocation
##                   96               1723
```

Recall that actual laser energy in each run was scaled by `EnergyScaleFactor`, but let's ignore this nuance for the time being. In stark contrast to our similar analysis on the field data, an ordinary least squares fit summarized in Table 2.4 indicates that all main effects which were varied in CE1 are statistically significant.

```
fit.ce1 <- lm(ShockLocation ~ ., data=ce1[,u.ce1 > 1])
kable(summary(fit.ce1)$coefficients,
  caption="Summary of linear regression fit to CE1.")
```

These results suggest that the computer simulation data, and subsequent fits, could usefully augment data and fitted dynamics from the field. Data from CE2 tell a similar story, but for a partially disjoint collection of inputs. Focusing back on CE1, consider Figure 2.11 which offers a view into how shock location varies with time and (scaled) laser energy. The heat plot in the figure is examining a linear interpolation of raw CE1 data, but alternatively we could extract a similar surface from `predict` applied to our `fit.ce1` object. It's apparent from the image that energy and time work together to determine how far/quickly shocks travel. That makes sense intuitively, but wasn't evident in our analysis of the field data alone. Some sort of hybrid modeling apparatus is needed in order to peruse the potential for further such synergies.

```
x <- ce1$Time
y <- ce1$LaserEnergy * ce1$EnergyScaleFactor
g <- interp(x/max(x), y/max(y), ce1$ShockLocation, dupl="mean")
image(g, col=heat.colors(128), xlab="scaled time", ylab="scaled energy")
```

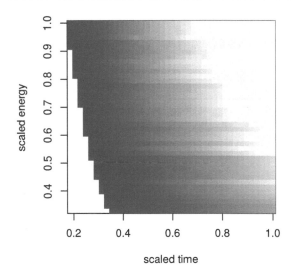

FIGURE 2.11: Energy and time interact to determine shock location.

Before delving headlong into that enterprise, it'll help to first get some of the base modeling components right. Linear modeling is likely insufficient: physical phenomena rarely covary linearly. With a wealth of simulation data we should be able to train a much more sophisticated meta-model. Plus even our linear model fits reveal that other variables matter besides energy and time. Chapter 8 details computer model calibration, combining a surrogate fit to a limited simulation campaign (Chapter 5) with a suite of methods for estimating bias between computer simulation and field observation, while at the same time determining the best setting of calibration parameters in order to "rein in" and correct for bias. With those elements in hand we'll be able to build a predictor which combines computer model surrogate with bias correction in order to develop a meta-model for the full suite of physical dynamics under study. Coping with a rather large simulation experiment in a modern surrogate modeling framework will require approximation. Chapter 9 revisits these data within an approximate surrogate–calibration framework.

2.3 Predicting satellite drag

Obtaining accurate estimates of satellite drag coefficients in low Earth orbit (LEO) is a crucial component in positioning (e.g., for scientists to plan experiments: what can be seen when?) and collision avoidance. Toward that end, researchers at Los Alamos National Laboratory (LANL) were tasked with predicting orbits for dozens of research satellites, e.g.:

- HST (Hubble space telescope)
- ISS (International space station)
- GRACE (Gravity Recovery and Climate Experiment), a NASA & German Aerospace Center collaboration
- CHAMP (Challenging Minisatellite Payload), a German satellite for atmospheric and ionospheric research

Drag coefficients are required to determine drag force, which plays a key role in predicting and maintaining orbit. The Committee for the Assessment of the U.S. Air Force's Astrodynamics Standards recently released a report citing atmospheric drag as the largest source of uncertainty for LEO objects, due in part to improper modeling of the interaction between atmosphere and object. Drag depends on geometry, orientation, ambient and surface temperatures, and atmospheric chemical composition in LEO, which depends on position: latitude, longitude, and altitude. Numerical simulations can produce accurate drag coefficient estimates as a function of these input coordinates, and up to uncertainties in atmospheric and gas–surface interaction (GSI) models. But the calculations are too slow for real-time applications.

Most of the input coordinates mentioned above are ordinary scalars. Satellite geometry however, is rather high dimensional. Geometry is specified in a mesh file: an ASCII representation of a picture like the one in Figure 2.12 for the Hubble space telescope (HST).

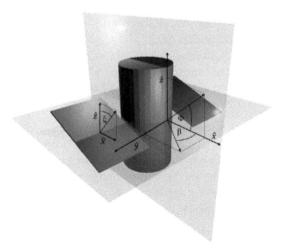

FIGURE 2.12: Rendering of the Hubble space telescope mesh. Reproduced from Mehta et al. (2014); used with permission from Elsevier.

A satellite's geometry is usually fixed; most don't change shape in orbit. The HST is an exception: its solar panels can rotate, which is a nuance I'll discuss in more detail later. For

TABLE 2.5: Satellite design variables (AC: accommodation coefficient).

Symbol [ascii]	Parameter [units]	Range
v_{rel} [Umag]	velocity [m/s]	[5500, 9500]
T_s [Ts]	surface temperature [K]	[100, 500]
T_a [Ta]	atmospheric temperature [K]	[200, 2000]
θ [theta]	yaw [radians]	$[-\pi/2, \pi/2]$
ϕ [phi]	pitch [radians]	$[-\pi, \pi]$
α_n [alphan]	normal energy AC [unitless]	[0, 1]
σ_t [sigmat]	tangential momentum AC [unitless]	[0, 1]

now, take geometry as a fixed input. Position and orientation inputs make up the *design variables*. These are listed alongside their units and ranges, determining the study region of interest, in Table 2.5.

Atmospheric chemical composition is an *environmental variable*. At LEO altitudes the atmosphere is primarily comprised of atomic oxygen (O), molecular oxygen (O_2), atomic nitrogen (N), molecular nitrogen (N_2), helium (He), and hydrogen (H) (Picone et al., 2002). Mixtures of these so-called chemical "species" vary with position, and there exist calculators, like this one from NASA[4], which can deliver mixture weights if provided position and time coordinates.

2.3.1 Simulating drag

Researchers at LANL developed the *Test Particle Monte Carlo (TPMC)* simulator in C, which I wrapped in an R interface called **tpm**. The entire library, packaging C internals and R wrapper, may be found on a public Bitbucket repository linked here[5]. The TPMC C backend simulates the environment encountered by satellites in LEO under free molecular flow (FMF) as modulated by coordinates in the set of three input categories (geometry, design and chemical composition) described above. Since the C simulations are time-consuming, but ultimately independent for each unique input configuration, the **tpm** R interface utilizes OpenMP[6] to facilitate symmetric multiprocessing (SMP)[7] parallelization of evaluations. A wrapper routine called **tpm.parallel**, utilizing an additional message passing interface (MPI) layer[8], is provided to further distribute parallel instances over nodes in a cluster.

The **tpm** R interface requires a pointer to a single mesh file, a single six-vector chemical mixture of environmental variables, and a design of as many seven-vector position/orientation configurations as desired, and over which parallel instances are partitioned. In other words the overall design, varying mesh file, mixture and configuration inputs, must be *blocked* over mesh and mixture. This setup eases distribution of configuration inputs, along which the strongest nonlinear spatial relationships manifest, over parallel instances.

To illustrate, let's set up an execution with two replicates on a design of eight runs. Note that **tpm** takes inputs on the nominal scale, with ranges indicated in Table 2.5. R code below

---

[4]https://ccmc.gsfc.nasa.gov/modelweb/models/nrlmsise00.php
[5]https://bitbucket.org/gramacylab/tpm
[6]https://en.wikipedia.org/wiki/OpenMP
[7]https://en.wikipedia.org/wiki/Symmetric_multiprocessing
[8]https://en.wikipedia.org/wiki/Message_Passing_Interface

builds the 7d design X and specifies one of the GRACE meshes. Meshes for over a dozen other satellites are provided in the "tpm-git/tpm/Mesh_Files" directory.

```
n <- 8
X <- data.frame(Umag=runif(n, 5500, 9500), Ts=runif(n, 100, 500),
   Ta=runif(n, 200, 2000), theta=runif(n, -pi, pi),
   phi=runif(n, -pi/2, pi/2), alphan=runif(n), sigmat=runif(n))
X <- rbind(X,X)
mesh <- "tpm-git/tpm/Mesh_Files/GRACE_A0_B0_ascii_redone.stl"
moles <- c(0,0,0,0,1,0)
```

The final line above sets up the atmospheric chemical composition as a unit-vector isolating pure helium (He). With input configurations in place, we are ready to run simulations. Code below loads the R interface and invokes a simulation instance which parallelizes evaluations over the full suite of cores on my machine.

```
source("tpm-git/tpm/R/tpm.R")
system.time(y <- tpm(X, moles=moles, stl=mesh, verb=0))
```

```
##      user   system  elapsed
## 5160.235    1.808  462.894
```

What do we see? It's not speedy despite OpenMP parallelization. I have eight cores on my machine, and I'm getting about a factor of 6× speedup. That efficiency improves with a larger design. For example, with n <- 800 runs it's close to parity at 8×. Also observe that the output is not deterministic.

```
mean((y[1:n] - y[(n + 1):length(y)])^2)
```

```
## [1] 3.445e-05
```

Each simulation involves pseudo-random numbers underlying trajectories of particles bombarding external facets of the satellite. Despite the stochastic response, variability in runs is quite low. Occasionally, in about two in every ten thousand runs, tpm fails to return a valid output (yielding NA) due to numerical issues. When that happens, a simple restart of the offending simulation usually suffices.

Since calculations underlying TPMC are implemented in C, compiling that code is a prerequisite to sourcing tpm-git/tpm/R/tpm.R. But this only needs to be done once per machine. On a Unix-based system, like Linux or Apple OSX, that's relatively easy with the Gnu C compiler gcc. Note the default compiler on OSX is clang from LLVM, which at the time of writing doesn't support OpenMP out of the box. The C code will still compile, but it won't SMP-parallelize. To obtain gcc compilers, visit the HPC for OSX page[9]. On Microsoft Windows, the Rtools[10] library is helpful, providing a Unix-like environment and gcc compilers, enabling commands similar to those below to be performed from the DOS command prompt.

The C code resides in tpm-git/tpm/src, and a shared library (for runtime linking with R)

---

[9]http://hpc.sourceforge.net/
[10]https://cran.r-project.org/bin/windows/Rtools/

can be built using R's `SHLIB` command from the Unix (or DOS command) prompt. Figure 2.13 provides a screen capture from a compile performed on my machine.

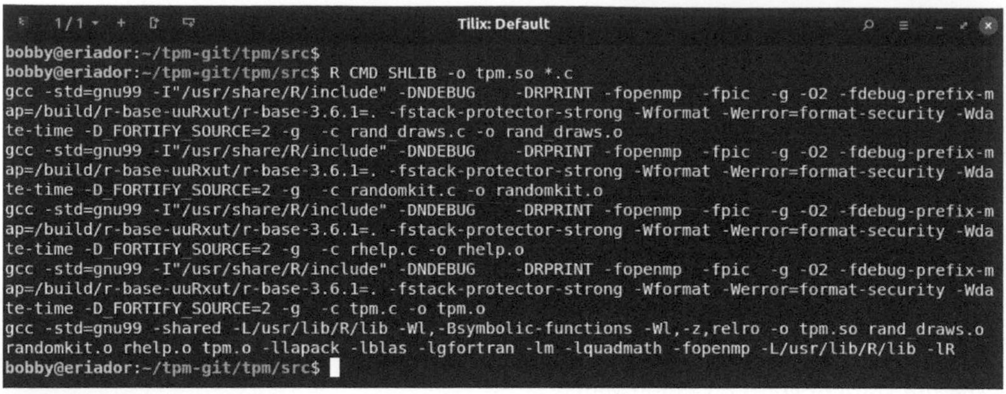

FIGURE 2.13: Compiling the C source code behind TPM.

The `tpm` library can be used to estimate new drag coefficients. However considering the substantial computational expense, it helps to have some pre-run datasets on hand. There are two suites of pre-run batches: one from a limited pilot experiment performed at LANL in order to demonstrate the potential value of surrogate modeling; and another far more extensive suite performed by me using UChicago's Research Computing Center (RCC)[11]. That second suite utilized seventy-thousand core hours (meaning hours you'd have to wait if only one core of one processor were being used, in serial), distributed over thousands of cores of hundreds of nodes of the RCC Midway supercomputer[12]. In fact, when RCC saw how much computing I was doing they decided to commission a puff piece[13] about it.

2.3.2   Surrogate drag

Consider the first suite of runs from LANL's pilot study. The goal of that study was to build a surrogate, via Gaussian processes (GPs), such that predictions from GP fits to `tpm` simulations were accurate to within 1% of actual simulation output, based on root mean-squared percentage error (RMSPE). That proved to be a difficult task considering some of the computational challenges behind GP inference and prediction. My LANL colleagues quickly realized that the data size they'd need, in the 7d input space described by Table 2.5, would be way bigger than what's conventionally tractable with GPs. In order to meet that 1% goal they had to dramatically reduce the input space for training, and consequently also reduce the domain on which reliable predictions could reasonably be expected (i.e., the testing set). The details and other reasoning behind the experiment they ultimately performed are provided by Mehta et al. (2014); a brief explanation and demonstration follow.

LANL researchers looked at data sets, containing TPMC simulations, that were about $N = 1000$ runs in size. You can handle slightly larger N with GPs without getting creative (e.g., special linear algebra subroutines; see Appendix A) using desktop workstations circa early 2010's, but not much. With such a limited number of runs, it was clear that they'd never achieve the 1% RMSPE goal in the full 7d space. Later in §9.3.6 this is verified empirically in

[11]https://rcc.uchicago.edu/
[12]https://rcc.uchicago.edu/docs/using-midway/index.html
[13]https://rcc.uchicago.edu/simulation-you-never-have-run

TABLE 2.6: Reduced angles for small GRACE drag simulation campaigns.

Symbol [ascii]	Parameter [units]	Ideal Range	Reduced Range	Percentage
θ [theta]	yaw [radians]	$[-\pi/2, \pi/2]$	[0, 0.06]	1.9
ϕ [phi]	pitch [radians]	$[-\pi, \pi]$	[-0.06, 0.06]	1.9

a Monte Carlo study. Yet it's easy to deduce that outcome from the simpler experiment our LANL colleagues ultimately decided to entertain instead. And that variation is as follows: they focused on a narrow, yet realistically representative, band of yaw (θ) and pitch (ϕ) angles. For the GRACE satellite, that reduction may be characterized approximately as outlined in Table 2.6.

A narrower range of angles, leading to an input space of smaller volume, has the effect of making the $N = 1000$ points closer together. This makes fitting and prediction in the reduced space easier, and more accurate. Covering the full range of angles, at the density implied by a space-filling set of $N = 1000$ inputs, would require about 4 million points: well beyond the capabilities that could be provided by a GP-based method – at least any known at the time.

Let's look at LANL's GRACE runs for pure He, reproducing the essence of their GP training and testing exercise. A $N = 1000$-sized training set, and $N' = 100$-sized testing set, was generated independently as a LHS (§4.1). GP predictions derived on the former for the latter were evaluated for out-of-sample accuracy with RMSPE.

```
train <- read.csv("tpm-git/data/GRACE/CD_GRACE_1000_He.csv")
test <- read.csv("tpm-git/data/GRACE/CD_GRACE_100_He.csv")
r <- apply(rbind(train, test)[,1:7], 2, range)
r
```

```
##       Umag    Ts      Ta     theta      phi     alphan     sigmat
## [1,] 5502 100.0   201.2 0.0000127 -0.06978 0.0008822 0.0007614
## [2,] 9498 499.8 2000.0 0.0697831  0.06971 0.9999078 0.9997902
```

As you can see above, the range of **theta** (yaw) and **phi** (pitch) are reduced compared to their ideal range. Exact ranges depend on the satellite and pure-species in question, and may not line up perfectly with either of the tables above. In particular notice that yaw- and pitch-angle reduced ranges slightly exceed those from Table 2.6. Columns eight and nine in the files provide estimated coefficients of drag, **Cd**. The ninth column, labeled **Cd_old**, provides a legacy estimate based on an older **tpm** simulation toolkit. These values are internally compatible with **Cd_old** values from other files in that **data/GRACE** directory, but may not match bespoke simulations from the latest **tpm** library. In the particular case of GRACE simulations, differences between **Cd** and **Cd_old** are slight.

Before fitting models, it helps to first convert to coded inputs.

```
X <- train[,1:7]
XX <- test[,1:7]
for(j in 1:ncol(X)) {
  X[,j] <- X[,j] - r[1,j]
  XX[,j] <- XX[,j] - r[1,j]
```

```
    X[,j] <- X[,j]/(r[2,j] - r[1,j])
    XX[,j] <- XX[,j]/(r[2,j] - r[1,j])
}
```

Now we can fit a GP to the training data and make predictions on a hold-out testing set. Suspend your disbelief for now; details of GP fitting and prediction are provided in gory detail in Chapter 5. The library used here, `laGP` (Gramacy and Sun, 2018) on CRAN, is the same as the one used in our introductory wing weight example from §1.2.1.

```
library(laGP)
```

The fitting code below is nearly cut-and-paste from that example.

```
fit.gp <- newGPsep(X, train[,8], 2, 1e-6, dK=TRUE)
mle <- mleGPsep(fit.gp)
p <- predGPsep(fit.gp, XX, lite=TRUE)
rmspe <- sqrt(mean((100*(p$mean - test[,8])/test[,8])^2))
rmspe
```

```
## [1] 0.672
```

Just as Mehta et al. found: better than 1%. Now that was for just one chemical species, pure He. In a real forecasting context there would be a mixture of elements in LEO depending on satellite position. LANL researchers address this by fitting six, separate GP surrogates, one for each pure species. The data directory provides these files:

```
list.files("tpm-git/data/GRACE", "1000*_[A-Z].csv")
```

```
## [1] "CD_GRACE_100_H.csv"   "CD_GRACE_100_N.csv"   "CD_GRACE_100_O.csv"
## [4] "CD_GRACE_1000_H.csv"  "CD_GRACE_1000_N.csv"  "CD_GRACE_1000_O.csv"
```

A similar suite of files with `.dat` extensions contain legacy data, duplicated in the `Cd_old` column of `*.csv` analogs. Designs in these files are identical up to a column reordering.

Once surrogates have been fit to pure-species data, their predictions may be combined for any mixture as

$$
C_D = \frac{\sum_{j=1}^{6} C_{D_j} \cdot \chi_j \cdot m_j}{\sum_{j=1}^{6} \chi_j \cdot m_j},
$$

where χ_j is the mole fraction at a particular LEO location in the atmosphere, and m_j are particle masses. For example, NASA's calculator[14] gives the following at 1/1/2000 0h at 0 deg L&L, and 550km altitude:

```
mf <- c(O=0.83575679477, O2=0.00004098807, N=0.01409589809,
    N2=0.00591827778, He=0.13795985368, H=0.00622818760)
```

---

[14]https://ccmc.gsfc.nasa.gov/modelweb/models/nrlmsise00.php

A periodic table provides the masses which, in relative terms, are proportional to the following.

```
pm <- c(O=2.65676, O2=5.31352, N=2.32586, N2=4.65173,
  He=0.665327, H=0.167372)
```

LANL went on to show that the six independently fit "pure emulators", when suitably combined, were still able to give RMSPEs out-of-sample that were within the desired 1% tolerance. A homework exercise (§2.6) asks the reader to duplicate this analysis by appropriately collating predictions for other species and comparing, out of sample, to results obtained directly under a mixture-of-species simulation.

Our LANL colleagues provided similar proof-of-concept runs and experiments for the Hubble Space Telescope (HST).

```
list.files("tpm-git/data/HST", "Satellite.*.csv")
```

```
## [1] "Satellite_H.csv"    "Satellite_He.csv"    "Satellite_N.csv"
## [4] "Satellite_N2.csv"   "Satellite_O.csv"     "Satellite_O2.csv"
## [7] "Satellite_TS_H.csv" "Satellite_TS_He.csv" "Satellite_TS_N.csv"
## [10] "Satellite_TS_N2.csv" "Satellite_TS_O.csv" "Satellite_TS_O2.csv"
```

A slightly different naming convention was used here compared to the GRACE files: "TS" means "test set". Analog files without the `.csv` extension contain legacy LANL simulation output. These legacy outputs differ substantially from the revised analog for reasons that have to do with special handling of its solar panels. HST is specified by a suite of mesh files, one for each of ten panel angles, resulting in an extra input column (i.e., 8d inputs). Efficient simulation requires additional blocking of `tpm` simulations, iterating over the ten meshes, each with 1/10th of the space-filling settings of the other inputs.

```
list.files("tpm-git/tpm/Mesh_Files", "HST")
```

```
## [1] "HST_0.stl"  "HST_10.stl" "HST_20.stl" "HST_30.stl" "HST_40.stl"
## [6] "HST_50.stl" "HST_60.stl" "HST_70.stl" "HST_80.stl" "HST_90.stl"
```

Finally, Mehta et al. reported on a similar experiment with the International Space Station (ISS). These files have not been updated from their original format and legacy output.

```
list.files("tpm-git/data/ISS", "_")
```

```
## [1] "ISS_H.dat"  "ISS_He.dat" "ISS_N.dat"  "ISS_N2.dat" "ISS_O.dat"
## [6] "ISS_O2.dat"
```

Although the ISS has many "moving parts", only one representative mesh file is provided.

```
list.files("tpm-git/tpm/Mesh_Files", "ISS")
```

```
## [1] "ISS_ascii.stl"
```

2.3.3 Big `tpm` runs

To address drawbacks of that initial pilot study, particularly the narrow range of input angles, I compiled a new suite of TPMC simulations using `tpm` for HST and GRACE. HST simulations were collected for $N = 1M$ and $N = 2M$ depending on the species. A testing set of size $N' = 1M$ was gathered under an ensemble of species for out-of-sample benchmarking. The designs were LHSs divided equally between the ten panel angles. For GRACE, which is easier to model, LHSs with $N = N' = 1M$ is sufficient throughout. This is fortunate since GRACE simulations are more than $10\times$ slower than HST.

```
c(list.files("tpm-git/data/HST", "hst.*dat"),
  list.files("tpm-git/data/GRACE", "grace.*dat"))
```

```
##   [1] "hstA_05.dat"   "hstA.dat"       "hstEns.dat"    "hstH.dat"
##   [5] "hstHe.dat"     "hstHe2.dat"     "hstN.dat"      "hstN2.dat"
##   [9] "hstO.dat"      "hstO2.dat"      "hstQ_05.dat"   "hstQ.dat"
##  [13] "graceA_05.dat" "graceEns.dat"   "graceH.dat"    "graceHe.dat"
##  [17] "graceN.dat"    "graceN2.dat"    "graceO.dat"    "graceO2.dat"
##  [21] "graceQ_05.dat" "graceQ.dat"
```

Files named with **Ens** correspond to chemical ensembles which were calculated using mole fractions quoted in §2.3.2. All together these took about 70K service units (SUs). An SU is equivalent to one CPU core-hour. Runs were distributed across dozens of batches farmed out to 32 16-core nodes for about 18 hours each, depending on the mesh being used and the size of the full design. Although the files have a `.dat` extension, matching the naming scheme of files containing legacy runs, this larger suite was performed with the latest `tpm`.

Divide-and-conquer is key to managing data of this size with GP surrogates. One option is hard partitioning, for example dividing up the input space by its angles, iterating the Mehta et al. idea. But soft partitioning works better in the sense that it's simultaneously more accurate, computationally more efficient, and utilizes a smaller training data set. (I needed only $N = 1$-$2M$ runs, as described above.) The method of local approximate Gaussian processes (LAGP), introduced in §9.3, facilitates one such approach to soft partitioning having the added benefit of being massively parallelizable, a key feature in the modern landscape of ubiquitous multicore SMP computing. Our examples above leverage full-GP features from the `laGP` package; local approximate enhancements will have to wait for §9.3.6.

2.4 Groundwater remediation

Worldwide there are more than 10,000 contaminated land sites (Ter Meer et al., 2007). Environmental cleanup at these sites has received increased attention in the last few decades. Preventing migration of contaminant plumes is vital to protect water supplies and prevent disease. One approach is pump-and-treat remediation, in which wells are strategically placed to pump out contaminated water, purify it, and inject treated water back into the system to prevent contaminant spread.

A case study of one such remediation effort is the 580-acre Lockwood Solvent Groundwater

Plume Site[15], an EPA Superfund site near Billings Montana. As a result of industrial practices, groundwater at this site is contaminated with volatile organic compounds that are hazardous to human health. Figure 2.14 shows the location of the site and provides a simplified illustration of the contaminant plumes that threaten a valuable water source: the Yellowstone River.

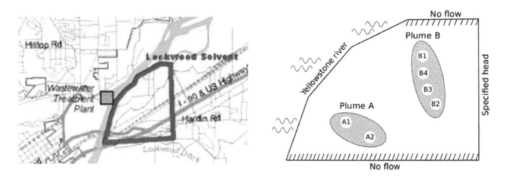

FIGURE 2.14: Lockwood site via map (left) and plume diagram (right). Captured from Gramacy et al. (2016) and used with permission from the American Statistical Association.

To prevent further expansion of these plumes, six pump and treat wells have been proposed. These are shown in sets of two and four in the right panel of Figure 2.14. The amount of contaminant exiting the boundaries of the system – entering the river in particular – depends on placement of the wells and their pumping rates. Here we treat placement as fixed, roughly at the locations shown in the diagram, and focus on determining appropriate pumping rates. An analytic element method (AEM)[16] groundwater model and solver was developed to simulate the amount of contaminant exiting the two boundaries under different pumping regimes (Matott et al., 2006).

Code implementing the solver takes a positive six-vector as input, specifying pumping rates for each of the six wells: $x \in [0, 2 \cdot 10^4]^6$. It returns two outputs: 1) a quantity proportional to the cost of pumping, which is just $\sum_j x_j$; and 2) a two-vector, indicating the amount of contaminant exiting the boundaries. The goal is to explore pumping rates where both entries of the contaminant vector are zero, indicating no contaminant spread. If the contaminant output vector is positive in either coordinate then that's bad (the larger the badder), because it means that some contaminant has escaped. Note that the contaminant vector output is never negative as long as input x is in the valid range, as described above.

The groundwater solver implementation is delicate, owing to the hodgepodge of C++ libraries that were weaved together to obtain the desired calculation(s).

- One was (at the time it was developed) called Bluebird, but now goes by VisualAEM[17].
- The other is called Ostrich[18].
- A shell script called RunLock acts as glue and provides the appropriate configuration files.
- An R wrapper (written by me) runlock.R enables Runlock to be invoked from R.

The underlying C++ programs, which read and write files with absolute paths, require runs be performed within the **runlock** directory (after you've run the build.sh script to compile

---

[15]https://cumulis.epa.gov/supercpad/cursites/csitinfo.cfm?id=0801709

[16]https://en.wikipedia.org/wiki/Analytic_element_method

[17]http://www.civil.uwaterloo.ca/jrcraig/visualaem/main.html

[18]http://www.eng.buffalo.edu/~lsmatott/Ostrich/OstrichMain.html

all the C++ code). Note that the makefiles used by the build scripts assume **g++** compilers, which is the default on most systems but not OSX. OSX uses LLVM/clang with aliases to **g++** and doesn't work with the **runlock** back-end. This is similar to the issue mentioned above for **tpm** nearby Figure 2.13, except in this case a true **g++** compiler is essential. See the HPC for OSX page[19] to obtain a GNU C/C++ compiler for OSX. At this time, this setup is not known to work on Windows systems. Binary Bluebird and Ostrich executables may be compiled for Windows with slight modification to the source code, but the shell scripts which glue them together assume a Unix environment.

```
setwd("runlock")
```

Here's how output looks on a random input vector.

```
source("runlock.R")
x <- runif(6, 0, 20000)
runlock(x)
```

```
## $obj
## [1] 60220
##
## $c
## [1] 0.3498 0.0000
```

Both outputs are derived from deterministic calculations. As described above, the $obj output is $\sum_j x_j$, the sum of the six pumping rates. The $c output is a two-vector indicating how much contaminant exited the system. About one in ten runs with random inputs, $x \sim \text{Unif}(0, 20000)^6$, yield output $c at zero for both boundaries. The second boundary is less likely to suffer contaminant breach. One "good" pumping rate is known, but it implies a fair amount of pumping.

```
runlock(rep(10000, 6))
```

```
## $obj
## [1] 60000
##
## $c
## [1] 0 0
```

Here's a run on one hundred random inputs.

```
runs <- matrix(NA, nrow=100, ncol=9)
runs[,1:6] <- matrix(runif(6*nrow(runs), 0, 20000), ncol=6)
tic <- proc.time()[3]
for(i in 1:nrow(runs)) {
  runs[i,7:9] <- unlist(runlock(runs[i,1:6]))
}
toc <- proc.time()[3]
toc - tic
```

---

[19]https://hpc.sourceforge.net/

```
## elapsed
##    115.5
```

As you can see, simulations are relatively quick (about 1.2s/run), but not instantaneous. More than a decade ago, when this problem was first studied, the computational cost was more prohibitive, being upwards of ten or so seconds per run. Improvements in processing speed and compiler optimizations have combined to provide about a tenfold speedup.

```
success <- sum(apply(runs, 1, function(x) { all(x[8:9] == 0) }))
success
```

```
## [1] 18
```

Above, 18% of the random inputs came back without contaminant breach. Before changing gears let's remember to back out of the **runlock** directory.

```
setwd("../")
```

2.4.1   Optimization and search

Mayer et al. (2002) proposed casting the pump-and-treat setting as a *constrained "blackbox" optimization*. For the version of the Lockwood problem considered here, pumping rates x can be varied to minimize the cost of operating wells subject to constraints on the contaminant staying within plume boundaries, whose evaluations require running groundwater simulations. This led to the following blackbox optimization problem, a so-called nonlinear mathematical program.

$$\min_{x} \left\{ f(x) = \sum_{j=1}^{6} x_j : c_1(x) \leq 0, \, c_2(x) \leq 0, \, x \in [0, 2 \cdot 10^4]^6 \right\}$$

The term *blackbox* means that inner-workings of the program are largely opaque to the optimizer. Objective f is linear and describes costs required to operate the wells; this matches up with output **\$obj** from **runlock** above. Absent constraints c (via **\$c** from **runlock**), which are satisfied when both components are zero, the solution is at the origin and corresponds to no pumping and no remediation. But this unconstrained solution is of little interest. We desire a pumping rate just low enough, minimizing costs of operating the wells, to accomplish the remediation goal: clean drinking water.

Matott et al. (2011) compared MATLAB® and Python optimizers, treating constraints with the *additive penalty method* (reviewed in more detail in §7.3.4), all initialized at the known-valid input $x_j^1 = 10^4$. These results are shown in Figure 2.15. Many of the optimizers, such as "Newton", "Nelder-Mead" and "BFGS" may be familiar. Several are implemented as options in the **optim** function for R and have their own Wikipedia pages. More detail will be provided in Chapter 7.

```
bvv <- read.csv("runlock/pato_results.csv")
cols <- 3:14
nc <- length(cols)
```

```
matplot(bvv[1:1000,1], bvv[1:1000,cols], xlim=c(1,1500), type="l", bty="n",
    xlab="number of evaluations", ylab="pump-and-treat cost to remediate")
legend("topright", names(bvv)[cols], lty=1:nc, col=1:nc, bty="n")
```

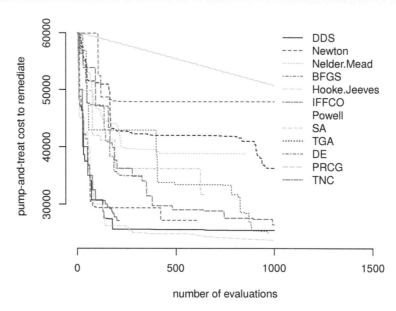

FIGURE 2.15: Best valid values found by MATLAB and Python APM optimizers over the course of 1000 expensive simulations; rebuild of a figure from Matott et al. (2011) using original data.

Observe from Figure 2.15 that there's great diversity in success of the methods deployed to solve the Lockwood constrained optimization. The goal is to obtain a value on the y-axis in that plot, indicating the best valid pumping rate, that's as low as possible as a function of the number of `runlock` evaluations, indicated on the x-axis. Better methods "hug the origin", with lines closer to the lower-left corner of the plot. The very best methods by this metric are "Hooke–Jeeves" and "DDS".

Looking at the results from that study, the following questions emerge. What makes good methods good, and why do bad methods fail so spectacularly? And by the way, how are statistics and surrogates involved? As a window to potential insight, consider the following random iterative search involving *objective improving candidates*, or OICs. Algorithm 2.1 shows how to obtain the next OIC given n runs of the simulator, collecting evaluations of f and c with rejection sampling[20]. Importantly, no expensive blackbox evaluations of $f(\cdot)$ are called for in the pseudocode. With x_{n+1} in hand, we can evaluate $f(x_{n+1})$ and $c(x_{n+1})$ and repeat, which might yield an improved $x^\star$ and thus result in a narrower subsequent search for the next OIC.

Figure 2.16 shows how such OICs fare on the Lockwood problem. Thin gray lines in the figure are extracted from the first 500 iterations from Matott et al. (2011)'s experiment. A thicker black line added to the plot shows average progress (best valid value, i.e., $f(x^\star)$ over the iterations, n) from thirty repeated runs of sequential selection of OICs, from $n = 2, \ldots, 500$, initialized with the same high pumping rate $x^1 = (10^4)^6$ used by all other methods. Note

---

[20]https://en.wikipedia.org/wiki/Rejection_sampling

Algorithm 2.1 Next Objective Improving Candidate

Assume the search region is $\mathcal{B}$, perhaps a unit hypercube.

Require input settings $x_1, \ldots, x_n$ at which the objective f and constraints c have already been evaluated.

Then

1. Find the current best valid input, $x^\star$, using discrete search over the existing runs $x^\star = \arg \min_{i=1,\ldots,n} \{ f(x_i) : c_j(x_i) \leq 0, \forall j \}$.
2. Draw x_{n+1} uniformly from $\{ x \in \mathcal{B} : f(x) < f(x^\star) \}$, for example by rejection sampling.

Return x_{n+1}, the next objective improving candidate.

---

that these are the first R results in the book which don't originate in Rmarkdown, owing to the substantial computational effort involved in evaluating **runlock** $500 \times 30 = 1500$ times. Output file `oic_prog.csv` is deliberately omitted from the supplementary material. Reproducing these results is the subject of a homework exercise in §2.6.

```
prog <- read.csv("oic_prog.csv")
matplot(bvv[,cols], col="gray", type="l", lty=1, xlim=c(1,500),
  xlab="evaluations (n)", ylab="best valid value")
lines(rowMeans(prog), lty=1, lwd=2)
legend("topright", "average OIC", lwd=2)
```

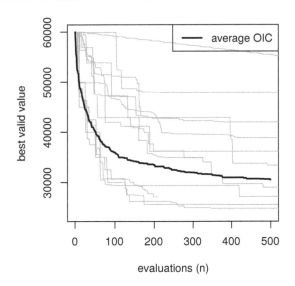

FIGURE 2.16: Objective improving candidates versus Matott optimizers. Rebuild of a figure from Gramacy and Lee (2011) with novel OIC simulations.

What do we notice from the OIC results in the figure? Half of the MATLAB/Python methods are not doing better (on average) than a slightly modified "random search", as represented by OICs. Those inferior methods are getting stuck in a local minima and failing to explore other opportunities. Now *stochastic search*, whether via OICs or purely at random, doesn't represent a compelling practical alternative in this context because of the chance of poor results in any particular run, even though average behavior out of thirty repetitions is pretty good. But those average results, which suggest that a more careful balance between exploration and exploitation could be beneficial, do indicate that a statistical decision process – where striking such balances is routine – could be advantageous. Statistical response surface methods, through suitably flexible surrogate models paired with searching criteria that can trade-off reward and uncertainty, have been shown to be quite effective in this context. These are the subject of Chapter 7.

As a taste of what's to come, surrogate Gaussian process (GP) models can be fitted to blackbox function evaluations, from the objective and/or from constraints, and sequential design criteria can be derived – *acquisition functions* in machine learning (ML) jargon – that leverage the full uncertainty of predictive equations (i.e., mean and variance) to decide where to sample next. These so-called Bayesian optimization (BO)[21] procedures have been shown empirically to lead to more global searches, finding better solutions with fewer evaluations, compared to conventional optimizers like those deployed in the Matott et al. (2011) study. The term "Bayesian optimization" is somewhat of a misnomer, because Bayesian methodology/thinking isn't essential. However the term has caught on owing to excitement in Bayesian updating, of which many sequential statistical calculations are an example. Although work on BO in ML is feverish at the moment – for example, to tune the hyperparameters of deep neural networks – the roots of these methods lie in industrial statistics under the moniker of statistical/surrogate-assisted optimization. In the mathematical programming world, surrogate assisted methods for blackbox optimization falls generically under the class of derivative-free methods[22] (Larson et al., 2019). They utilize only function evaluations, requiring little knowledge of how codes implementing those functions are comprised, in particular not requiring derivatives. However use of derivative-like information, often through a surrogate, can be both explicit and implicit in such enterprises.

2.4.2   Where from here?

Once we nail down surrogate models (e.g., via GPs) we'll be able to address sequential design and optimization problems on both more general and more specific terms. The process of using surrogate model fits to further data collection, updating the design to maximize information or reduce variance, say, has become fundamental to computer simulation experiments. It's also popular in ML where it goes by the name *active learning*, where the learner gets to choose the examples it's trained on. Extending that idea to more general optimization problems, with appropriate design/acquisition criteria – the most popular of which is called *expected improvement* (EI) – is relatively straightforward. The setting becomes somewhat more challenging when constraints are involved, or when targets are more nuanced: a search for level sets or contours, classification boundaries, etc. We'll expound upon blackbox constrained optimization in some detail. The others are hot areas at the moment and I shall refer to the literature for many of those in order to keep the exposition relatively self-contained.

Everyone knows that modern statistical learning benefits from optimization methodology. Just think about the myriad numerical schemes for maximizing log likelihoods, or convex optimization in penalized least squares. Throughout the text we'll make liberal use of optimization libraries as subroutines, ones which over the years have been engineered to

---

[21]https://en.wikipedia.org/wiki/Bayesian_optimization
[22]https://en.wikipedia.org/wiki/Derivative-free_optimization

the extent that they've become so robust that we take their good behavior for granted. But in the spectrum of problems from the mathematical programming literature, which is where optimization experts play, these represent relatively easy examples. The harder ones, like with blackbox objectives and constraints, could benefit from a fresh perspective. Mathematical programmers are learning from statisticians in a big way, porting robust surrogate modeling and decision theoretic criteria to aid in search for global optima. Perhaps a virtuous cycle is forming.

2.5 Data out there

The examples in this chapter are interesting because each involves several facets of modern response surface methods, via surrogates, and sequential design. Yet it'll be useful to draw from a cache of simpler – you might say toy – examples for our early illustrations, isolating particular features common to real data and real-world simulation experiments and methods which have been designed to suit. Synthetic experiments on simple data, as a first pass on demonstrating and benchmarking new methodology, abound in the literature.

There are several good outposts offering libraries of toy problems, curated and regularly updated. My new favorite is the Virtual Library of Simulation Experiments (VLSE)[23] out of Simon Fraser University. Several homework exercises throughout the book feature data pulled from that site. Good optimization-oriented virtual libraries can be found on the world wide web, although for our purposes the VLSE is sufficient – containing a specifically optimization-oriented tab of problems. An impressive exception is the Decision Tree for Optimization Software[24] out of Arizona State University which not only provides examples, but also sets of live benchmarks offering a kind of unit testing[25] of updated implementations submitted to the site. Two other cool examples get their own heading.

Fire modeling and bottle rockets

I came across a page on Fire Modeling Software[26] when browsing Santner et al. (2003), first edition. That text contains an example in the introductory chapter involving an old FORTRAN/BASIC program called ASET, which is one of the ones listed on that page. Others from that page could offer a source of pseudo-synthetic examples with real code (i.e., real computer model simulations like TPMC), assuming they could be made to compile and run. I couldn't get ASET-B to compile, but it's been such a long time since I've looked at BASIC. Some of the newer codes are more involved, but potentially more exciting.

Ever do a bottle rocket project in middle school? I certainly did. Folks have written calculators that estimate the height and distance traveled, and other outputs, as a function of bottle geometry, water pressure, etc. There's an entire web page dedicated to simulators for flying bottle rockets[27], but sadly the site is old and many links therein are now broken. NASA thinks bottle rockets are pretty cool too, in addition to the real kind. They offer two

---

[23]http://www.sfu.ca/~ssurjano/
[24]http://plato.asu.edu/guide.html
[25]https://en.wikipedia.org/wiki/Unit_testing
[26]http://www.nist.gov/el/fire_protection/buildings/fire-modeling-programs.cfm
[27]http://cjh.polyplex.org/rockets/simulation/

simulators: a simpler one called BottleRocketSim[28] and a more complex alternative called RocketModeler III[29]. Both have somewhat dated web interfaces but could easily work in a lab/field experiment or "field trip" setting. Students could collect data on real rockets and calibrate simulations (§8.1) in order to improve accuracy of predictions for future runs.

2.6   Homework exercises

These exercises give the reader an opportunity to play with data sets and simulation codes introduced in this chapter and to help head off any system dependency issues in compiling codes required for those simulations.

#1: The other five LGBB outputs

Our rocket booster example in §2.1 emphasized the `lift` output. Repeat similar slice visuals, for example like Figure 2.6, for the other five outputs. In each case you'll need to choose a value of the third input, `beta`, to hold fixed for the visualization.

a. Begin with the choice of `beta=1` following the `lift` example. Comment on any trends or variations across the five (or six including `lift`) outputs.
b. Experiment with other `beta` choices. In particular what happens when `beta=0` for the latter three outputs: `side`, `yaw` and `roll`? How about with larger `beta` settings for those outputs? Explain what you think might be going on.

#2: Exploring CRASH with feature expanded linear models

Revisit the CRASH simulation linear model/curve fitting analysis nearby Figure 2.10 by expanding the data and the linear basis.

a. Form a `data.frame` combing CE1 data (`ce1`) and scale factor expanded CE2 data (`ce2.sf`), and don't forget to drop the `FileNumber` column.
b. Fit a linear model with `ShockLocation` as the response and the other columns as predictors. Which predictors load (i.e., have estimated slope coefficients which are statistically different from zero)?
c. Consider interactions among the predictors. Which load? Are there any collinearity concerns? Fix those if so. You might try `step`wise regression.
d. Consider quadratic feature expansion (i.e., augment columns with squared terms) with and without interactions. Again watch out for collinearity; try `step` and comment on what loads.
e. Contemplate higher-order polynomial terms as features. Does this represent a sensible, parsimonious approach to nonlinear surrogate modeling?

---

[28]http://www.grc.nasa.gov/WWW/K-12/bottlerocket/
[29]http://www.grc.nasa.gov/WWW/K-12/rocket/rktsim.html

#3: Ensemble satellite drag benchmarking

Combine surrogates for pure species TPMC simulations to furnish forecasts of satellite drag under a realistic mixture of chemical species in LEO.

a. Get familiar with using `tpm` to calculate satellite drag coefficients. Begin by compiling the C code to produce a shared object for use with the R wrapper in `tpm-git/tpm/R/tpm.R`. Run `R CMD SHLIB -o tpm.so *.c` in the `tpm-git/tpm/src` directory. Then double-check you have a working `tpm` library by trying some of the examples in §2.3.2. *You only need to do this once per machine; see note nearby Figure 2.13 about OSX/OpenMP.*

b. Generate a random one-hundred run, 7d testing design uniformly in the ranges provided by Table 2.5 with restricted angle inputs as described in Table 2.6. Evaluate these inputs for GRACE under the mole fractions provided in §2.3.2 for a particular LEO position.

c. Train six GP surrogates on the pure species data provided in the directory `tpm-git/data/GRACE`. *A few helpful notes: pure He was done already (see §2.3.2), so there are only five left to do; be careful to use the same coding scheme for all six sets of inputs; you may wish to double-check your pure species predictors on the pure species testing sets provided.*

d. Collect predictions from all six GP surrogates at the testing locations from #a. Combine them into a single prediction for that LEO position and calculate RMSPE to the true simulation outputs from #a. Is the RMSPE close to 1%?

e. Repeat #b–d for HST with an identical setup except that you'll need to augment your design with 100 random panel angles in $\{0, 10, 20, \ldots, 80, 90\}$ and utilize the appropriate mesh files in your simulations.

#4: Objective improving candidates

Reproduce the OIC comparison for the Lockwood problem summarized in Figure 2.16.

a. Compile the `runlock` back-end using the `build.sh` script provided in the root `lockwood` directory. Double-check that the compiled library works with the R `runlock` interface by trying the code in §2.4.

b. Implement a rejection sampler for generating OICs or figure out how to use `laGP:::rbetter` from `laGP` on CRAN.

c. Starting with the known valid setting, $x^* \equiv x^1 = (10^4)^6$, implement 100 iterations of constrained optimization with OICs as described in Algorithm 2.1. Be sure to save your progress in terms of the best valid value found over the iterations. Plot that progress against the MATLAB/Python optimizers.

d. Repeat #c 30 times and plot the average progress against the MATLAB/Python optimizers.

3

Steepest Ascent and Ridge Analysis

This chapter offers a fly-by of classical methods for response surfaces, focusing on local linear modeling. Development closely follows Chapters 5-6 of Myers et al. (2016), however with rather less on design and greater emphasis on implementation in R. Some of the examples/data are taken verbatim, although with less back-story. Readers who are keen on a more in-depth treatment of the subjects in this chapter will find the Myers text far more satisfying. Of course neither treatment is complete. There are a wealth of texts and modern surveys on the topic.

The goal here is to offer some historical context, in order to facilitate comparisons and contrasts to more modern methodology introduced in subsequent chapters. That said, the methods herein are far better understood, and more widely deployed in practice, particularly in industrial settings. One can argue that they offer more control and potential for insight to the experienced modeler, who deeply understands potential limitations and pitfalls. However, as will become clear as topics progress, many rigid assumptions are embedded in this framework, primarily for analytical tractability. These may be difficult to justify in practice, especially in settings where the goal is to produce an autonomous framework that can operate without constant (expert) human oversight.

What is this chapter about, more specifically? At a high level it's about collecting and learning from data. Since that's perhaps too generic, a better question entails not what but why. One reason for data collection and modeling is to develop understanding or to make predictions under novel conditions. Such understanding and predictions could facilitate many aims. But sometimes that's too ambitious, especially when experimentation (obtaining the "runs" that make up the data) is expensive, and it usually is, or when the ideal modeling apparatus is data hungry. A more humble goal is to make things a little better, which may require less data and where cruder models may be sufficient. That's what this chapter is about. We'll return to bigger models and bigger data in subsequent chapters.

The choice of data – the experimental design – and models is supremely important. The two are intimately linked. Yet simple choices work well when focused locally in the input domain, appealing to Taylor's theorem[1] from calculus. Tools outlined in this chapter are canonical in process optimization or process improvement. Although the goal is to find a configuration of inputs providing a point of optimum response, one more pragmatically settles for simply making an improvement on a previously utilized input setting. Imagine a manufacturing process whose current operating conditions are good, but could potentially be better. Some data could be collected nearby the current regime which, through model fitting and analysis, leads to a potentially new operating regime that may be more efficient. Then the process might iterate, potentially leading to even greater efficiencies, if experimental budgets allow.

The underlying methodology deployed in each iteration is compartmentalized by whether a first-order or second-order model is being entertained. First-order methods fall under the heading of *steepest ascent*, and these are rather straightforward to anyone with experience in

---

[1]https://en.wikipedia.org/wiki/Taylor's_theorem

statistical modeling and optimization. Second-order methods are more involved owing to the many shapes a second-order surface can take on (see, e.g., §1.1.2), leading to so-called *ridge analysis*. Here decisions about how to refine or improve the experiment are nuanced, leaning on careful analysis of standard errors of coefficients whose values determine the nature of the surface (local maxima, rising ridge, saddle, etc.) and thereby what to do in the next stage. Both steepest ascent and ridge analysis compartments share an intimate relationship with experimental design and benefit from a human-in-the-loop throughout stages of iterative refinement. That's in stark contrast to Bayesian optimization (BO) methods (Chapter 7) which are rather more hands-off, intended for autonomous/automatic implementation.

3.1   Path of steepest ascent

Our presentation here privileges maximization of input–output relationships – in keeping with classical RSM tradition – as measured by data collected on a certain process. Adaptations for minimization, which is more common in modern BO settings, are straightforward. Emphasis is more on modeling and decision making and less on appropriate choices of design. For more details on design(s) such as orthogonal, factorial or central composite design, see Myers et al. (2016). Throughout we shall work with coded variables, unless otherwise prefaced, and assume designs centered at $x_j = 0$ in $[-1, +1]$, for $j = 1, \ldots, m$. Some designs, such as those named above, would have x_j take on *only* these values $\{-1, 0, 1\}$. Finally, we presume to be working in a highly localized subset of the input domain on the natural scale, perhaps centering around a setting of inputs representing current operating conditions on which we plan to improve.

The method of *steepest ascent* involves a first-order model, sometimes fitted with interactions. Predictive equations emitted from a fitted first-order model depend upon estimated regression coefficients $\hat{\beta}_j \equiv b_j$. Therefore, ignoring their standard errors s_{b_j} for the moment (see §3.2.3), it's perhaps not surprising that coordinates along the path of steepest ascent depend on those values. Coordinate directions are determined by the sign of b_j; step sizes depend proportionally on their magnitude $|b_j|$.

3.1.1   Signs and magnitudes

To begin with an example, suppose an experiment in two input variables produces the following fitted first-order model. So we're skipping design and fitting stages here and going right to equations defining fitted values and predictive means.

$$\hat{y} = 20 + 3x_1 - 1.5x_2 \tag{3.1}$$

In R:

```
first.order <- function(x1, x2)
  {
    20 + 3*x1 - 1.5*x2
  }
```

Signs and magnitudes of estimated coefficients, $b_1 = 3$ and $b_2 = -1.5$, determine the path of steepest ascent and result in x_1 moving in a positive direction, twice as fast as x_2 moving in a negative direction. Evaluating `first.order` as `yhat` ($\hat{y}$) over a grid `x1` $\times$ `x2` in the input space ...

```
x1 <- x2 <- seq(-2, 3.5, length=1000)
g <- expand.grid(x1, x2)
yhat <- matrix(first.order(g[,1], g[,2]), ncol=length(x2))
D <- rbind(c(-1,-1), c(-1,1), c(1,-1), c(1,1))
```

... facilitates the graphical depiction provided in Figure 3.1: contours of `yhat` values overlayed on a heat plot. Hypothetical design locations are shown as filled dots outlining a square in the input space.

```
cols <- heat.colors(128)
image(x1, x2, yhat, col=cols)
contour(x1, x2, yhat, add=TRUE)
points(D, pch=19)
points(0, 0)
arrows(0, 0, 3, -1.5)
```

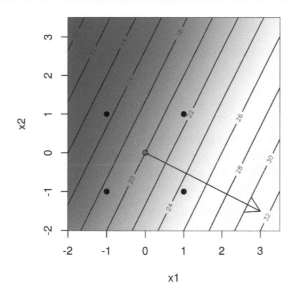

FIGURE 3.1: Example first-order response surface and direction of steepest ascent.

The arrow in the figure, emitting from the center of the design at $(0, 0)$, gets its slope from (b_1, b_2). Theoretically the arrow, outlining the path of steepest ascent, runs perpendicular to the contours. However peculiarities in Rmarkdown's handling of graphical device specifications can introduce skew. If we're careful, acknowledging that the fitted surface isn't a perfect representation of the actual surface (presumably getting worse away from the design center), and take baby steps along the path, we may expect to find inputs that improve upon the actual output of the system. So an important question is: how far to explore along the path?

Backing up a bit from the example, let's now examine the setup more generically in order to formally establish the basic calculations involved in the method of steepest ascent. Consider the fitted first-order regression model

$$\hat{y}(x) = b_0 + b_1 x_1 + b_2 x_2 + \cdots + b_m x_m.$$

The *path of steepest ascent* is one which produces a maximum estimated response, with the constraint that all coordinates are a fixed distance, say radius r, away from the center of the design. That definition is captured mathematically by the program

$$\max_x \hat{y}(x) \quad \text{subject to} \quad \sum_{j=1}^{m} x_j^2 = r^2.$$

The constraint serves two purposes. One: it's clearly folly to venture too far away from the origin of the original design, as the fitted surface can only be trusted nearby. Two: in this first-order linear formulation the unconstrained fitted surface is maximized out at infinity, which is an impractical input setting for most purposes.

Solving this optimization problem proceeds by straightforward application of Lagrange multipliers[2].

$$L = b_0 + b_1 x_1 + b_2 x_2 + \cdots + b_m x_m - \lambda \left(\sum_{j=1}^{m} x_j^2 - r^2 \right)$$

Differentiating and setting to zero leads to specifications for each x_j on the path of steepest ascent,

$$x_j = \frac{b_j}{2\lambda} \equiv \rho b_j, \quad \text{where} \quad \rho = \frac{1}{2\lambda}$$

may be viewed as a constant of proportionality. For ascent, our default, ρ is positive; for descent it's negative. The parameter ρ, which is related to λ and thus to radius r, determines the distance from the design center where the resulting (new) point would reside, which ultimately must be specified by the practitioner.

Extending Figure 3.1, Figure 3.2 shows two potential radiuses $r_1 = 1$ and $r_2 = 1.75$ as circles in the plane.

```
image(x1, x2, yhat, col=cols)
contour(x1, x2, yhat, add=TRUE)
points(D, pch=19)
points(0, 0)
arrows(0, 0, 3, -1.5)
library(plotrix)
draw.circle(0, 0, 1)
text(1, 0, "r1")
draw.circle(0, 0, 1.75)
text(1.75, 0, "r2 > r1")
```

Input coordinates of new runs can be determined by the intersection between the circle(s)

---

[2] purposely/unremarkably using/misusing Lagrange multiplier

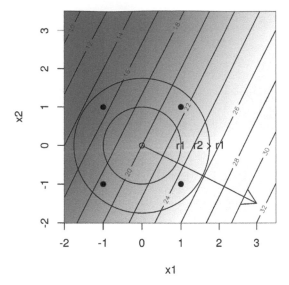

FIGURE 3.2: Radii along paths of steepest ascent.

and the path, depicted by the arrow. That enterprise is more simply carried out via choice of ρ and estimated coefficients b_j. An example follows.

Example: plasma etch process

Plasma etching[3] is a process involved in the fabrication of integrated circuits where a high-speed stream of plasma, the source of which is called an etch species, is shot at a sample. Here we're interested in `etch` rate, our response, measured in units of Å/min, as a function of two inputs describing the process: anode-cathode `gap` (ξ_1), with natural units of centimeters, and the `power` applied to the cathode (ξ_2) in watts. The data in plasma.txt[4], quoted in Table 3.1, are derived from a 2^2 factorial design with four center points.

```
plasma <- read.table("plasma.txt", header=TRUE)
kable(plasma, caption="Plasma etching data.")
```

Observe that the data file contains both natural and coded inputs. We wish to move to a region of the input space where `etch` rate is increased. A goal of the experiment from which these data were derived was to move to a region where `etch` rate is close to 1000 Å/min. Begin by fitting a first-order model. The code below, with `summary.lm` output in Table 3.2, first tries a variation with interactions, mostly with the aim of illustrating that the interaction term is unnecessary.

```
fit.int <- lm(etch ~ x1*x2, data=plasma)
s <- summary(fit.int)$coefficients
kable(s, caption="Summary of first-order model fit with interactions.")
```

---

[3]https://en.wikipedia.org/wiki/Plasma_etching
[4]http://bobby.gramacy.com/surrogates/plasma.txt

TABLE 3.1: Plasma etching data.

gap	power	x1	x2	etch
1.2	275	-1	-1	775
1.6	275	1	-1	670
1.2	325	-1	1	890
1.6	325	1	1	730
1.4	300	0	0	745
1.4	300	0	0	760
1.4	300	0	0	780
1.4	300	0	0	720

TABLE 3.2: Summary of first-order model fit with interactions.

	Estimate	Std. Error	t value	Pr($>$|t|)
(Intercept)	758.75	8.604	88.189	0.0000
x1	-66.25	12.168	-5.445	0.0055
x2	43.75	12.168	3.596	0.0228
x1:x2	-13.75	12.168	-1.130	0.3216

Since the `x1:x2` coefficient isn't statistically significant, re-fit with an ordinary first-order model as follows ...

```
fit <- lm(etch ~ x1 + x2, data=plasma)
coef(fit)
```

```
## (Intercept)            x1            x2
##      758.75        -66.25         43.75
```

... yielding the following fit which we shall use as the basis for constructing a path of steepest ascent.

$$\hat{y} = 758.75 - 66.25x_1 + 43.75x_2$$

The sign of x_1 is negative and the sign of x_2 is positive, so we shall seek improvements in `etch` rate by decreasing `gap` and increasing `power`. For every unit change in gap (x_1), the corresponding change in `power` (x_2) may be calculated as follows.

```
b1 <- coef(fit)[2]
b2 <- coef(fit)[3]
delta2 <- abs(b2/b1)
delta2
```

```
##      x2
## 0.6604
```

If we choose the gap step size (in coded units) to be $\Delta x_1 = -1$, then the power step size is `delta2 = 0.6604`. All right, let's consider three potential new runs along that path with $\Delta x_1 \in \{-1, -2, -3\}$, stored in an R `data.frame` as follows.

```
Dnew <- data.frame(x1=(-1):(-3), x2=(1:3)*delta2)
```

To aid in visualization, code below first evaluates the fitted model, through its predictive equations, on a grid for image/contour plotting.

```
x1 <- seq(-4, 1.5, length=100)
x2 <- seq(-1.5, 4*delta2, length=100)
g <- expand.grid(x1=x1, x2=x2)
yhat <- matrix(predict(fit, newdata=g), ncol=length(x2))
```

Figure 3.3 shows that surface, again via contours overlayed on a heat plot and design indicated as filled dots. An arrow outlines the path of steepest ascent, and three open circles along that path are derived from Dnew, calculated above.

```
image(x1, x2, yhat, col=cols)
contour(x1, x2, yhat, add=TRUE)
points(plasma$x1, plasma$x2, pch=19)
points(0, 0)
arrows(0, 0, -3.5, 3.5*delta2)
points(Dnew)
```

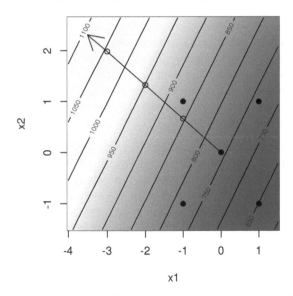

FIGURE 3.3: Steps along the path of steepest ascent for plasma etch data.

Steepest ascent path in hand, including particular input settings, the next step is to perform new runs at those settings. Doing that requires inputs in natural units, which can be derived one point at a time or for the path at large. The latter in our case is $(\Delta \text{gap}, \Delta \text{power}) = (-0.2\text{cm}, 16.5W)$, and the file plasma_delta.txt[5] contains inputs and responses for three runs along this path, i.e., via $\Delta = \{1, 2, 3\}$. These are shown in Table 3.3.

---

[5]http://bobby.gramacy.com/surrogates/plasma_delta.txt

TABLE 3.3: Plasma observations along the path of steepest ascent.

Delta	gap	power	x1	x2	etch	p.etch
1	1.2	316.5	-1	0.66	845	853.9
2	1.0	333.0	-2	1.32	950	949.0
3	0.8	349.5	-3	1.98	1040	1044.1

```
plasma.delta <- read.table("plasma_delta.txt", header=TRUE)
plasma.delta$p.etch <- predict(fit, newdata=plasma.delta[,4:5])
kable(plasma.delta,
  caption="Plasma observations along the path of steepest ascent.")
```

In the jargon, the first run, corresponding to $\Delta = 1$, is called a *confirmation test*. It's reassuring to observe that the output (about 845) is close to our prediction (854). This output, assuming we haven't seen/performed the other two runs yet, suggests we're moving in the right direction. However, 845 and 854 are not better than the best run on our original design.

```
max(plasma$etch)
```

```
## [1] 890
```

Therefore it stands to reason (with caution) that we could go a little farther along the path of steepest ascent and expect further gains. Indeed that is the case, and again our predicted and observed outputs farther along the path are in remarkable agreement. The response from the last run, corresponding to $\Delta = 3$, indicates that the experimental objective of an `etch` rate of 1000 Å/min has been met. Note that the actual `etch` response (1040) is slightly lower than our predicted one (1044.125).

Considering our success, it might be tempting to explore farther along the path. Actually, the conventional wisdom here – once substantially outside the original experimental region – would be to re-design an experiment centered at the new best value in order to allow for a potential *mid-course correction*, i.e., to better align the estimated and actual path of steepest ascent. Before turning to another, bigger example, let's take a moment to codify the method in terms of an easy-to-find (and follow) algorithm, say for help on homework exercises (§3.3). The description in Algorithm 3.1 is liberally cribbed from Myers et al. (2016).

A bigger example: shrinkage

To accommodate for shrinkage[6] in injection-molding processes it's common for dies for parts to be built larger than their nominal, or desired size. In order to minimize the amount of shrinkage for a particular part, an experiment was conducted varying four factors known to impact shrinkage. The experiment consisted of an un-replicated 2^4 factorial design using the factors outlined in Table 3.4.

Let's extract those ranges as an R `matrix` so that they may be used by the implementation to follow.

---

[6]http://www.dc.engr.scu.edu/cmdoc/dg_doc/develop/process/physics/b3500001.htm

---

Algorithm 3.1 Steepest Ascent for First-Order Models

Assume that the point $x_1 = x_2 = \cdots = x_m = 0$ is the base or origin point.

Require fitted first-order model with coefficients $b_1, \ldots, b_m$.

Then

1. Choose a reference variable index $j \in \{1, \ldots, m\}$. It doesn't matter which; common choices include:
   a. the variable with the largest absolute coefficient $|b_j|$, possibly adjusted by standard error $|b_j|/s_{b_j}$;
   b. the input most familiar to the practitioner;
   c. the variable with the widest (relative) range as measured in natural units.
2. Choose a step size Δx_j in coded units.
3. Calculate the step size in the other variables relative to x_j.

$$\Delta x_k = \frac{b_k}{b_j/\Delta x_j}, \quad k \neq j$$

Return Δx_k, for $k = 1, \ldots, m$, possibly after first mapping from coded back to natural variables.

TABLE 3.4: Mapping of shrinkage inputs.

Factor	Name (natural units)	-1	+1
x_1	Injection velocity (ft/sec)	1.0	2.0
x_2	Mold temperature (°C)	100	150
x_3	Mold pressure (psi)	500	1000
x_4	Back pressure (psi)	75	120

```
r <- cbind(c(1,2), c(100, 150), c(500,1000), c(75,120))
colnames(r) <- c("vel", "temp", "mpress", "bpress")
```

The center of the space, representing our baseline setting, is ...

```
base <- r[1,] + (r[2,] - r[1,])/2
base
```

```
##     vel    temp mpress bpress
##     1.5   125.0  750.0   97.5
```

... in natural units. This corresponds to the zero vector on the coded scale. Unfortunately we don't have access to the actual observed responses (which are measured in units of 10^{-4} as deviations from nominal) at each input combination, but we do have a first-order model fit to that data (in coded units):

$$\hat{y} = 80 - 5.28x_1 - 6.22x_2 - 1.21x_3 - 1.07x_4.$$

Following Step 1 in Algorithm 3.1, suppose that we were to select x_1 as the reference variable

TABLE 3.5: Steps along path of steepest ascent for shrinkage example.

		x1	x2	x3	x4	ft/sec	°C	psi	psi
Base +0Δ	0	0.000	0.0000	0.0000	1.5	125.0	750.0	97.5
Base +1Δ	1	1.178	0.2292	0.2026	2.0	154.5	807.3	102.1
Base +2Δ	2	2.356	0.4583	0.4053	2.5	183.9	864.6	106.6
Base +3Δ	3	3.534	0.6875	0.6079	3.0	213.4	921.9	111.2
Base +4Δ	4	4.712	0.9167	0.8106	3.5	242.8	979.2	115.7

defining the step size. In Step 2 choose $\Delta x_1 = 1$, corresponding to a 0.5 ft/sec injection velocity. Then, relative to that choice we have the following adjustments Δ.

```
b <- c(-5.28, -6.22, -1.21, -1.07)
delta <- b/b[1]
delta
```

```
## [1] 1.0000 1.1780 0.2292 0.2027
```

R code below gathers input settings, collected into a `data.frame`, along the path of steepest descent up to 4Δ in coded units.

```
path <- rbind(0,
  apply(matrix(rep(delta, 4), ncol=4, byrow=TRUE), 2, cumsum))
colnames(path) <- paste0("x", 1:4)
rownames(path) <- paste0("Base +", 0:4, "Δ")
```

Before displaying coordinates on the path, let's convert to natural units. The corresponding changes on that scale are ...

```
dnat <- delta*(r[2,] - r[1,])/2
dnat
```

```
##     vel    temp mpress bpress
##    0.50   29.45  57.29   4.56
```

... augmenting our `data.frame` as shown in Table 3.5.

```
pnat <- rbind(base, matrix(rep(base, 4), ncol=4, byrow=TRUE) +
  apply(matrix(rep(dnat, 4), ncol=4, byrow=TRUE), 2, cumsum))
colnames(pnat) <- c("ft/sec", "°C", "psi", "psi")
rownames(pnat) <- rownames(path)
kable(cbind(path, pnat), digits=5,
  caption="Steps along path of steepest ascent for shrinkage example.")
```

Then it's simply a matter of convincing whomever manages the process to tinker with settings along that schedule. Sometimes that's easier said than done. It could help to give them an inkling of the likelihood of success, especially when it comes to exploring 1Δ away from the center point, which may be well outside the bounding box of the original experiment. Besides eliminating statistically useless predictors, so far the procedure lacks a fundamental

– some may say hallmark – aspect of statistical decision making: an acknowledgment of sampling variability. So far it's just least squares and calculus.

3.1.2 Confidence regions

Understanding uncertainty is key to effective analysis, and incorporating that uncertainty into decision making is a recurring theme in this text. Doing so isn't always an easy task, sometimes involving many shortcuts or conversely requiring Monte Carlo to obtain a rough accounting of prevailing variabilities. Fortunately, in the case of linear/first-order models the process is rather straightforward thanks to a high degree of analytical tractability in requisite calculations. The resulting uncertainty set, which is actually a range of angles tracing out a cone around the path of steepest ascent, can be of great value to practitioners. A tight region around the path (narrow set of angles) indicates promise for success; a looser set may suggest any new runs are speculative and may not be worth the cost.

An underlying theme in our presentation here is one of propagating uncertainty. The path of steepest ascent is based on estimated regression coefficients, which in turn have sampling distributions whose standard errors are readily available in output summaries from software. Introductory regression texts would have a section explaining how that variability propagates to predictive equations, leading to predictive intervals. Here we show how they may be propagated to the path of steepest ascent.

We have seen how m least squares estimated slope coefficients $b_1, b_2, \ldots, b_m$ determine the path of steepest ascent in an m-dimensional design space, via movement relative to a reference coordinate x_j. Standard errors for those coefficients are derived, in a classical linear modeling setup, by comparing estimated b_j coefficients to their true but unknown values β_j. Specifically, if t_{ν_b} is the standard (mean zero, scale one) Student-t distribution[7] with ν_b degrees of freedom (DoF), then

$$\frac{b_j - \beta_j}{s_{b_j}} \sim t_{n-m-1}, \quad \text{for} \ \ j = 1, \ldots, m. \tag{3.2}$$

The shorthand $b_j \sim t_{\nu_b}(\beta_j, s_{b_j}^2)$, mimicking the parameterization of a Gaussian distribution, is common. Here DoF $\nu_b = n - m - 1$ is the same for all j. Sampling distribution scales $s_{b_j}^2$, whose formulas can be found in most regression texts (more in §3.2.3), are usually unique to each coordinate j. However it turns out that in the case of a standard two-level orthogonal design on coded inputs they're constant across j. That is, $s_{b_j}^2 \equiv s_b^2$ for all $j \in \{1, \ldots, m\}$, a fact which we shall leverage shortly in order to simplify calculations.

Now recall that one moves along the path of steepest ascent as $x_j = \rho b_j$, so reversing that logic a bit and assuming the first-order model is correct (at least locally), we have

$$\beta_j = \gamma X_j, \quad j = 1, \ldots, m.$$

In other words, the true path may be traced via true but unknown coefficients β_j and unobserved constants X_j, called *direction cosines*. Direction cosines are merely a mathematical device that allows one to ask about the chance, according to the sampling distribution, that particular coordinates are on the true path. Now γ, another unknown quantity, is the analog

---

[7]https://en.wikipedia.org/wiki/Student's_t-distribution

of ρ for true coefficients β_j and direction cosines X_j. This setup suggests a mechanism for pinning down γ through the following regression model without an intercept:

$$b_j = \gamma X_j + \varepsilon_j, \quad j = 1, \ldots, m,$$

leading to ordinary least squares solution

$$\hat{\gamma} = \frac{\sum_{j=1}^m b_j X_j}{\sum_{j=1}^m X_j^2}.$$

The usual residual sum of squares estimate of the variance of ε_j is

$$s_b^{2*} = \frac{1}{m-1} \sum_{j=1}^m (b_j - \hat{\gamma} X_j)^2 \quad \text{on } m-1 \text{ DoF.}$$

Since s_b^2 (our common variance for b_j from the sampling distribution under a certain two-level orthogonal design) and s_b^{2*} are scaled residual sums of squares, they're χ^2 distributed. Moreover they're independent, so their ratio has an F distribution[8],

$$\frac{s_b^{2*}}{s_b^2} \sim F_{m-1, \nu_b}$$

which can be used to outline a confidence region for direction cosines X_j. A $100(1-\alpha)\%$ confidence region is defined as the central set of $X_1, \ldots, X_m$-values for which

$$\sum_{j=1}^m \frac{(b_j - \hat{\gamma} X_j)^2 / (m-1)}{s_b^2} \leq F_{m-1, \nu_b}^\alpha,$$

where F_{m-1, ν_b}^α is the $100(1-\alpha)\%$ point of the F distribution with $m-1$ numerator and ν_b denominator DoF.

This region turns out to be a cone, or a hypercone in more than three variables. The apex of the hypercone is at the design origin. (Remember these are coded variables!) After expanding out $\hat{\gamma}$ and considering, say, all points at unit distance from the origin, i.e., satisfying $\sum_j X_j^2 = 1$, one obtains the following confidence hyper-ring:

$$\sum_{j=1}^m b_j^2 - \frac{\left(\sum_{j=1}^m b_j X_j \right)^2}{(m-1) \sum_{j=1}^m X_j^2} \leq s_b^2 F_{m-1, \nu_b}^\alpha.$$

To see how to use that result, consider our earlier example (3.1) where $b_1 = 3$ and $b_2 = -1.5$. Since that example didn't use actual data, we don't have estimates of standard error to work with, so let's further suppose here that the variances of the coefficients were $s_b^2 = \frac{1}{4}$, under $\nu_b = 4$ DoF. Given those values and . . .

```
qf(0.95, 1, 4)
```

```
## [1] 7.709
```

---

[8]https://en.wikipedia.org/wiki/F-distribution

... the 95% confidence region for the path of steepest ascent at a fixed distance $X_1^2 + X_2^2 = 1.0$ is determined by solutions (X_1, X_2) to

$$9 + 2.25 - (3X_1 - 1.5X_2)^2 \leq \frac{1}{4}(7.71),$$

$$\text{or} \quad (3X_1 - 1.5X_2)^2 \geq 9.3225.$$

Well that's neat, but what can we do with that? One option is to simply plug-in X_j-values and see if they're "in there". If they are, then we can (loosely) say that we have a high confidence that they could be on the true path of steepest ascent. The correct interpretation is that we can't reject the null hypothesis, at the 5% level, that they're on the path of steepest ascent. In 2d and maybe 3d you can plot, but such graphics are prettier than they are useful. Figure 3.4 captures the confidence set above visually by extending Figure 3.1. Green dots are inside and red dots, as well as all points in the left quadrants, are outside.

```
x1 <- seq(0, 1, length=1000)
x2 <- sqrt(1 - x1^2)
x1 <- c(x1, x1)
x2 <- c(x2, -x2)
ci95 <- (3*x1 - 1.5*x2)^2 >= 9.3225
plot(0, type="n", xlim=c(-2,3.5), ylim=c(-2,3.5), xlab="x1", ylab="x2")
points(D, pch=19)
points(0, 0)
arrows(0, 0, 3, -1.5)
points(x1, x2, col=2 + ci95, pch=19, cex=0.5)
```

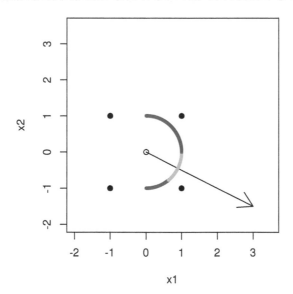

FIGURE 3.4: Confidence region along the path of steepest ascent, extending Figure 3.1.

Tracing out a full confidence hypercone, i.e., for non-unit distances from the origin, is left to an exercise in §3.3. The width of the hypercone – size of the green area in Figure 3.4 – increases as distance $(X_1 + X_2)^2$ increases from the center of the design region. The rate of that increase can be depicted by an angle θ indicating directions included in (or excluded from) the 95% confidence region, giving a sense of the tightness of the region. It turns out that

$$\theta = \arcsin \left\{ \frac{(m-1)s_b^2 F_{m-1,\nu_b}^\alpha}{\sum_{j=1}^m b_j^2} \right\}^{1/2}.$$

Returning back to our simple example (3.1), we have the following.

```
theta <- asin(sqrt(0.25*7.71*(1/11.25)))
theta
```

```
## [1] 0.4268
```

So the interpretation is that angles lower than around 0.43 radians, or 24°, are within the confidence region for any particular fixed distance from the center of the design. Larger angles are outside. Since this simple example is two-dimensional, this means that the green arc in Figure 3.4 traces out about 48° which is $100 \times 48/360 = 13.33\%$ of the input space, i.e., excluding 86.67 degrees. Obviously the larger the amount excluded the better, indicating greater knowledge about the computed path of steepest ascent.

3.1.3   Constrained ascent

It can happen that a steepest ascent path moves into an impermissible region of the design space, for one or more variables. The simplest example of this is where an input exceeds the practical limits of the apparatus involved in the process or experiment. In such cases, we desire a modified path of ascent with a constraint imposed. Here focus is on constraints which may be coded in linear combination, nesting bound constraints as a special case. Extensions for multiple linear constraints are a matter of repeated application, which may be easier said than done. The presentation here is provided primarily as a precursor to our later, far more generic treatment of constrained Bayesian optimization in §7.3.1 where multiple (even nonlinear and unknown) constraints are handled in a unifying framework.

Consider a boundary constraint of the form

$$c_0 + \sum_{j=1}^m c_j x_j \leq 0,$$

where possibly some $c_j = 0$, indicating that variable j is unconstrained. Usually the constraint is formulated in the natural (uncoded) variable, in which case it must first be written in coded form for manipulation.

How do we go about finding the path of steepest ascent subject to a constraint? The recipe is pretty simple if the starting point is within the constraint satisfaction region. Begin by proceeding along the path of steepest ascent, starting at the center of the design, until the path meets the constraint boundary, which is a line when $m = 2$ or a (hyper) plane for $m > 2$. Let O denote the intersection point between steepest ascent path and constraint boundary. From O follow a modified path that modulates steepest ascent in light of the constraint, the description of which will be provided shortly.

As a warm-up example, consider the following setup in $m = 2$ input dimensions. Suppose the unconstrained path of steepest ascent is defined as $x_2 = 2x_1$ and the constraint boundary follows the formula above, with equality, using constants $(c_0, c_1, c_2) = (4, 1, -2)$ such that

$$0 = c_0 + c_1 x_1 + c_2 x_2$$
$$x_2 = -\frac{c_0}{c_2} - \frac{c_1}{c_2} x_1$$
$$= 2 + 0.5x_1.$$

Those two lines intersect when

$$2x_1 = 2 + 0.5x_1 \quad \rightarrow \quad (x_1, x_2) = (4/3, 8/3),$$

defining the point $O = (4/3, 8/3)$.

```
O <- c(4/3, 8/3)
```

Figure 3.5 shows the path of steepest ascent in red, originating from $(0, 0)$ and proceeding until O is reached. Once at O a modified path proceeds along the constraint boundary shown in green, following the direction best aligned with that of steepest ascent.

```
plot(0,0, type="n", xlab="x1", ylab="x2", xlim=c(0,4), ylim=c(0,5))
arrows(0, 0, O[1], O[2], col=2, lwd=3)
arrows(O[1], O[2], 3, 2 + 3/2, col=3, lwd=3)
text(O[1] - 0.1, O[2] + 0.2, "O")
abline(0, 2, lty=2)
abline(2, 0.5, lty=3)
text(1.5, 1, "steepest ascent", col=2)
text(2.5, 2.7, "constrained", col=3)
```

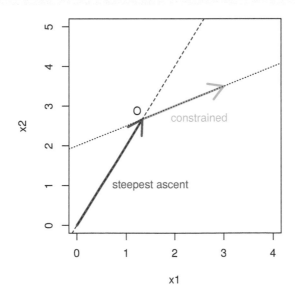

FIGURE 3.5: Modified path under constrained steepest ascent.

Although visually intuitive in two dimensions, the scheme requires a bit of math to operationalize more generally. That is, to determine the vector (in particular the direction) of ascent once the intersection point O is reached. Observe that the modified path can be parameterized by the vector $b_j - dc_j$, for $j = 1, \ldots, m$, for which

$$\sum_{j=1}^{m} (b_j - dc_j)^2 \quad \text{is minimized.}$$

In other words, the direction along the constraint line (or plane) is taken so as to be the closest to the original path. Just like with our confidence region calculations above, this is another "funny regression" situation in which b_j are being regressed against c_j, without an intercept, through which we calculate a coefficient d, playing here the role of regression slope. Ordinary least squares calculations give

$$d = \frac{\sum_{j=1}^{m} b_j c_j}{\sum_{j=1}^{m} c_j^2}.$$

Using that value, the modified portion of the path would begin at O and proceed in the direction described by an m-vector with components $b_j - dc_j$, for $j = 1, \ldots, m$. How to find the intersection point O? In our simple example we expressed the steepest ascent and constraint relationships as x_2 in terms of x_1. Setting the two x_2 expressions equal to one another allowed us to solve for x_1. That's made more general as follows. We know that a steepest ascent path is given by $x_j = \rho b_j$, for $j = 1, \ldots, m$ and that $c_0 + \sum_{j=1}^{m} c_j x_j = 0$ must also hold. Therefore, they collide at ρ_0 satisfying

$$c_0 + \left(\sum_{j=1}^{m} c_j b_j \right) \rho_o = 0. \qquad \text{Solving gives} \qquad \rho_o = \frac{-c_0}{\sum_{j=1}^{m} c_j b_j}.$$

Using that calculation, coordinates of the intersection point are given as $O = (x_1^o, \ldots, x_m^o)$ where $x_j^o = \rho_o b_j$, for $j = 1, \ldots, m$. One may then move along the modified portion of the path, starting at O, following

$$x_j = x_j^o + \lambda(b_j - dc_j),$$

where d is determined by the least squares solution, above, and $\lambda \geq 0$ is chosen by the practitioner. In this way the process for following the modified path is identical to that in Algorithm 3.1 except that the starting point is O rather than the zero-vector, and coefficient b_j is replaced by $b_j - dc_j$.

Example: fabric strength

This example concerns the breaking strength in grams per square inch of a certain type of fabric as a function of three components ξ_1, ξ_2, ξ_3. Raw data is unavailable, but Table 3.6 provides ranges from an experiment involving three input variables which are measured in grams.

Consider a constraint on the first two variables, ξ_1 and ξ_2, as follows:

$$\xi_1 + \xi_2 \leq 500.$$

The first step is to transform to coded variables, in particular mapping the constraint to coded variables. R code below derives constants involved in that enterprise.

TABLE 3.6: Summarizing fabric strength data.

Material	-1	+1
1	100	150
2	50	100
3	20	40

```
upper <- c(150, 100, 40)
lower <- c(100, 50, 20)
scale <- (upper - lower)/2
shift <- scale + lower
toxi <- data.frame(scale=scale, shift=shift)
toxi
```

```
##   scale shift
## 1    25   125
## 2    25    75
## 3    10    30
```

That results in the following mapping,

$$x_1 = \frac{\xi_1 - 125}{25} \quad x_2 = \frac{\xi_2 - 75}{25} \quad x_3 = \frac{\xi_3 - 30}{10},$$

whose inverse is straightforward and will be provided below. Under this mapping the constraint reduces to

$$25x_1 + 25x_2 \leq 300,$$

which leads to the following constants c_0, c_1, c_2, c_3:

```
c <- c(-300, 25, 25, 0)
```

Since we don't have the data we can't perform a fit, but suppose we have the following from an off-line analysis,

$$\hat{y} = 150 + 1.7x_1 + 0.8x_2 + 0.5x_3,$$

represented in R as b below.

```
b <- c(150, 1.7, 0.8, 0.5)
```

Now ρ_o can be calculated as ...

```
rhoo <- -c[1]/sum(c[-1]*b[-1])
rhoo
```

```
## [1] 4.8
```

... and as a result the modified path starts at:

```
xo <- rhoo*b[-1]
xo
```

```
## [1] 8.16 3.84 2.40
```

Ordinary least squares provides d.

```
d <- as.numeric(coef(lm(b[-1] ~ c[-1] - 1)))
d
```

```
## [1] 0.05
```

Then the coordinates of the hybrid path, including ordinary steepest ascent up to O and modified thereafter, are given by the following function of the quantities above. Step sizes are controlled with scale parameter (vector) λ specified by the practitioner.

```
hpath <- function(lambda, b, c, rhoo, d)
  {
  ## steepest ascent up to one step past the constraint boundary
  delta <- b/b[1]
  path <- matrix(0, ncol=length(b), nrow=1)
  while(1) {
    lpath <- nrow(path)
    path <- rbind(path, path[lpath,] + delta)
    if(c[1] + sum(path[lpath + 1,]*c[-1]) > 0) break
  }

  ## intersection point plus steps along the modified portion
  cpath <- rhoo*b
  for(i in 1:length(lambda)) {
    cpath <- rbind(cpath, rhoo*b + lambda[i]*(b - d*c[-1]))
  }

  ## pasting the hybrid path together and naming the rows and columns
  path <- rbind(path[1:lpath,], cpath)
  colnames(path) <- paste("x", 1:length(b), sep="")
  rownames(path) <- c(rep("u", lpath), "o", rep("c", length(lambda)))
  return(path)
  }
```

Notice that the function makes several assumptions, including

- `lambda` should not contain zero, but the zero setting (corresponding to intersection O) is automatically calculated and included in the hybrid path;
- the steepest ascent path must eventually violate the constraint, otherwise there's an infinite loop;
- the baseline variable is x_1, corresponding to $j = 1$ in Algorithm 3.1.

Evaluating `hpath` with a sequence of `lambda` values, chosen somewhat arbitrarily, provides coordinates along the hybrid steepest ascent and modified path.

TABLE 3.7: Coordinates on hybrid path.

	xi1	xi2	xi3
u	125.0	75.00	30.00
u	150.0	86.76	32.94
u	175.0	98.53	35.88
u	200.0	110.29	38.82
u	225.0	122.06	41.76
u	250.0	133.82	44.71
u	275.0	145.59	47.65
u	300.0	157.35	50.59
u	325.0	169.12	53.53
o	329.0	171.00	54.00
c	340.2	159.75	59.00
c	351.5	148.50	64.00
c	362.8	137.25	69.00
c	374.0	126.00	74.00

```
lambda <- c(1,2,3,4)
path <- hpath(lambda, b[-1], c, rhoo, d)
```

Finally, we may use `toxi` to map the hybrid path to the natural scale before formatting it for display in Table 3.7.

```
A <- matrix(rep(toxi[,1], nrow(path)), ncol=ncol(path), byrow=TRUE)
B <- matrix(rep(toxi[,2], nrow(path)), ncol=ncol(path), byrow=TRUE)
pathxi <- A * path + B
colnames(pathxi) <- paste("xi", 1:3, sep="")
kable(pathxi, caption="Coordinates on hybrid path.")
```

The first row is the center of the design region, and the tenth records point O, both on the natural scale. Observe that most of the coordinates on the hybrid path are well outside of the bounding box of the original design. Perhaps it's unwise to venture out into that region before performing a new experiment – potentially leading to a course correction – part way between the edge of that box and O. With that logic, a good candidate for the center point of the new design may be the fourth or fifth step along the hybrid path:

```
pathxi[4:5,]
```

```
##   xi1   xi2   xi3
## u 200 110.3 38.82
## u 225 122.1 41.76
```

What to do if we encounter runs in a follow-up experiment that don't show improvement in the response? This may indicate that the first-order model isn't a good approximation. A notion of "locale" might not be appropriate in this wider design space. Higher-order modeling could help, motivating the methods in the next section.

3.2   Second-order response surfaces

We can only get so far with steepest ascent on first-order fits, even with interactions. Linear models only offer decent approximation locally. That local scope may in practice be exceedingly small, especially near the optimal input configuration. Eventually observed responses will systematically diverge from estimated ones along the steepest ascent path, at least for most interesting problems. Divergence could be substantial. When that happens it's worth entertaining a richer class of models, the simplest of which is a second-order (linear) model.

As a reminder, the *second-order model* is characterized by the following linear equation.

$$y = \beta_0 + \sum_{j=1}^{m} \beta_j x_j + \sum_{j=1}^{m} \beta_{jj} x_j^2 + \sum \sum_{j<k} \beta_{jk} x_j x_k + \varepsilon$$

I won't spend much time on appropriate designs for second-order models. However it's worth noting that, since the model contains $1 + 2m + m(m-1)/2$ parameters, the design must therefore contain at least as many distinct locations in order for all unknown regression coefficients β to be estimable via least squares/likelihood-based methods. Moreover the design must contain at least three levels of each variable to estimate any pure quadratic terms. For more details on design for second-order models see Chapters 8–9 of Myers et al. (2016).

3.2.1   Canonical analysis

All second-order response surface models have a *stationary point* which, as shown visually in Figures 1.5–1.8, may be a maximum, minimum, or a saddle point. When the stationary point is a saddle point, the model is sometimes called a *saddle* or *minimax system*. Detecting the nature of the system with a design focused around nearby operating conditions, and determining the location of the stationary point, represent an integral first step in second-order analysis. As in the direction of the path of steepest ascent, the nature and location of a stationary point depends on the signs and magnitudes of estimated coefficients. In particular, estimated interaction/pure quadratic effects play a vital role, and their standard errors convey uncertainty about local behavior of the system.

As a simple example to fix ideas before delving deeper, consider a fitted second-order model given by

$$\hat{y} = 100 + 5x_1 + 10x_2 - 8x_1^2 - 12x_2^2 - 12x_1 x_2. \tag{3.3}$$

Because $m = 2$, simple graphics can help determine the location and nature of the stationary point. The code chunk below defines the fit, and then evaluates it on a grid in a rectangular input domain.

```
second.order <- function(x1, x2)
 {
   100 + 5*x1 + 10*x2 - 8*x1^2 - 12*x2^2 - 12*x1*x2
```

```
 }

x1 <- x2 <- seq(-4, 4, length=100)
g <- expand.grid(x1, x2)
y <- matrix(second.order(g[,1], g[,2]), ncol=length(x2))
```

Figure 3.6 shows the fit in perspective (left) and image/contour (right) formats. Recall that lighter colors correspond to higher values.

```
par(mfrow=c(1,2))
persp(x1, x2, y, theta=30, phi=30, zlab="eta", expand=0.75, lwd=0.25)
image(x1, x2, y, col=cols)
contour(x1, x2, y, add=TRUE)
```

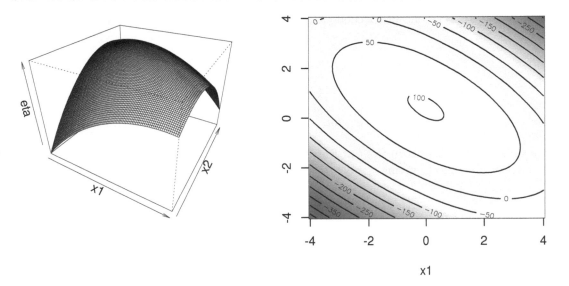

FIGURE 3.6: Simple maximum second-order response surface following Eq. (3.3).

By inspection, the response surface is concave down, a *simple maximum* like in Figure 1.5. Evidently it's maximized near the origin, perhaps with x_2 a little above zero. No need to ballpark it when we can find coordinates precisely. Calculus says that the stationary point is the solution to

$$\frac{\partial \hat{y}}{\partial x_1} = 0 \quad \text{and} \quad \frac{\partial \hat{y}}{\partial x_2} = 0.$$

This results in a system of linear equations ...

$$16x_1 + 12x_2 = 5$$
$$12x_2 + 24x_2 = 10,$$

... which can be solved in R as follows.

```
dy <- rbind(c(16, 12), c(12, 24))
xh <- solve(dy, c(5,10))
yh <- 100 + 5*xh[1] + 10*xh[2] - 8*xh[1]^2 - 12*xh[2]^2 - 12*xh[1]*xh[2]
c(x1=xh[1], x2=xh[2], y=yh)
```

```
##        x1       x2        y
##    0.0000   0.4167 102.0833
```

Therefore, the stationary point is at $(\hat{x}_1, \hat{x}_2) = (0, 0.42)$, and the response value at that location is $\hat{y}(\hat{x}) = 102.1$. To abstract a bit, towards obtaining a more generic recipe, it helps to write the fitted model in matrix notation as follows.

$$\hat{y} = b_0 + x^\top b + x^\top B x$$

Above, the scalar b_0, m-vector b, and $m \times m$ matrix B are estimates of intercept, linear (or main effects), and second-order coefficients, respectively. Specifically, B is the symmetric matrix

$$B = \begin{bmatrix} b_{11} & b_{12}/2 & \cdots & b_{1m}/2 \\ & b_{22} & \cdots & b_{2m}/2 \\ & & \ddots & \vdots \\ \text{sym.} & & & b_{mm} \end{bmatrix}$$

containing all coefficients in front of features which are derived from original inputs (§3.2) through products $x_j x_k$ giving

$$x^\top B x = \sum_{j=1}^{m} b_{jj} x_j^2 + \sum \sum_{j<k}^{m} b_{jk} x_j x_k.$$

The $1/2$ term arises due to the symmetry of B, so that two contributions add up to one in the final sum. Setting up those quantities for our toy 2d example visualized in Figure 3.6 proceeds as follows in R.

```
b0 <- 100
b <- c(5, 10)
B <- matrix(c(-8, -12/2, -12/2, -12), ncol=2, byrow=TRUE)
B
```

```
##        [,1] [,2]
## [1,]    -8   -6
## [2,]    -6  -12
```

At first this representation (especially the B matrix) seems too clever by half. Are double-sums really that bad? But the investment is worth its weight in implementation simplicity, and in off-loading to matrix properties for interpretation. For example, in this formulation it's straightforward to give a general expression for the location of the stationary point, x_s. One can differentiate $\hat{y}$ with respect to x to obtain

$$\frac{\partial \hat{y}}{\partial x} = b + 2Bx,$$

$$\text{giving} \quad x_s = -\frac{1}{2}B^{-1}b$$

after setting equal to zero and solving. The predicted response at the stationary point is obtained by plugging x_s back into the quadratic form for $\hat{y}$:

$$\hat{y}_s = b_0 + x_s^\top b + x_s^\top B x_s$$
$$= b_0 + x_s^\top b - \frac{1}{2}b^\top B^{-1} B x_s$$
$$= b_0 + \frac{1}{2}x_s^\top b.$$

Double-checking those formulas in R with our results above for the toy 2d example (3.3) indicates success.

```
xs <- - 0.5*solve(B) %*% b
ys <- b0 + 0.5*t(xs) %*% b
sols <- rbind(h=c(xh, yh), s=c(xs, ys))
colnames(sols) <- c("x1", "x2", "y")
sols
```

```
##    x1     x2      y
## h  0 0.4167 102.1
## s  0 0.4167 102.1
```

Now the nature of the stationary point is determined from the signs of the eigenvalues of the matrix B, the relative magnitudes of which are key to interpretation. Such is the real value in investing in a matrix representation. Technically, location of the stationary point and its predicted response are not material to analysis of B. However re-centering the system at x_s does aid interpretation. Translation and rotation of the axes, and inspection of eigenvalues described below, is referred to as the *canonical analysis* of the response system.

Toward that end, let $z = x - x_s$ so that

$$\hat{y} = b_0 + (z+x_s)^\top b + (z+x_s)^\top B(z+x_s)$$
$$= [b_0 + x_s^\top b + x_s^\top B x_s] + z^\top b + z^\top B z + 2x_s^\top B z$$
$$= \hat{y}_s + z^\top B z.$$

Again, $\hat{y}_s$ is the estimated response at the stationary point, and thus the final step comes from $x_s = -\frac{1}{2}B^{-1}b$, giving $2x_s^\top B z = -z^\top b$.

Once the system is centered at x_s, the next step is to rotate axes according to their principal components – in other words, to determine the principal axes of contours of the response surface. Let P be an $m \times m$ matrix whose columns contain normalized eigenvectors associated with eigenvalues of B. Denote $\Lambda = P^\top B P$ as the diagonal matrix of eigenvalues $\lambda_1, \ldots, \lambda_m$ corresponding to that system. Now, if $w = P^\top z$, then

$$\hat{y} = \hat{y}_s + w^\top P^\top B P w$$

$$= \hat{y}_s + w^\top \Lambda w$$

$$= \hat{y}_s + \sum_{j=1}^m \lambda_j w_j^2. \tag{3.4}$$

These $w_1, w_2, \ldots, w_m$ are called *canonical variables*, and form the basis for principal axes. The final line in the equations above nicely describes the nature of the stationary point x_s via signs and relative magnitudes of eigenvalues $\lambda_1, \ldots, \lambda_m$.

1. If the λ_j are negative for all j, the stationary point corresponds to a local maxima.
2. If λ_j are positive for all j, we have a local minima.
3. If any λ_j and λ_k for $j \neq k$ have *mixed* sign, we have a saddle point.

The development below explores some of these quantities on our running toy 2d example (Eq. (3.3) and Figure 3.6). To start with, eigenvalues may be calculated as follows.

```
E <- eigen(B)
lambda <- E$values
o <- order(abs(lambda), decreasing=TRUE)
lambda <- lambda[o]
lambda
```

```
## [1] -16.325  -3.675
```

Notice that both are negative, suggesting the stationary point x_s is a maximum. (We already knew that, but it doesn't hurt to check understanding.) The code stores eigenvalues in decreasing order. Observe that the first principal axis is elongated relative to the other by nearly a factor of four. Actually those numbers are in units of squared distance in the input space, so the true scaling factor is more like two. Code below extracts eigenvectors, taking their ordering from the eigenvalues.

```
V <- E$vectors[,o]
```

These eigenvectors have unit norm, so it can help to scale them a bit when visualizing. Figure 3.7 shows principal axes estimated for this response surface, upscaling eigenvectors by a factor of ten so that they cover the entire range of x_1 and x_2.

```
image(x1, x2, y, col=cols)
contour(x1, x2, y, add=TRUE)
lines(c(-V[1,1], V[1,1])*10 + xs[1], c(-V[2,1], V[2,1])*10 + xs[2], lty=2)
lines(c(-V[1,2], V[1,2])*10 + xs[1], c(-V[2,2], V[2,2])*10 + xs[2], lty=2)
```

When we work with the system in canonical variables through

$$\hat{y} = 102.0833 - 16.3246 w_1^2 - 3.6754 w_2^2$$

we're essentially working on the original system pictured above, although there's rarely a good reason to do so in practice. Eigenvalue analysis gives us all the information we need to proceed with further calculations, directly on original or coded inputs. Before outlining next

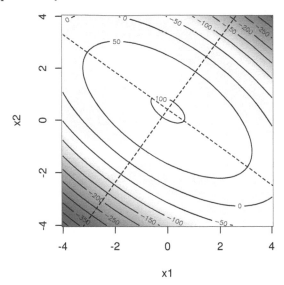

FIGURE 3.7: Principal axes for the response surface in Eq. (3.3).

steps, the example below provides a second look, but in a somewhat more realistic setting with a response surface estimated from actual data.

Example: chemical process

Consider an investigation into the effect of two variables, reaction temperature (ξ_1) and reactant concentration (ξ_2), on the percentage conversion of a chemical process (y). By way of a bit of back-story, we arrived at the current setup after an initial screening experiment was conducted involving several factors, with temperature and concentration being isolated as the two most important variables. Since experimenters believed that the process was operating in the vicinity of the optimum, a quadratic model is appropriate for the next stage of analysis. Measurements of the response variables were subsequently collected on a central composite design whose center point was replicated four times.

```
chem <- read.table("chemical.txt", header=TRUE)
uchem <- unique(chem[,1:2])
reps <- apply(uchem, 1,
  function(x) { sum(apply(chem[,1:2], 1, function(y) { all(y == x) })) })
```

Figure 3.8 provides a visualization of that design.

```
plot(uchem, type="n")
text(uchem, labels=reps)
```

Note that the data file contains measurements in both natural and coded units. After expanding out into squared and interaction features, the second-order model may be fit to these data (in coded units) with the following code.

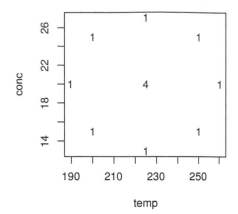

FIGURE 3.8: Central composite design for an experiment involving a chemical process.

TABLE 3.8: Chemical process data.

	Estimate	Std. Error	t value	Pr(>|t|)
(Intercept)	79.750	1.2462	63.996	0.0000
x1	10.178	0.8812	11.551	0.0000
x2	4.216	0.8812	4.785	0.0030
x11	-8.500	0.9852	-8.628	0.0001
x22	-5.250	0.9852	-5.329	0.0018
x12	-7.750	1.2462	-6.219	0.0008

```
X <- data.frame(x1=chem$x1, x2=chem$x2, x11=chem$x1^2, x22=chem$x2^2,
  x12=chem$x1*chem$x2)
y <- chem$y
fit <- lm(y ~ ., data=X)
kable(summary(fit)$coefficients, caption="Chemical process data.")
```

Observe in Table 3.8 that all estimated coefficients are statistically significant at the typical 5% level. The fitted model may be summarized as

$$\hat{y} = 79.75 + 10.18x_1 + 4.22x_2 - 8.50x_1^2 - 5.25x_2^2 - 7.75x_1x_2.$$

To obtain some visuals, code below builds a predictive grid in natural inputs, and converts these into coded units for prediction.

```
r <- cbind(c(200, 250), c(15,25))
d <- (r[2,] - r[1,])/2
xi1 <- seq(min(chem$temp), max(chem$temp), length=100)
xi2 <- seq(min(chem$conc), max(chem$conc), length=100)
xi <- expand.grid(xi1, xi2)
x <- cbind((xi[,1] - r[2,1] + d[1])/d[1], (xi[,2] - r[2,2] + d[2])/d[2])
```

Using that grid in coded units, a `data.frame` of features is built, and fed into the `predict.lm` method.

```
XX <- data.frame(x1=x[,1], x2=x[,2],
  x11=x[,1]^2, x22=x[,2]^2, x12=x[,1]*x[,2])
p <- predict(fit, newdata=XX)
```

Figure 3.9 shows an image/contour view into those predictions on natural axes.

```
xlab <- "Temperature (°C)"
ylab <- "Concentration (%)"
image(xi1, xi2, matrix(p, nrow=length(xi1)),
  col=cols, xlab=xlab, ylab=ylab)
contour(xi1, xi2, matrix(p, nrow=length(xi1)), add=TRUE)
```

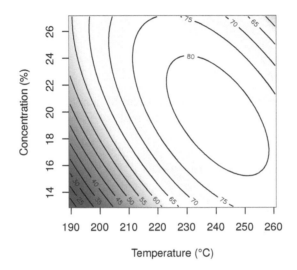

FIGURE 3.9: Fitted response surface for chemical process data whose fit is summarized in Table 3.8.

Canonical analysis begins by determining the location of stationary point x_s. Essentially cutting-and-pasting from above ...

```
b <- coef(fit)[2:3]
B <- matrix(NA, nrow=2, ncol=2)
diag(B) <- coef(fit)[4:5]
B[1,2] <- B[2,1] <- coef(fit)[6]/2
xs <- -(1/2)*solve(B, b)
xs
```

```
## [1]  0.62648 -0.06088
```

... which is converted back to natural units as follows.

```
xis <- xs*d + (r[2,] - d)
xis
```

```
## [1] 240.7  19.7
```

A check that this is correct is deferred until after we calculate canonical axes, below. Both eigenvalues of B are negative, so the stationary point is a maximum (as is obvious by inspecting the surface in Figure 3.9).

```
E <- eigen(B)
E
```

```
## eigen() decomposition
## $values
## [1]  -2.673 -11.077
##
## $vectors
##          [,1]    [,2]
## [1,]   0.5537 -0.8327
## [2,]  -0.8327 -0.5537
```

The code below saves those values, re-ordering them by magnitude, extracts eigenvectors (in that order), and converts them to their natural scale for visualization.

```
lambda <- E$values
o <- order(abs(lambda), decreasing=TRUE)
lambda <- lambda[o]
V <- E$vectors[,o]
Vxi <- V
for(j in 1:ncol(Vxi)) Vxi[,j] <- Vxi[,j]*d*10
```

We may then depict those axes, which intersect at stationary point x_s, as in Figure 3.10.

```
image(xi1, xi2, matrix(p, nrow=length(xi1)), col=cols, xlab=xlab, ylab=ylab)
contour(xi1, xi2, matrix(p, nrow=length(xi1)), add=TRUE)
lines(c(-Vxi[1,1], Vxi[1,1])+xis[1], c(-Vxi[2,1], Vxi[2,1])+xis[2], lty=2)
lines(c(-Vxi[1,2], Vxi[1,2])+xis[1], c(-Vxi[2,2], Vxi[2,2])+xis[2], lty=2)
points(xis[1], xis[2])
text(xis[1], xis[2], "xs", pos=4)
```

The canonical form of the second-order model is

$$\hat{y} = \hat{y}_s + \lambda_1 w_1^2 + \lambda_2 w_2^2,$$

where $\hat{y}_s$ is computed as follows.

```
ys <- as.numeric(coef(fit)[1]) + drop(0.5*t(xs) %*% b)
ys
```

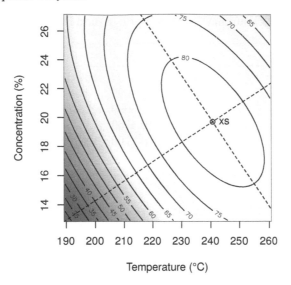

FIGURE 3.10: Principal axes for response surface in Figure 3.9.

```
## [1] 82.81
```

So we have

$$\hat{y} = 82.8099 - 11.0769w_1^2 - 2.6731w_2^2.$$

The input corresponding to $(w_1 = 0, w_2 = 0)$, which is x_s in coded units, is an obvious candidate for a confirmation test or as the center of a new central composite design over a narrower range of inputs. One potential criticism of this approach is that subsequent iterations retain little memory of previous (expensive) experimental work. Local second-order Taylor expansions are not well-suited to data collected at multiple – even somewhat nearby – locales. This will remain a motivating theme for much of the more global, nonparametric surrogate-based methods in later chapters. In the meantime, what happens when the eigenvalues aren't all of the same sign?

3.2.2 Ridge analysis

When the system involves a pure minimum or maximum the procedure outlined above is fairly straightforward. But what happens when eigenvalues are mixed, or near zero? It's perhaps more common for data on real response surfaces to lead to estimated models implying a degree of ambiguity. Saddle points are one possibility, although these are rather less common in practice, suggesting a bifurcation in the input space. A far more involved study may be required to choose between two or more competing locally optimal operating regimes on submanifolds of the study region. The murky spaces in-between, when some eigenvalues are approximately zero (and the others are of generally the same sign) are rather more common. This scenario indicates an elongation of the surface in the canonical direction corresponding to that eigenvalue, resulting in what's called a *ridge system*. An expanded toolset is required, not only to deal with uncertainties regarding the nature of the system – to test for and detect the nature of ridges – but subsequently to decide how to iterate towards improved operating conditions: to "rise the ridge".

TABLE 3.9: Linear model summary for rising ridge data.

	Estimate	Std. Error	t value	Pr($>$|t|)
(Intercept)	50.263	0.3894	129.079	0e+00
A	-12.417	0.3099	-40.068	0e+00
B	8.283	0.3099	26.730	0e+00
A2	-4.108	0.4769	-8.614	3e-04
B2	-9.108	0.4769	-19.098	0e+00
AB	11.125	0.3795	29.312	0e+00

Ridge systems are classified into two types depending on the estimated location of the stationary point x_s, relative to the experimental design used to fit the second-order model. If x_s is within the region of the design then we have a *stationary ridge system*. This is a fortunate circumstance since it means that many inputs, along the ridge, provide nearly the same, almost optimal result. Practitioners therefore have some freedom to choose among operating conditions along that ridge, perhaps according to other criteria. For example if one setting along the ridge is easier to implement, or represents the smallest divergence from previous operating conditions, that setting might be preferred over others. On the other hand, if x_s isn't inside the design region, this suggests that additional experimentation may be in order. We have a *rising* or *falling ridge*. Remote x_s may be spurious, so small steps in its direction are preferred over big jumps into a new regime. Any tests failing to reject a hypothesis about one or more zero eigenvalues, leading to a ridge local to the current design, will likely need to be re-cvaluated in the new experimental region.

As a warm-up example to make things a little more concrete, consider data from an experiment on a square region with two factors, A and B, using a central composite design with three center runs. Code supporting the second-order summary provided in Table 3.9 reads in data, expands into second-order features and performs the least squares fit.

```
rr <- read.table("risingridge.txt", header=TRUE)
rr$A2 <- rr$A^2
rr$B2 <- rr$B^2
rr$AB <- rr$A*rr$B
fit <- lm(y ~ ., data=rr)
kable(summary(fit)$coefficients,
  caption="Linear model summary for rising ridge data.")
```

Observe that t-tests indicate that all estimated coefficients are statistically significant. To perform the canonical analysis, let's extract the coefficients into vector b and matrix B, and use those values to estimate x_s.

```
b <- coef(fit)[2:3]
B <- matrix(NA, nrow=2, ncol=2)
diag(B) <- coef(fit)[4:5]
B[1,2] <- B[2,1] <- coef(fit)[6]/2
xs <- -(1/2)*solve(B, b)
xs
```

```
## [1] -5.177 -2.707
```

Now, the bounding box of the design is ...

```
apply(rr[,1:2], 2, range)
```

```
##       A  B
## [1,] -1 -1
## [2,]  1  1
```

... so x_s is well outside the experimental region. Eigen-analysis indicates that x_s is a maximum, as both estimated eigenvalues are negative. However one of them is quite close to zero.

```
E <- eigen(B)
lambda <- E$values
o <- order(abs(lambda), decreasing=TRUE)
V <- E$vectors[,o]*20
lambda <- lambda[o]
lambda
```

```
## [1] -12.7064  -0.5094
```

We'll have to wait until §3.2.5 to say how close is close in statistically meaningful terms. Here the point is that statistically significant regression coefficients don't necessarily imply that eigenvalues are (statistically) non-zero and vice versa. Coefficients can "cancel each other out" when they work in concert to represent the estimated response surface. The canonical form offers another view. Combining those eigenvectors and a calculation of $\hat{y}_s$ ...

```
ys <- coef(fit)[1] + 0.5*t(xs) %*% b
ys
```

```
##         [,1]
## [1,] 71.19
```

... leads to the representation

$$\hat{y} = 71.1902 - 12.7064w_1^2 - 0.5094w_2^2.$$

So the question is: should we ignore the second canonical axis by taking $\lambda_2^2 = 0$? Although testing for $\lambda_2^2 = 0$ is technically a matter of statistical inference, the outcome of that test is somewhat of a moot point since x_s is so far outside the design region, and the first principal axis has more than 24× the "weight" of the second-one. For all practical purposes, we have a rising ridge scenario.

If that doesn't make sense, perhaps the following visualization will help. Using the eigenvectors and gathering predictions on a grid of inputs (A and B) chosen large enough to show all relevant features, we obtain the picture provided by Figure 3.11.

```
x <- seq(-6, 6, length=100)
xx <- expand.grid(x, x)
XX <- data.frame(A=xx[,1], B=xx[,2], A2=xx[,1]^2,
```

```
  B2=xx[,2]^2, AB=xx[,1]*xx[,2])
p <- predict(fit, newdata=XX)
image(x, x, matrix(p, nrow=length(x)), col=cols, xlab="A", ylab="B")
contour(x, x, matrix(p, nrow=length(x)), add=TRUE)
lines(c(-V[1,1], V[1,1]) + xs[1], c(-V[2,1], V[2,1]) + xs[2], lty=2)
lines(c(-V[1,2], V[1,2]) + xs[1], c(-V[2,2], V[2,2]) + xs[2], lty=2)
polygon(c(1,1,-1,-1), c(1,-1,-1,1), lty=3)
text(0,-0.5, "design region", cex=0.5)
points(xs[1], xs[2])
text(xs[1], xs[2], "xs", pos=4)
```

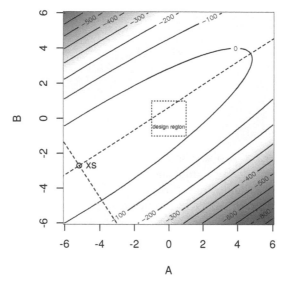

FIGURE 3.11: Visual of the rising ridge on principal axes whose stationary point x_s is far from the design region.

Regardless of whether or not λ_2^2 is statistically zero, it's both clear which way to move the experimental region (toward x_s) and foolish to trust x_s and move the experimental design all the way over there. It's never a good idea to extrapolate too far outside the design region, especially with linear models. It'll be beneficial to develop a more formal framework for deciding on next steps. When x_s is closer to or within the design region, and/or when λs are similar in magnitude, a more nuanced analysis may be warranted. In such cases it'll be crucial to link outcomes of statistical tests to knobs in our framework in order to cautiously but expediently rise the ridge.

Ridge analysis is steepest ascent applied to second-order models. Since second-order models are generally undertaken when the practitioner believes that s/he is quite near the region of the optimum, ridge analysis is typically entertained only in such settings. However, as we saw above, results from experiments may reveal a stationary point well outside the design region, contradicting that belief. Most often this is an artifact of the local nature of analysis via low-order polynomial (Taylor) approximation. Nevertheless, it's important for practitioners to keep an open mind, and let evidence in the data be suggestive, if not entirely authoritative, nearby the experimental regime. But enough with disclaimers ...

Consider the fitted second-order response surface model

$$\hat{y} = b_0 + x^\top b + x^\top B x.$$

Without loss of generality, presume that the center of the design region is $x_1 = \cdots = x_m = 0$, likely in coded units. In steepest ascent with first-order models we optimized the fitted response surface under a constraint, lest the solution be out at infinity. While second-order response surfaces may not as acutely suffer from that pathology, it's nevertheless a good idea to explore paths of steepest ascent with "baby steps". In a ridge analysis, the custom is to maximize (or minimize) $\hat{y}$ subject to the constraint

$$x^\top x = R^2.$$

Lagrange multipliers[9] facilitate optimization by differentiating

$$L = b_0 + x^\top b + x^\top B x - \mu(x^\top x - R^2)$$

with respect to the vector x.

$$\frac{\partial L}{\partial x} = b + 2Bx - 2\mu x$$

The constrained optimal stationary point is determined by setting that expression to zero and then solving for x.

$$(B - \mu I_m)x = -\frac{1}{2}b$$

$$x = -\frac{1}{2}(B - \mu I_m)^{-1}b$$

Notice how that solution looks like a ridge regression[10] estimator. Although a coincidence in terminology, that analogy serves as an easy way to remember the role of μ in a ridge analysis.

Appropriate choices of μ depend upon eigenvalues of B, and whether ascent or descent is desired. If a) $\mu > \lambda_{\max}$, the largest eigenvalue of B, then x will be an absolute maximum for $\hat{y}$ subject to the constraint; otherwise if b) $\mu < \lambda_{\min}$, then the solution will be an absolute minimum. To get better intuition on why and how, consider the orthogonal matrix P that diagonalizes B, i.e., $P^\top B P = \Lambda$, where $\Lambda = \mathrm{diag}(\lambda_1, \ldots, \lambda_m)$. Now for $(B - \mu I)$, which must be inverted to find x, we find that

$$P^\top (B - \mu I)P = \Lambda - \mu I_m, \quad \text{since} \quad P^\top P = I_m.$$

Observe that situation a) thus results in a negative definite $B - \mu I_m$; and b) a positive definite one. This result suggests, but doesn't guarantee that x corresponds to an absolute minima or maxima in $\hat{y}$; see Draper (1963) for more details.

Before turning to an example, let's codify the steps in the form of pseudocode for easy reference. Although applications of steepest ascent on ridge systems vary widely in practice, the essence is often a variation on themes outlined in Algorithm 3.2. Step 5 of the algorithm might benefit from elaboration. One typical approach here is to perform a bisection-style or

---

[9]https://en.wikipedia.org/wiki/Lagrange_multiplier
[10]https://en.wikipedia.org/wiki/Tikhonov_regularization

Algorithm 3.2 Steepest Ascent for Second-Order Models

Assume we wish to maximize a second-order response surface (ascent); let x_0 be the design center, e.g., the origin in coded inputs.

Require fitted first-order model via intercept b_0; m-vector of main effects b; and matrix of second-order terms B. The bounding box of the design region must also be on hand.

Then

1. Calculate eigenvalues of B and let $\lambda_{\max}$ be the largest such value.
2. Choose $\mu \geq \lambda_{\max}$, perhaps with equality to start out with.
3. Solve for x, the constrained optimal stationary point

$$x = -\frac{1}{2}(B - \mu I_m)^{-1}b.$$

4. Using that x, calculate the implied radius $R = \sqrt{x^\top x}$.
5. With (μ, R) pair calculated in Steps 3–4, initialize an iterative (perhaps interactive) exploration of x and $\hat{y}(x) = b_0 + x^\top b + x^\top B x$ values and standard errors near the boundary of the design.

   - Increasing μ will result in coordinates x nearer to x_0, the design center.
   - Decreasing μ, maintaining $\mu > \lambda_{\max}$, will push x outside of the design region.

Return the resulting collection of $(\mu, x, \hat{y})$ tuples and standard errors, possibly after first mapping back from coded to natural units.

---

root-finding[11] search for (μ, R) pairs which yield an x as close to the boundary of the design as possible. Often the initial μ-value, $\mu^{(1)} = \lambda_{\max}$, leads to an x which is well outside the design area, so a sensible search area may be in the range $(\mu^{(1)}, 10\mu^{(1)})$, with larger orders of magnitude on the upper end possibly being required if an initial search fails to converge on a solution away from the boundary. Another variation, which is actually an embellishment on the previous theme, is to use root-finding to derive an adaptive grid of μ-values based on a regular grid of R-values, with evenly spaced radius from $R = 0$ (implying $\mu = \infty$) to $R = 2$ in coded units. Both variations will be explored in some detail in our example below. Having a collection of $(\mu, x, \hat{y})$ tuples for a range of Rs may be beneficial in determining where to center the next experimental design. In particular, an inspection of prediction intervals obtained for $\hat{y}$ could help determine the point near, or just beyond the design boundary where quadratic growth in variance begins to dominate the nature of their spread.

Example: ridge analysis of a saddle

Consider a chemical process that converts 1,2-propanediol to 2,5-dimethylpiperazine, where a maximum conversion is sought as a function of the following coded factors:

$$x_1 = \frac{\text{NH}_3 \text{ level} - 102}{51} \qquad\qquad x_2 = \frac{\text{temp.} - 250}{200}$$

$$x_3 = \frac{\text{H}_2\text{O level} - 300}{200} \qquad\qquad x_4 = \frac{\text{H press.} - 850}{350}. \qquad (3.5)$$

---

[11]https://en.wikipedia.org/wiki/Root-finding_algorithm

TABLE 3.10: Summary of second-order fit to saddle data (3.5).

	Estimate	Std. Error	t value	Pr($>$|t|)
(Intercept)	40.1982	8.322	4.8305	0.0007
x1	-1.5110	3.152	-0.4794	0.6420
x2	1.2841	3.152	0.4074	0.6923
x3	-8.7390	3.152	-2.7725	0.0197
x4	4.9548	3.152	1.5720	0.1470
x11	-6.3324	5.035	-1.2575	0.2371
x22	-4.2916	5.035	-0.8523	0.4140
x33	0.0196	5.035	0.0039	0.9970
x44	-2.5059	5.035	-0.4976	0.6295
x12	2.1938	3.517	0.6238	0.5467
x13	-0.1437	3.517	-0.0409	0.9682
x14	1.5812	3.517	0.4496	0.6626
x23	8.0063	3.517	2.2765	0.0461
x24	2.8062	3.517	0.7979	0.4435
x34	0.2937	3.517	0.0835	0.9351

The data file contains measurements of the response on inputs (in those coded units) following a central composite design. R code below reads in the data and expands the resulting **data.frame** to include features of a second-order model. A short-hand is used to avoid the tedium of writing out all terms by hand. Some of the renaming and reordering of columns isn't strictly necessary, but helps here to maintain a degree of consistency across analyses.

```
saddle <- read.table("saddle.txt", header=TRUE)
saddle <- cbind(saddle[,-5]^2,
  model.matrix(~ .^2 - 1, saddle[,-5]), y=saddle[,5])
names(saddle)[1:4] <- paste("x", 1:4, 1:4, sep="")
names(saddle)[9:14] <- sub(":x", "", names(saddle)[9:14])
saddle <- saddle[c(5:8,1:4,9:15)]
```

A summary of the fit, provided in Table 3.10, suggests that perhaps we don't have enough data ($n = 25$) for all of the estimated quantities ($m = 14$). By independent t-tests at the 5% level – a crude inspection to be sure – there are far fewer useful than useless predictors.

```
fit <- lm(y ~ ., data=saddle)
kable(summary(fit)$coefficients,
  caption="Summary of second-order fit to saddle data(3.5).")
```

Recall that a representation on canonical axes hinges on fewer ($m + 1 = 5$) derived quantities. So nevertheless we proceed, prudently with caution. First, extract coefficient main effects vector b and matrix B of second-order terms.

```
b <- coef(fit)[2:5]
B <- matrix(NA, nrow=4, ncol=4)
```

```
diag(B) <- coef(fit)[6:9]
i <- 10
for(j in 1:3) {
  for(k in (j+1):4) {
    B[j,k] <- B[k,j] <- coef(fit)[i]/2
    i <- i + 1
  }
}
```

Using those coefficients, stationary point x_s may be calculated as follows.

```
xs <- -(1/2)*solve(B, b)
xs
```

```
## [1] 0.2647 1.0336 0.2906 1.6680
```

Observe that this is within the vicinity of the design region.

```
apply(saddle[,1:4], 2, range)
```

```
##         x1   x2   x3   x4
## [1,] -1.4 -1.4 -1.4 -1.4
## [2,]  1.4  1.4  1.4  1.4
```

Visualization is somewhat more challenging in four input dimensions, making eigen-analysis crucial from both interpretive and algorithmic perspectives.

```
E <- eigen(B)
lambda <- E$values
o <- order(abs(lambda), decreasing=TRUE)
lambda <- lambda[o]
lambda
```

```
## [1] -7.547 -6.008  2.604 -2.159
```

Evidently we have a saddle point: these coefficients are far from zero but don't agree on sign. Of course, we don't have their standard errors so we don't know if they're statistically non-zero. (Chances are not good, since so many of the estimated coordinates of b and B are likely insignificant. More later.) Stationary point x_s, being centered on the trough of the saddle, is well-within the design region, so we're in a somewhat different situation compared to our warm-up example from Table 3.9, where the center of the design was clearly along a ridge far from the estimated mode.

On the other hand, when choosing initial $\mu^{(1)} = \lambda_{\max}$ and calculating the x this implies, following Steps 2–3 in Algorithm 3.2 ...

```
mul <- max(lambda)
x <- solve(B - mul*diag(4), -b/2)
x
```

```
## [1] 1.163e+13 8.290e+13 1.297e+14 2.829e+13
```

... we get a location which is well outside of the design region. So a bit of work will be required to iterate on larger μ-values, placing us closer to the boundary of the design region. To automate that process, code below sets up function f to serve as an objective in the search for that boundary via the root (i.e., zero-crossing) of $R^2 - x^\top x$, with x derived as a function of μ. By default, f is set up to target $R^2 = 1$ unless otherwise specified.

```
f <- function(mu, R2=1)
 {
   x <- solve(B - mu*diag(4), -b/2)
   R2 - t(x) %*% x
 }
```

The boundary of our search region is at 1.4 in absolute value. We know that we need a $\mu > \lambda_{\max}$, so it's reasonable to set $\mu = \lambda_{\max}$, stored as mul in the code, as the left-hand side of the search interval. For starters, we'll search up to 10λ on the right.

```
mu <- uniroot(f, c(mul, 10*mul), R2=1.4^2)$root
mu
```

```
## [1] 4.834
```

Having located a μ-value in the interior of the search region $[\lambda_{\max}, 10\lambda_{\max}]$, we can be confident that there's no need to widen the range. It makes sense to check that input x associated with that μ-value lies near one of the boundaries.

```
x <- solve(B - mu*diag(4), -b/2)
x
```

```
## [1] -0.09124 -0.47679 -1.29616  0.21061
```

Indeed, the third coordinate is quite close to 1.4. Double-checking the R-value ...

```
drop(sqrt(t(x) %*% x))
```

```
## [1] 1.4
```

... verifies that the desired solution has been found. Since our x is within the design region, i.e., where predictions from the fitted second-order model are most reliable, this may represent a decent location for a confirmation test. Or it may serve as the center of a new design in an iteration along a path of ascent. But before doing that, it's a good idea to inspect predictive standard errors, and corresponding error-bars. To avoid benchmarking in a vacuum, code below considers a range of R-values in $[0, 2]$, and the μ's and x's they imply, so that ultimately predictive equations can be derived at those locations, and compared relative to one another.

```
mus <- rs <- seq(0.1, 2, length=20)
xp <- matrix(NA, nrow=length(rs), ncol=4)
colnames(xp) <- c("x1", "x2", "x3", "x4")
```

```
for(i in 1:length(rs)) {
  mus[i] <- uniroot(f, c(mul, 100*mul), R2=rs[i]^2)$root
  xp[i,] <- solve(B - mus[i]*diag(4), -b/2)
}
xp <- rbind(rep(0,4), xp)
rs <- c(0, rs)
mus <- c(Inf, mus)
```

Obtaining predictions with those x-values requires expanding out into second-order features. Code below builds up a `data.frame` that can be passed into the `predict.lm` method.

```
Xp <- data.frame(xp)
Xp <- cbind(Xp^2, model.matrix(~ .^2 - 1, Xp))
names(Xp)[1:4] <- paste("x", 1:4, 1:4, sep="")
names(Xp)[9:14] <- sub(":x", "", names(Xp)[9:14])
Xp <- Xp[c(5:8,1:4,9:14)]
```

We're ready to predict.

```
p <- predict(fit, newdata=Xp, se.fit=TRUE)
```

An inspection of the sequence(s) of numbers thus calculated, as collated in Table 3.11, reveals that predictive uncertainly grows very quickly away from the design boundary. It certainly seems risky to trust predictions derived from $R > 1.4$.

```
kable(cbind(R=rs, mu=mus, data.frame(pred=p$fit, se=p$se.fit), round(xp,6)),
  caption="Predictions for the saddle experiment calculated along a path
  leading away from the center of the design region.")
```

Perhaps the case is better made visually. Figure 3.12 plots confidence intervals (CIs) derived from these means and standard errors.

```
plot(rs, p$fit, type="b", ylim=c(20,100), xlab="radius (R)",
  ylab="y.hat(x) & 95% CIs")
lines(rs, p$fit + 2*p$se, col=2, lty=2)
lines(rs, p$fit - 2*p$se, col=2, lty=2)
```

Not surprisingly, predictive CIs begin to rapidly diverge from one another as we leave the design region, with $R > 1.4$. Observe that the amount of predicted improvement (from the predictive mean), even at $R = 2$ (compared to $R = 0$), is apparently dwarfed by estimation uncertainty. This suggests we need more runs inside/nearby the design region before venturing farther afield. It may be sensible to augment with a design centered near the $R = 1.4$ solution, collecting new responses in a more limited range of inputs. Combining these with the original design should yield improved predictions and further inform on statistical relevance for coefficients and eigenvectors/values underpinning the analysis. Assessing that relevance will require a bit more scaffolding, making for a nice segue into our final classical response surfaces segment(s), on sampling properties.

TABLE 3.11: Predictions for the saddle experiment calculated along a path leading away from the center of the design region.

R	mu	pred	se	x1	x2	x3	x4
0.0	Inf	40.20	8.322	0.0000	0.0000	0.0000	0.0000
0.1	49.811	41.21	8.305	-0.0126	0.0064	-0.0871	0.0471
0.2	24.557	42.20	8.255	-0.0217	0.0012	-0.1773	0.0900
0.3	16.343	43.18	8.175	-0.0287	-0.0141	-0.2699	0.1271
0.4	12.368	44.16	8.074	-0.0346	-0.0379	-0.3641	0.1575
0.5	10.071	45.16	7.960	-0.0399	-0.0686	-0.4591	0.1815
0.6	8.598	46.18	7.849	-0.0451	-0.1045	-0.5543	0.1994
0.7	7.585	47.22	7.757	-0.0503	-0.1444	-0.6494	0.2120
0.8	6.853	48.30	7.708	-0.0556	-0.1873	-0.7438	0.2203
0.9	6.302	49.42	7.724	-0.0612	-0.2325	-0.8377	0.2248
1.0	5.875	50.57	7.832	-0.0669	-0.2793	-0.9308	0.2262
1.1	5.534	51.77	8.056	-0.0727	-0.3275	-1.0231	0.2251
1.2	5.257	53.01	8.414	-0.0788	-0.3765	-1.1148	0.2220
1.3	5.027	54.29	8.921	-0.0849	-0.4264	-1.2058	0.2170
1.4	4.834	55.62	9.581	-0.0912	-0.4768	-1.2961	0.2106
1.5	4.670	57.00	10.395	-0.0976	-0.5276	-1.3860	0.2030
1.6	4.528	58.42	11.357	-0.1041	-0.5789	-1.4752	0.1942
1.7	4.404	59.90	12.461	-0.1107	-0.6304	-1.5641	0.1846
1.8	4.296	61.42	13.699	-0.1173	-0.6821	-1.6525	0.1742
1.9	4.199	62.99	15.062	-0.1241	-0.7340	-1.7405	0.1631
2.0	4.114	64.61	16.543	-0.1308	-0.7861	-1.8281	0.1514

3.2.3 Sampling properties

The stationary point x_s, or its constrained analog, are only estimates. Any point on the contour of a response surface, as well as the contour itself, possesses sampling variability. These quantities depend on estimated coefficients b_0, b and B, which have standard errors. When uncertainties are propagated through predictive equations to build the fitted response surface, and its constrained optima, those derived quantities inherit a sampling uncertainty which is, as yet, unexplored in our development. Below we review that predictive distribution, borrowing highlights from a first course in linear models, as a first step in understanding how uncertainty propagates to the stationary point(s) of the fitted surface, and steps along the path of steepest ascent.

Let $y = (y_1, \ldots, y_n)$ be an n-vector of responses and write X as the $n \times (1 + 2m + \binom{m}{2})$ model matrix comprised of rows x_i built, e.g., as

$$x_i^\top = [1, x_{i1}, x_{i2}, x_{i1}^2, x_{i2}^2, x_{i1}x_{i2}], \quad \text{for the special case of } m = 2.$$

Then we have $\quad b = (X^\top X)^{-1} X^\top y, \quad$ generally for any X.

For the time being, note that we're not utilizing the (b_0, b, B) representation, but rather a flattened p-vector of regression coefficients b, arising from a feature expanded X encoding second-order model structure. Solution $b = (X^\top X)^{-1} X^\top y$ comes from solving least

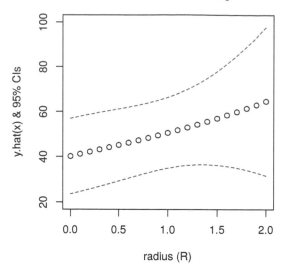

FIGURE 3.12: Visualizing predictive means and quantiles from Table 3.11.

squares equations, or as maximum likelihood estimator (MLE) $b \equiv \hat{\beta}$ under the model $Y \sim \mathcal{N}_n(X\beta, \sigma^2 \mathbb{I}_n)$, where $\mathcal{N}_n$ is an n-variate multivariate normal (MVN) distribution.

Now a useful property of projections of MVN random vectors is that if $Y \sim \mathcal{N}_n(\mu, \Sigma)$ and P is a $p \times n$ matrix, then $PY \sim \mathcal{N}_p(P\mu, P\Sigma P^\top)$. Notice that b involves one such projection. In particular, let $P = (X^\top X)^{-1} X^\top$ so that $\hat{\beta} = Py$. Observe that

- $PX\beta = (X^\top X)^{-1} X^\top X\beta = \beta$, and
- $PP^\top = (X^\top X)^{-1} X^\top X (X^\top X)^{-1} = (X^\top X)^{-1}$

since $X^\top X$ and its inverse are symmetric. Using the MVN result above, we obtain

$$\hat{\beta} = PY \sim \mathcal{N}_p(\beta, \sigma^2 (X^\top X)^{-1}), \qquad (3.6)$$

giving the sampling distribution of $b \equiv \hat{\beta}$.

Predictions, provided as $\hat{y}(x) \equiv x^\top b$, are a function of estimated b and thus inherit its sampling distribution. That distribution may be derived through a second projection, this time onto the (one-dimensional) real line. We obtain that $\hat{y}(x)$ is univariate Gaussian with mean $x^\top \beta$, which is the true (but unknown) response modulo the local nature of the approximation of the second-order response surface, with variance equal to $\sigma^2 x^\top (X^\top X)^{-1} x$.

Usually σ^2 is unknown, so it too must be estimated from data. MLE $\hat{\sigma}^2$ may be derived as a mean sum of squares

$$\hat{\sigma}^2 = \frac{1}{n} \sum_{i=1}^{n} (y_i - x_i^\top b)^2, \quad \text{however} \quad s^2 = \frac{1}{n-p} \sum_{i=1}^{n} (y_i - \hat{y}_i)^2,$$

which corrects for bias in $\hat{\sigma}^2$ by subtracting off p DoF spent in estimating b, is more often used in practice. An application of Cochran's Theorem[12] gives that $s^2 \sim \frac{\sigma^2}{n-p}\chi^2_{n-p}$. The Gaussian nature of $Y(x) \mid \sigma^2$ and χ^2 relationship between s^2 and σ^2 combines to give a Student-t sampling distribution for the prediction

---

[12]https://en.wikipedia.org/wiki/Cochran's_theorem

$$\hat{y}(x) \sim t_{n-p}(x^\top \beta, s_{\hat{y}}^2(x)), \quad \text{as shorthand for} \quad \frac{\hat{y}(x) - x^\top \beta}{s_{\hat{y}}(x)} \sim t_{n-p},$$

where $s_{\hat{y}}(x)$ is the standard error of $\hat{y}(x) \equiv x^\top b$:

$$s_{\hat{y}(x)} = s\sqrt{x^\top (X^\top X)^{-1} x}.$$

When using `predict.lm` with fitted `lm` objects in R, argument `se.fit=TRUE` causes estimates of $s_{\hat{y}(x)}$ to be returned on output. Given that standard error, a CI for $\hat{y}(x)$ is $\hat{y}(x) \pm t_{\alpha/2,n-p} s_{\hat{y}(x)}$. In R this is what you get when you provide `interval="confidence"` to `predict` with `level=1 - alpha`. Note that this is different from a prediction interval whose standard error includes an extra factor of s:

$$\sqrt{s^2 + s_{\hat{y}(x)}^2}.$$

You get intervals based on this estimate with `interval="prediction"`. For completeness, let me remark that a similar argument, paired with the Gaussian sampling distribution (3.6) leads to the quantities reported for $b_j \equiv \hat{\beta}_j$ in `summary.lm` output, e.g. most recently in Table 3.9. The square-root of the diagonal of $s(X^\top X)^{-1}$ are the standard errors s_{b_j} from Eq. (3.2), also on $n - p$ DoF.

To illustrate a predictive application, consider again our chemical conversion "saddle" example (3.5) on four input variables. Since visualization is challenging in 4d, here we explore how predictive standard error varies as a function of (x_1, x_2) under fixed settings of (x_3^0, x_4^0) values. First, set up a 2d grid in the first two inputs to aid in visualization via slices.

```
x12 <- seq(-2,2,length=100)
g <- expand.grid(x12, x12)
Xp <- data.frame(x1=g[,1], x2=g[,2],
  x11=g[,1]^2, x22=g[,2]^2, x12=g[,1]*g[,2])
```

Code below completes the `data.frame` with features derived from two settings of the latter two inputs, $(x_3, x_4) = (0, 0)$ and $(x_3, x_4) = (1, 1)$, and collects predictive quantities under the response surface fit earlier.

```
## x3 = x4 = 0
Xp$x3 <- Xp$x33 <- Xp$x4 <- Xp$x44 <- Xp$x34 <- 0
Xp$x13 <- Xp$x14 <- Xp$x23 <- Xp$x24 <- Xp$x34 <- 0
p0 <- predict(fit, newdata=Xp, se.fit=TRUE)
## x3 = x4 = 1
Xp$x3 <- Xp$x33 <- Xp$x4 <- Xp$x44 <- Xp$x34 <- 1
Xp$x13 <- Xp$x14 <- Xp$x1; Xp$x23 <- Xp$x24 <- Xp$x2
p1 <- predict(fit, newdata=Xp, se.fit=TRUE)
```

Surfaces showing the predictive standard error $s_{\hat{y}(x)}$ in those two cases are provided in Figure 3.13. Projected experimental design coordinates are overlaid as open circles; the subset of inputs lying in the corresponding (x_3, x_4) slice are shown as closed circles.

```
par(mfrow=c(1,2))
bs <- seq(min(p0$se.fit), max(p0$se.fit), length=129)
image(x12, x12, matrix(p0$se.fit, nrow=length(x12)), col=cols, breaks=bs,
    xlab="x1", ylab="x2", main="se.fit, x3 = x4 = 0")
contour(x12, x12, matrix(p0$se.fit, nrow=length(x12)), add=TRUE)
points(saddle[,1:2])
points(saddle[apply(saddle[,3:4] == c(0,0), 1, all),1:2], pch=19)
image(x12, x12, matrix(p1$se.fit, nrow=length(x12)), col=cols, breaks=bs,
    xlab="x1", ylab="x2", main="se.fit, x3 = x4 = 1")
contour(x12, x12, matrix(p1$se.fit, nrow=length(x12)), add=TRUE)
points(saddle[,1:2])
points(saddle[apply(saddle[,3:4] == c(1,1), 1, all), 1:2], pch=19)
```

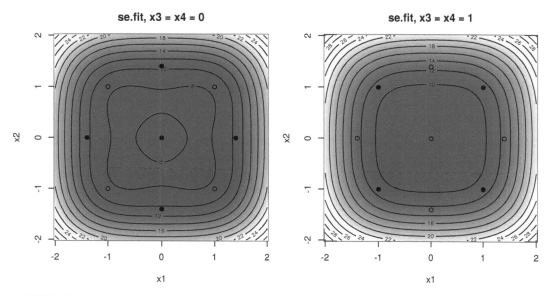

FIGURE 3.13: Predictive variance slices for the chemical conversion "saddle" example (3.5).

What do we take away from these plots? Prediction with this second-order response surface becomes worse as one gets near the design perimeter. If the predicted value of the optimum isn't inside the design region, searching outside comes with substantial risk. Observe that $(X^\top X)^{-1}$ plays a prominent role in $s_{\hat{y}(x)}$, so the design X matters! What might be less obvious is that X (and y) also affect the location of the (estimated) stationary point x_s and its nature. So if we're wondering about uncertainty in $\hat{y}(x_s)$, say, we're missing an important source of variability.

3.2.4 Confidence in the stationary point

Clearly the location of the stationary point x_s is of considerable interest in its own right, beyond its out-sized role in a ridge analysis. But how good can an estimate based on local second-order modelling and limited experimentation possibly be? Can we get a sense of confidence in our estimates? Could uncertainty so swamp our estimates that another set of locations offers about the same improvement at far lower risk? Answering these questions requires propagating predictive standard error through the derivative-based optimization.

Recall that

$$\hat{y}(x) = b_0 + x^\top b + 2x^\top Bx, \quad \text{where we calculated} \quad \frac{\partial \hat{y}(x)}{\partial x} = b + 2Bx.$$

So let $d(x) = b + 2Bx$ be the vector of derivatives (the gradient) with each component being $d_j(x) = b_j + 2B_j x$, where B_j is the jth row of B. Observe that $d(x)$ is comprised of linear functions of $x_1, \ldots, x_m$. Now let $t = (t_1, t_2, \ldots, t_m)$ denote the true but unknown stationary point of the system. The Gaussian error structure of our linear modeling framework, and the fact that t is a critical point of the true response surface, implies that

$$d(t) \sim \mathcal{N}_m(0, \mathbb{V}\text{ar}\{d(t)\}).$$

This is useful because, if we can obtain an estimate of that variance–covariance matrix, we may use the following result as a means of developing a CI for t, the true location of the stationary point:

$$\frac{d^\top(t)[\widehat{\mathbb{V}\text{ar}}\{d(t)\}]^{-1} d(t)}{m} \sim F_{m,n-p}, \tag{3.7}$$

where $\widehat{\mathbb{V}\text{ar}}\{d(t)\}$ is $\mathbb{V}\text{ar}\{d(t)\}$ with σ^2 estimated by s^2 on $n - p$ DoF. In particular, a $100(1-\alpha)\%$ confidence region for the stationary point includes those t which evaluate, under that quadratic form (3.7), to a number less than an α quantile of the $F_{m,n-p}$ distribution:

$$d^\top(t)[\widehat{\mathbb{V}\text{ar}}\{d(t)\}]^{-1} d(t) \leq m F^\alpha_{m,n-p}. \tag{3.8}$$

The utility of this result is, however, unfortunately limited to graphical analysis in two or three dimensions at most. Also, the devil is in the details of estimating the variance of the derivatives, which is greatly simplified if the design is chosen fortuitously. That's best illustrated by example.

Consider data on a process in two coded input dimensions under a central composite design whose center is replicated four times.

```
crdat <- read.table("confreg.txt", header=TRUE)
```

R code below combines several steps: expanding to second-order features, model fitting and extracting b and B, differentiating and solving to estimate stationary point x_s.

```
crdat$x11 <- crdat$x1^2
crdat$x22 <- crdat$x2^2
crdat$x12 <- crdat$x1 * crdat$x2
fit <- lm(y ~ ., data=crdat)
b <- coef(fit)[2:3]
B <- matrix(NA, nrow=2, ncol=2)
diag(B) <- coef(fit)[4:5]
B[1,2] <- B[2,1] <- coef(fit)[6]/2
```

```
xs <- -(1/2)*solve(B, b)
xs
```

```
## [1] -0.1716 -0.1806
```

Stationary point x_s is a maximum, as indicated by the eigenvalues.

```
eigen(B)$values
```

```
## [1] -2.244 -3.061
```

Next evaluate predictive equations on a grid in order to visualize the response surface, design and stationary point, which is well within the interior of the experimental region. See Figure 3.14.

```
xx <- xx <- seq(-1.6, 1.6, length=200)
g <- expand.grid(xx, xx)
XX <- data.frame(x1=g[,1], x2=g[,2],
  x11=g[,1]^2, x22=g[,2]^2, x12=g[,1]*g[,2])
p <- as.numeric(predict(fit, newdata=XX))
image(xx, xx, matrix(p, nrow=length(xx)), col=cols, xlab="x1", ylab="x2")
contour(xx, xx, matrix(p, nrow=length(xx)), add=TRUE)
points(crdat$x1, crdat$x2, pch=20)
points(xs[1], xs[2])
text(xs[1], xs[2], "xs", pos=4)
```

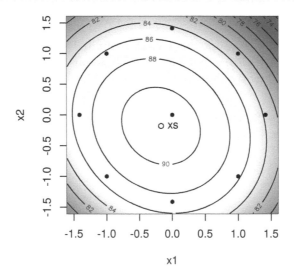

FIGURE 3.14: Simple maximum response surface.

Combining $d(t) = b + 2Bt$ with calculations of b and B ...

```
d <- rbind(c(b[1], 2*B[,1]), c(b[2], 2*B[,2]))
colnames(d) <- c("(Intercept)", "+1", "t2")
d
```

```
##          (Intercept)      t1      t2
## [1,]        -1.095  -5.562  -0.775
## [2,]        -1.045  -0.775  -5.047
```

... gives the following expressions for derivatives of the response surface as a function of the true unknown stationary point t.

$$d_1(t) = -1.095 - 5.562t_1 - 0.775t_2$$
$$d_2(t) = -1.045 - 0.775t_1 - 5.048t_2$$

Now the variances of $d(t)$, via $\hat{\beta} \sim \mathcal{N}_p(\beta, \sigma^2(X^\top X)^{-1})$, are obtained from our estimate of the residual variance s^2 and the matrix $(X^\top X)^{-1}$. The former may be extracted from `fit` ...

```
s2 <- summary(fit)$sigma^2
s2
```

```
## [1] 3.164
```

... on $n - p = 7$ DoF. The latter is most easily calculated "by hand" as follows.

```
X <- cbind(1, as.matrix(crdat[,-3]))
XtXi <- solve(t(X) %*% X)
XtXi
```

```
##                  x1      x2       x11       x22    x12
##          0.3333  0.000  0.000  -0.16669  -0.16669  0.00
## x1       0.0000  0.125  0.000   0.00000   0.00000  0.00
## x2       0.0000  0.000  0.125   0.00000   0.00000  0.00
## x11     -0.1667  0.000  0.000   0.17716   0.05208  0.00
## x22     -0.1667  0.000  0.000   0.05208   0.17716  0.00
## x12      0.0000  0.000  0.000   0.00000   0.00000  0.25
```

Observe that $(X^\top X)^{-1}$ is sparse, owing to the orthogonal structure of our central composite design. We finally have all necessary ingredients to build up the covariance matrix of $d(t)$, as a function of t. We shall proceed one component at a time. First: diagonal elements. By linearity of variances, and reading off entries of $(X^\top X)^{-1}$, we obtain

$$\widehat{\mathrm{Var}}\{d_1(t)\} = \frac{s^2}{\sigma^2}[\mathrm{Var}b_1 + 4t_1^2\mathrm{Var}b_{11} + t_2^2\mathrm{Var}b_{12}]$$
$$= s^2[1/8 + 4t_1^2(0.1772) + t_2^2/4],$$

and similarly $\quad \widehat{\mathrm{Var}}\{d_2(t)\} = \frac{s^2}{\sigma^2}[\mathrm{Var}b_2 + t_1^2\mathrm{Var}b_{12} + 4t_2^2\mathrm{Var}b_{22}]$
$$= s^2[1/8 + t_1^2/4 + 4t_2^2(0.1772)].$$

Now for the off-diagonals. From $(X^\top X)^{-1}$ we can see that all covariances involving b_1 are zero and $\mathbb{Cov}(b_{11}, b_{12}) = \mathbb{Cov}(b_{12}, b_{22}) = 0$. Therefore,

$$\widehat{\mathbb{Cov}}(d_1(t), d_2(t)) = \frac{s^2}{\sigma^2}[4t_1t_2\mathbb{Cov}(b_{11}, b_{22}) + t_1t_2\mathrm{Var}b_{12}]$$
$$= s^2[4(0.0521)t_1t_2 + t_1t_2/4].$$

To help visualize the resulting confidence region, code below captures that covariance matrix as a function of its elements, and coordinates t.

```
Vard <- function(t1, t2, s2, XtXi)
 {
   v11 <- XtXi[2,2] + 4*t1^2*XtXi[4,4] + t2^2*XtXi[6,6]
   v22 <- XtXi[3,3] + t1^2*XtXi[6,6] + 4*t2^2*XtXi[5,5]
   v12 <- v21 <- 4*t1*t2*XtXi[4,5] + t1*t2*XtXi[6,6]
   v <- s2 * matrix(c(v11,v12,v21,v22), ncol=2, byrow=TRUE)
   return(v)
 }
```

This implementation isn't generic; it leverages the particular structure of sparsity in `XtXi` from above. Developing a general purpose version could be a good exercise for the interested reader. To continue, we must combine that variance with a function that evaluates the quadratic form in Eq. (3.7).

```
CIqf <- function(t1, t2, s2, XtXi, b, B)
 {
   dt <- b + 2*B %*% c(t1, t2)
   V <- Vard(t1, t2, s2, XtXi)
   Vi <- solve(V)
   t(dt) %*% Vi %*% dt
 }
```

With that we've assembled all building blocks necessary to evaluate Eq. (3.8) on our (x_1, x_2) predictive grid `g` from before.

```
quadform <- rep(NA, nrow(g))
for(i in 1:nrow(g)) quadform[i] <- CIqf(g[i,1], g[i,2], s2, XtXi, b, B)
```

Code supporting Figure 3.15 completes Eq. (3.8) by converting evaluations of that quadratic form into logical vectors under quantiles obtained from an $F_{2,5}$ distribution at levels $\alpha = 0.1$ and $\alpha = 0.05$, respectively. These are then plotted as grayscale images.

```
ci90 <- quadform <= 2*qf(0.9, 2, nrow(X)-ncol(X))
ci95 <- quadform <= 2*qf(0.95, 2, nrow(X)-ncol(X))
image(xx, xx, matrix(ci90 + ci95, ncol=length(xx)), xlab="x1", ylab="x2",
   col=c("white", "lightgray", "darkgray"))
text(c(-0.2,-1), c(-0.5,1), c("90%","95%"))
```

Unfortunately, these confidence regions present a rather grim picture of the utility of our estimated stationary point. Although the smaller 90% region is clearly closed, and mostly contained with the study area, the larger 95% set doesn't close inside the plot window which extends to cover the entirety of the experimental region. Recall the interpretation of a confidence set: that the true response surface maximum, which could reside at any location inside the (e.g., 95%) set, could readily have produced the data that were observed. Apparently, at the 95% level, that could be nearly any location nearby the design, (I leave it as an exercise to the curious reader to expand the grid and re-calculate the 95% region.)

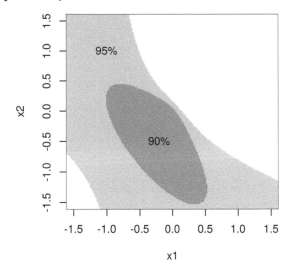

FIGURE 3.15: 90% and 95% confidence regions for stationary point x_s.

It's perhaps surprising that we have such a high degree of uncertainty, even though our fit explains 84% of the variability in the response.

```
summary(fit)$r.squared
```

```
## [1] 0.8389
```

So as not to end on such a bleak message, consider a second dataset with the same design (so $X^\top X$ is the same), but with a considerably better fit, explaining 99% of variability.

```
crdat$y <- c(87.6, 86.5, 85.7, 86.9, 86.7, 86.8, 87.4, 86.7, 90.3,
  91.0, 90.8)
fit2 <- lm(y ~ ., data=crdat)
summary(fit2)$r.squared
```

```
## [1] 0.9891
```

Updating the calculations from above with the new estimated coefficients ...

```
b <- coef(fit)[2:3]
B <- matrix(NA, nrow=2, ncol=2)
diag(B) <- coef(fit)[4:5]
B[1,2] <- B[2,1] <- coef(fit)[6]/2
xs <- -(1/2)*solve(B, b)
s2 <- summary(fit2)$sigma^2
for(i in 1:nrow(g)) quadform[i] <- CIqf(g[i,1], g[i,2], s2, XtXi, b, B)
```

... leads to the 95% confidence region shown in Figure 3.16: a very compact set indeed.

```
ci95 <- quadform <= 2*qf(0.95, 2, nrow(X)-ncol(X))
```

```
image(xx, xx, matrix(ci95, ncol=length(xx)), xlab="x1", ylab="x2",
    col=c("white", "lightgray"))
```

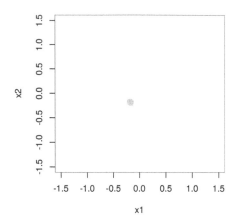

FIGURE 3.16: 95% confidence regions for stationary point x_s under higher R^2; compare to Figure 3.15.

A smaller confidence region for the same level α is a welcome result. It's nice to have low ambiguity about where a local fit thinks best operating conditions lie, especially when the estimated x_s value is located within the design region. On the other hand, larger confidence regions are not always bad news. As in Figure 3.15, it's possible for a model to fit quite well and yet generate a large confidence region. This state of affairs is merely a reflection of the reality that the response surface is locally flat. A positive spin may be that such situations imply a degree of flexibility in choosing the optimum. Communicating that effectively to stakeholders requires care, of course, but is definitely of interest.

3.2.5   Intervals on eigenvalues

In our discussion of the canonical analysis (§3.2.1) we noted that when one or more eigenvalues of B are close to zero a ridge system of some type is present. In judging the size of an eigenvalue it can be helpful to have a notion of standard error – to put some statistical heft behind just how small is basically zero. Standard errors can then be converted into CIs to ease reporting and interpretation.

A convenient procedure for developing standard errors for λ, and subsequently CIs, is the *double linear regression (DLR)* method (Bisgaard and Ankenman, 1996). As the name implies, it entails fitting two linear models. The first is a second-order model fit to data, as in any ridge analysis, providing the matrix B and its eigenvalues and vectors. The second is framed in the canonical model, whose so-called B-canonical form (3.4) we've used extensively throughout the latter half of this chapter. The DLR development instead utilizes an equivalent "A-canonical" form which is somewhat less common when studying ridge systems. The A-canonical form is based upon the representation $\hat{y} = a_0 + u^\top a + u^\top \Lambda u$, where $a = P^\top b$ and $u = P^\top x$ with eigenvectors stacked in P. Observe that the A-form utilizes axes rotated as in the B-form, but without translating from the stationary point x_s to the origin.

Rather than directly using a and Λ-values extracted from the eigen-analysis, they're "re-inferred" through least squares on a design matrix of u and u^2 predictors. Fitted values of that regression may be expressed as

$$\hat{y} = \hat{a}_0 + \sum_{j=1}^{m} \hat{a}_j u_j + \sum_{j=1}^{m} \hat{\lambda}_j u_j^2.$$

Standard errors for $\hat{\lambda}_j$ obtained in this way can be used to make $100(1 - \alpha)\%$ CIs.

To illustrate, consider again the chemical processes experiment whose design is depicted in Figure 3.8. Code below duplicates some of our earlier effort to recreate variables pertinent to a ridge analysis, comprising the first regression in the DLR method.

```
chem <- read.table("chemical.txt", header=TRUE)
X <- data.frame(x1=chem$x1, x2=chem$x2, x11=chem$x1^2, x22=chem$x2^2,
  x12=chem$x1*chem$x2)
y <- chem$y
fit <- lm(y ~ ., data=X)
b <- coef(fit)[2:3]
B <- matrix(NA, nrow=2, ncol=2)
diag(B) <- coef(fit)[4:5]
B[1,2] <- B[2,1] <- coef(fit)[6]/2
E <- eigen(B)
lambda <- E$values
o <- order(abs(lambda), decreasing=TRUE)
P <- E$vectors[,o]
print(lambda <- lambda[o])
```

```
## [1] -11.077  -2.673
```

Next derive u coordinates, following $u = P^\top x$, and then expand out into second-order features (without interactions, since the system is already orthogonalized). Then fit the second regression involved in the DLR method.

```
U <- data.frame(cbind(chem$x1, chem$x2) %*% P)
names(U) <- c("u1", "u2")
U$u11 <- U$u1^2
U$u22 <- U$u2^2
fitU <- lm(y ~ ., data=U)
```

It's not a coincidence that fitted coefficients from our canonical analysis match projected coefficients from the original model, and the eigenvalues.

```
rbind(dlr2=coef(fitU)[-1], eigen=c(b %*% P, lambda))
```

```
##               u1    u2     u11     u22
## dlr2   -10.81 2.126 -11.08 -2.673
## eigen -10.81 2.126 -11.08 -2.673
```

But now we can use summary.lm on the output of the second regression to extract standard errors on those eigenvalues.

```
summary(fitU)$coefficients[4:5,]
```

```
##      Estimate Std. Error t value  Pr(>|t|)
## u11   -11.077    0.9121 -12.144 5.868e-06
## u22    -2.673    0.9121  -2.931 2.200e-02
```

Observe that the t-tests provided in the table above, taken separately, reject the null hypothesis that $\lambda_{4:5} = 0$. Therefore we conclude that both are indeed negative: the true response is a maximum. Alternatively, these standard errors can be used to construct CIs for the λ_j's, or we can ask R to do it for us.

```
confint(fitU)[4:5,]
```

```
##       2.5 %  97.5 %
## u11 -13.23 -8.9201
## u22  -4.83 -0.5163
```

As expected, neither includes zero: both are squarely in the negative. We now have statistical evidence that the response surface is (locally) concave down, and can be reasonably confident that baby steps toward ascent will bear fruit.

Summarizing remarks

This concludes our chapter on classical response surface methods. Although barely scratching the surface, many of the underlying themes are present in abundance. In this careful enterprise there's potential to learn a great deal, at least locally, from appropriately planned experiments. When conditions are right, a cautious ascent will likely lead to improvements. When they're not, a statistical explanation can justify staying "right where you are" when little scope is apparent for incremental refinements to the process under study.

On the surface, it may seem that the biggest downside is the local nature of analysis. Another, perhaps more modern, perspective might suggest another drawback: reproducibility. Two different statistical experts might obtain dramatically different results or conclusions due to small changes or different choices (of design, size of ascent steps, etc.) along the way, despite being largely faithful to similar underlying principles. The process is far from automatable, which makes a meta-analysis, about what happens in the long run after repeatedly applying sequential procedures such as these, nearly impossible even in the abstract. Choice of design for convenience of analytical calculation, say in the calculation of confidence regions, may not be ideal either. Nevertheless, precedence for such tactics is well-established throughout academic statistics.

These drawbacks, local emphasis and removing expert variability toward machine automation, motivate much of what's presented in subsequent chapters. Optimization and analysis of computer simulation experiments, nonlinear regression in spatial statistics and machine learning, are increasingly nonparametric. At first blush nonparametric methods seem more complex, and thus are often dismissed as black boxes. But that vastly oversimplifies a wide class of estimators. One clear positive side effect of non-parameter-"ism" is fewer choices requiring expert judgement, and therefore greater potential for automation and reproducibility. The canonical nonparametric apparatus in our setting is the Gaussian process. While no panacea, these remarkably agile beasts have already revolutionized scientific inquiry and optimization in the engineering and physical sciences, and deserve to be entertained as replacements to many well-established methods. Classical response surface methods are just one example.

3.3   Homework exercises

These exercises are designed to check your understanding of the method of steepest ascent, ridge analysis, and assessments of uncertainty thereupon. Data and parts of several questions are borrowed from Chapters 5–6 of Myers et al. (2016), as detailed parenthetically below. Other exercises from those chapters come highly recommended.

#1: Steepest ascent

The file sadat.txt[13] contains runs of an experiment on two input variables. (Synthesizes exercises 5.12, 5.24, and 5.25 from Myers et al. (2016) using data from Table E5.602.)

a. Apply the method of steepest ascent and construct an appropriate path based on a first-order model. Report the path on both natural and coded variables.
b. Show graphically a confidence region for the path of steepest ascent. What fraction of the possible directions from the design origin are excluded by the path you computed in #a?
c. Perform tests for interaction and curvature. From these tests, do you feel comfortable engaging in the method of steepest ascent? Explain why or why not. Would you suggest any new runs besides those which are on the path from #a?

#2: Metallurgy

In a metallurgy experiment it's desired to test the effect of four factors and their interactions on the concentration (percent by weight) of a particular phosphorus compound in casting material. The variables are: x_1, percent phosphorus in the refinement; x_2, percent remelted material; x_3, fluxing time; and x_4, holding time. The four factors are varied in a 2^4 factorial design with two castings taken at each factor combination. The 32 castings were made in random order, and are provided in metallurgy.txt[14], where the factors are presented in coded form. (Reproduced almost verbatim from exercise 5.5 in Myers et al. (2016) using data from Table E5.601.)

a. Build a first-order response function.
b. Construct a table summarizing the path of steepest ascent in the coded design variables.
c. It's important to constrain the percentages of phosphorus and remelted material. In fact, in coded variables we obtain $x_1 + x_2 \leq 2.7$, where x_1 is percent phosphorus and x_2 is percent remelted material. Recalculate the path of steepest ascent subject to the above constraint.

---

[13]http://bobby.gramacy.com/surrogates/sadat.txt
[14]http://bobby.gramacy.com/surrogates/metallurgy.txt

#3: Heat transfer

In a chemical engineering experiment dealing with heat transfer in a shallow fluidized bed, data (in heat.txt[15]) are collected on the following four regressor variables: fluidizing gas flow rate, lb/hr (**fluid**); supernatant gas flow rate, lb/hr (**gas**); supernatant gas inlet nozzle opening, mm (**open**); supernatant gas inlet temperature, °F (**temp**). The responses measured are heat transfer efficiency (**heat**) and thermal efficiency (**therm**). (Reproduced almost verbatim from exercise 6.2 in Myers et al. (2016) using data from Table E6.1.)

This is a good example of what often happens in practice. An attempt was made to use a particular second-order design. However, errors in controlling the variables produced a design that's only an approximation of the standard design.

a. Center and scale the design variables. That is, create a design matrix in coded form.
b. Fit a second-order model separately for both responses.
c. In the case of transfer efficiency (**heat**), do a canonical analysis and determine the nature of the stationary point. Do the same for thermal efficiency (**therm**).
d. For the case of transfer efficiency, what (natural) levels of the design variables would you recommend if maximum transfer efficiency is sought?
e. Do the same for thermal efficiency; that is, find levels of the design variables that you recommend for maximization of thermal efficiency.

#4: Bumper plating

A computer program simulates an auto-bumper plating process using thickness as the response with **time**, temperature (**temp**), and **nickel** pH as design variables. An experiment was conducted so that a response surface optimization could be entertained. A coding for the design variables is given by

$$x_1 = \frac{\text{time} - 8}{4} \qquad x_2 = \frac{\text{temp} - 24}{8} \qquad x_3 = \frac{\text{nickel} - 14}{4}.$$

Using the data in bumper.txt[16], perform the following steps. (Inspired by exercise 6.9 in Myers et al. (2016) using data from Table E6.6 which can be traced back to Schmidt and Launsby (1989).)

a. With the conversion above, write out the design matrix in coded form.
b. Fit and edit a second-order response surface model. That is, fit and eliminate insignificant terms.
c. Find conditions that maximize thickness, with the constraint that the condition falls inside the design region. In addition, compute the standard error for prediction at the location of optimum conditions.

#5: Gas turbine generators

The file turbine.txt[17] contains runs of an experiment on two input variables, blade **speed** (in/sec) and voltage measuring sensor **extension** (in) describing the configuration of a gas turbine generator, and measuring the **volts** (voltage) output by the system. (Stripped down version of exercise 6.4 in Myers et al. (2016) using data from Table E6.4.)

---

[15]http://bobby.gramacy.com/surrogates/heat.txt
[16]http://bobby.gramacy.com/surrogates/bumper.txt
[17]http://bobby.gramacy.com/surrogates/turbine.txt

a. Write the design matrix in coded form.
b. Fit a second-order model, find the stationary point and interpret and visualize the fitted response surface.
c. Calculate and visualize a confidence region for the stationary point at the 90% and 95% levels. Approximately what proportion of those regions lie inside the design region?

#6: Viscosity

Consider an experiment summarized in viscosity.txt[18] that studied a response triple (`yield`, `viscosity` and molecular weight `molewt`) as a function of two inputs, `time` and temperature (`temp`). For this question focus only on the `viscosity` output, ignoring the other two (`yield` and `molewt`). (Adapted from exercise 6.15 in Myers et al. (2016) using data from Table E7.4.)

a. Perform a canonical analysis on the second-order model for these data.
b. Use the double linear regression method to find confidence intervals on the eigenvalues.
c. Interpret the fitted surface.

---

[18]http://bobby.gramacy.com/surrogates/viscosity.txt

4

Space-filling Design

This segment puts the cart before the horse a little. Nonparametric spatial regression, emphasizing Gaussian processes in Chapter 5, benefits from a more agnostic approach to design compared to classical, linear modeling-based, response surface methods. One of the goals here is pragmatic from an organizational perspective: to have some simple, good designs for illustrations and comparisons in later chapters. Designs here are model-free, meaning that we don't need to know anything about the (Gaussian process) models we intend to use with them, except in the loose sense that those are highly flexible models which impose limited structure on the underlying data they're are trained on. Later in Chapter 6 we'll develop model-specific analogs and find striking similarity, and in some sense inferiority – a bizarre result considering their optimality – compared to these model-free analogs.

Here we seek so-called *space-filling designs*, ones which spread out points with the aim of encouraging a diversity of data once responses are observed. A spread of training examples, the thinking goes, will ultimately yield fitted models which smooth/interpolate/extrapolate best, leading to more accurate predictors at out-of-sample testing locations. Our development will focus on variations between, and combinations of, two of the most popular space-filling schemes: Latin hypercube sampling (LHS), and maximin distance designs. Both are based on geometric criteria but offer optimal spread in different senses. LHSs are random, so they disperse in a probabilistic sense, targeting a certain uniformity property. Maximin designs are more deterministic, even if many solvers for such designs deploy stochastic search. They seek spread in terms of relative distance.

Plenty of other model-free, space-filling designs enjoy wide popularity, but they're mostly variations on a theme. It's a lot of splitting hairs about subtly different geometric criteria, coupled with vastly different highly customized solving algorithms, producing designs that are visually and practically indistinguishable from one another. The aim here is twofold: 1) to have a nice conceptual base to jump off from, and contrast with the more interesting topic of model-based sequential design (§6.2 and Chapter 7); 2) to have an intuitive default from which to initialize a more ambitious such sequential search.

The entirety of this chapter, and indeed much of the rest of the text, presumes coded inputs in $[0, 1]^m$ unless otherwise stated. So the goal is a space-filling design of a desired run size n in the unit hypercube. As an auxiliary consideration, we'll explore the possibility of "growing" a design to size $n' + n$ from an initial fixed design of size n when additional experimental resources allow augmenting the training dataset. There the goal will be more modest: to best respect, or at least not drastically offend, an underlying space-filling criteria as regards the choice of n' new locations.

4.1 Latin hypercube sample

One simple option for design is random uniform. That is, fill a design matrix X_n in $[0,1]^m$ with n runs via `runif`.

```
m <- 2
n <- 10
X <- matrix(runif(n*m), ncol=m)
colnames(X) <- paste0("x", 1:m)
```

The trouble is, randomness is clumpy. With n of any reasonable size relative to input dimension m you're almost guaranteed to get design locations right next to one another, and thus gaps in other parts of the input space. Figure 4.1 illustrates this with the single random example generated above.

```
plot(X, xlim=c(0,1), ylim=c(0,1))
```

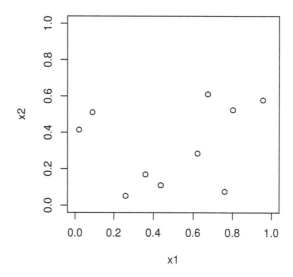

FIGURE 4.1: Random designs can be clumpy.

It can be useful to repeat that example, visualizing new random designs in several replicates. A takeaway from such an exercise is undoubtedly that designs offering reasonable "fill" in the 2d input space are quite rare indeed. Although we don't bother to describe the formalism, nor undertake calculations to make the following statement precise, it's not a difficult exercise for one so-inclined. The probability of observing at least two points close to one other, as a relative share of the total input space (with $n = 10$ runs in $m = 2$ input dimensions), is quite high. Therefore the chance of adequately filling the space, at least from an aesthetic perspective, is quite low. Now clumpiness in a design for computer surrogate modeling is not necessarily a bad thing. In fact, purely random designs can be good in some cases (Zhang et al., 2020). But let's table that discussion for a moment.

Latin hypercube samples (LHSs) were created, or perhaps borrowed (more details below) from another literature, to alleviate this problem. The goal is to guarantee a certain degree of spread in the design, while otherwise enjoying the properties of a random uniform sample. LHSs accomplish that by divvying the design region evenly into cubes, by partitioning coordinates marginally into equal-sized segments, and ensuring that the sample contains just one point in each such segment. In 2d, which is perhaps over-emphasized due to ease of visualization, the pattern of selected cubes (i.e., those containing a sample) resemble Latin squares[1].

Since the location of the point within the selected cube is allowed to be random, the LHS doesn't preclude two points located nearby one another. For example, two points may reside near a corner in common between a cube from an adjacent row and column. This will be easy to see in our visualizations shortly. However the LHS does limit the number of such adjacent cases, guaranteeing a certain amount of spread. But we're getting ahead of ourselves; how do we construct an LHS? For the description below, I'd like to acknowledge a chapter by Lin and Tang (2015) as a primary source of material for this presentation.

A *Latin hypercube (LH)* of n runs for m factors is represented by an $n \times m$ matrix $L = (l_{ij})$. Each column l_j, for $j = 1, \ldots, m$, contains a permutation of n equally spaced levels. For convenience, the n levels may be taken to be

$$-(n-1)/2, -(n-3)/2, \ldots, (n-3)/2, (n-1)/2.$$

Right now levels span from $-(n-1)/2$ to $(n-1)/2$, so L spans $[-(n-1)/2, (n-1)/2]^m$. This is a mismatch to our design region $[0, 1]$ in m coordinates. But don't worry – we'll fix that with a normalization step momentarily.

A *Latin hypercube sample* (LHS), or sometimes design (LHD), X_n in the design space $[0, 1]^m$ is an $n \times m$ matrix with $(i, j)^{\text{th}}$ entry

$$x_{ij} = \frac{l_{ij} + (n-1)/2 + u_{ij}}{n}, \quad i = 1, \ldots, n, \; j = 1, \ldots, m, \tag{4.1}$$

where the u_{ij}'s are independent uniform random deviates in $[0, 1]$. Observe that the denominator n normalizes so that x_{ij} is mapped into $[0, 1]^m$. The $(n-1)/2$ term chooses a point in the corner of the selected square or (hyper) cube. Random adjustments u_{ij} place the jth coordinate of run i elsewhere in the cube. If instead each u_{ij} is taken to be 0.5 the result is called a *Latin sample*, which will put x_{ij} right in the middle of one of the squares or (hyper) cubes.

If that verbal description seems complicated, its codification is relatively straightforward. The first step is to create the levels, which depend on the desired number of runs in the design, n.

```
l <- (-(n - 1)/2):((n - 1)/2)
```

Next, put m randomly permuted versions of the level vector l into a $n \times m$ matrix L.

```
L <- matrix(NA, nrow=n, ncol=m)
for(j in 1:m) L[,j] <- sample(l, n)
```

---

[1]https://en.wikipedia.org/wiki/Latin_square

Finally, create a uniform jitter matrix U and combine with the level matrix L following a vectorized version of Eq. (4.1).

```
U <- matrix(runif(n*m), ncol=m)
X <- (L + (n - 1)/2 + U)/n
```

What do we get? Figure 4.2 shows the design X as open circles, and uses the vector of levels l to demarcate candidate Latin squares with gray-dashed lines.

```
plot(X, xlim=c(0,1), ylim=c(0,1), xlab="x1", ylab="x2")
abline(h=c((1 + (n - 1)/2)/n,1), col="grey", lty=2)
abline(v=c((1 + (n - 1)/2)/n,1), col="grey", lty=2)
```

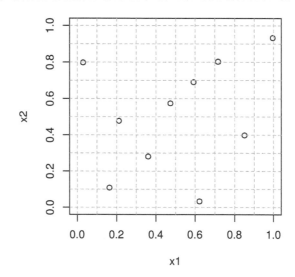

FIGURE 4.2: Latin hypercube sample (LHS) in 2d.

If we were to instead shade the squares containing the dots gray, dropping the dots, we'd have a Latin hypercube; forcing U=1/2 and placing the dot in the middle of the square yields the Latin sample. Both represent useful exercises for the curious reader hoping to gain better insight into what's going on. Permuting l, comprising columns of L, ensures that each margin, $j \in \{1, 2\}$ representing x_1 and x_2 in this $m = 2$ dimensional example, contains just one dot in each segment of the grid in that dimension. Said another way, each column and each row (demarcated by the gray-dashed lines) in the plot has just one dot. Consequently each margin, when projecting over the other input, is uniformly distributed in that coordinate. More detail on these features is provided momentarily, and in greater generality in higher dimension m.

But first, let's write a function that does it so we don't have to keep cutting-and-pasting the code above. That code is essentially ready-to-go for arbitrary $m \equiv m$, so all we're really doing here is wrapping those lines in an R function. Think of this as a more directly useful alternative to the formal environment used elsewhere in the book to demarcate pseudocode for higher-level algorithmics.

```
mylhs <- function(n, m)
 {
   ## generate the Latin hypercube
   l <- (-(n - 1)/2):((n - 1)/2)
   L <- matrix(NA, nrow=n, ncol=m)
   for(j in 1:m) L[,j] <- sample(l, n)

   ## draw the random uniforms and turn the hypercube into a sample
   U <- matrix(runif(n*m), ncol=m)
   X <- (L + (n - 1)/2 + U)/n
   colnames(X) <- paste0("x", 1:m)

   ## return the design and the grid it lives on for visualization
   return(list(X=X, g=c((1 + (n - 1)/2)/n,1)))
 }
```

The LHS output resides in the `$X` field of the list returned. To aid in visualization, grid points are also returned via `$g`.

4.1.1 LHS properties

When using the new "library routine", what can be expected in higher dimension $m \equiv$ m? As we noticed in Figure 4.2, LHSs are constructed so that there's exactly one point in each of the n intervals

$$[0, 1/n), [1/n, 2/n), \dots, [(n - 1)/n, 1),$$

partitioning up each input coordinate, $j = 1, \dots, m$. This property is referred to as *one-dimensional uniformity*. As a consequence of that property, any projection into lower dimensions that can be obtained by dropping some of the coordinates will also be distributed uniformly. Therefore that lower m'-dimensional design will also be an LHS in the $m' < m$ dimensional unit hypercube. This property is perhaps apparent by inspecting the algorithmic description above, or its implementation in code. What happens for a column of L or X is independent of, yet identical to, calculations for other columns.

As an illustration, consider an $m = 3$ dimensional LHS with $n = 10$ runs using our library routine.

```
Dlist <- mylhs(10, 3)
```

Figure 4.3 shows the first two coordinates of that sample.

```
plot(Dlist$X[,1:2], xlim=c(0,1), ylim=c(0,1), xlab="x1", ylab="x2")
abline(h=Dlist$g, col="grey", lty=2)
abline(v=Dlist$g, col="grey", lty=2)
```

Observe that this sample is a perfectly good $m = 2$ LHS, having only one point in each row and column of the 2d grid, despite being generated in 3d. The other two pairs of inputs show

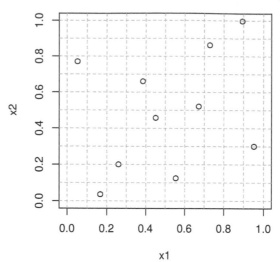

FIGURE 4.3: Two-dimensional projection of a 3d LHS.

the same property in Figure 4.4. Any of these three (or $\binom{m}{2}$ more generally) two-dimensional projections would be a perfectly good 2d LHS.

```
Is <- as.list(as.data.frame(combn(ncol(Dlist$X),2)))[-1]
par(mfrow=c(1,length(Is)))
for(i in Is) {
  plot(Dlist$X[,i], xlim=c(0,1), ylim=c(0,1),
    xlab=paste0("x", i[1]), ylab=paste0("x", i[2]))
  abline(h=Dlist$g, col="grey", lty=2)
  abline(v=Dlist$g, col="grey", lty=2)
}
```

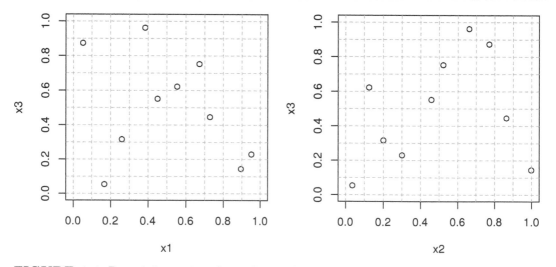

FIGURE 4.4: Remaining pairs of coordinates beyond the one shown in Figure 4.3.

Projecting down into one dimension reveals uniform distributions on the margins. Figure 4.5 utilizes a much larger LHS so that the resulting histograms look more convincingly uniform.

An analog using the previous $n = 10$ sample, potentially under replication, also provides an instructive suite of visualizations.

```
X <- mylhs(1000, 3)$X
par(mfrow=c(1,ncol(X)))
for(i in 1:ncol(X)) hist(X[,i], main="", xlab=paste0("x", i))
```

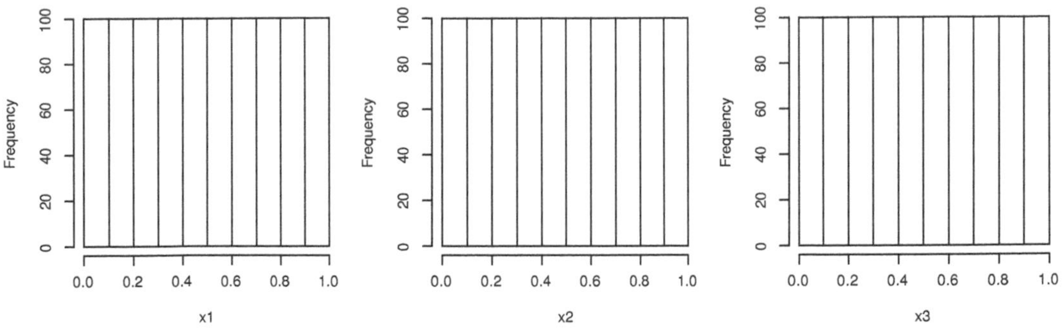

FIGURE 4.5: Histograms of one-dimensional LHS margins.

Popularity of LHSs as designs can largely be attributed to their role in reducing variance in numerical integration. Consider a function $y = f(x)$ where f is known, and x has a uniform distribution on $[0,1)^m$, and where we wish to estimate $\mathbb{E}\{Y\}$ via

$$\hat{\mu} = \frac{1}{n} \sum_{i=1}^{n} f(x_i), \quad x_i \overset{iid}{\sim} U[0,1].$$

It turns out that if $f(x)$ is monotone in each coordinate of x then the variance of $\hat{\mu}$ is lower if an LHS is used rather than simple random sampling.

One drawback of LHSs is that, although a degree of space-fillingness is guaranteed at the margins, there's no guarantee that the resulting design won't otherwise be somehow *aliased*, i.e., revealing patterns or clumpiness in other aspects of the variables' joint distribution. For example, the code chunk below repeatedly draws LHSs with $(n, m) = (10, 2)$, stopping when one is found satisfying a loosely specified monotonicity condition: where the second coordinate is roughly increasing in the first, basically lining up along a jittered diagonal. A counter is used to tally the number of attempts before that criterion is met.

```
count <- 0
while(1) {
  count <- count + 1
  Dlist <- mylhs(10, 2)
  o <- order(Dlist$X[,1])
  x <- Dlist$X[o,2]
  if(all(x[1:9] < x[2:10] + 1/20)) break
}
```

Figure 4.6 shows the design that caused the code's execution to break out of the while loop, finding an LHS that met that loose monotonicity criterion.

```
plot(Dlist$X, xlim=c(0,1), ylim=c(0,1), xlab="x1", ylab="x2")
text(0.2, 0.85, paste("count =", count))
abline(v=Dlist$g, col="grey", lty=2)
abline(h=Dlist$g, col="grey", lty=2)
```

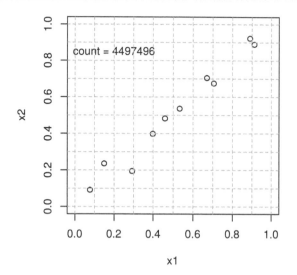

FIGURE 4.6: Potential for aliasing in LHSs.

Clearly this is an undesirable space-filling design in 2d, even though the 1d marginals are nice and uniform. Admittedly, it takes a ton of iterations (approximately 4500 thousand in this run) to find the pattern we were looking for. But that's one of many patterns out of perhaps an enormous collection of dots that could have yielded an aesthetically undesirable design. Despite many attractive properties, besides being easy to compute, a condition on marginals is a crude way to spread things out, and one which has diminishing returns in increasing dimension. In high dimensions (big m relative to n) everything looks like some kind of pattern, so perhaps the distinction between (totally) uniform random and LHSs is academic in that setting.

Suggesting potential remedies will require extra scaffolding which is offered, in part, in §4.2 on maximin design. In the meantime – and to end the LHS description on a high rather than low note – it's worth discussing several simple-yet-powerful enhancements unique to the LHS setting.

4.1.2   LHS variations and extensions

LHSs may be extended to marginals other than uniform through a simple inverse cumulative distribution function (CDF) transformation[2]. The steps are:

1. Generate an ordinary LHS, then
2. apply an inverse CDF (quantile function) of your choice on each of the marginals.

This will cause the marginals to take on distributions with those CDFs, but at the same time ensure that there's not "more clumpiness than necessary" in the joint distribution.

---

[2]https://en.wikipedia.org/wiki/Inverse_transform_sampling

The simplest application of this is to re-scale to a custom hyperrectangle to accommodate another coding of the inputs, or directly in the natural scale.

Again in lieu of a formal algorithm, R code below generalizes our `mylhs` library routine to utilize beta distributed[3] marginals, of which uniform is a special case. Parameters α and β are specified as m-vectors `shape1` and `shape2` following the nomenclature used by `qbeta`, R's built-in quantile function for beta distributions. Besides returning a matrix of design elements, the implementation also provides a matrix of beta-mapped grid lines `$g`. Vector `$g` no longer suffices since grid elements are now input-coordinate dependent.

```
mylhs.beta <- function(n, m, shape1, shape2)
 {
  ## generate the Latin Hypercube and turn it into a sample
  l <- (-(n - 1)/2):((n - 1)/2)
  L <- matrix(NA, nrow=n, ncol=m)
  for(j in 1:m) L[,j] <- sample(l, n)
  U <- matrix(runif(n*m), ncol=m)
  X <- (L + (n - 1)/2 + U)/n

  ## calculate the grid for that design
  g <- (L + (n - 1)/2)/n
  g <- rbind(g, 1)

  for(j in 1:m) { ## redistrbute according to beta quantiles
     X[,j] <- qbeta(X[,j], shape1[j], shape2[j])
     g[,j] <- qbeta(g[,j], shape1[j], shape2[j])
  }
  colnames(X) <- paste0("x", 1:m)

  ## return the design and the grid it lives on for visualization
  return(list(X=X, g=g))
}
```

By way of illustration, consider a 2d LHS with $n = 10$ under $\text{Beta}(3, 2)$ and $\text{Beta}(1/2, 1/2)$ marginals.

```
Dlist <- mylhs.beta(10, 2, shape1=c(3,1/2), shape2=c(2,1/2))
```

Figure 4.7 shows behavior matching what was prescribed: density of grid locations (and samples) is highest to the right of middle in x_1, targeting a mean of $3/(2+3) = 3/5$, and at the edges of the x_2 coordinate owing to the "U-shape" of $\text{Beta}(1/2, 1/2)$.

```
plot(Dlist$X, xlim=c(0,1), ylim=c(0,1), xlab="x1", ylab="x2")
abline(v=Dlist$g[,1], col="grey", lty=2)
abline(h=Dlist$g[,2], col="grey", lty=2)
```

To look closer at the marginals, code below generates a much larger LHS ...

---

[3]https://en.wikipedia.org/wiki/Beta_distribution

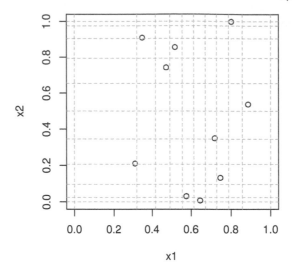

FIGURE 4.7: Two-dimensional LHS with beta marginals.

```
X <- mylhs.beta(1000, 2, shape1=c(3,1/2), shape2=c(2,1/2))$X
```

... and Figure 4.8 presents histograms of samples obtained in each coordinate against PDF evaluations under the desired beta distributions.

```
par(mfrow=c(1,2))
x <- seq(0,1,length=100)
hist(X[,1], main="", xlab="x1", freq=FALSE)
lines(x, dbeta(x, 3, 2), col=2)
hist(X[,2], main="", xlab="x2", freq=FALSE)
lines(x, dbeta(x, 1/2, 1/2), col=2)
```

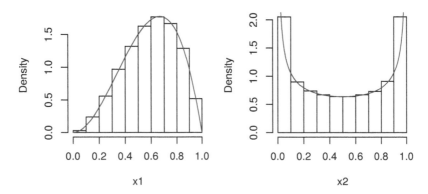

FIGURE 4.8: Marginals of LHS with beta distributions.

Designs generated in this way are useful for input sensitivity analyses, discussed in detail in §8.2, and related exercises common in the uncertainty quantification (UQ) literature. An important aspect in both settings involves exploring how a distribution on inputs propagates to a distribution of outputs through an opaque blackbox apparatus. Random sampling can facilitate a study of main effects and second-order uncertainty indices for nominal inputs under random jitter, or for inputs concentrated in a typical regime. LHSs offer similar functionality while requiring many fewer samples be pushed through a computationally intensive blackbox, say one constructed as a daisy-chain of numerical solvers, surrogates, and so on.

LHSs are also handy in synthetic predictive benchmarking exercises emphasizing assessment of generalization error. Sometimes these are called "bakeoffs" or "horse races". The idea is to generate training and out-of-sample testing sets where input–output pairs from the former are used to build fitted models of several varieties, and ones from the latter are held out to subsequently compare predictors derived from those fitted models. Inputs from an LHS offer certain guarantees on spread, and when collected simultaneously for training and testing can yield a partition where out-of-sample measurements are made on novel (but not too novel) input settings. Outputs for such exercises could come from any of the synthetic data generating functions offered by the VLSE[4], for example.

To illustrate, consider the following partition of an $n = 20$ LHS in 2d where training inputs come from the first ten, and testing from the last ten.

```
Dlist <- mylhs(20, 2)
Xtrain <- Dlist$X[1:10,]
Xtest <- Dlist$X[11:20,]
```

Figure 4.9 visualizes the two sets on the combined LH grid.

```
plot(Xtrain, xlim=c(0,1), ylim=c(0,1), xlab="x1", ylab="x2")
points(Xtest, pch=20)
abline(v=Dlist$g, col="grey", lty=2)
abline(h=Dlist$g, col="grey", lty=2)
```

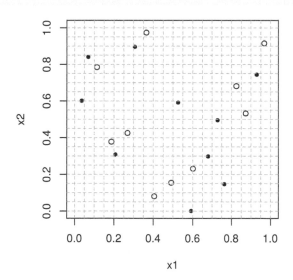

FIGURE 4.9: LHS training and testing partition.

Observe that testing and training locations are spaced out both relative to themselves and

---

to those from the other set. The thinking is that a prediction exercise based upon these inputs would reduce variance in assessments of generalization error, i.e., when measuring the quality of predictions made for inputs somewhat different than ones the fitted models were trained on. It's common to base horse races or bakeoffs on summaries obtained from repetitions on this theme, as codified by Algorithm 4.1.

Algorithm 4.1 LHS Bakeoff

Assume input dimension (m) is a match for test function f, and that the K competing methods are appropriate for f.

Require test function f, K fitting/prediction methods, training size n, testing size n', and number of repetitions R.

Repeat the following **for** $r = 1, \ldots, R$:

1. Generate an LHS of size $n + n'$ creating X.
2. Evaluate the true response at all $n + n'$ sites, yielding $Y \sim f(X)$, and combine them with inputs to obtain $n + n'$ data pairs $D = (X, Y)$.
3. Randomly partition into n training pairs $T_n = (X_n^t, Y_n^t)$ and n' testing or validation pairs $V_{n'} = (X_{n'}^v, Y_{n'}^v)$.
   - Since rows of the LHS are random, as from `mylhs`, it's usually sufficient to take the first n and last n' pairs, respectively.
4. Train K competing methods on T_n yielding predictors $\hat{y}^{(k)}(\cdot) \equiv \hat{y}^{(k)}(\cdot) \mid T_n$, for $k = 1, \ldots, K$.
5. Test out-of-sample by making predictions on the validation set $V_{n'}$, saving them as $\hat{Y}_{n'}^{(k)} = \hat{y}^{(k)}(X_{n'}^v)$.
6. Calculate metrics $m_r^{(k)}$ for each method $k = 1, \ldots, K$ comparing $\hat{Y}_{n'}^{(k)}$ to the held-out values $Y_{n'}^v$. Examples include
   - proper scoring rule (See §5.2 and Gneiting and Raftery, 2007),
   - *root mean-squared (prediction) error* (RMSE) or *mean absolute error* (MAE), respectively

$$m_r^{(k)} = \sqrt{\frac{1}{n'} \sum_{i=1}^{n'} (y_i^v - \hat{y}_i^{(k)})^2}, \quad \text{or} \quad m_r^{(k)} = \frac{1}{n'} \sum_{i=1}^{n'} |y_i^v - \hat{y}_i^{(k)}|.$$

Return one or more $R \times K$ matrices M comprised of $m_r^{(k)}$-values calculated in Step 6, above, and/or more compact summaries of their empirical distribution in order to determine the winner of the bakeoff.

Some notes about typical use follow. Perhaps the most common setting is $n' = n$ so that testing size is commensurate with training. However in situations where computation involved in fitting limits the training size, n, but prediction is much faster (i.e., as is the case for GPs in Chapter 5), an $n' \gg n$ setup is sometimes used. It's typical for y-values from f to be observed with noise, e.g., $y = f(x) + \varepsilon$, where $\varepsilon \sim \mathcal{N}(0, \sigma^2)$, yet deterministic settings are common when studying computer experiments. Notation $Y \sim f(X)$ above is intended to cover both cases. Sometimes one is interested in training on noisy data, but making predictive comparisons to de-noised versions, say via RMSE or MAE. This may be implemented with a simple adaptation to the algorithm. However note that proper scores must use a noisy testing set.

One reason to prefer returning a raw $R \times K$ matrix of metric evaluations, M, as opposed to a summary (say column-wise averages, quantiles or boxplots), is that the former makes it easier to change your mind later. If the relative comparison turns out to be a "close call", in terms of average values say, then pairwise t-tests (possibly on the log of the metric) can be helpful to adjudicate in order to determine a winner.

But we're getting a little ahead of ourselves here. We haven't really introduced any predictors to compare yet, so it's hard to make things concrete with an example at this time. The curious reader may wish to "fast forward" to a comparison between GP variations and MARS in §5.2.7 styled in the form of Algorithm 4.1.

While the testing set is spread out relative to the training set in the partition in Figure 4.9, these partition elements are not themselves LHSs. That is, they don't make up a *nested LHS* (Qian, 2009), which might be a desirable property to have. Moreover, unless the plan is to randomize over the partition, as in the GP comparison referenced above, the random nature of an LHS may be undesirable. Or at the very least it's a double-edged sword: randomization necessitates devoting valuable computational resources to repetition and can lead to insecurities about how much is enough. Many of the drawbacks of LHSs, like that spread is not guaranteed (aliasing is always present to a degree), are a consequence of its inherent randomness. Getting points with maximal spread, whether for one-shot design, train–test partitions, or for sequential analysis, requires swapping a measure of randomization for optimization.

4.2   Maximin designs

If what we want is points spread out, then perhaps it makes sense to design a criteria that deliberately spreads points out! For that we need a notion of distance by which to measure spread. The development here takes the simple choice of (squared) Euclidean distance

$$d(x, x') = ||x - x'||^2 = \sum_{j=1}^{m}(x_j - x'_j)^2,$$

but the framework is otherwise quite general if other choices, such as say Mahalanobis distance[5], were desired instead. A design $X_n = \{x_1, \ldots, x_n\}$ which maximizes the minimum distance between all pairs of points,

$$X_n = \text{argmax}_{X_n} \min\{d(x_i, x_k) : i \neq k = 1, \ldots, n\},$$

is called a *maximin design*.

The opposite way around, so-called *minimax designs*, also leads to points all spread out. By limiting how far apart pairs of points can be, you end up encouraging spread in a circuitous manner. Many of the considerations are similar, so we won't get into further detail here on minimax designs. It turns out they're somewhat harder to calculate than the more conceptually straightforward maximin in practice. See Johnson et al. (1990) for more discussion. Tan (2013) is also an excellent resource.

---

[5]https://en.wikipedia.org/wiki/Mahalanobis_distance

4.2.1 Calculating maximin designs

Maximin designs: easy to say, possibly not so easy to do. The details are all in the algorithm. The leading $\max_X$ is effectively a maximization over $n \times m$ "dimensions". Potential design sites $x_1, \ldots, x_n$ making up X_n each occupy m coordinates in the input space. Every X_n entertained must be evaluated under a criterion that entails at best $\mathcal{O}(n^2)$ operations, at least at first glance. Although the criterion suggests a deterministic solution for a particular choice of n and m, in fact most solvers are stochastic, owing to the challenge of the optimization setting. A naïve but simple iterative algorithm entails proposing a random change to one of the n sites x_i, and accepting or rejecting that change by consulting the criterion. Since only $n - 1$ distances change under such a swap, a clever implementation can get away with $\mathcal{O}(n)$ rather than $\mathcal{O}(n^2)$ calculations.

Having a fast (Euclidean) distance calculation, not just for one pair of points but potentially for many, is crucial to creating maximin designs efficiently. R has a `dist` function which is appropriate. For slightly simpler implementation and to connect better to GP code in later chapters (particularly Chapter 5), we shall instead borrow `distance` from the `plgp` package (Gramacy, 2014) on CRAN. The `distance` function allows distances to be calculated between pairs of coordinates, from one coordinate to many, and from many to many, where the `for` loops are optimized in compiled C code.

```
library(plgp)
distance(X[1:2,], X[3:5,])
```

```
##          [,1]    [,2]     [,3]
## [1,]  0.5317  1.0479  0.09896
## [2,]  0.1169  0.4173  0.04746
```

If you give `distance` just one matrix it'll calculate all pairwise distances between the rows in that matrix. Although the output records redundant lower and upper triangles, as complete $n \times n$ matrices are required for many GP calculations, the implementation doesn't double-up effort when calculating duplicate entries. To use that output in a maximin calculation we must ignore the diagonal, which is zero for all $i = 1, \ldots, n$.

Consider two random uniform designs, and a comparison between them based on the maximin criterion.

```
X1 <- matrix(runif(n*m), ncol=m)
dX1 <- distance(X1)
dX1 <- dX1[upper.tri(dX1)]
md1 <- min(dX1)
X2 <- matrix(runif(n*m), ncol=m)
dX2 <- distance(X2)
dX2 <- dX2[upper.tri(dX2)]
md2 <- min(dX2)
```

Now we can visualize the two and see if we think that the one with larger distance "looks" better as a space-filling design. Figure 4.10 shows the two designs, one with open circles and the other closed, with maximin distances quoted in the legend.

```
plot(X1, xlim=c(0,1.25), ylim=c(0,1.25), xlab="x1", ylab="x2")
points(X2, pch=20)
legend("topright", paste("md =", round(c(md1, md2), 5)), pch=c(21,20))
```

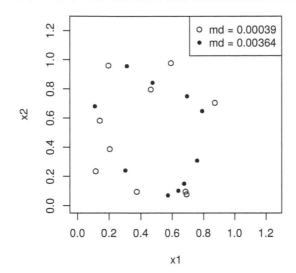

FIGURE 4.10: Comparing two designs via the maximin distance criterion.

Rmarkdown builds are sensitive to random whim, so it's hard to write text here to anticipate the outcome except to say that usually it's pretty obvious which of the two designs is better. It's highly likely that at least one of the random uniform designs has a pair of points very close together. Typically, one has an order of magnitude smaller `md` than the other. However, a single pair of points doesn't a holistic judgment make. It could well be that the design with the smallest minimum distance pair is actually the far better one when taking the distances of all other pairs into account. It's tempting to consider average minimum distance or some such more aggregate alternative, but that turns out to be a poor criteria for refinement through iterative search. That is, unless aggregate criteria are developed with care. See Morris and Mitchell (1995)'s ϕ_p criteria and related homework exercise (§4.4). More on computational strategies is peppered throughout the discussion below; specific libraries/implementations and off-shoots will be provided in §4.3.

Iterative pairwise comparisons of designs of size n, as in the example above, would be quite cumbersome computationally. Yet the simplicity of this idea makes it attractive from an implementation perspective. The coding effort is just as easy, but more efficient computationally, if modified to consider pairs `X1` and `X2` which differ by just one point. Along those lines, code below sets up a simple stochastic exchange algorithm to maximize minimum distance, starting from a random uniform design (`X1` above), proposing a random row to swap out, and a new set of random coordinates to swap back in (thus implicitly creating `X2`). If the resulting maximin distance is not improved, then the proposed exchange is undone: a rejection. Otherwise the implicit `X2` is accepted as the new `X1`. After accepting or rejecting 10000 such proposals, the algorithm terminates.

```
T <- 10000
for(t in 1:T) {
  row <- sample(1:n, 1)
```

```
  xold <- X1[row,]        ## random row selection
  X1[row,] <- runif(m)    ## random new row
  d <- distance(X1)
  d <- d[upper.tri(d)]
  mdprime <- min(d)
  if(mdprime > md1) {     ## accept
    md1 <- mdprime
  } else {                ## reject
    X1[row,] <- xold
  }
}
```

There are clearly many inefficiencies which are easily remedied, and addressing these will be left to exercises in §4.4. For example, any swapped-out `row` not corresponding to a minimum distance pair will result in rejection. So choosing `row` completely at random is clumsy. Replacing completely at random is also inefficient, although relatively cheap compared to other alternatives. Secondly, `distance` is applied to the entirety of the modified `X1`, an $\mathcal{O}(n^2)$ operation, when only one row and column (of $\mathcal{O}(n)$) has changed. Finally, no convergence is monitored. It could be that stopping much sooner than `T=10000` would yield a nearly identical design.

Despite gross inefficiency, execution is reasonably fast – because n is pretty small – and results are quite good, as shown in Figure 4.11.

```
plot(X1, xlim=c(0,1), ylim=c(0,1), xlab="x1", ylab="x2")
```

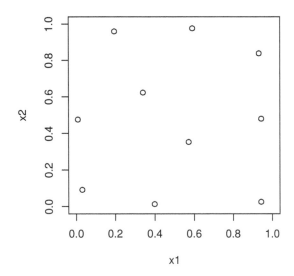

FIGURE 4.11: Maximin design in two input dimensions.

What do we notice? Design sites are nice and spread out, but they're pushed to boundaries which may be undesirable. Upon reflection, perhaps that's an obvious consequence of the search criteria. Points are pushed to the edge of the input space because they're repelled away from other ones nearby. As input dimension m increases this boundary effect becomes more extreme. Surface area will eventually dwarf interior volume in relative terms. Also notice a lack of one-dimensional uniformity here. Marginally, projecting over either of x_1 or x_2, there appears to be just three or four truly unique settings. Whereas with an LHS using $n = 10$ there were exactly ten.

Before exploring further, particularly in higher dimension, let's make a "library" function just as we did with `mylhs`.

```
mymaximin <- function(n, m, T=100000)
  {
    X <- matrix(runif(n*m), ncol=m)      ## initial design
    d <- distance(X)
    d <- d[upper.tri(d)]
    md <- min(d)

    for(t in 1:T) {
      row <- sample(1:n, 1)
      xold <- X[row,]                     ## random row selection
      X[row,] <- runif(m)                 ## random new row
      d <- distance(X)
      d <- d[upper.tri(d)]
      mdprime <- min(d)
      if(mdprime > md) { md <- mdprime    ## accept
      } else { X[row,] <- xold }          ## reject
    }

    return(X)
}
```

Now how about generating a maximin design with $n = 10$ and $m = 3$?

```
X <- mymaximin(10, 3)
```

Figure 4.12 offers 2d projections of that design for all pairs of inputs. Numbers plotted spot indices of each unique design element in order to help link points across panels.

```
Is <- as.list(as.data.frame(combn(ncol(X),2)))
par(mfrow=c(1,length(Is)))
for(i in Is) {
  plot(X[,i], xlim=c(0,1), ylim=c(0,1), type="n", xlab=paste0("x", i[1]),
    ylab=paste0("x", i[2]))
  text(X[,i], labels=1:nrow(X))
}
```

Things don't look so good here, at least compared to similar plots for the 3d LHS in Figures 4.3–4.4. Of course, it's harder to visualize a 3d design via 2d projections. An alternative is provided at the end of §4.2.2, but similar features emerge here: a push to boundaries, near non-uniqueness of values in the input coordinates, etc. Some 2d projections may have close pairs, but other panels indicate that those points are far apart in their other coordinates. In fact, those which are among the closest in one input pair tend to be among the farthest in

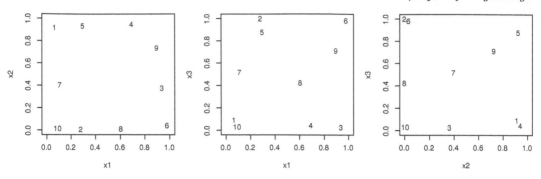

FIGURE 4.12: Three 2d projections of a maximin design in three inputs.

the others. It's abundantly clear that a projection of a maximin design is not itself maximin in the lower dimension. There's no one-dimensional uniformity.

Many of these drawbacks are addressed by variations available in library implementations, including the gross computational inefficiencies in `mymaximin` discussed above. We'll get to some of these, including hybridizations, in §4.3. First comes one of my favorite features of maximin designs.

4.2.2   Sequential maximin design

Maximin selection naturally extends to *sequential design*. That is, we may condition on an existing design $X_n^{\text{orig}} \equiv$ `Xorig` and ask what new runs could augment that design to produce $X_{n'} \equiv$ `X` where the new runs optimize the maximin criterion. Only elements of `X` are being chosen, but the desired settings of its coordinates are spaced out relative to themselves, and to existing locations in `Xorig`. Mathematically,

$$X_{n'} = \text{argmax}_{X_{n'}} \min\{d(x_i, x_k) : i = 1, \ldots, n'; k \neq i = 1, \ldots, n' + n\},$$

where X_n^{orig} is comprised of n locations indexed as $x_{n'+1}, \ldots, x_{n'+n}$. Sometimes $X_{n'}$ in this context is called an *augmenting design*.

Adapting our `mymaximin` function (§4.2.1) to work in this way isn't hard. Again, this implementation lacks solutions to many of the same inefficiencies pointed out for our earlier version, but benefits from simplicity. New code is highlighted by "`## new code`" comments below, comprised of two `if` statements that are active only when `Xorig` is defined.

```
mymaximin <- function(n, m, T=100000, Xorig=NULL)
 {
   X <- matrix(runif(n*m), ncol=m)        ## initial design
   d <- distance(X)
   d <- d[upper.tri(d)]
   md <- min(d)
   if(!is.null(Xorig)) {                  ## new code
     md2 <- min(distance(X, Xorig))
     if(md2 < md) md <- md2
   }
```

```
for(t in 1:T) {
  row <- sample(1:n, 1)
  xold <- X[row,]                    ## random row selection
  X[row,] <- runif(m)                ## random new row
  d <- distance(X)
  d <- d[upper.tri(d)]
  mdprime <- min(d)
  if(!is.null(Xorig)) {              ## new code
    mdprime2 <- min(distance(X, Xorig))
    if(mdprime2 < mdprime) mdprime <- mdprime2
  }
  if(mdprime > md) { md <- mdprime   ## accept
  } else { X[row,] <- xold }         ## reject
}

return(X)
}
```

Let's see how it works. Below we ask for $n' = 5$ new runs, augmenting our earlier $n = 10$ sites stored in X1.

```
X2 <- mymaximin(5, 2, Xorig=X1)
```

Figure 4.13 shows how new runs (closed dots) are spaced out relative to themselves and to the previously obtained maximin design (open circles).

```
plot(X1, xlim=c(0,1), ylim=c(0,1), xlab="x1", ylab="x2")
points(X2, pch=20)
```

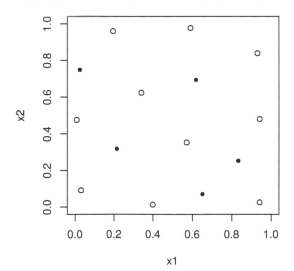

FIGURE 4.13: Sequential maximin design with augmenting points as closed dots.

The chosen locations look sensible, with spacing being optimized. Observe however that very few new settings are chosen marginally, when mentally projecting down onto coordinate axes for x_1 and x_2. The random nature of search makes it hard to comment on specific patterns in this Rmarkdown document. Yet in repeated applications it's quite typical for new design elements to have just one new unique coordinate value, up to reasonable relative tolerance. Lack of one-dimensional uniformity remains a drawback of maximin relative to LHS, even in the sequential or augmenting setting.

Despite that downside, this example demonstrates that a sequential maximin design could be used to create a training and testing partition much like our LHS analog in §4.1.2 and Algorithm 4.1. For settings where a large degree of training–testing repetition is too expensive, a single partition obtained via sequential maximin design might represent a sensible alternative. First select an n-sized maximin training design X_n; then create a sequential maximin testing set $X_{n'}$ given X_n.

The story is similar in higher input dimension. For example, consider expanding our 3d design to include five more runs.

```
X2 <- mymaximin(5, 3, Xorig=X)
X <- rbind(X2, X)
```

To ease visualization in Figure 4.14, new runs from X2 are subsumed into X. In addition to occupying indices 1:5, new design elements are colored in red.

```
Is <- as.list(as.data.frame(combn(ncol(X),2)))
par(mfrow=c(1,length(Is)))
for(i in Is) {
  plot(X[,i], xlim=c(0,1), ylim=c(0,1), type="n", xlab=paste0("x", i[1]),
    ylab=paste0("x", i[2]))
  text(X[,i], labels=1:nrow(X), col=c(rep(2,5), rep(1,10)))
}
```

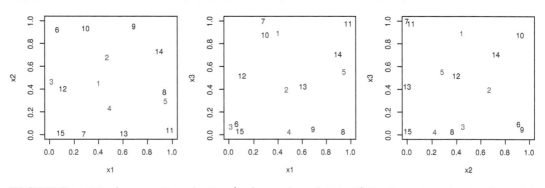

FIGURE 4.14: Augmenting design (red, numbered 1 to 5) in three inputs via three 2d projections.

As before, any particular projection might not be perfect, but the ensemble of panel views indicates that new runs are indeed spaced out relative to themselves and to the previous locations (numbered 6:15 and colored in black). Sometimes the illusion of perspective offered by the `scatterplot3d` library is helpful in such inspections. Some code is provided below, but graphical output is suppressed due to the excessive whitespace it creates in the Rmarkdown environment.

```
library(scatterplot3d)
scatterplot3d(X, type="h", color=c(rep(2,5), rep(1,10)),
  pch=c(rep(20,5), rep(21,10)), xlab="x1", ylab="x2", zlab="x3")
```

The color and point scheme indicates new locations $X_{n'}$ as red filled circles and X_n^{orig} as black open circles. Vertical lines under the dots, as provided by `type="h"`, are essential to visually track locations in 3d space and see that the design is truly space-filling. Unfortunately, it's not obvious how to use text to plot indices rather than points in `scatterplot3d`.

4.3 Libraries and hybrids

Our `mylhs` function from §4.1, and its extension `mylhs.beta` (§4.1.2) are hard to improve upon in terms of computational efficiency. Nevertheless it's simpler to rely on canned alternatives. Similar functionality is offered by **randomLHS** in the **lhs** library (Carnell, 2018) on CRAN. Although tailored to uniform marginals, that library does offer a sequential alternative in **augmentLHS**.

The **lhs** package also provides a nice hybrid between maximin and LHS in **maximinLHS** which helps avoid aliasing problems of both ordinary LHS and maximin designs. Think of this as entertaining candidate LHSs and preferring ones whose minimum distance between design element pairs is large (Morris and Mitchell, 1995). Typical search methods proceed similarly to our sequential maximin method, described above, usually involving stochastic exchange of pairs of levels, separately (and possibly also randomly) in each coordinate, in order to preserve a Latin square structure. To compare/contrast with some of our earlier examples, code below builds one of these hybrid designs.

```
library(lhs)
X <- maximinLHS(10, 2)
```

As shown in Figure 4.15, the resulting points are spread out, but perhaps not as spread out as a fully maximin design. By default, `maximinLHS` performs a *greedy*[6], stepwise search. A homework exercise (§4.4) demonstrates that it's possible to get better maximin LHS hybrid performance with a more exhaustive search. The library doesn't provide a Latin square grid on output, so code supporting Figure 4.15 offers an educated guess as to what that grid might look like.

```
plot(X, xlim=c(0,1), ylim=c(0,1), xlab="x1", ylab="x2")
abline(h=seq(0,1,length=11), col="gray", lty=2)
abline(v=seq(0,1,length=11), col="gray", lty=2)
```

Since it's still an LHS, points are not pushed to the boundaries. One-dimensional uniformity guarantees at most one point near each edge. Consequently such designs are often more desirable than an LHS or maximin on their own, although there are reasons to prefer these

---

[6]https://en.wikipedia.org/wiki/Greedy_algorithm

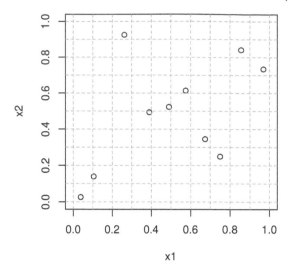

FIGURE 4.15: Maximin/LHS hybrid design.

original, raw (un-hybridized), options. For example, I'm not aware of any software offering a sequential update to a hybrid maximin LHS design. Related, so-called maximum projection design (Joseph et al., 2015), extends the LHS one-dimensional uniformity property to larger subspaces. Building a *maxpro* design entails calculations similar to those involved in maximin, and like maximin they emit a natural sequential analog. A package is available on CRAN called `MaxPro` (Ba and Joseph, 2018). Sometimes computer experiments contain factor-valued inputs – i.e., categorical rather than real-valued ones. Maxpro designs have recently been extended to accommodate this case. Maximin "sliced" LHD (SLHD) hybrids (Ba et al., 2015) also represent an attractive option in this setting. See `SLHD` (Ba, 2015) on CRAN.

A package called `maximin` (Sun and Gramacy, 2019) on CRAN offers batch and sequential (ordinary) maximin design. By optionally including distance to boundaries in search criteria it can avoid sites selected along extremities of the input space. Its implementation addresses many of the computational inefficiencies highlighted for `mymaximin` in §4.2.1, and which are the subject of an exercise in §4.4. Additionally `maximin` replaces random search with a heuristically motivated derivative-based optimization, leveraging local continuity in the objective criteria. Sometimes, however, a more discrete alternative is handy. A function `maximin.cand` allows search to be restricted to a candidate grid, which can be helpful when simulators only accept certain input locations. A homework exercise in §5.5 requires such a feature in order to work with surrogate LGBB data from §2.1. Sun et al. (2019b) used candidate-restricted sequential maximin designs to combine field and simulation data for a solar irradiance forecasting project. Obtaining accurate IBM PAIRS[7] simulations of solar irradiance, emphasizing ones geographically far from sporadically spaced weather station data, required limiting future simulation runs to a particular grid.

There are many further options when it comes to sensible, and computationally efficient, model-free space-filling designs. `DiceDesign` on CRAN (Franco et al., 2018; Dupuy et al., 2015) covers several others in addition ones I've introduced/pointed to above. We now have everything we need to move on to GP modeling in Chapter 5, in particular to generate data for illustrative examples provided therein. As mentioned at the outset of this chapter, topics in model-based design (using GPs) in Chapter 6 are in many ways twinned with the

---

[7] https://ibmpairs.mybluemix.net/

model-free analogs here. Despite conditioning on the model, resulting designs are quite similar to more agnostic alternatives, except when you have something more particular in mind. Perhaps that explains the popularity of a model-free option: all of the effect of "knowing" the model for the purposes of learning more about it, without potential for pathology as inherent in any presumptive choice before learning actually takes place.

Unfortunately, that characterization is not quite right because choosing a space-filling design can indeed influence what you learn. In fact Zhang et al. (2020) very specifically demonstrate pathological behavior when using space-filling designs like maximin and LHS with GP surrogates. Poor behavior can be particularly acute when using maximin or LHS as a small *seed design* in order to initiate an inherently sequential analysis, such as in Bayesian optimization (Chapter 7). Surprisingly, simple completely random design can protect against such pathologies. Going further, Zhang et al. show how designs offering more control on the distribution of pairwise distances – as opposed to attempting to maximize the smallest of those – fare better too. The reason has to do with mechanisms behind how GPs are typically specified, and fit to data. But we're getting well ahead of ourselves. I must first introduce the GP.

4.4 Homework exercises

These exercises give the reader an opportunity to explore space-filling design properties and algorithms. Throughout, use up to T=10000 iterations for your searches.

#1: Faster maximin

This question targets improvements to our initial, inefficient `mymaximin` implementation in §4.2.1. In each case, measure improvement in terms of the rate of increase in minimum distance over search iterations.

a. Fix a random design `Xinit` of size n=100 in m=2 dimensions. Modify `mymaximin` to use `Xinit` as its starting design, and to keep track (and return) `md` progress over the T iterations of search. (It might help to change the default number of iterations to T=10000.) Generate a maximin design using your new `mymaximin` implementation. Provide a visual of the final design and progress in `md` over the iterations. Report execution time.
b. Update `mymaximin` to more cleverly choose the `row` swapped out in each iteration. Argue that only two rows are worth potentially swapping out if progress is to be measured by increasing `md` over stochastic exchange iterations. How would you choose among these two? After you implement your new scheme, compare it to results from #a via visuals of the final design and `md` progress on common axes. Compare the execution time to #a.
c. Update `mymaximin` so that not all pairwise distances are calculated in each iteration. Calculate and update only those distances, and minimum pairwise distances, which change. Verify that your new implementation gives identical results (final design and `md` progress) to those which you obtained in #b, but show that the calculation is now much faster in terms of execution time. If your algorithm is randomized, verifying identical results will require identical RNG seeds.
d. Repeat #a–c thirty times with novel `Xinit` and compare average progress in `md` in terms

of means and 90% quantiles for each of the three methods. Also compare distributions of compute times.

#2: Hybrid space-filling design

Latin hypercube samples (LHSs) (§4.1) and maximin distance (§4.2) are both space-filling in a certain sense. Consider a hybrid of the two methods, a so-called maximin LHS (§4.3) as introduced by Morris and Mitchell (1995). That is, among LHSs we desire ones that maximize the minimum distance between design points. Such a hybrid can offer the best of both words: nice margins (whether uniform or otherwise), no clumping at the boundaries, and no diagonal aliasing!

There are lots of ways to make a precise definition for maximim LHS, but the spirit is conveyed in a simple algorithm for how it might be (approximately) calculated.

1. Generate an initializing LHS `Xinit` of appropriate size.
2. Randomly choose a pair of design points, randomly choose a column, and propose swapping that pair's coordinates in that column.
3. Keep the exchange if the minimum distances between those two points and all others is improved; otherwise swap back.
4. Repeat 2–3.

Your task is the following.

a. Convince yourself (and your instructor) that the result is still an LHS.
b. Implement the method and try it out in $m \in \{2, 3, 4\}$ dimensional design spaces, with design sizes $n \in \{100, 1000\}$. How do your maximin LHS designs compare to maximin? In particular, how do 2d projections for $m \in \{3, 4\}$ compare? Can you improve upon the numerical performance of your algorithm by deploying some of the strategies from exercises #1b–c?
c. Report on average progress of the algorithm, in terms of its ability to maximize minimum distance(s), over iterations of steps 2–3 above. (Use Monte Carlo; e.g., see #1d above.)
d. How does your algorithm compare to the `maximinLHS` method in the `lhs` library for R?

(The attentive reader will recognize that the choice of random jitter in the Latin square, i.e., the U step in generating `Xinit`, is at odds with the maximin criterion. But let's not get distracted by such details here; if you prefer choose U=0.)

#3: ϕ_p designs

In addition to maximin LHS hybridization (#2 above), Morris and Mitchell (1995) suggested the following search criteria for the purpose of space-filling design. Let d_k, for $k = 1, \ldots, K$, denote the K unique pairwise distances in a design X_n, and let J_k denote the number of rows of X_n who share pairwise distance d_k. In most applications, $K = \binom{n}{2}$ and all $J_k = 1$. Then, one may search for a design X_n by minimizing $\phi_p(X_n)$ where

$$\phi_p = \left[\sum_{k=1}^{K} J_k d_k^{-p} \right]^{1/p}.$$

At $p = \infty$, minimizing the criterion is equivalent to solving for a maximin design. Interestingly, Morris & Mitchell show how all p lead to maximin designs in the sense that the smallest distance, $\min_k d_k$ is maximized – but resulting designs differ in the distribution of other distances $d_{(-k)}$. Behavior of numerical methods for optimizing ϕ_p also differ from maximin due to the smoother nature of the criteria for smaller p.

Your task is to port `mymaximin` to `myminphi`, entertain enhancements similar to those of exercises #1a–c (i.e., think about search and computational efficiency), and to make a Monte Carlo comparison to those methods by extending #1d. Limit your study to $p \in \{2, 5\}$. Then hybridize with LHS following exercise #2b–c. How do the most efficient variations on the four methods: maximin, ϕ_p, maximin LHS, and ϕ_p LHS compare?

#4: Candidate-based augmenting maximin design

For a solar irradiance forecasting project, Sun et al. (2019b) worked with data from a network of weather stations scattered throughout the continental United States. Monitoring sites are heterogeneously dispersed, with heavy concentration in national/state parks and sparse coverage in the Great Plains and lower Midwest. Code below reads in the data ...

```
lola <- read.csv("lola.csv")
```

... and Figure 4.16 offers a visual of the sites where irradiance measurements have been gathered.

```
library(maps)
myUS <- map("state", fill=TRUE, plot=FALSE)
omit <- c("massachusetts:martha's vineyard", "massachusetts:nantucket",
  "new york:manhattan", "new york:staten island", "new york:long island",
  "north carolina:knotts", "north carolina:spit", "virginia:chesapeake",
  "virginia:chincoteague", "washington:san juan island",
  "washington:lopez island", "washington:orcas island",
  "washington:whidbey island")
myNames <- myUS$names[!(myUS$names %in% omit)]
myUS <- map("state", myNames, fill=TRUE, col="gray", plot=TRUE)
points(lola$lon, lola$lat, pch=19, cex=0.3, col="blue")
legend("bottomleft", legend=c("Original/Weather Stations"), col="blue",
  pch=19, pt.cex=0.3, bty="n")
```

To extrapolate better to sparsely covered regions of the country, Sun et al. (2019b) augmented their data with computer model irradiance evaluations. For reasons that are somewhat more thoroughly explained in §4.3 and in their manuscript, such runs could only be obtained at the grid of locations stored in the file below.

```
lola.cands <- read.csv("lola_cands.csv")
```

a. Demonstrate visually that the grid uniformly covers the continental USA except where it doubles-up on original `lola` locations. For brownie points, how would you automate the process of creating `lola.cands` from a uniform 0.1-spaced grid in longitude and latitude throughout `myUS$range`; that is, so that there are no grid points in the oceans/Great Lakes and none overlapping with `lola`?

b. Design an algorithm which delivers 100 new sites whose locations are confined to `lola.cands` and whose minimum distance to themselves and to `lola` is maximized.

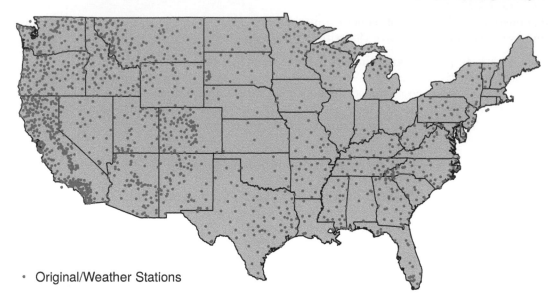

• Original/Weather Stations

FIGURE 4.16: Weather station sites in the continental USA.

Augment the visualization above to include the newly selected sites. Inspect `md` progress over the iterations. You may wish to adapt your method from exercise #1.

c. Using your implementation from #b, or with `maximin.cands` in the `maximin` package on CRAN, select 900 more (for 1000 total) new sites and provide visuals of the results. Plot `md` progress over the iterations.

5

Gaussian Process Regression

Here the goal is humble on theoretical fronts, but fundamental in application. Our aim is to understand the Gaussian process (GP) as a prior over random functions, a posterior over functions given observed data, as a tool for spatial data modeling and surrogate modeling for computer experiments, and simply as a flexible nonparametric regression. We'll see that, almost in spite of a technical over-analysis of its properties, and sometimes strange vocabulary used to describe its features, GP regression is a simple extension of linear modeling. Knowing that is all it takes to make use of it as a nearly unbeatable regression tool when input–output relationships are relatively smooth, and signal-to-noise ratios relatively high. And even sometimes when they're not.

The subject of this chapter goes by many names and acronyms. Some call it *kriging*, which is a term that comes from geostatistics (Matheron, 1963); some call it Gaussian spatial modeling or a Gaussian stochastic process. Both, if you squint at them the right way, have the acronym GaSP. Machine learning (ML) researchers like Gaussian process regression (GPR). All of these instances are about regression: training on inputs and outputs with the ultimate goal of prediction and uncertainty quantification (UQ), and ancillary goals that are either tantamount to, or at least crucially depend upon, qualities and quantities derived from a predictive distribution. Although the chapter is titled "Gaussian process regression", and we'll talk lots about Gaussian process surrogate modeling throughout this book, we'll typically shorten that mouthful to Gaussian process (GP), or use "GP surrogate" for short. GPS[1] would be confusing and GPSM[2] is too scary. I'll try to make this as painless as possible.

After understanding how it all works, we'll see how GPs excel in several common response surface tasks: as a sequential design tool in Chapter 6; as the workhorse in modern (Bayesian) optimization of blackbox functions in Chapter 7; and all that with a "hands off" approach. Classical RSMs of Chapter 3 have many attractive features, but most of that technology was designed specifically, and creatively, to cope with limitations arising in first- and second-order linear modeling. Once in the more flexible framework that GPs provide, one can think big without compromising finer detail on smaller things.

Of course GPs are no panacea. Specialized tools can work better in less generic contexts. And GPs have their limitations. We'll have the opportunity to explore just what they are, through practical examples. And down the line in Chapter 9 we'll see that most of those are easy to sweep away with a bit of cunning. These days it's hard to make the case that a GP shouldn't be involved as a component in a larger analysis, or at least attempted as such, where ultimately limited knowledge of the modeling context can be met by a touch of flexibility, taking us that much closer to human-free statistical "thinking" – a fundamental desire in ML and thus, increasingly, in tools developed for modern analytics.

---

[1]https://en.wikipedia.org/wiki/Global_Positioning_System
[2]https://en.wikipedia.org/wiki/S&M_(disambiguation)

5.1   Gaussian process prior

Gaussian process is a generic term that pops up, taking on disparate but quite specific meanings, in various statistical and probabilistic modeling enterprises. As a generic term, all it means is that any finite collection of realizations (i.e., n observations) is modeled as having a multivariate normal (MVN)[3] distribution. That, in turn, means that characteristics of those realizations are completely described by their mean n-vector μ and $n \times n$ covariance matrix Σ. With interest in modeling functions, we'll sometimes use the term *mean function*, thinking of $\mu(x)$, and *covariance function*, thinking of $\Sigma(x, x')$. But ultimately we'll end up with vectors μ and matrices Σ after evaluating those functions at specific input locations $x_1, \dots, x_n$.

You'll hear people talk about function spaces, reproducing kernel Hilbert spaces, and so on, in the context of GP modeling of functions. Sometimes thinking about those aspects/properties is important, depending on context. For most purposes that makes things seem fancier than they really need to be.

The action, at least the part that's interesting, in a GP treatment of functions is all in the covariance. Consider a covariance function defined by inverse exponentiated squared Euclidean distance:

$$\Sigma(x, x') = \exp\{-||x - x'||^2\}.$$

Here covariance decays exponentially fast as x and x' become farther apart in the input, or x-space. In this specification, observe that $\Sigma(x, x) = 1$ and $\Sigma(x, x') < 1$ for $x' \neq x$. The function $\Sigma(x, x')$ must be *positive definite*[4]. For us this means that if we define a covariance matrix Σ_n, based on evaluating $\Sigma(x_i, x_j)$ at pairs of n x-values $x_1, \dots, x_n$, we must have that

$$x^\top \Sigma_n x > 0 \quad \text{for all } x \neq 0.$$

We intend to use Σ_n as a covariance matrix in an MVN, and a positive (semi-) definite covariance matrix is required for MVN analysis. In that context, positive definiteness is the multivariate extension of requiring that a univariate Gaussian have positive variance parameter, σ^2.

To ultimately see how a GP with that simple choice of covariance Σ_n can be used to perform regression, let's first see how GPs can be used to generate random data following a smooth functional relationship. Suppose we take a bunch of x-values: $x_1, \dots, x_n$, define Σ_n via $\Sigma_n^{ij} = \Sigma(x_i, x_j)$, for $i, j = 1, \dots, n$, then draw an n-variate realization

$$Y \sim \mathcal{N}_n(0, \Sigma_n),$$

and plot the result in the x-y plane. That was a mouthful, but don't worry: we'll see it in code momentarily. First note that the mean of this MVN is zero; this need not be but it's quite surprising how well things work even in this special case. Location invariant zero-mean GP modeling, sometimes after subtracting off a middle value of the response (e.g., $\bar{y}$), is the

---

[3] https://en.wikipedia.org/wiki/Multivariate_normal_distribution
[4] https://en.wikipedia.org/wiki/Positive-definite_matrix

default in computer surrogate modeling and (ML) literatures. We'll talk about generalizing this later.

Here's a version of that verbal description with x-values in 1d. First create an input grid with 100 elements.

```
n <- 100
X <- matrix(seq(0, 10, length=n), ncol=1)
```

Next calculate pairwise squared Euclidean distances between those inputs. I like the `distance` function from the **plgp** package (Gramacy, 2014) in R because it was designed exactly for this purpose (i.e., for use with GPs), however `dist` in base R provides similar functionality.

```
library(plgp)
D <- distance(X)
```

Then build up covariance matrix Σ_n as inverse exponentiated squared Euclidean distances. Notice that the code below augments the diagonal with a small number `eps` $\equiv \epsilon$. Although inverse exponentiated distances guarantee a positive definite matrix in theory, sometimes in practice the matrix is numerically ill-conditioned[5]. Augmenting the diagonal a tiny bit prevents that. Neal (1998), a GP vanguard in the statistical/ML literature, calls ϵ the *jitter* in this context.

```
eps <- sqrt(.Machine$double.eps)
Sigma <- exp(-D) + diag(eps, n)
```

Finally, plug that covariance matrix into an MVN random generator; below I use one from the **mvtnorm** package (Genz et al., 2018) on CRAN.

```
library(mvtnorm)
Y <- rmvnorm(1, sigma=Sigma)
```

That's it! We've generated a finite realization of a random function under a GP prior with a particular covariance structure. Now all that's left is visualization. Figure 5.1 plots those X and Y pairs as tiny connected line segments on an x-y plane.

```
plot(X, Y, type="l")
```

Because the Y-values are random, you'll get a different curve when you try this on your own. We'll generate some more below in a moment. But first, what are the properties of this function, or more precisely of a random function generated in this way? Several are easy to deduce from the form of the covariance structure. We'll get a range of about $[-2, 2]$, with 95% probability, because the scale of the covariance is 1, ignoring the jitter ϵ added to the diagonal. We'll get several bumps in the x-range of $[0, 10]$ because short distances are highly correlated (about 37%) and long distances are essentially uncorrelated ($1e^{-7}$).

---

[5]https://en.wikipedia.org/wiki/Condition_number

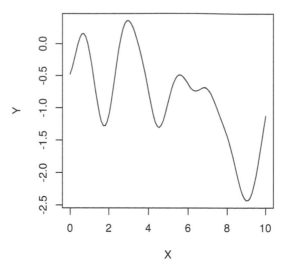

FIGURE 5.1: A random function under a GP prior.

```
c(exp(-1^2), exp(-4^2))
```

```
## [1] 3.679e-01 1.125e-07
```

Now the function plotted above is only a finite realization, meaning that we really only have 100 pairs of points. Those points look smooth, in a tactile sense, because they're close together and because the `plot` function is "connecting the dots" with lines. The full surface, which you might conceptually extend to an infinite realization over a compact domain, is extremely smooth in a calculus sense because the covariance function is infinitely differentiable, a discussion we'll table for a little bit later.

Besides those three things – scale of two, several bumps, smooth look – we won't be able to anticipate much else about the nature of a particular realization. Figure 5.2 shows three new random draws obtained in a similar way, which will again look different when you run the code on your own.

```
Y <- rmvnorm(3, sigma=Sigma)
matplot(X, t(Y), type="l", ylab="Y")
```

Each random finite collection is different than the next. They all have similar range, about the same number of bumps, and are smooth. That's what it means to have function realizations under a GP prior: $Y(x) \sim \mathcal{GP}$.

5.1.1 Gaussian process posterior

Of course, we're not in the business of generating random functions. I'm not sure what that would be useful for. Instead, we ask: given examples of a function in pairs $(x_1, y_1), \ldots, (x_n, y_n)$, comprising data $D_n = (X_n, Y_n)$, what random function realizations could explain – could have generated – those observed values? That is, we want to know about the conditional

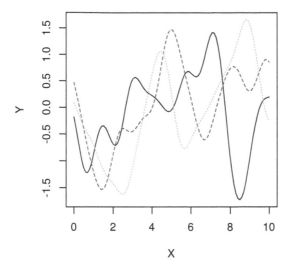

FIGURE 5.2: Three more random functions under a GP prior.

distribution of $Y(x) \mid D_n$. If we call $Y(x) \sim \mathcal{GP}$ the prior, then $Y(x) \mid D_n$ must be the posterior.

Fortunately, you don't need to be a card-carrying Bayesian to appreciate what's going on, although that perspective has really taken hold in ML. That conditional distribution, $Y(x) \mid D_n$, which one might more simply call a *predictive distribution*, is a familiar quantity in regression analysis. Forget for the moment that when regressing one is often interested in other aspects, like relevance of predictors through estimates of parameter standard errors, etc., and that so far our random functions look like they have no noise. The somewhat strange, and certainly most noteworthy, thing is that so far there are no parameters!

Let's shelve interpretation (Bayesian updating or a twist on simple regression) for a moment and focus on conditional distributions, because that's what it's really all about. Deriving that predictive distribution is a simple application of deducing a conditional from a (joint) MVN. From Wikipedia[6], if an N-dimensional random vector X is partitioned as

$$X = \begin{pmatrix} X_1 \\ X_2 \end{pmatrix} \quad \text{with sizes} \quad \begin{pmatrix} q \times 1 \\ (N-q) \times 1 \end{pmatrix},$$

and accordingly μ and Σ are partitioned as,

$$\mu = \begin{pmatrix} \mu_1 \\ \mu_2 \end{pmatrix} \quad \text{with sizes} \quad \begin{pmatrix} q \times 1 \\ (N-q) \times 1 \end{pmatrix}$$

and

$$\Sigma = \begin{pmatrix} \Sigma_{11} & \Sigma_{12} \\ \Sigma_{21} & \Sigma_{22} \end{pmatrix} \quad \text{with sizes} \quad \begin{pmatrix} q \times q & q \times (N-q) \\ (N-q) \times q & (N-q) \times (N-q) \end{pmatrix},$$

then the distribution of X_1 conditional on $X_2 = x_2$ is MVN $X_1 \mid x_2 \sim \mathcal{N}_q(\bar{\mu}, \bar{\Sigma})$, where

---

[6]https://en.wikipedia.org/wiki/Multivariate_normal_distribution#Conditional_distributions

$$\bar{\mu} = \mu_1 + \Sigma_{12}\Sigma_{22}^{-1}(x_2 - \mu_2) \tag{5.1}$$

$$\text{and} \quad \bar{\Sigma} = \Sigma_{11} - \Sigma_{12}\Sigma_{22}^{-1}\Sigma_{21}.$$

An interesting feature of this result is that conditioning upon x_2 alters the variance of X_1. Observe that $\bar{\Sigma}$ above is reduced compared to its marginal analog Σ_{11}. Reduction in variance when conditioning on data is a hallmark of statistical learning. We know more – have less uncertainty – after incorporating data. Curiously, the amount by which variance is decreased doesn't depend on the value of x_2. Observe that the mean is also altered, comparing μ_1 to $\bar{\mu}$. In fact, the equation for $\bar{\mu}$ is a linear mapping, i.e., of the form $ax + b$ for vectors a and b. Finally, note that $\Sigma_{12} = \Sigma_{21}^{\top}$ so that $\bar{\Sigma}$ is symmetric.

Ok, how do we deploy that fundamental MVN result towards deriving the GP predictive distribution $Y(x) \mid D_n$? Consider an $n + 1^{\text{st}}$ observation $Y(x)$. Allow $Y(x)$ and Y_n to have a joint MVN distribution with mean zero and covariance function $\Sigma(x, x')$. That is, stack

$$\begin{pmatrix} Y(x) \\ Y_n \end{pmatrix} \quad \text{with sizes} \quad \begin{pmatrix} 1 \times 1 \\ n \times 1 \end{pmatrix},$$

and if $\Sigma(X_n, x)$ is the $n \times 1$ matrix comprised of $\Sigma(x_1, x), \ldots, \Sigma(x_n, x)$, its covariance structure can be partitioned as follows:

$$\begin{pmatrix} \Sigma(x, x) & \Sigma(x, X_n) \\ \Sigma(X_n, x) & \Sigma_n \end{pmatrix} \quad \text{with sizes} \quad \begin{pmatrix} 1 \times 1 & 1 \times n \\ n \times 1 & n \times n \end{pmatrix}.$$

Recall that $\Sigma(x, x) = 1$ with our simple choice of covariance function, and that symmetry provides $\Sigma(x, X_n) = \Sigma(X_n, x)^{\top}$.

Applying Eq. (5.1) yields the following predictive distribution

$$Y(x) \mid D_n \sim \mathcal{N}(\mu(x), \sigma^2(x))$$

with

$$\text{mean} \quad \mu(x) = \Sigma(x, X_n)\Sigma_n^{-1}Y_n \tag{5.2}$$

$$\text{and variance} \quad \sigma^2(x) = \Sigma(x, x) - \Sigma(x, X_n)\Sigma_n^{-1}\Sigma(X_n, x).$$

Observe that $\mu(x)$ is linear in observations Y_n, so we have a linear predictor! In fact it's the best linear unbiased predictor (BLUP), an argument we'll leave to other texts (e.g., Santner et al., 2018). Also notice that $\sigma^2(x)$ is lower than the marginal variance. So we learn something from data Y_n; in fact the amount that variance goes down is a quadratic function of distance between x and X_n. Learning is most efficient for x that are close to training data locations X_n. However the amount learned doesn't depend upon Y_n. We'll return to that later.

The derivation above is for "pointwise" GP predictive calculations. These are sometimes called the *kriging[7] equations*, especially in geospatial contexts. We can apply them, separately, for many predictive/testing locations x, one x at a time, but that would ignore the obvious correlation they'd experience in a big MVN analysis. Alternatively, we may consider a bunch of x locations jointly, in a testing design $\mathcal{X}$ of n' rows, say, all at once:

---

[7]https://en.wikipedia.org/wiki/Kriging

$$Y(\mathcal{X}) \mid D_n \sim \mathcal{N}_{n'}(\mu(\mathcal{X}), \Sigma(\mathcal{X}))$$

with

$$\text{mean} \quad \mu(\mathcal{X}) = \Sigma(\mathcal{X}, X_n)\Sigma_n^{-1}Y_n \tag{5.3}$$
$$\text{and variance} \quad \Sigma(\mathcal{X}) = \Sigma(\mathcal{X}, \mathcal{X}) - \Sigma(\mathcal{X}, X_n)\Sigma_n^{-1}\Sigma(\mathcal{X}, X_n)^\top,$$

where $\Sigma(\mathcal{X}, X_n)$ is an $n' \times n$ matrix. Having a full covariance structure offers a more complete picture of the random functions which explain data under a GP posterior, but also more computation. The $n' \times n'$ matrix $\Sigma(\mathcal{X})$ could be enormous even for seemingly moderate n'.

Simple 1d GP prediction example

Consider a toy example in 1d where the response is a simple sinusoid measured at eight equally spaced x-locations in the span of a single period of oscillation. R code below provides relevant data quantities, including pairwise squared distances between the input locations collected in the matrix D, and its inverse exponentiation in `Sigma`.

```
n <- 8
X <- matrix(seq(0,2*pi,length=n), ncol=1)
y <- sin(X)
D <- distance(X)
Sigma <- exp(-D) + diag(eps, ncol(D))
```

Now this is where the example diverges from our earlier one, where we used such quantities to generate data from a GP prior. Applying MVN conditioning equations requires similar calculations on a testing design $\mathcal{X}$, coded as XX below. We need inverse exponentiated squared distances between those XX locations ...

```
XX <- matrix(seq(-0.5, 2*pi+0.5, length=100), ncol=1)
DXX <- distance(XX)
SXX <- exp(-DXX) + diag(eps, ncol(DXX))
```

... as well as between testing locations $\mathcal{X}$ and training data locations X_n.

```
DX <- distance(XX, X)
SX <- exp(-DX)
```

Note that an ϵ jitter adjustment is not required for SX because it need not be decomposed in the conditioning calculations (and SX is anyways not square). We do need jitter on the diagonal of SXX though, because this matrix is directly involved in calculation of the predictive covariance which we shall feed into an MVN generator below.

Now simply follow Eq. (5.3) to derive joint predictive equations for XX $\equiv \mathcal{X}$: invert Σ_n, apply the linear predictor, and calculate reduction in covariance.

```
Si <- solve(Sigma)
mup <- SX %*% Si %*% y
Sigmap <- SXX - SX %*% Si %*% t(SX)
```

Above `mup` maps to $\mu(\mathcal{X})$ evaluated at our testing grid $\mathcal{X} \equiv$ `XX`, and `Sigmap` similarly for $\Sigma(\mathcal{X})$ via pairs in `XX`. As a computational note, observe that `Siy <- Si %*% y` may be pre-computed in time quadratic in $n =$ `length(y)` so that `mup` may subsequently be calculated for any `XX` in time linear in n, without redoing `Siy`; for example, as `solve(Sigma, y)`. There are two reasons we're not doing that here. One is to establish a clean link between code and mathematical formulae. The other is a presumption that the variance calculation, which remains quadratic in n no matter what, is at least as important as the mean.

Mean vector and covariance matrix in hand, we may generate Y-values from the posterior/predictive distribution $Y(\mathcal{X}) \mid D_n$ in the same manner as we did from the prior.

```
YY <- rmvnorm(100, mup, Sigmap)
```

Those $Y(\mathcal{X}) \equiv$ `YY` samples may then be plotted as a function of predictive input $\mathcal{X} \equiv$ `XX` locations. Before doing that, extract some pointwise quantile-based error-bars from the diagonal of $\Sigma(\mathcal{X})$ to aid in visualization.

```
q1 <- mup + qnorm(0.05, 0, sqrt(diag(Sigmap)))
q2 <- mup + qnorm(0.95, 0, sqrt(diag(Sigmap)))
```

Figure 5.3 plots each of the random predictive, finite realizations as gray curves. Training data points are overlayed, along with true response at the $\mathcal{X}$ locations as a thin blue line. Predictive mean $\mu(\mathcal{X})$ in black, and 90% quantiles in dashed-red, are added as thicker lines.

```
matplot(XX, t(YY), type="l", col="gray", lty=1, xlab="x", ylab="y")
points(X, y, pch=20, cex=2)
lines(XX, sin(XX), col="blue")
lines(XX, mup, lwd=2)
lines(XX, q1, lwd=2, lty=2, col=2)
lines(XX, q2, lwd=2, lty=2, col=2)
```

What do we observe in the figure? Notice how the predictive surface interpolates the data. That's because $\Sigma(x, x) = 1$ and $\Sigma(x, x') \to 1^-$ as $x' \to x$. Error-bars take on a "football" shape, or some say a "sausage" shape, being widest at locations farthest from x_i-values in the data. Error-bars get really big outside the range of the data, a typical feature in ordinary linear regression settings. But the predictive mean behaves rather differently than under an ordinary linear model. For GPs it's mean-reverting, eventually leveling off to zero as $x \in \mathcal{X}$ gets far away from X_n. Predictive variance, as exemplified by those error-bars, is also reverting to something: a prior variance of 1. In particular, variance won't continue to increase as x gets farther and farther from X_n. Together those two "reversions" imply that although we can't trust extrapolations too far outside of the data range, at least their behavior isn't unpredictable, as can sometimes happen in linear regression contexts, for example when based upon feature-expanded (e.g., polynomial basis) covariates.

These characteristics, especially the football/sausage shape, is what makes GPs popular

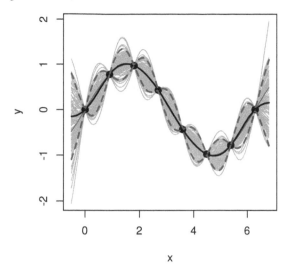

FIGURE 5.3: Posterior predictive distribution in terms of means (solid black), quantiles (dashed-red), and draws (gray). The truth is shown as a thin blue line.

as surrogates for computer simulation experiments. That literature, which historically emphasized study of deterministic computer simulators, drew comfort from interpolation-plus-expansion of variance away from training simulations. Perhaps more importantly, they liked that out-of-sample prediction was highly accurate. Come to think of it, that's why spatial statisticians and machine learners like them too. But hold that thought; there are a few more things to do before we get to predictive comparisons.

5.1.2 Higher dimension?

There's nothing particularly special about the presentation above that would preclude application in higher input dimension. Except perhaps that visualization is a lot simpler in 1d or 2d. We'll get to even higher dimensions with some of our later examples. For now, consider a random function in 2d sampled from a GP prior. The plan is to go back through the process above: first prior, then (posterior) predictive, etc.

Begin by creating an input set, X_n, in two dimensions. Here we'll use a regular 20×20 grid.

```
nx <- 20
x <- seq(0, 2, length=nx)
X <- expand.grid(x, x)
```

Then calculate pairwise distances and evaluate covariances under inverse exponentiated squared Euclidean distances, plus jitter.

```
D <- distance(X)
Sigma <- exp(-D) + diag(eps, nrow(X))
```

Finally make random MVN draws in exactly the same way as before. Below we save two such draws.

```
Y <- rmvnorm(2, sigma=Sigma)
```

For visualization in Figure 5.4, `persp` is used to stretch each $20 \times 20 = 400$-variate draw over a mesh with a fortuitously chosen viewing angle.

```
par(mfrow=c(1,2))
persp(x, x, matrix(Y[1,], ncol=nx), theta=-30, phi=30, xlab="x1",
  ylab="x2", zlab="y")
persp(x, x, matrix(Y[2,], ncol=nx), theta=-30, phi=30, xlab="x1",
  ylab="x2", zlab="y")
```

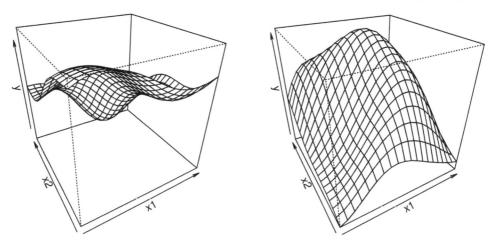

FIGURE 5.4: Two random functions under a GP prior in 2d.

So drawing from a GP prior in 2d is identical to the 1d case, except with a 2d input grid. All other code is "cut-and-paste". Visualization is more cumbersome, but that's a cosmetic detail. Learning from training data, i.e., calculating the predictive distribution for observed (x_i, y_i) pairs, is no different: more cut-and-paste.

To try it out we need to cook up some toy data from which to learn. Consider the 2d function $y(x) = x_1 \exp\{-x_1^2 - x_2^2\}$ which is highly nonlinear near the origin, but flat (zero) as inputs get large. This function has become a benchmark 2d problem in the literature for reasons that we'll get more into in Chapter 9. Suffice it to say that thinking up simple-yet-challenging toy problems is a great way to get noticed[8] in the community, even when you borrow a common example in vector calculus textbooks or one used to demonstrate 3d plotting features in MATLAB®.

```
library(lhs)
X <- randomLHS(40, 2)
X[,1] <- (X[,1] - 0.5)*6 + 1
X[,2] <- (X[,2] - 0.5)*6 + 1
y <- X[,1]*exp(-X[,1]^2 - X[,2]^2)
```

Above, a Latin hypercube sample (LHS; §4.1) is used to generate forty (coded) input

---

[8]https://www.sfu.ca/~ssurjano/grlee08.html

locations in lieu of a regular grid in order to create a space-filling input design. A regular grid with 400 elements would have been overkill, but a uniform random design of size forty or so would have worked equally well. Coded inputs are mapped onto a scale of $[-2, 4]^2$ in order to include both bumpy and flat regions.

Let's suppose that we wish to interpolate those forty points onto a regular 40×40 grid, say for stretching over a mesh. Here's code that creates such testing locations XX $\equiv \mathcal{X}$ in natural units.

```
xx <- seq(-2, 4, length=40)
XX <- expand.grid(xx, xx)
```

Now that we have inputs and outputs, X and y, and predictive locations XX we can start cutting-and-pasting. Start with the relevant training data quantities ...

```
D <- distance(X)
Sigma <- exp(-D)
```

... and follow with similar calculations between input sets X and XX.

```
DXX <- distance(XX)
SXX <- exp(-DXX) + diag(eps, ncol(DXX))
DX <- distance(XX, X)
SX <- exp(-DX)
```

Then apply Eq. (5.3). Code wise, these lines are identical to what we did in the 1d case.

```
Si <- solve(Sigma)
mup <- SX %*% Si %*% y
Sigmap <- SXX - SX %*% Si %*% t(SX)
```

It's hard to visualize a multitude of sample paths in 2d – two was plenty when generating from the prior – but if desired, we may obtain them with the same **rmvnorm** commands as in §5.1.1. Instead focus on plotting pointwise summaries, namely predictive mean $\mu(x) \equiv$ mup and predictive standard deviation $\sigma(x)$:

```
sdp <- sqrt(diag(Sigmap))
```

The left panel in Figure 5.5 provides an image plot of the mean over our regularly-gridded inputs XX; the right panel shows standard deviation.

```
par(mfrow=c(1,2))
cols <- heat.colors(128)
image(xx, xx, matrix(mup, ncol=length(xx)), xlab="x1", ylab="x2", col=cols)
points(X[,1], X[,2])
image(xx, xx, matrix(sdp, ncol=length(xx)), xlab="x1", ylab="x2", col=cols)
points(X[,1], X[,2])
```

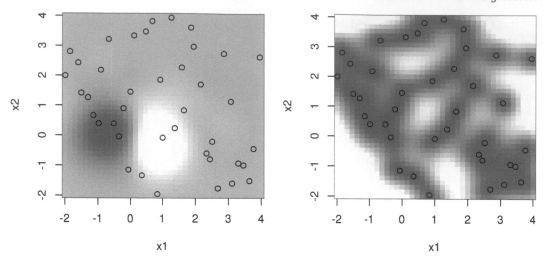

FIGURE 5.5: Posterior predictive for a two-dimensional example, via mean (left) and standard deviation (right) surfaces. Training data input locations are indicated by open circles.

What do we observe? Pretty much the same thing as in the 1d case. We can't see it, but the predictive surface interpolates. Predictive uncertainty, here as standard deviation $\sigma(x)$, is highest away from x_i-values in the training data. Predictive intervals don't look as much like footballs or sausages, yet somehow that analogy still works. Training data locations act as anchors to smooth variation between points with an organic rise in uncertainty as we imagine predictive inputs moving away from one toward the next.

Figure 5.6 provides another look, obtained by stretching the predictive mean over a mesh. Bumps near the origin are clearly visible, with a flat region emerging for larger x_1 and x_2 settings.

```
persp(xx, xx, matrix(mup, ncol=40), theta=-30, phi=30, xlab="x1",
  ylab="x2", zlab="y")
```

Well that's basically it! Now you know GP regression. Where to go from here? Hopefully I've convinced you that GPs hold great potential as a nonlinear regression tool. It's kinda-cool that they perform so well – that they "learn" – without having to tune anything. In statistics, we're so used to seeking out optimal settings of parameters that a GP predictive surface might seem like voodoo. Simple MVN conditioning is able to capture input–output dynamics without having to "fit" anything, or without trying to minimize a loss criteria. That flexibility, without any tuning knobs, is what people think of when they call GPs a nonparametric regression tool. All we did was define covariance in terms of (inverse exponentiated squared Euclidean) distance, condition, and voilà.

But when you think about it a little bit, there are lots of (hidden) assumptions which are going to be violated by most real-data contexts. Data is noisy. The amplitude of all functions we might hope to learn will not be 2. Correlation won't decay uniformly in all directions, i.e., radially. Even the most ideally smooth physical relationships are rarely infinitely smooth.

Yet we'll see that even gross violations of those assumptions are easy to address, or "fix up". At the same time GPs are relatively robust to transgressions between assumptions and

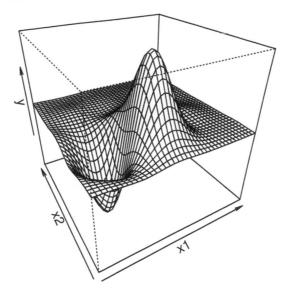

FIGURE 5.6: Perspective view on the posterior mean surface from the left panel of Figure 5.5.

reality. In other words, sometimes it works well even when it ought not. As I see it – once we clean things up – there are really only two serious problems that GPs face in practice: stationarity of covariance (§5.3.3), and computational burden, which in most contexts go hand-in-hand. Remedies for both will have to wait for Chapter 9. For now, let's keep the message upbeat. There's lots that can be accomplished with the canonical setup, whose description continues below.

5.2 GP hyperparameters

All this business about nonparametric regression and here we are introducing parameters, passive–aggressively you might say: refusing to call them parameters. How can one have hyperparameters without parameters to start with, or at least to somehow distinguish from? To make things even more confusing, we go about learning those hyperparameters in the usual way, by optimizing something, just like parameters. I guess it's all to remind you that the real power – the real flexibility – comes from MVN conditioning. These hyperparameters are more of a fine tuning. There's something to that mindset, as we shall see. Below we revisit the drawbacks alluded to above – scale, noise, and decay of correlation – with a (fitted) hyperparameter targeting each one.

5.2.1 Scale

Suppose you want your GP prior to generate random functions with an amplitude larger than two. You could introduce a scale parameter τ^2 and then take $\Sigma_n = \tau^2 C_n$. Here C is basically the same as our Σ from before: a correlation function for which $C(x,x) = 1$ and $C(x,x') < 1$ for $x \neq x'$, and positive definite; for example

$$C(x,x') = \exp\{-||x - x'||^2\}.$$

But we need a more nuanced notion of covariance to allow more flexibility on scale, so we're re-parameterizing a bit. Now our MVN generator looks like

$$Y \sim \mathcal{N}_n(0, \tau^2 C_n).$$

Let's check that that does the trick. First rebuild X_n-locations, e.g., a sequence of one hundred from zero to ten, and then calculate pairwise distances. Nothing different yet compared to our earlier illustration in §5.1.

```
n <- 100
X <- matrix(seq(0, 10, length=n), ncol=1)
D <- distance(X)
```

Now amplitude, via 95% of the range of function realizations, is approximately $2\sigma(x)$ where $\sigma^2 \equiv \text{diag}(\Sigma_n)$. So for an amplitude of 10, say, choose $\tau^2 = 5^2 = 25$. The code below calculates inverse exponentiated squared Euclidean distances in C_n and makes ten draws from an MVN whose covariance is obtained by pre-multiplying C_n by τ^2.

```
C <- exp(-D) + diag(eps, n)
tau2 <- 25
Y <- rmvnorm(10, sigma=tau2*C)
```

As Figure 5.7 shows, amplitude has increased. Not all draws completely lie between -10 and 10, but most are in the ballpark.

```
matplot(X, t(Y), type="l")
```

But again, who cares about generating random functions? We want to be able to learn about functions on any scale from training data. What would happen if we had some data with an amplitude of 5, say, but we used a GP with a built-in scale of 1 (amplitude of 2). In other words, what would happen if we did things the "old-fashioned way", with code cut-and-pasted directly from §5.1.1?

First generate some data with that property. Here we're revisiting sinusoidal data from §5.1.1, but multiplying by 5 on the way out of the sin call.

```
n <- 8
X <- matrix(seq(0, 2*pi, length=n), ncol=1)
y <- 5*sin(X)
```

Next cut-and-paste code from earlier, including our predictive grid of 100 equally spaced locations.

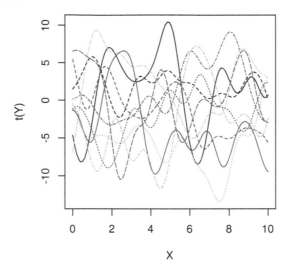

FIGURE 5.7: Higher amplitude draws from a GP prior.

```
D <- distance(X)
Sigma <- exp(-D)
XX <- matrix(seq(-0.5, 2*pi + 0.5, length=100), ncol=1)
DXX <- distance(XX)
SXX <- exp(-DXX) + diag(eps, ncol(DXX))
DX <- distance(XX, X)
SX <- exp(-DX)
Si <- solve(Sigma);
mup <- SX %*% Si %*% y
Sigmap <- SXX - SX %*% Si %*% t(SX)
```

Now we have everything we need to visualize the resulting predictive surface, which is shown in Figure 5.8 using plotting code identical to that behind Figure 5.3.

```
YY <- rmvnorm(100, mup, Sigmap)
q1 <- mup + qnorm(0.05, 0, sqrt(diag(Sigmap)))
q2 <- mup + qnorm(0.95, 0, sqrt(diag(Sigmap)))
matplot(XX, t(YY), type="l", col="gray", lty=1, xlab="x", ylab="y")
points(X, y, pch=20, cex=2)
lines(XX, mup, lwd=2)
lines(XX, 5*sin(XX), col="blue")
lines(XX, q1, lwd=2, lty=2, col=2)
lines(XX, q2, lwd=2, lty=2, col=2)
```

What happened? In fact the "scale 1" GP is pretty robust. It gets the predictive mean almost perfectly, despite using the "wrong prior" relative to the actual data generating mechanism, at least as regards scale. But it's over-confident. Besides a change of scale, the new training data exhibit no change in relative error, nor any other changes for that matter, compared to the example we did above where the scale was actually 1. So we must now be under-estimating predictive uncertainty, which is obvious by visually comparing the

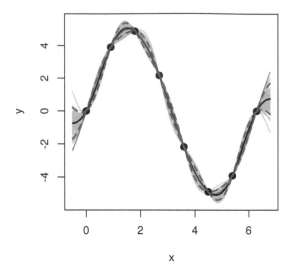

FIGURE 5.8: GP fit to higher amplitude sinusoid.

error-bars to those obtained from our earlier fit (Figure 5.3). Looking closely, notice that the true function goes well outside of our predictive interval at the edges of the input space. That didn't happen before.

How to estimate the right scale? Well for starters, admit that scale may be captured by a parameter, τ^2, even though we're going to call it a hyperparameter to remind ourselves that its impact on the overall estimation procedure is really more of a fine-tuning. The analysis above lends some credence to that perspective, since our results weren't too bad even though we assumed an amplitude that was off by a factor of five. Whether benevolently gifted the right scale or not, GPs clearly retain a great deal of flexibility to adapt to the dynamics at play in data. Decent predictive surfaces often materialize, as we have seen, in spite of less than ideal parametric specifications.

As with any "parameter", there are many choices when it comes to estimation: method of moments (MoM), likelihood (maximum likelihood, Bayesian inference), cross validation (CV), the "eyeball norm". Some, such as those based on (semi-) variograms[9], are preferred in the spatial statistics literature. All of those are legitimate, except maybe the eyeball norm which isn't very easily automated and challenges reproducibility. I'm not aware of any MoM approaches to GP inference for hyperparameters. Stochastic kriging[10] (Ankenman et al., 2010) utilizes MoM in a slightly more ambitious, latent variable setting which is the subject of Chapter 10. Whereas CV is common in some circles, such frameworks generalize rather less well to higher dimensional hyperparameter spaces, which we're going to get to momentarily. I prefer likelihood-based inferential schemes for GPs, partly because they're the most common and, especially in the case of maximizing (MLE/MAP) solutions, they're also relatively hands-off (easy automation), and nicely generalize to higher dimensional hyperparameter spaces.

But wait a minute, what's the likelihood in this context? It's a bit bizarre that we've been talking about priors and posteriors without ever talking about likelihood. Both prior and likelihood are needed to form a posterior. We'll get into finer detail later. For now, recognize

---

[9]http://petrowiki.org/Spatial_statistics#Semivariograms_and_covariance
[10]http://users.iems.northwestern.edu/~nelsonb/SK/

that our data-generating process is $Y \sim \mathcal{N}_n(0, \tau^2 C_n)$, so the relevant quantity, which we'll call the likelihood now (but was our prior earlier), comes from an MVN PDF:

$$L \equiv L(\tau^2, C_n) = (2\pi\tau^2)^{-\frac{n}{2}} |C_n|^{-\frac{1}{2}} \exp\left\{-\frac{1}{2\tau^2} Y_n^\top C_n^{-1} Y_n\right\}.$$

Taking the log of that is easy, and we get

$$\ell = \log L = -\frac{n}{2} \log 2\pi - \frac{n}{2} \log \tau^2 - \frac{1}{2} \log |C_n| - \frac{1}{2\tau^2} Y_n^\top C_n^{-1} Y_n. \tag{5.4}$$

To maximize that (log) likelihood with respect to τ^2, just differentiate and solve.

$$0 \stackrel{\text{set}}{=} \ell' = -\frac{n}{2\tau^2} + \frac{1}{2(\tau^2)^2} Y_n^\top C_n^{-1} Y_n,$$

$$\text{so } \hat{\tau}^2 = \frac{Y_n^\top C_n^{-1} Y_n}{n}. \tag{5.5}$$

In other words, we get that the MLE for scale τ^2 is a mean residual sum of squares under the quadratic form obtained from an MVN PDF with a mean of $\mu(x) = 0$: $(Y_n - 0)^\top C_n^{-1}(Y_n - 0)$.

How would this analysis change if we were to take a Bayesian approach? A homework exercise (§5.5) invites the curious reader to investigate the form of the posterior under prior $\tau^2 \sim \text{IG}(a/2, b/2)$. For example, what happens when $a = b = 0$ which is equivalent to $p(\tau^2) \propto 1/\tau^2$, a so-called *reference prior* in this context (Berger et al., 2001, 2009)?

Estimate of scale $\hat{\tau}^2$ in hand, we may simply "plug it in" to the predictive equations (5.2)–(5.3). Now technically, when you estimate a variance and plug it into a (multivariate) Gaussian, you're turning that Gaussian into a (multivariate) Student-t, in this case with n degrees of freedom (DoF). (There's no loss of DoF when the mean is assumed to be zero.) For details, see for example Gramacy and Polson (2011). For now, presume that n is large enough so that this distinction doesn't matter. As we generalize to more hyperparameters, DoF correction could indeed matter but we still obtain a decent approximation, which is so common in practice that the word "approximation" is often dropped from the description – a transgression I shall be guilty of as well.

So to summarize, we have the following scale-adjusted (approximately) MVN predictive equations:

$$Y(\mathcal{X}) \mid D_n \sim \mathcal{N}_{n'}(\mu(\mathcal{X}), \Sigma(\mathcal{X}))$$
$$\text{with mean} \quad \mu(\mathcal{X}) = C(\mathcal{X}, X_n) C_n^{-1} Y_n$$
$$\text{and variance} \quad \Sigma(\mathcal{X}) = \hat{\tau}^2 [C(\mathcal{X}, \mathcal{X}) - C(\mathcal{X}, X_n) C_n^{-1} C(\mathcal{X}, X_n)^\top].$$

Notice how $\hat{\tau}^2$ doesn't factor into the predictive mean, but it does figure into predictive variance. That's important because it means that Y_n-values are finally involved in assessment of predictive uncertainty, whereas previously (5.2)–(5.3) only X_n-values were involved.

To see it all in action, let's return to our simple 1d sinusoidal example, continuing from Figure 5.8. Start by performing calculations for $\hat{\tau}^2$.

```
CX <- SX
Ci <- Si
CXX <- SXX
tau2hat <- drop(t(y) %*% Ci %*% y / length(y))
```

Checking that we get something reasonable, consider ...

```
2*sqrt(tau2hat)
```

```
## [1] 5.487
```

... which is quite close to what we know to be the true value of five in this case. Next plug $\hat{\tau}^2$ into the MVN conditioning equations to obtain a predictive mean vector and covariance matrix.

```
mup2 <- CX %*% Ci %*% y
Sigmap2 <- tau2hat*(CXX - CX %*% Ci %*% t(CX))
```

Finally gather some sample paths using MVN draws and summarize predictive quantiles by cutting-and-pasting from above.

```
YY <- rmvnorm(100, mup2, Sigmap2)
q1 <- mup + qnorm(0.05, 0, sqrt(diag(Sigmap2)))
q2 <- mup + qnorm(0.95, 0, sqrt(diag(Sigmap2)))
```

Figure 5.9 shows a much better surface compared to Figure 5.8.

```
matplot(XX, t(YY), type="l", col="gray", lty=1, xlab="x", ylab="y")
points(X, y, pch=20, cex=2)
lines(XX, mup, lwd=2)
lines(XX, 5*sin(XX), col="blue")
lines(XX, q1, lwd=2, lty=2, col=2); lines(XX, q2, lwd=2, lty=2, col=2)
```

Excepting the appropriately expanded scale of the y-axis, the view in Figure 5.9 looks nearly identical to Figure 5.3 with data back on the two-unit scale. Besides that this last fit (with $\hat{\tau}^2$) looks better (particularly the variance) than the one before it (with implicit $\tau^2 = 1$ when the observed scale was really much bigger), how can one be more objective about which is best out-of-sample?

A great paper by Gneiting and Raftery (2007) offers *proper scoring rules* that facilitate comparisons between predictors in a number of different situations, basically depending on what common distribution characterizes predictors being compared. These are a great resource when comparing apples and oranges, even though we're about to use them to compare apples to apples: two GPs under different scales.

We have the first two moments, so Eq. (25) from Gneiting and Raftery (2007) may be used. Given $Y(\mathcal{X})$-values observed out of sample, the proper scoring rule is given by

$$\text{score}(Y, \mu, \Sigma; \mathcal{X}) = -\log|\Sigma(\mathcal{X})| - (Y(\mathcal{X}) - \mu(\mathcal{X}))^{\top}(\Sigma(\mathcal{X}))^{-1}(Y(\mathcal{X}) - \mu(\mathcal{X})). \quad (5.6)$$

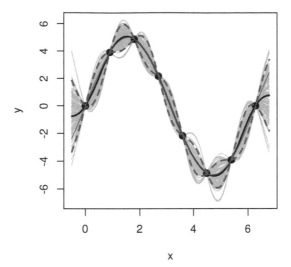

FIGURE 5.9: Sinusoidal GP predictive surface with estimated scale $\hat{\tau}^2$. Compare to Figure 5.8.

In the case where predictors are actually MVN, which they aren't quite in our case (they're Student-t), this is within an additive constant of what's called *predictive log likelihood*. Higher scores, or higher predictive log likelihoods, are better. The first term $-\log|\Sigma(\mathcal{X})|$ measures magnitude of uncertainty. Smaller uncertainty is better, all things considered, so larger is better here. The second term $(Y(\mathcal{X}) - \mu(\mathcal{X}))^\top (\Sigma(\mathcal{X}))^{-1}(Y(\mathcal{X}) - \mu(\mathcal{X}))$ is mean-squared error (MSE) adjusted for covariance. Smaller MSE is better, but when predictions are inaccurate it's also important to capture that uncertainty through $\Sigma(\mathcal{X})$. Score compensates for that second-order consideration: it's ok to mispredict as long as you know you're mispredicting.

A more recent paper by Bastos and O'Hagan (2009) tailors the scoring discussion to deterministic computer experiments, which better suits our current setting: interpolating function observations without noise. They recommend using Mahalanobis distance[11], which for the multivariate Gaussian is the same as the (negative of the) formula above, except without the determinant of $\Sigma(\mathcal{X})$, and square-rooted.

$$\text{mah}(y, \mu, \Sigma; \mathcal{X}) = \sqrt{(y(\mathcal{X}) - \mu(\mathcal{X}))^\top (\Sigma(\mathcal{X}))^{-1}(y(\mathcal{X}) - \mu(\mathcal{X}))} \tag{5.7}$$

Smaller distances are otherwise equivalent to higher scores. Here's code that calculates both in one function.

```
score <- function(Y, mu, Sigma, mah=FALSE)
  {
  Ymmu <- Y - mu
  Sigmai <- solve(Sigma)
  mahdist <- t(Ymmu) %*% Sigmai %*% Ymmu
  if(mah) return(sqrt(mahdist))
  return (- determinant(Sigma, logarithm=TRUE)$modulus - mahdist)
  }
```

---

[11]https://en.wikipedia.org/wiki/Mahalanobis_distance

How about using Mahalanobis distance (Mah for short) to make a comparison between the quality of predictions from our two most recent fits ($\tau^2 = 1$ versus $\hat{\tau}^2$)?

```
Ytrue <- 5*sin(XX)
df <- data.frame(score(Ytrue, mup, Sigmap, mah=TRUE),
  score(Ytrue, mup2, Sigmap2, mah=TRUE))
colnames(df) <- c("tau2=1", "tau2hat")
df
```

```
##    tau2=1 tau2hat
## 1  6.259   2.282
```

Estimated scale wins! Actually if you do `score` without `mah=TRUE` you come to the opposite conclusion, as Bastos and O'Hagan (2009) caution. Knowledge that the true response is deterministic is important to coming to the correct conclusion about estimates of accuracy as regards variations in scale, in this case, with signal and (lack of) noise contributing to the range of observed measurements. Now what about when there's noise?

5.2.2 Noise and nuggets

We've been saying "regression" for a while, but actually interpolation is a more apt description. Regression is about extracting signal from noise, or about smoothing over noisy data, and so far our example training data have no noise. By inspecting a GP prior, in particular its correlation structure $C(x, x')$, it's clear that the current setup precludes idiosyncratic behavior because correlation decays smoothly as a function of distance. Observe that $C(x, x') \to 1^-$ as $x \to x'$, implying that the closer x is to x' the higher the correlation, until correlation is perfect, which is what "connects the dots" when conditioning on data and deriving the predictive distribution.

Moving from GP interpolation to smoothing over noise is all about breaking interpolation, or about breaking continuity in $C(x, x')$ as $x \to x'$. Said another way, we must introduce a discontinuity between diagonal and off-diagonal entries in the correlation matrix C_n to smooth over noise. There are a lot of ways to skin this cat, and a lot of storytelling that goes with it, but the simplest way to "break it" is with something like

$$K(x, x') = C(x, x') + g\delta_{x,x'}.$$

Above, $g > 0$ is a new hyperparameter called the *nugget* (or sometimes nugget effect[12]), which determines the size of the discontinuity as $x' \to x$. The function δ is more like the Kronecker delta[13], although the way it's written above makes it look like the Dirac delta[14]. Observe that g generalizes Neal's ϵ jitter.

Neither delta is perfect in terms of describing what to do in practice. The simplest, correct description, of how to break continuity is to only add g on a diagonal – when indices of x are the same, not simply for identical values – and nowhere else. Never add g to an off-diagonal correlation even if that correlation is based on zero distances: i.e., identical x and x'-values. Specifically,

---

[12]http://petrowiki.org/Spatial_statistics#Nugget_effect
[13]https://en.wikipedia.org/wiki/Kronecker_delta
[14]https://en.wikipedia.org/wiki/Dirac_delta_function

- $K(x_i, x_j) = C(x_i, x_j)$ when $i \neq j$, even if $x_i = x_j$;
- only $K(x_i, x_i) = C(x_i, x_i) + g$.

This leads to the following representation of the data-generating mechanism.

$$Y \sim \mathcal{N}_n(0, \tau^2 K_n)$$

Unfolding terms, covariance matrix Σ_n contains entries

$$\Sigma_n^{ij} = \tau^2(C(x_i, x_j) + g\delta_{ij}),$$

or in other words $\Sigma_n = \tau^2 K_n = \tau^2(C_n + g\mathbb{I}_n)$. This all looks like a hack, but it's operationally equivalent to positing the following model.

$$Y(x) = w(x) + \varepsilon,$$

where $w(x) \sim \mathcal{GP}$ with scale τ^2, i.e., $W \sim \mathcal{N}_n(0, \tau^2 C_n)$, and ε is independent Gaussian noise with variance $\tau^2 g$, i.e., $\varepsilon \overset{\text{iid}}{\sim} \mathcal{N}(0, \tau^2 g)$.

A more aesthetically pleasing model might instead use $w(x) \sim \mathcal{GP}$ with scale τ^2, i.e., $W \sim \mathcal{N}_n(0, \tau^2 C_n)$, and where $\varepsilon(x)$ is iid Gaussian noise with variance σ^2, i.e., $\varepsilon(x) \overset{\text{iid}}{\sim} \mathcal{N}(0, \sigma^2)$. An advantage of this representation is two totally "separate" hyperparameters, with one acting to scale noiseless spatial correlations, and another determining the magnitude of white noise. Those two formulations are actually equivalent. There's a 1:1 mapping between the two. Many researchers prefer the latter to the former on intuition grounds. But inference in the latter is harder. Conditional on g, $\hat{\tau}^2$ is available in closed form, which we'll show momentarily. Conditional on σ^2, numerical methods are required for $\hat{\tau}^2$.

Ok, so back to plan-A with $Y \sim \mathcal{N}(0, \Sigma_n)$, where $\Sigma_n = \tau^2 K_n = \tau^2(C_n + g\mathbb{I}_n)$. Recall that C_n is an $n \times n$ matrix of inverse exponentiated pairwise squared Euclidean distances. How, then, to estimate two hyperparameters: scale τ^2 and nugget g? Again, we have all the usual suspects (MoM, likelihood, CV, variogram) but likelihood-based methods are by far most common. First, suppose that g is known.

MLE $\hat{\tau}^2$ given a fixed g is

$$\hat{\tau}^2 = \frac{Y_n^\top K_n^{-1} Y_n}{n} = \frac{Y_n^\top (C_n + g\mathbb{I}_n)^{-1} Y_n}{n}.$$

The derivation involves an identical application of Eq. (5.5), except with K_n instead of C_n.

Plug $\hat{\tau}^2$ back into our log likelihood to get a *concentrated* (or *profile*) log likelihood involving just the remaining parameter g.

$$\begin{aligned}
\ell(g) &= -\frac{n}{2} \log 2\pi - \frac{n}{2} \log \hat{\tau}^2 - \frac{1}{2} \log |K_n| - \frac{1}{2\hat{\tau}^2} Y_n^\top K_n^{-1} Y_n \\
&= c - \frac{n}{2} \log Y_n^\top K_n^{-1} Y_n - \frac{1}{2} \log |K_n|
\end{aligned} \tag{5.8}$$

Unfortunately taking a derivative and setting to zero doesn't lead to a closed form solution. Calculating the derivative is analytic, which we show below momentarily, but solving is not. Maximizing $\ell(g)$ requires numerical methods. The simplest thing to do is throw it into `optimize` and let a polished library do all the work. Since most optimization libraries prefer to minimize, we'll code up $-\ell(g)$ in R. The `nlg` function below doesn't directly work on X inputs, rather through distances D. This is slightly more efficient since distances can be pre-calculated, rather than re-calculated in each evaluation for new g.

```
nlg <- function(g, D, Y)
 {
  n <- length(Y)
  K <- exp(-D) + diag(g, n)
  Ki <- solve(K)
  ldetK <- determinant(K, logarithm=TRUE)$modulus
  ll <- - - (n/2)*log(t(Y) %*% Ki %*% Y) - (1/2)*ldetK
  counter <<- counter + 1
  return(-ll)
 }
```

Observe a direct correspondence between `nlg` and $-\ell(g)$ with the exception of a `counter` increment (accessing a global variable). This variable is not required, but we'll find it handy later when comparing alternatives on efficiency grounds in numerical optimization, via the number of times our likelihood objective function is evaluated. Although optimization libraries often provide iteration counts on output, sometimes that report can misrepresent the actual number of objective function calls. So I've jerry-rigged my own counter here to fill in.

Example: noisy 1d sinusoid

Before illustrating numerical nugget (and scale) optimization towards the MLE, we need some example data. Let's return to our running sinusoid example from §5.1.1, picking up where we left off but augmented with standard Gaussian noise. Code below utilizes the same uniform Xs from earlier, but doubles them up. Adding replication into a design is recommended in noisy data contexts, as discussed in more detail in Chapter 10. Replication is not essential for this example, but it helps guarantee predictable outcomes which is important for a randomly seeded, fully reproducible Rmarkdown build.

```
X <- rbind(X, X)
n <- nrow(X)
y <- 5*sin(X) + rnorm(n, sd=1)
D <- distance(X)
```

Everything is in place to estimate the optimal nugget. The `optimize` function in R is ideal in 1d derivative-free contexts. It doesn't require an initial value for g, but it does demand a search interval. A sensible yet conservative range for g-values is from `eps` to `var(y)`. The former corresponds to the noise-free/jitter-only case we entertained earlier. The latter is the observed marginal variance of Y, or in other words about as big as variance could be if these data were all noise and no signal.

```
counter <- 0
g <- optimize(nlg, interval=c(eps, var(y)), D=D, Y=y)$minimum
g
```

```
## [1] 0.2878
```

Now the value of that estimate isn't directly useful to us, at least on an intuitive level. We need $\hat{\tau}^2$ to understand the full decomposition of variance. But backing out those quantities is relatively straightforward.

```
K <- exp(-D) + diag(g, n)
Ki <- solve(K)
tau2hat <- drop(t(y) %*% Ki %*% y / n)
c(tau=sqrt(tau2hat), sigma=sqrt(tau2hat*g))
```

```
##    tau sigma
## 2.304 1.236
```

Both are close to their true values of $5/2 = 2.5$ and 1, respectively. Estimated hyperparameters in hand, prediction is a straightforward application of MVN conditionals. First calculate quantities involved in covariance between testing and training locations, and between testing locations and themselves.

```
DX <- distance(XX, X)
KX <- exp(-DX)
KXX <- exp(-DXX) + diag(g, nrow(DXX))
```

Notice that only KXX is augmented with g on the diagonal. KX is not a square symmetric matrix calculated from identically indexed x-values. Even if it were coincidentally square, or if DX contained zero distances because elements of XX and X coincide, still no nugget augmentation is deployed. Only with KXX, which is identically indexed with respect to itself, does a nugget augment the diagonal.

Covariance matrices in hand, we may then calculate the predictive mean vector and covariance matrix.

```
mup <- KX %*% Ki %*% y
Sigmap <- tau2hat*(KXX - KX %*% Ki %*% t(KX))
q1 <- mup + qnorm(0.05, 0, sqrt(diag(Sigmap)))
q2 <- mup + qnorm(0.95, 0, sqrt(diag(Sigmap)))
```

Showing sample predictive realizations that look pretty requires "subtracting" out idiosyncratic noise, i.e., the part due to nugget g. Otherwise sample paths will be "jagged" and hard to interpret.

```
Sigma.int <- tau2hat*(exp(-DXX) + diag(eps, nrow(DXX))
  - KX %*% Ki %*% t(KX))
YY <- rmvnorm(100, mup, Sigma.int)
```

§5.3.2 explains how this maneuver makes sense in a latent function-space view of GP posterior updating, and again when we delve into a deeper signal-to-noise discussion in Chapter 10. For now this is just a trick to get a prettier picture, only affecting gray lines plotted in Figure 5.10.

```
matplot(XX, t(YY), type="l", lty=1, col="gray", xlab="x", ylab="y")
points(X, y, pch=20, cex=2)
lines(XX, mup, lwd=2)
lines(XX, 5*sin(XX), col="blue")
lines(XX, q1, lwd=2, lty=2, col=2)
lines(XX, q2, lwd=2, lty=2, col=2)
```

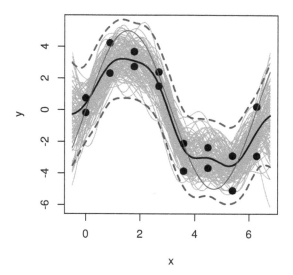

FIGURE 5.10: GP fit to sinusoidal data with estimated nugget.

Notice how the error-bars, which do provide a full accounting of predictive uncertainty, lie mostly outside of the gray lines and appropriately capture variability in the training data observations, shown as filled black dots. That's it: now we can fit noisy data with GPs using a simple library-based numerical optimizer and about twenty lines of code.

5.2.3 Derivative-based hyperparameter optimization

It can be unsatisfying to brute-force an optimization for a hyperparameter like g, even though 1d solving with `optimize` is often superior to cleverer methods. Can we improve upon the number of evaluations?

```
nlg.count <- counter
nlg.count
```

```
## [1] 16
```

Actually, that's pretty good. If you can already optimize numerically in fewer than twenty or so evaluations there isn't much scope for improvement. Yet we're leaving information on the table: closed-form derivatives. Differentiating $\ell(g)$ involves pushing the chain rule through the inverse of covariance matrix K_n and its determinant, which is where hyperparameter g is involved. The following identities, which are framed for an arbitrary parameter ϕ, will come in handy.

$$\frac{\partial K_n^{-1}}{\partial \phi} = -K_n^{-1}\frac{\partial K_n}{\partial \phi}K_n^{-1} \quad \text{and} \quad \frac{\partial \log |K_n|}{\partial \phi} = \text{tr}\left\{K_n^{-1}\frac{\partial K_n}{\partial \phi}\right\} \tag{5.9}$$

The chain rule, and a single application of each of the identities above, gives

$$\begin{aligned}
\ell'(g) &= -\frac{n}{2}\frac{Y_n^\top \frac{\partial K_n^{-1}}{\partial g}Y_n}{Y_n^\top K_n^{-1}Y_n} - \frac{1}{2}\frac{\partial \log |K_n|}{\partial g} \\
&= \frac{n}{2}\frac{Y_n^\top K_n^{-1}\frac{\partial K_n}{\partial g}K_n^{-1}Y_n}{Y_n^\top K_n^{-1}Y_n} - \frac{1}{2}\text{tr}\left\{K_n^{-1}\frac{\partial K_n}{\partial g}\right\}.
\end{aligned} \tag{5.10}$$

Off-diagonal elements of K_n don't depend on g. The diagonal is simply $1+g$. Therefore $\frac{\partial K_n}{\partial g}$ is an n-dimensional identity matrix. Putting it all together:

$$\ell'(g) = \frac{n}{2}\frac{Y_n^\top (K_n^{-1})^2 Y_n}{Y_n^\top K_n^{-1}Y_n} - \frac{1}{2}\text{tr}\left\{K_n^{-1}\right\}.$$

Here's an implementation of the negative of that derivative for the purpose of minimization. The letter "g" for gradient in the function name is overkill in this scalar context, but I'm thinking ahead to where yet more hyperparameters will be optimized.

```
gnlg <- function(g, D, Y)
 {
  n <- length(Y)
  K <- exp(-D) + diag(g, n)
  Ki <- solve(K)
  KiY <- Ki %*% Y
  dll <- (n/2) * t(KiY) %*% KiY / (t(Y) %*% KiY) - (1/2)*sum(diag(Ki))
  return(-dll)
 }
```

Objective (negative concentrated log likelihood, `nlg`) and gradient (`gnlg`) in hand, we're ready to numerically optimize using derivative information. The `optimize` function doesn't support derivatives, so we'll use `optim` instead. The `optim` function supports many optimization methods, and not all accommodate derivatives. I've chosen to illustrate `method="L-BFGS-B"` here because it supports derivatives and allows bound constraints (Byrd et al., 1995). As above, we know we don't want a nugget lower than `eps` for numerical reasons, and it seems unlikely that g will be bigger than the marginal variance.

Here we go ... first reinitializing the evaluation `counter` and choosing 10% of marginal variance as a starting value.

```
counter <- 0
out <- optim(0.1*var(y), nlg, gnlg, method="L-BFGS-B", lower=eps,
  upper=var(y), D=D, Y=y)
c(g, out$par)
```

```
## [1] 0.2878 0.2879
```

Output is similar to what we obtained from `optimize`, which is reassuring. How many iterations?

```
c(out$counts, actual=counter)
```

```
## function gradient    actual
##        8        8        8
```

Notice that in this scalar case our internal, manual counter agrees with `optim`'s. Just 8 evaluations to optimize something is pretty excellent, but possibly not noteworthy compared to `optimize`'s 16, especially when you consider that an extra 8 gradient evaluations (with similar computational complexity) are also required. When you put it that way, our new derivative-based version is potentially no better, requiring 16 combined evaluations of commensurate computational complexity. Hold that thought. We shall return to counting iterations after introducing more hyperparameters.

5.2.4   Lengthscale: rate of decay of correlation

How about modulating the rate of decay of spatial correlation in terms of distance? Surely unadulterated Euclidean distance isn't equally suited to all data. Consider the following generalization, known as the *isotropic Gaussian* family.

$$C_\theta(x, x') = \exp\left\{ -\frac{||x - x'||^2}{\theta} \right\}$$

Isotropic Gaussian correlation functions are indexed by a scalar hyperparameter θ, called the *characteristic lengthscale*. Sometimes this is shortened to *lengthscale*, or θ may be referred to as a *range* parameter, especially in geostatistics. When $\theta = 1$ we get back our inverse exponentiated squared Euclidean distance-based correlation as a special case. *Isotropy* means that correlation decays radially; Gaussian suggests inverse exponentiated squared Euclidean distance. Gaussian processes should not be confused with Gaussian-family correlation or *kernel* functions, which appear in many contexts. GPs get their name from their connection with the MVN, not because they often feature Gaussian kernels as a component of the covariance structure. Further discussion of kernel variations and properties is deferred until later in §5.3.3.

How to perform inference for θ? Should our GP have a slow decay of correlation in space, leading to visually smooth/slowly changing surfaces, or a fast one looking more wiggly? Like with nugget g, embedding θ deep within coordinates of a covariance matrix thwarts analytic maximization of log likelihood. Yet again like g, numerical methods are rather straightforward. In fact the setup is identical except now we have two unknown hyperparameters.

Consider brute-force optimization without derivatives. The R function `nl` is identical to `nlg` except argument `par` takes in a two-vector whose first coordinate is θ and second is g. Only two lines differ, and those are indicated by comments in the code below.

```
nl <- function(par, D, Y)
  {
    theta <- par[1]                              ## change 1
    g <- par[2]
    n <- length(Y)
    K <- exp(-D/theta) + diag(g, n)              ## change 2
```

```
    Ki <- solve(K)
    ldetK <- determinant(K, logarithm=TRUE)$modulus
    ll <- - (n/2)*log(t(Y) %*% Ki %*% Y) - (1/2)*ldetK
    counter <<- counter + 1
    return(-ll)
  }
```

That's it: just shove it into `optim`. Note that `optimize` isn't an option here as that routine only optimizes in 1d. But first we'll need an example. For variety, consider again our 2d exponential data from §5.1.2 and Figure 5.5, this time observed with noise and entertaining non-unit lengthscales.

```
library(lhs)
X2 <- randomLHS(40, 2)
X2 <- rbind(X2, X2)
X2[,1] <- (X2[,1] - 0.5)*6 + 1
X2[,2] <- (X2[,2] - 0.5)*6 + 1
y2 <- X2[,1]*exp(-X2[,1]^2 - X2[,2]^2) + rnorm(nrow(X2), sd=0.01)
```

Again, replication is helpful for stability in reproduction, but is not absolutely necessary. Estimating lengthscale and nugget simultaneously represents an attempt to strike balance between signal and noise (Chapter 10). Once we get more experience, we'll see that long lengthscales are more common when noise/nugget is high, whereas short lengthscales offer the potential to explain away noise as quickly changing dynamics in the data. Sometimes choosing between those two can be a difficult enterprise.

With `optim` it helps to think a little about starting values and search ranges. The nugget is rather straightforward, and we'll copy ranges and starting values from our earlier example: from ϵ to $\mathbb{V}\text{ar}\{Y\}$. The lengthscale is a little harder. Sensible choices for θ follow the following rationale, leveraging x-values in coded units ($\in [0,1]^2$). A lengthscale of 0.1, which is about $\sqrt{0.1} = 0.32$ in units of x, biases towards surfaces three times more wiggly than in our earlier setup, with implicit $\theta = 1$, in a certain loose sense. More precise assessments are quoted later after learning more about kernel properties (§5.3.3) and upcrossings (5.17). Initializing in a more signal, less noise regime seems prudent. If we thought the response was "really straight", perhaps an ordinary linear model would suffice. A lower bound of `eps` allows the optimizer to find even wigglier surfaces, however it might be sensible to view solutions close to `eps` as suspect. A value of $\theta = 10$, or $\sqrt{10} = 3.16$ is commensurately (3x) less wiggly than our earlier analysis. If we find a $\hat{\theta}$ on this upper boundary we can always re-run with a new, bigger upper bound. For a more in-depth discussion of suitable lengthscale and nugget ranges, and even priors for regularization, see Appendix A of the tutorial (Gramacy, 2016) for the `laGP` library (Gramacy and Sun, 2018) introduced in more detail in §5.2.6.

Ok, here we go. (With new X we must first refresh D.)

```
D <- distance(X2)
counter <- 0
out <- optim(c(0.1, 0.1*var(y2)), nl, method="L-BFGS-B", lower=eps,
   upper=c(10, var(y2)), D=D, Y=y2)
out$par
```

```
## [1] 0.902791 0.009972
```

Actually the outcome, as regards the first coordinate $\hat{\theta}$, is pretty close to our initial version with implied $\theta = 1$. Since `"L-BFGS-B"` is calculating a gradient numerically through finite differences, the reported count of evaluations in the output doesn't match the number of actual evaluations.

```
brute <- c(out$counts, actual=counter)
brute
```

```
## function gradient    actual
##       14       14        70
```

We're searching in two input dimensions, and a rule of thumb is that it takes two evaluations in each dimension to build a tangent plane to approximate a derivative. So if 14 `function` evaluations are reported, it'd take about $2 \times 2 \to 4 \times 14 = 56$ additional runs to approximate derivatives, which agrees with our "by-hand" `counter`.

How can we improve upon those counts? Reducing the number of evaluations should speed up computation time. It might not be a big deal now, but as n gets bigger the repeated cubic cost of matrix inverses and determinants really adds up. What if we take derivatives with respect to θ and combine with those for g to form a gradient? That requires $\dot{K}_n \equiv \frac{\partial K_n}{\partial \theta}$, to plug into inverse and determinant derivative identities (5.9). The diagonal is zero because the exponent is zero no matter what θ is. Off-diagonal entries of $\dot{K}_n$ work out as follows. Since

$$K_\theta(x, x') = \exp\left\{ -\frac{\|x - x'\|^2}{\theta} \right\}, \quad \text{we have} \quad \frac{\partial K_\theta(x_i, x_j)}{\partial \theta} = K_\theta(x_i, x_j) \frac{\|x_i - x_j\|^2}{\theta^2}.$$

A slightly more compact way to write the same thing would be $\dot{K}_n = K_n \circ \mathrm{Dist}_n / \theta^2$ where $\circ$ is a component-wise, Hadamard product[15], and Dist_n contains a matrix of squared Euclidean distances – our `D` in the code. An identical application of the chain rule for the nugget (5.10), but this time for θ, gives

$$\ell'(\theta) \equiv \frac{\partial}{\partial \theta} \ell(\theta, g) = \frac{n}{2} \frac{Y_n^\top K_n^{-1} \dot{K}_n K_n^{-1} Y_n}{Y_n^\top K_n^{-1} Y_n} - \frac{1}{2} \mathrm{tr}\left\{ K_n^{-1} \dot{K}_n \right\}. \tag{5.11}$$

A vector collecting the two sets of derivatives forms the gradient of $\ell(\theta, g)$, a joint log likelihood with τ^2 concentrated out. R code below implements the negative of that gradient for the purposes of MLE calculation with `optim` minimization. Comments therein help explain the steps involved.

```
gradnl <- function(par, D, Y)
 {
  ## extract parameters
  theta <- par[1]
  g <- par[2]

  ## calculate covariance quantities from data and parameters
  n <- length(Y)
```

---

[15]https://en.wikipedia.org/wiki/Hadamard_product_(matrices)

```
K <- exp(-D/theta) + diag(g, n)
Ki <- solve(K)
dotK <- K*D/theta^2
KiY <- Ki %*% Y

## theta component
dlltheta <- (n/2) * t(KiY) %*% dotK %*% KiY / (t(Y) %*% KiY) -
  (1/2)*sum(diag(Ki %*% dotK))

## g component
dllg <- (n/2) * t(KiY) %*% KiY / (t(Y) %*% KiY) - (1/2)*sum(diag(Ki))

## combine the components into a gradient vector
return(-c(dlltheta, dllg))
}
```

How well does `optim` work when it has access to actual gradient evaluations? Observe here that we're otherwise using exactly the same calls as earlier.

```
counter <- 0
outg <- optim(c(0.1, 0.1*var(y2)), nl, gradnl, method="L-BFGS-B",
  lower=eps, upper=c(10, var(y2)), D=D, Y=y2)
rbind(grad=outg$par, brute=out$par)
```

```
##            [,1]     [,2]
## grad    0.9028 0.009972
## brute   0.9028 0.009972
```

Parameter estimates are nearly identical. Availability of a true gradient evaluation changes the steps of the algorithm slightly, often leading to a different end-result even when identical convergence criteria are applied. What about the number of evaluations?

```
rbind(grad=c(outg$counts, actual=counter), brute)
```

```
##        function gradient actual
## grad         11       11     11
## brute        14       14     70
```

Woah! That's way better. No only does our `actual` "by-hand" count of evaluations match what's reported on output from `optim`, but it can be an order of magnitude lower, roughly, compared to what we had before. (Variations depend on the random data used to generate this Rmarkdown document.) A factor of five-to-ten savings is definitely worth the extra effort to derive and code up a gradient. As you can imagine, and we'll show shortly, gradients are commensurately more valuable when there are even more hyperparameters. "But what other hyperparameters?", you ask. Hold that thought.

Optimized hyperparameters in hand, we can go about rebuilding quantities required for prediction. Begin with training quantities ...

```
K <- exp(- D/outg$par[1]) + diag(outg$par[2], nrow(X2))
Ki <- solve(K)
tau2hat <- drop(t(y2) %*% Ki %*% y2 / nrow(X2))
```

... then predictive/testing ones ...

```
gn <- 40
xx <- seq(-2, 4, length=gn)
XX <- expand.grid(xx, xx)
DXX <- distance(XX)
KXX <- exp(-DXX/outg$par[1]) + diag(outg$par[2], ncol(DXX))
DX <- distance(XX, X2)
KX <- exp(-DX/outg$par[1])
```

... and finally kriging equations.

```
mup <- KX %*% Ki %*% y2
Sigmap <- tau2hat*(KXX - KX %*% Ki %*% t(KX))
sdp <- sqrt(diag(Sigmap))
```

The resulting predictive surfaces look pretty much the same as before, as shown in Figure 5.11.

```
par(mfrow=c(1,2))
image(xx, xx, matrix(mup, ncol=gn), main="mean", xlab="x1",
  ylab="x2", col=cols)
points(X2)
image(xx, xx, matrix(sdp, ncol=gn), main="sd", xlab="x1",
  ylab="x2", col=cols)
points(X2)
```

This is perhaps not an exciting way to end the example, but it serves to illustrate the basic idea of estimating unknown quantities and plugging them into predictive equations. I've only illustrated 1d and 2d so far, but the principle is no different in higher dimensions.

5.2.5 Anisotropic modeling

It's time to expand input dimension a bit, and get ambitious. Visualization will be challenging, but there are other metrics of success. Consider the Friedman function[16], a popular toy problem from the seminal multivariate adaptive regression splines (MARS[17]; Friedman, 1991) paper. Splines are a popular alternative to GPs in low input dimension. The idea is to "stitch" together low-order polynomials. The "stitching boundary" becomes exponentially huge as dimension increases, which challenges computation. For more details, see the splines supplement linked here[18] which is based on Hastie et al. (2009), Chapters 5, 7 and 8.

---

[16]https://www.sfu.ca/~ssurjano/filed.html
[17]https://en.wikipedia.org/wiki/Multivariate_adaptive_regression_splines
[18]http://bobby.gramacy.com/surrogates/splines.html

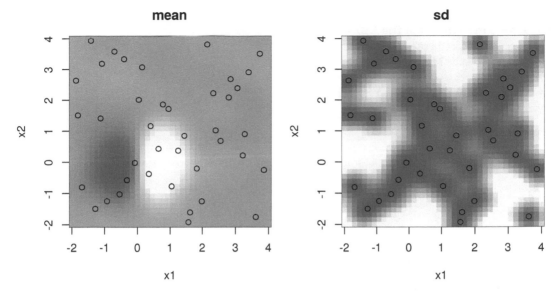

FIGURE 5.11: Predictive mean (left) and standard deviation (right) after estimating a lengthscale $\hat{\theta}$.

MARS circumvents many of those computational challenges by simplifying basis elements (to piecewise linear) on main effects and limiting (to two-way) interactions. Over-fitting is mitigated by aggressively pruning useless basis elements with a generalized CV scheme.

```
fried <- function(n=50, m=6)
 {
  if(m < 5) stop("must have at least 5 cols")
  X <- randomLHS(n, m)
  Ytrue <- 10*sin(pi*X[,1]*X[,2]) + 20*(X[,3] - 0.5)^2 + 10*X[,4] + 5*X[,5]
  Y <- Ytrue + rnorm(n, 0, 1)
  return(data.frame(X, Y, Ytrue))
 }
```

The surface is nonlinear in five input coordinates,

$$\mathbb{E}\{Y(x)\} = 10\sin(\pi x_1 x_2) + 20(x_3 - 0.5)^2 + 10x_4 - 5x_5, \qquad (5.12)$$

combining periodic, quadratic and linear effects. Notice that you can ask for more (useless) coordinates if you want: inputs $x_6, x_7, \ldots$ The **fried** function, as written above, generates both the X-values, via LHS (§4.1) in $[0,1]^m$, and Y-values. Let's create training and testing sets in seven input dimensions, i.e., with two irrelevant inputs x_6 and x_7. Code below uses **fried** to generate an LHS training–testing partition (see, e.g., Figure 4.9) with $n = 200$ and $n' = 1000$ observations, respectively. Such a partition could represent one instance in the "bakeoff" described by Algorithm 4.1. See §5.2.7 for iteration on that theme.

```
m <- 7
n <- 200
nprime <- 1000
```

```
data <- fried(n + nprime, m)
X <- as.matrix(data[1:n,1:m])
y <- drop(data$Y[1:n])
XX <- as.matrix(data[(n + 1):nprime,1:m])
yy <- drop(data$Y[(n + 1):nprime])
yytrue <- drop(data$Ytrue[(n + 1):nprime])
```

The code above extracts two types of Y-values for use in out-of-sample testing. De-noised yytrue values facilitate comparison with root mean-squared error (RMSE),

$$\sqrt{\frac{1}{n'}\sum_{i=1}^{n'}(y_i - \mu(x_i))^2}. \tag{5.13}$$

Notice that RMSE is square-root Mahalanobis distance (5.7) calculated with an identity covariance matrix. Noisy out-of-sample evaluations yy can be used for comparison by proper score (5.6), combining both mean accuracy and estimates of covariance.

First learning. Inputs X and outputs y are re-defined, overwriting those from earlier examples. After re-calculating pairwise distances D, we may cut-and-paste gradient-based optim on objective nl and gradient gnl.

```
D <- distance(X)
out <- optim(c(0.1, 0.1*var(y)), nl, gradnl, method="L-BFGS-B", lower=eps,
  upper=c(10, var(y)), D=D, Y=y)
out
```

```
## $par
## [1] 2.534239 0.005208
##
## $value
## [1] 683.6
##
## $counts
## function gradient
##       33       33
##
## $convergence
## [1] 0
##
## $message
## [1] "CONVERGENCE: REL_REDUCTION_OF_F <= FACTR*EPSMCH"
```

Output indicates convergence has been achieved. Based on estimated $\hat{\theta} = 2.534$ and $\hat{g} = 0.0052$, we may rebuild the data covariance quantities ...

```
K <- exp(- D/out$par[1]) + diag(out$par[2], nrow(D))
Ki <- solve(K)
tau2hat <- drop(t(y) %*% Ki %*% y / nrow(D))
```

... as well as those involved in predicting at XX testing locations.

```
DXX <- distance(XX)
KXX <- exp(-DXX/out$par[1]) + diag(out$par[2], ncol(DXX))
DX <- distance(XX, X)
KX <- exp(-DX/out$par[1])
```

Kriging equations are then derived as follows.

```
mup <- KX %*% Ki %*% y
Sigmap <- tau2hat*(KXX - KX %*% Ki %*% t(KX))
```

Notice how not a single line in the code above, pasted directly from identical lines used in earlier examples, requires tweaking to accommodate the novel 7d setting. Our previous examples were in 1d and 2d, but the code works verbatim in 7d. However the number of evaluations required to maximize is greater now than in previous examples. Here we have 33 compared to 11 previously in 2d.

How accurate are predictions? RMSE on the testing set is calculated below, but we don't yet have a benchmark to compare this to.

```
rmse <- c(gpiso=sqrt(mean((yytrue - mup)^2)))
rmse
```

```
## gpiso
## 1.073
```

How about comparing to MARS? That seems natural considering these data were created as a showcase for that very method. MARS implementations can be found in the **mda** (Leisch et al., 2017) and **earth** (Milborrow, 2019) packages on CRAN.

```
library(mda)
fit.mars <- mars(X, y)
p.mars <- predict(fit.mars, XX)
```

Which wins between the isotropic GP and MARS based on RMSE to the truth?

```
rmse <- c(rmse, mars=sqrt(mean((yytrue - p.mars)^2)))
rmse
```

```
## gpiso  mars
## 1.073 1.529
```

Usually the GP wins in this comparison. In about one time out of twenty random Rmarkdown rebuilds MARS wins. Unfortunately MARS doesn't natively provide a notion of predictive variance. That is, not without an extra bootstrap layer or a Bayesian treatment; e.g., see **BASS** (Francom, 2017) on CRAN. So a comparison to MARS by proper score isn't readily available. Some may argue that this comparison isn't fair. MARS software has lots of tuning parameters that we aren't exploring. Results from **mars** improve with argument **degree=2** and, for reasons that aren't immediately clear to me at this time, they're even better with

earth after the same **degree=2** modification. I've deliberately put up a relatively "vanilla" straw man in this comparison. This is in part because our GP setup is itself relatively vanilla. An exercise in §5.5 invites the reader to explore a wider range of alternatives on both fronts.

How can we add more flavor? If that was vanilla GP regression, what does rocky road look like? To help motivate, recall that the Friedman function involved a diverse combination of effects on the input variables: trigonometric, quadratic and linear. Although we wouldn't generally know that much detail in a new application – and GPs excel in settings where little is known about input–output relationships, except perhaps that it might be worth trying methods beyond the familiar linear model – it's worth wondering if our modeling apparatus is not at odds with typically encountered dynamics. More to the point, GP modeling flexibility comes from the MVN covariance structure which is based on scaled (by θ) inverse exponentiated squared Euclidean distance. That structure implies uniform decay in correlation in each input direction. Is such radial symmetry reasonable? Probably not in general, and definitely not in the case of the Friedman function.

How about the following generalization?

$$C_\theta(x, x') = \exp\left\{ -\sum_{k=1}^{m} \frac{(x_k - x'_k)^2}{\theta_k} \right\}$$

Here we're using a vectorized lengthscale parameter $\theta = (\theta_1, \ldots, \theta_m)$, allowing strength of correlation to be modulated separately by distance in each input coordinate. This family of correlation functions is called the *separable* or *anisotropic Gaussian*. Separable because the sum is a product when taken outside the exponent, implying independence in each coordinate direction. Anisotopic because, except in the special case where all θ_k are equal, decay of correlation is not radial.

How does one perform inference for such a vectorized parameter? Simple; just expand log likelihood and derivative functions to work with vectorized θ. Thinking about implementation: a **for** loop in the gradient function can iterate over coordinates, wherein each iteration we plug

$$\frac{\partial K_n^{ij}}{\partial \theta_k} = K_n^{ij} \frac{(x_{ik} - x_{jk})^2}{\theta_k^2} \tag{5.14}$$

into our formula for $\ell'(\theta_k)$ in Eq. (5.11), which is otherwise unchanged.

Each coordinate has a different θ_k, so pre-computing a distance matrix isn't helpful. Instead we'll use the **covar.sep** function from the **plgp** package which takes vectorized **d** $\equiv \theta$ and scalar **g** arguments, combing distance and inverse-scaling into one step. Rather than going derivative crazy immediately, let's focus on the likelihood first, which we'll need anyways before going "whole hog". The function below is nearly identical to **nl** from §5.2.4 except the first **ncol(X)** components of argument **par** are sectioned off for **theta**, and **covar.sep** is used directly on **X** inputs rather than operating on pre-calculated **D**.

```
nlsep <- function(par, X, Y)
  {
   theta <- par[1:ncol(X)]
   g <- par[ncol(X)+1]
   n <- length(Y)
   K <- covar.sep(X, d=theta, g=g)
```

```
Ki <- solve(K)
ldetK <- determinant(K, logarithm=TRUE)$modulus
ll <- - (n/2)*log(t(Y) %*% Ki %*% Y) - (1/2)*ldetK
counter <<- counter + 1
return(-ll)
}
```

As a testament to how easy it is to optimize that likelihood, at least in terms of coding, below we port our `optim` on `nl` above to `nlsep` below with the only change being to repeat `upper` and `lower` arguments, and supply `X` instead of `D`. (Extra commands for timing will be discussed momentarily.)

```
tic <- proc.time()[3]
counter <- 0
out <- optim(c(rep(0.1, ncol(X)), 0.1*var(y)), nlsep, method="L-BFGS-B",
  X=X, Y=y, lower=eps, upper=c(rep(10, ncol(X)), var(y)))
toc <- proc.time()[3]
out$par
```

```
## [1]  1.046068  1.156524  1.792535  9.036107  9.979581 10.000000
## [7]  9.207463  0.008191
```

What can be seen on output? Notice how $\hat{\theta}_k$-values track what we know about the Friedman function. The first three inputs have relatively shorter lengthscales compared to inputs four and five. Recall that shorter lengthscale means "more wiggly", which is appropriate for those nonlinear terms; longer lengthscale corresponds to linearly contributing inputs. Finally, the last two (save g in the final position of `out$par`) also have long lengthscales, which is similarly reasonable for inputs which aren't contributing.

But how about the number of evaluations?

```
brute <- c(out$counts, actual=counter)
brute
```

```
## function gradient   actual
##       71       71     1207
```

Woah, lots! Although only 71 optimization steps were required, in 8d (including nugget `g` in `par`) that amounts to evaluating the objective function more than one-thousand-odd times, plus-or-minus depending on the random Rmarkdown build. When $n = 200$, and with cubic matrix decompositions, that can be quite a slog time-wise: about 9 seconds.

```
toc - tic
```

```
## elapsed
##   9.341
```

To attempt to improve on that slow state of affairs, code below implements a gradient (5.14) for vectorized θ.

```
gradnlsep <- function(par, X, Y)
 {
  theta <- par[1:ncol(X)]
  g <- par[ncol(X)+1]
  n <- length(Y)
  K <- covar.sep(X, d=theta, g=g)
  Ki <- solve(K)
  KiY <- Ki %*% Y

  ## loop over theta components
  dlltheta <- rep(NA, length(theta))
  for(k in 1:length(dlltheta)) {
    dotK <- K * distance(X[,k])/(theta[k]^2)
    dlltheta[k] <- (n/2) * t(KiY) %*% dotK %*% KiY / (t(Y) %*% KiY) -
      (1/2)*sum(diag(Ki %*% dotK))
  }

  ## for g
  dllg <- (n/2) * t(KiY) %*% KiY / (t(Y) %*% KiY) - (1/2)*sum(diag(Ki))

  return(-c(dlltheta, dllg))
 }
```

Here's what you get when you feed `gradnlsep` into `optim`, otherwise with the same calls as before.

```
tic <- proc.time()[3]
counter <- 0
outg <- optim(c(rep(0.1, ncol(X)), 0.1*var(y)), nlsep, gradnlsep,
  method="L-BFGS-B", lower=eps, upper=c(rep(10, ncol(X)), var(y)), X=X, Y=y)
toc <- proc.time()[3]
thetahat <- rbind(grad=outg$par, brute=out$par)
colnames(thetahat) <- c(paste0("d", 1:ncol(X)), "g")
thetahat
```

```
##            d1    d2    d3    d4     d5 d6    d7        g
## grad   1.111 1.116 1.755 7.457 10.00 10 8.910 0.008419
## brute  1.046 1.157 1.793 9.036  9.98 10 9.207 0.008191
```

First, observe the similar, but not always identical result in terms of optimized parameter(s). Derivatives enhance accuracy and alter convergence criteria compared to tangent-based approximations which sometimes leads to small discrepancies. How about number of evaluations?

```
rbind(grad=c(outg$counts, actual=counter), brute)
```

```
##        function gradient actual
## grad        138      130    130
## brute        71       71   1207
```

Far fewer; an order of magnitude fewer actually, and that pays dividends in time.

```
toc - tic
```

```
## elapsed
##   5.818
```

Unfortunately, it's not 10× faster with 10× fewer evaluations because gradient evaluation takes time. Evaluating each derivative component – each iteration of the `for` loop in `gradnlsep` – involves a matrix multiplication quadratic in n. So that's eight more quadratic-n-cost operations per evaluation compared to one for `nlsep` alone. Consequently, we see a 2–3× speedup. There are some inefficiencies in this implementation. For example, notice that `nlsep` and `gradnlsep` repeat some calculations. Also the matrix trace implementation, `sum(diag(Ki %*% dotK)` is wasteful. Yet again I'll ask you to hold that thought for when we get to library-based implementations, momentarily.

Ok, we got onto this tangent after wondering if GPs could do much better, in terms of prediction, on the Friedman data. So how does a separable GP compare against the isotropic one and MARS? First, take MLE hyperparameters and plug them into the predictive equations.

```
K <- covar.sep(X, d=outg$par[1:ncol(X)], g=outg$par[ncol(X)+1])
Ki <- solve(K)
tau2hat <- drop(t(y) %*% Ki %*% y / nrow(X))
KXX <- covar.sep(XX, d=outg$par[1:ncol(X)], g=outg$par[ncol(X)+1])
KX <- covar.sep(XX, X, d=outg$par[1:ncol(X)], g=0)
mup2 <- KX %*% Ki %*% y
Sigmap2 <- tau2hat*(KXX - KX %*% Ki %*% t(KX))
```

A 2 is tacked onto the variable names above so as not to trample on isotropic analogs. We'll need both sets of variables to make a comparison based on score shortly. But first, here are RMSEs.

```
rmse <- c(rmse, gpsep=sqrt(mean((yytrue - mup2)^2)))
rmse
```

```
## gpiso   mars  gpsep
## 1.0732 1.5288 0.6441
```

The separable covariance structure performs much better. Whereas the isotropic GP only beats MARS 19/20 times in random Rmarkdown builds, the separable GP is never worse than MARS, and it's also never worse than its isotropic cousin. It pays to learn separate lengthscales for each input coordinate.

Since GPs emit full covariance structures we can also make a comparison by proper score (5.6). Mahalanobis distance is not appropriate here because training responses are not deterministic. Score calculations should commence on `yy` here, i.e., with noise, not on `yytrue` which is deterministic.

```
scores <- c(gp=score(yy, mup, Sigmap), mars=NA,
```

```
  gpsep=score(yy, mup2, Sigmap2))
scores
```

```
##        gp    mars   gpsep
## -1093.4      NA  -932.4
```

Recall that larger scores are better; so again the separable GP wins.

5.2.6 Library

All this cutting-and-pasting is getting a bit repetitive. Isn't there a library for that? Yes, several! But first, this might be a good opportunity to pin down the steps for GP regression in a formal algorithm. Think of it as a capstone. Some steps in Algorithm 5.1 are a little informal since the equations are long, and provided earlier. There are many variations/choices on exactly how to proceed, especially to do with MVN correlation structure, or kernel. More options follow, i.e., beyond isotropic and separable Gaussian variations, later in the chapter.

Algorithm 5.1 Gaussian Process Regression

Assume correlation structure $K(\cdot, \cdot)$ has been chosen, which may include hyperparameter lengthscale (vector) θ and nugget g; we simply refer to a combined $\theta \equiv (\theta, g)$ below.

Require $n \times m$ matrix of inputs X_n and n-vector of outputs Y_n; optionally an $n' \times m$ matrix of predictive locations $\mathcal{X}$.

Then

1. Derive the concentrated log likelihood $\ell(\theta)$ following Eq. (5.8) under MVN sampling model with hyperparameters θ and develop code to evaluate that likelihood as a function of θ.

   - Variations may depend on choice of $K(\cdot, \cdot)$, otherwise the referenced equations can be applied directly.

2. Optionally, differentiate that log likelihood (5.11) with respect to θ, forming a gradient $\ell'(\theta) \equiv \nabla \ell(\theta)$, and implement it too as a code which can be evaluated as a function of θ.

   - Referenced equations apply directly so long as $\dot{K}_n$, the derivative of the covariance matrix with respect to the components of θ, may be evaluated.

3. Choose initial values and search ranges for the components of θ being optimized.
4. Plug log likelihood and (optionally) gradient code into your favorite optimizer (e.g., `optim` with `method="L-BFGS-B"`), along with initial values and ranges, obtaining $\hat{\theta}$.

   - If any components of $\hat{\theta}$ are on the boundary of the chosen search range, consider expanding those ranges and repeat step 3.

5. If $\mathcal{X}$ is provided, plug $\hat{\theta}$ and $\mathcal{X}$ into either pointwise (5.2) or joint (5.3) predictive equations.

Return MLE $\hat{\theta}$, which can be used later for predictions; mean vector $\mu(\mathcal{X})$ and covariance matrix $\Sigma(\mathcal{X})$ or variance vector $\sigma^2(\mathcal{X})$ if $\mathcal{X}$ provided.

Referenced equations in the algorithm are meant as examples. Text surrounding those links offers more context about how such equations are intended to be applied. Observe that the description treats predictive/testing locations $\mathcal{X}$ as optional. It's quite common in implementation to separate inference and prediction, however Algorithm 5.1 combines them. If new $\mathcal{X}$ comes along, steps 1–4 can be skipped if $\hat{\theta}$ has been saved. If $\hat{\tau}^2$ and K_n^{-1}, which depend on $\hat{\theta}$, have also been saved then pointwise prediction is quadratic in n. They're quadratic in n' when a full predictive covariance $\Sigma(\mathcal{X})$ is desired, which may be problematic for large grids. Evaluating those equations, say to obtain draws, necessitates decomposition and is thus cubic in n'.

There are many libraries automating the process outlined by Algorithm 5.1, providing several choices of families of covariance functions and variations in hyperparameterization. For R these include `mlegp` (Dancik, 2018), `GPfit` (MacDonald et al., 2019), `spatial` (Ripley, 2015) `fields` (Nychka et al., 2019), `RobustGaSP` (Gu et al., 2018), and `kernlab` (Karatzoglou et al., 2018) – all performing maximum likelihood (or maximum *a posteriori*/Bayesian regularized) point inference; or `tgp` (Gramacy and Taddy, 2016), `emulator` (Hankin, 2019), `plgp`, and `spBayes` (Finley and Banerjee, 2019) – performing fully Bayesian inference. There are a few more that will be of greater interest later, in Chapters 9 and 10. For Python see GPy[19], and for MATLAB/Octave see gpstuff[20] (Vanhatalo et al., 2012). Erickson et al. (2018) provide a nice review and comparison of several libraries.

Here we shall demonstrate the implementation in `laGP` (Gramacy and Sun, 2018), in part due to my intimate familiarity. It's the fastest GP regression library that I'm aware of, being almost entirely implemented in C. We'll say a little more about speed momentarily. The main reason for highlighting `laGP` here is because of its more advanced features, and other convenient add-ons for sequential design and Bayesian optimization, which will come in handy in later chapters. The basic GP interface in the `laGP` package works a little differently than other packages do, for example compared to those above. But it's the considerations behind those peculiarities from which `laGP` draws its unmatched speed.

Ok, now for `laGP`'s basic GP functionality on the Friedman data introduced in §5.2.5. After loading the package, the first step is to initialize a GP fit. This is where we provide the training data, and choose initial values for lengthscale θ and nugget g. It's a bit like a constructor function, for readers familiar with C or C++. Code below also checks a clock so we can compare to earlier timings.

```
library(laGP)
tic <- proc.time()[3]
gpi <- newGPsep(X, y, d=0.1, g=0.1*var(y), dK=TRUE)
```

The "`sep`" in `newGPsep` indicates a separable/anisotropic Gaussian formulation. An isotropic version is available from `newGP`. At this time, the `laGP` package only implements Gaussian families (and we haven't talked about any others yet anyways).

After initialization, an MLE subroutine may be invoked. Rather than maximizing a concentrated log likelihood, `laGP` actually maximizes a Bayesian integrated log likelihood. But that's not an important detail. In fact, the software deliberately obscures that nuance with its `mle...` naming convention, rather than `mbile...` or something similar, which would probably look strange to the average practitioner.

---

[19]https://sheffieldml.github.io/GPy/
[20]https://research.cs.aalto.fi/pml/software/gpstuff/

```
mle <- mleGPsep(gpi, param="both", tmin=c(eps, eps), tmax=c(10, var(y)))
toc <- proc.time()[3]
```

Notice that we don't need to provide training data (X, y) again. Everything passed to newGPsep, and all data quantities derived therefrom, is stored internally by the gpi object. Once MLE calculation is finished, that object is updated to reflect the new, optimal hyperparameter setting. More on implementation details is provided below. Outputs from mleGPsep report hyperparameters and convergence diagnostics primarily for the purposes of inspection.

```
thetahat <- rbind(grad=outg$par, brute=out$par, laGP=mle$theta)
colnames(thetahat) <- c(paste0("d", 1:ncol(X)), "g")
thetahat
```

```
##            d1    d2    d3    d4    d5 d6     d7         g
## grad   1.111 1.116 1.755 7.457 10.00 10  8.910 0.008419
## brute  1.046 1.157 1.793 9.036  9.98 10  9.207 0.008191
## laGP   1.100 1.071 1.732 8.099 10.00 10 10.000 0.008527
```

Not exactly the same estimates as we had before, but pretty close. Since it's not the same objective being optimized, we shouldn't expect exactly the same estimate. And how long did it take?

```
toc - tic
```

```
## elapsed
##    1.05
```

Now that *is* faster! Almost five times faster than our bespoke gradient-based version, and ten times faster than our earlier non-gradient-based one. What makes it so fast? The answer is not that it performs fewer optimization iterations, although sometimes that is the case, ...

```
rbind(grad=c(outg$counts, actual=counter), brute,
  laGP=c(mle$its, mle$its, NA))
```

```
##        function gradient actual
## grad        138      138    138
## brute        71       71   1207
## laGP        139      139     NA
```

... or that it uses a different optimization library. In fact, laGP's C backend borrows the C subroutines behind L-BFGS-B optimization provided with R. One explanation for that speed boost is the compiled (and optimized) C code, but that's only part of the story. The implementation is very careful not to re-calculate anything. Matrices and decompositions are shared between objective and gradient, which involve many of the same operations. Inverses are based on Cholesky decompositions, which can be re-used to calculate determinants without new decompositions. (Note that this can be done in R too, with chol and chol2inv, but it's quite a bit faster in C, where pointers and pass-by-reference save on automatic copies necessitated by an R-only implementation.)

TABLE 5.1: RMSEs and proper scores on the Friedman data.

	gpiso	mars	gpsep	laGP
rmse	1.073	1.529	0.6441	0.6355
scores	-1093.429	NA	-932.3916	-929.8908

Although `mle` output reports estimated hyperparameter values, those are mostly for information purposes. That `mle` object is not intended for direct use in subsequent calculations, such as to make predictions. The `gpi` output reference from `newGPsep`, which is passed to `mleGPsep`, is where the real information lies. In fact, the `gpi` variable is merely an index – a unique integer – pointing to a GP object stored by backend C data structures, containing updated K_n and K_n^{-1}, and related derivative quantities, and everything else that's needed to do more calculations: more MLE iterations if needed, predictions, quick updates if new training data arrive (more in Chapters 6–7). These are modified as a "side effect" of the `mle` calculation. That means nothing needs to be "rebuilt" to make predictions. No copying of matrices back and forth. The C-side GP object is ready for whatever, behind the scenes.

```
p <- predGPsep(gpi, XX)
```

How good are these predictions compared to what we had before? Let's complete the table, fancy this time because we're done with this experiment. See Table 5.1.

```
rmse <- c(rmse, laGP=sqrt(mean((yytrue - p$mean)^2)))
scores <- c(scores, laGP=score(yy, p$mean, p$Sigma))
kable(rbind(rmse, scores),
  caption="RMSEs and proper scores on the Friedman data.")
```

About the same as before; we'll take a closer look at potential differences momentarily. When finished using the data structures stored for a GP fit in C, we must remember to call the destructor function otherwise memory will leak. The stored GP object referenced by `gpi` is not under R's memory management. (Calling `rm(gpi)` would free the integer reference, but not the matrices it refers to as C data structures otherwise hidden to R and to the user.)

```
deleteGPsep(gpi)
```

5.2.7 A bakeoff

As a capstone on the example above, and to connect to a dangling thread from Chapter 4, code below performs an LHS Bakeoff, in the style of Algorithm 4.1, over $R = 30$ Monte Carlo (MC) repetitions with the four comparators above. Begin by setting up matrices to store our two metrics, RMSE and proper score, and one new one: execution time.

```
R <- 30
scores <- rmses <- times <- matrix(NA, nrow=R, ncol=4)
colnames(scores) <- colnames(rmses) <- colnames(times) <- names(rmse)
```

Then loop over replicate data with each comparator applied to the same LHS-generated training and testing partition, each of size $n = n' = 200$. Note that this implementation discards the one MC replicate we already performed above, which was anyways slightly different under $n' = 1000$ testing runs. Since we're repeating thirty times here, a smaller testing set suffices. As in our previous example, out-of-sample RMSEs are calculated against the true (no noise) response, and scores against a noisy version. Times recorded encapsulate both fitting and prediction calculations.

```
for(r in 1:R) {

  ## train-test partition and application of f(x) on both
  data <- fried(2*n, m)
  train <- data[1:n,]
  test <- data[(n + 1):(2*n),]

  ## extract data elements from both train and test
  X <- as.matrix(train[,1:m])
  y <- drop(train$Y)
  XX <- as.matrix(test[,1:m])
  yy <- drop(test$Y)            ## for score
  yytrue <- drop(test$Ytrue)    ## for RMSE

  ## isotropic GP fit and predict by hand
  tic <- proc.time()[3]
  D <- distance(X)
  out <- optim(c(0.1, 0.1*var(y)), nl, gradnl, method="L-BFGS-B",
    lower=eps, upper=c(10, var(y)), D=D, Y=y)
  K <- exp(-D/out$par[1]) + diag(out$par[2], nrow(D))
  Ki <- solve(K)
  tau2hat <- drop(t(y) %*% Ki %*% y / nrow(D))
  DXX <- distance(XX)
  KXX <- exp(-DXX/out$par[1]) + diag(out$par[2], ncol(DXX))
  DX <- distance(XX, X)
  KX <- exp(-DX/out$par[1])
  mup <- KX %*% Ki %*% y
  Sigmap <- tau2hat*(KXX - KX %*% Ki %*% t(KX))
  toc <- proc.time()[3]

  ## calculation of metrics for GP by hand
  rmses[r,1] <- sqrt(mean((yytrue - mup)^2))
  scores[r,1] <- score(yy, mup, Sigmap)
  times[r,1] <- toc - tic

  ## MARS fit, predict, and RMSE calculation (no score)
  tic <- proc.time()[3]
  fit.mars <- mars(X, y)
  p.mars <- predict(fit.mars, XX)
  toc <- proc.time()[3]
  rmses[r,2] <- sqrt(mean((yytrue - p.mars)^2))
  times[r,2] <- toc - tic
```

```
## separable GP fit and predict by hand
tic <- proc.time()[3]
outg <- optim(c(rep(0.1, ncol(X)), 0.1*var(y)), nlsep, gradnlsep,
  method="L-BFGS-B", lower=eps, upper=c(rep(10, m), var(y)), X=X, Y=y)
K <- covar.sep(X, d=outg$par[1:m], g=outg$par[m+1])
Ki <- solve(K)
tau2hat <- drop(t(y) %*% Ki %*% y / nrow(X))
KXX <- covar.sep(XX, d=outg$par[1:m], g=outg$par[m+1])
KX <- covar.sep(XX, X, d=outg$par[1:m], g=0)
mup2 <- KX %*% Ki %*% y
Sigmap2 <- tau2hat*(KXX - KX %*% Ki %*% t(KX))
toc <- proc.time()[3]

## calculation of metrics for separable GP by hand
rmses[r,3] <- sqrt(mean((yytrue - mup2)^2))
scores[r,3] <- score(yy, mup2, Sigmap2)
times[r,3] <- toc - tic

## laGP based separable GP
tic <- proc.time()[3]
gpi <- newGPsep(X, y, d=0.1, g=0.1*var(y), dK=TRUE)
mle <- mleGPsep(gpi, param="both", tmin=c(eps, eps), tmax=c(10, var(y)))
p <- predGPsep(gpi, XX)
deleteGPsep(gpi)
toc <- proc.time()[3]

## calculation of metrics for laGP based separable GP
rmses[r,4] <- sqrt(mean((yytrue - p$mean)^2))
scores[r,4] <- score(yy, p$mean, p$Sigma)
times[r,4] <- toc - tic
}
```

Three sets of boxplots in Figure 5.12 show the outcome of the experiment in terms of RMSE, proper score and time, respectively. Smaller is better in the case of the first and last, whereas larger scores are preferred.

```
par(mfrow=c(1,3))
boxplot(rmses, ylab="rmse")
boxplot(scores, ylab="score")
boxplot(times, ylab="execution time (seconds)")
```

MARS is the fastest but least accurate. Library-based GP prediction with laGP is substantially faster than its by-hand analog. Variability in timings is largely due to differing numbers of iterations to convergence when calculating MLEs. Our by-hand separable GP and laGP analog are nearly the same in terms of RMSE and score. However boxplots only summarize marginal results, masking any systematic (if subtle) patterns between comparators over the thirty trials. One advantage to having a randomized experiment where each replicate is trained and tested on the same data is that a paired *t*-test can be used to check for systematic differences between pairs of competitors.

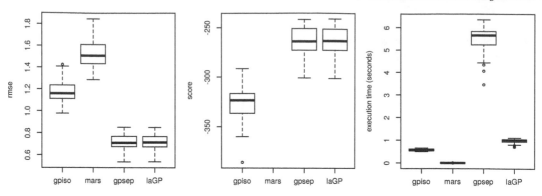

FIGURE 5.12: RMSEs (left), proper scores (middle) and execution times (right) for a bakeoff based on the Friedman data.

```
t.test(scores[,4], scores[,3], paired=TRUE)
```

```
##
##   Paired t-test
##
## data:  scores[, 4] and scores[, 3]
## t = -0.26, df = 29, p-value = 0.8
## alternative hypothesis: true difference in means is not equal to 0
## 95 percent confidence interval:
##   -0.3280  0.2549
## sample estimates:
## mean of the differences
##                  -0.03659
```

The outcome of the test reveals that there's no difference between these two predictors in terms of accuracy (*p*-value greater than the usual 5%). When comparing RMSEs in this way, it may be appropriate to take the log. (Scores are already on a log scale.) So that's it: now you can try GP regression on data of your own! Of course, there are natural variations we could add to the bakeoff above, such as laGP's isotropic implementation, and MARS with degree=2. I'll leave those to the curious reader as a homework exercise (§5.5). You could stop reading now and be satisfied knowing almost all there is to know about deploying GPs in practice.

Or you could let me get a few more words in, since I have your attention. What else is there? We've been alluding to "other covariance structures", so it might be a good idea to be a little more concrete on that. We've talked about "properties", but that's been vague. Same thing with limitations. We'll scratch the surface now, and spend whole chapters on that later on. And then there's the matter of perspective on what a GP is really doing when it's giving you predictions? Is it really Bayesian?

5.3   Some interpretation and perspective

So we fit some hyperparameters and used MVN conditioning identities to predict. But what are we doing really? Well that depends upon your perspective. I was raised a Bayesian, but have come to view things more pragmatically over the years. There's a lot you can do with ordinary least squares (OLS) regression, e.g., with the lm command in R. A bespoke fully Bayesian analog might feel like the right thing to do, but represents a huge investment with uncertain reward. Yet there's value in a clear (Bayesian) chain of reasoning, and the almost automatic regularization that a posterior distribution provides. That is, assuming all calculations are doable, both in terms of coder effort and execution time.

Those "pragmatic Bayesian" perspectives have heavily influenced my view of GP regression, which I'll attempt to summarize below through a sequence of loosely-connected musings. For those of you who like magic more than mystery, feel free to skip ahead to the next chapter. For those who want lots of theory, try another text. I'm afraid you'll find the presentation below to be woefully inadequate. For those who are curious, read on.

5.3.1   Bayesian linear regression?

Recall the standard multiple linear regression model, perhaps from your first or second class in statistics.

$$Y(x) = x^\top \beta + \varepsilon, \quad \varepsilon \overset{\text{iid}}{\sim} \mathcal{N}(0, \sigma^2).$$

For n inputs $x_1, \ldots, x_n$, stacked row-wise into a design matrix X_n, the sampling model may compactly be written as

$$Y \sim \mathcal{N}_n(X_n\beta, \sigma^2 \mathbb{I}_n).$$

That data-generating mechanism leads to the following likelihood, using observed data values $Y_n = (y_1, \ldots, y_n)^\top$ or RV analogs (for studying sampling distributions):

$$L(\beta, \sigma^2) = \left(\frac{1}{2\pi\sigma^2}\right)^{-\frac{n}{2}} \exp\left\{-\frac{1}{2\sigma^2}(Y_n - X_n\beta)^\top(Y_n - X_n\beta)\right\}.$$

Using those equations alone, i.e., only the likelihood, nobody would claim to be positing a Bayesian model for anything. A Bayesian version additionally requires priors on unknown parameters, β and σ^2.

So why is it that a slight generalization, obtained by replacing the identity with a covariance matrix parameterized by θ and g, and mapping $\sigma^2 \equiv \tau^2$,

$$Y \sim \mathcal{N}_n(X_n\beta, \tau^2 K_n), \quad K_n = C_n + g\mathbb{I}_n, \quad C_n^{ij} = C_\theta(x_i, x_j),$$

causes everyone to suddenly start talking about priors over function spaces and calling the whole thing Bayesian? Surely this setup implies a likelihood just as before

$$L(\beta, \tau^2, \theta, g) = \left(\frac{1}{2\pi\tau^2}\right)^{-\frac{n}{2}} |K_n|^{-\frac{1}{2}} \exp\left\{-\frac{1}{2\tau^2}(Y_n - X_n\beta)^\top K_n^{-1}(Y_n - X\beta)\right\}.$$

We've already seen how to infer τ^2, θ and g hyperparameters – no priors required. Conditional on $\hat{\theta}$ and $\hat{g}$-values, defining K_n, the same MLE calculations for $\hat{\beta}$ from your first stats class are easy with calculus. See exercises in §5.5. There's really nothing Bayesian about it until you start putting priors on all unknowns: θ, g, τ^2 and β. Right?

Not so fast! Here's my cynical take, which I only halfheartedly believe and will readily admit has as many grains of truth as it has innuendo. Nevertheless, this narrative has some merit as an instructional device, even if a caricature of a (possibly revisionist) history.

First set the stage. The ML community of the mid–late 1990s was enamored with Bayesian learning as an alternative to the prevailing computational learning theory (CoLT) and probably approximately correct (PAC) trends at the time. ML has a habit of appropriating methodology from related disciplines (statistics, operations research, mathematical programming, computer science), usually making substantial enhancements – which are of great value to all, don't get me wrong – but at the same time re-branding them to the extent that pedigree is all but obscured. One great example is the single-layer perceptron[21] versus logistic regression. Another is active learning[22] versus sequential design, which we shall discuss in more detail in Chapter 6. You have to admit, they're better at naming things!

Here's the claim. A core of innovators in ML were getting excited about GPs because they yielded great results for prediction problems in robotics, reinforcement learning, and e-commerce to name a few. Their work led to many fantastic publications, including the text by Rasmussen and Williams (2006) which has greatly influenced my own work, and many of the passages coming shortly. (In many ways, over the years machine learners have made more – and more accessible – advances to the GP arsenal and corpus of related codes than any other community.) I believe that those GP-ML vanguards calculated, possibly subconsciously, that they'd be better able to promote their work to the ML community by dressing it in a fancy Bayesian framework. That description encourages reading with a pejorative tone, but I think this maneuver was largely successful, and those of us who have benefited from their subsequent work owe them a debt of gratitude, because it all could've flopped. At that time, in the 1990s, they were hawking humble extensions to old ideas from 1960s spatial statistics, an enterprise that was already in full swing in the computer modeling literature. The decision to push GPs as nonparametric Bayesian learning machines served both as catalyst for enthusiasm in a community that essentially equated "Bayesian" with "better", and as a feature distinguishing their work from that of other communities who were somewhat ahead of the game, but for whom emphasizing Bayesian perspectives was not as advantageous. That foundation blossomed into a vibrant literature that nearly twenty years later is still churning out some of the best ideas in GP modeling, far beyond regression.

The Bayesian reinterpretation of more classical procedures ended up subsequently becoming a fad in other areas. A great example is the connection between lasso/L1-penalized regression and Bayesian linear modeling with independent double-exponential (Laplace) priors on β: the so-called Bayesian lasso (Park and Casella, 2008). Such procedures had been exposed decades earlier, in less fashionable times (Carlin and Polson, 1991) and had been all but forgotten. Discovering that a classical procedure had a Bayesian interpretation, and attaching to it a catchy name, breathed new life into old ideas and facilitated a great many practical extensions.

For GPs that turned out to be a particularly straightforward endeavor. We've already seen how to generate $Y(x)$'s independent of data, so why not call that a prior? And then we saw how MVN identities could help condition on data, generating $Y(x) \mid D_n$, so why not call

---

[21]https://en.wikipedia.org/wiki/Perceptron

[22]https://en.wikipedia.org/wiki/Active_learning_(machine_learning)

that a posterior? Never mind that we could've derived those predictive equations (and we essentially did just that in §5.1) under the "generalized" linear regression model.

But the fact that we can do it, i.e., force a Bayesian GP interpretation, begs three questions.

1. What is the prior over?
2. What likelihood is being paired with that prior (if not the MVN we've been using, which doesn't need a prior to work magic)?
3. What useful "thing" do we extract from this Bayesian enterprise that wasn't there before?

The answers to those questions lie in the concept of a latent random field.

5.3.2 Latent random field

Another apparatus that machine learners like is graphical models, or diagrammatic depictions of Bayesian hierarchical models. The prevailing such representation for GPs is similar to schematics used to represent hidden Markov models (HMMs)[23]. A major difference, however, is that HMMs obey the Markov property, so there's a direction of flow: the next time step is independent of the past given the current time step. For GPs there's potential for everything to depend upon (i.e., be correlated with) everything else.

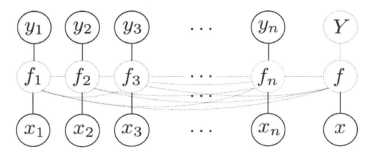

FIGURE 5.13: Diagram of the latent random field; similar to one from Rasmussen and Williams (2006).

The diagram in Figure 5.13 resembles one from Rasmussen and Williams (2006) except it's more explicit about this interconnectivity. The f_i-values in the middle row represent a *latent* Gaussian *random field*, sitting between inputs x_i and outputs y_i. They're latent because we don't observe them directly, and they're interdependent on one another. Instead we observe noise-corrupted y-values, but these are conditionally independent of one another given latent f-values. The Rasmussen & Williams analog shows these latent fs residing along a bus[24], to borrow a computer engineering term, in lieu of the many interconnections shown in Figure 5.13. To emphasize that these latents are unobserved, they reside in dashed/gray rather than solid circles. Interconnections are also dashed/gray, representing unobserved correlations and unobserved noises connecting latents together and linking them with their observed y-values. The rightmost $(x, f, Y) = (x, f(x), Y(x))$ set of nodes in the figure represent prediction at a new x location, exhibiting the same degree of interconnectedness and independence in noisy output.

---

[23]https://en.wikipedia.org/wiki/Hidden_Markov_model
[24]https://en.wikipedia.org/wiki/Bus_(computing)

The Bayesian interpretation inherited by the posterior predictive distribution hinges on placing a GP prior on latent f_i-values, not directly on measured y_i's.

$$F \sim \mathcal{N}_n(0, \tau^2 C_n) \tag{5.15}$$

Above I'm using C_n rather than K_n which is our notation for a correlation matrix without nugget. But otherwise this uses our preferred setup from earlier. Since there are no data (Y_n) values in Eq. (5.15), we can be (more) comfortable about calling this a prior. It's a prior over latent functions.

Now here's how the likelihood comes in, so that we'll have all of the ingredients comprising a posterior: likelihood + prior. Take an iid Gaussian sampling model for $Y_n \equiv (y_1, \ldots, y_n)$ around F:

$$Y \sim \mathcal{N}_n(F, \sigma^2 \mathbb{I}_n).$$

To map to our earlier notation with a nugget, choose $\sigma^2 = g\tau^2$. Embellishments may include adding a linear component $X_n \beta$ into the mean, as $F + X_n \beta$.

Now I don't know about you, but to me this looks a little contrived. With prior $F \mid X_n$ and likelihood $Y_n \mid F, X_n$, notation for the posterior is $F \mid D_n$. It's algebraic form isn't interesting in and of itself, but an application of Bayes' rule does reveal a connection to our earlier development.

$$p(F \mid D_n) = \frac{p(Y_n \mid F, X_n)p(F \mid X_n)}{p(Y_n \mid X_n)}$$

The denominator $p(Y_n \mid X_n)$ above, for which we have an expression (5.8) in log form, is sometimes called a *marginal likelihood* or *evidence*. The former moniker arises from the law of total probability[25], which allows us to interpret this quantity as arising after integrating out the latent random field F from the likelihood $Y_n \mid F, X_n$:

$$p(Y_n \mid X_n) = \int p(Y_n \mid f, X_n) \cdot p(f \mid X_n) \, df.$$

We only get what we really want, a posterior predictive $Y(x) \mid D_n = (X_n, Y_n)$, through a different marginalization, this time over the posterior of F:

$$p(Y(x) \mid D_n) = \int p(Y(x) \mid f) \cdot p(f \mid D_n) \, df.$$

But of course we already have expressions for that as well (5.2). Similar logic, through joint modeling of Y_n with $f(x)$, instead of with $Y(x)$, and subsequent MVN conditioning can be applied to derive $f(x) \mid D_n$. The result is Gaussian with identical mean $\mu(x)$, from Eq. (5.2), and variance

$$\breve{\sigma}^2(x) = \sigma^2(x) - \hat{\tau}^2 \hat{g} = \hat{\tau}^2(1 - k_n^\top(x)K_n^{-1}k_n(x)). \tag{5.16}$$

Thus, this latent function space interpretation justifies a trick we performed in §5.2.2, when exploring smoothed sinusoid visuals in Figure 5.10. Recall that we obtained de-noised sample paths (gray lines) by omitting the nugget in the predictive variance calculation. This is the

---

[25]https://en.wikipedia.org/wiki/Law_of_total_probability

same as plugging in $g = 0$ when calculating $K(\mathcal{X}, \mathcal{X})$, or equivalently using $C(\mathcal{X}, \mathcal{X})$. At the time we described that maneuver as yielding the epistemic uncertainty[26] – that due to model uncertainty – motivated by a desire for pretty-looking gray lines. In fact, sample paths resulting from that calculation are precisely draws from $f(\mathcal{X}) \mid D_n$, i.e., posterior samples of the latent function.

Error-bars calculated from that predictive variance are tighter, and characterize uncertainty in predictive mean, i.e., without additional uncertainty coming from a residual sum of squares. Note that $\hat{g}$ is only removed from the predictive calculation, not from the diagonal of K_n leading to K_n^{-1}. To make an analogy to linear modeling in R with `lm`, the former based on $\sigma^2(x)$ corresponds to `interval="prediction"`, and $\breve{\sigma}^2(x)$ to `interval="confidence"` when using `predict.lm`. (See §3.2.3.) As $n \to \infty$, we'll have that $\breve{\sigma}^2(x) \to 0$ for all x, meaning that eventually we'll learn the latent functions without uncertainty. On the other hand, $\sigma^2(x)$ could be no smaller than $\hat{\tau}^2 \hat{g}$, meaning that when it comes to making predictions, they'll offer an imperfect forecast of the noisy (held-out) response value no matter how much data, n, is available for training.

Was all that worth it? We learned something: predictive distribution $Y(x) \mid D_n$ involves integrating over a latent function space, at least notionally, even if the requisite calculations and their derivation don't require working that way. An examination of the properties of those latent functions, or rather correlation functions $C(\cdot, \cdot)$ which generate those f's, could provide insight into the nature of our nonparametric regression. I think that's the biggest feature of this interpretation.

5.3.3 Stationary kernels

Properties of the correlation function, $C(x, x')$, or covariance function $\Sigma(x, x')$, or generically *kernel* $k(x, x')$ as preferred in ML, dictate properties of latent functions f. For example, a *stationary* kernel $k(x, x')$ is a function only of $r = x - x'$,

$$k(x, x') = k(r).$$

More commonly, $r = |x - x'|$. Our isotropic and separable Gaussian families are both stationary kernels. The most commonly used kernels are stationary, despite the strong restrictions they place on the nature of the underlying process. Characteristics of the function f, via a stationary kernel k, are global since they're determined only by displacement between coordinates, not positions of the coordinates themselves; they must exhibit the same dynamics everywhere. What, more specifically, are some of the properties of a stationary kernel?

Consider wiggliness, which is easiest to characterize in a single input dimension $x \in [0, 1]$. Define the number of level-u *upcrossings* N_u to be the number of times a random realization of f crosses level u, from below to above on the y-axis, when traversing from left to right on the x-axis. It can be shown that the expected number of level-u upcrossings under a stationary kernel $k(\cdot)$ is

$$\mathbb{E}\{N_u\} = \frac{1}{2\pi} \sqrt{\frac{-k''(0)}{k(0)}} \exp\left(-\frac{u^2}{2k(0)}\right). \tag{5.17}$$

For a Gaussian kernel, the expected number of zero-crossings (i.e., $u = 0$) works out to be

[26]https://en.wikipedia.org/wiki/Uncertainty_quantification#Aleatoric_and_epistemic_uncertainty

$$\mathbb{E}\{N_0\} \propto \theta^{-1/2}.$$

See Adler (2010) for more details. Hyperparameter θ appears in the denominator of squared distance, so on the scale of inputs x, i.e., "un-squared" distance, θ is proportional to the expected length in the input space before crossing zero: hence the name *characteristic lengthscale*. Therefore θ directly controls how often latent functions change direction, which they must do in order to cross zero more than once (from below). A function with high $\mathbb{E}\{N_u\}$, which comes from small θ, would have more bumps, which we might colloquially describe as more wiggly. The same result can be applied separately in each coordinate direction of a separable Gaussian kernel using θ_k, for all $k = 1, \ldots, m$.

Consider smoothness, in the calculus sense – not as the opposite of wiggliness. A Gaussian kernel is infinitely differentiable, which in turn means that its latent f process has mean-square derivatives of all orders, and this is very smooth. Not many physical phenomena – think back to your results from Physics 101 or a first class in differential equations – are infinitely smooth. Despite this not being realistic for most real-world processes, (separable) Gaussian kernels are still the most widely used. Why? Because they're easy to code up; GPs are relatively robust to misspecifications, within reason (remember what happened when we forced $\tau^2 = 1$); faults of infinite smoothness are easily masked by "fudges" already present in the model for other reasons, e.g., with nugget or jitter (Andrianakis and Challenor, 2012; Gramacy and Lee, 2012; Peng and Wu, 2014). And finally there's the practical matter of guaranteeing a proper, positive definite covariance regardless of input dimension.

If a Gaussian kernel isn't working well, or if you want finer control on mean-square differentiability, then consider a *Matèrn family* member, with

$$k_\nu(r) = \frac{2^{1-\nu}}{\Gamma(\nu)} \left(r\sqrt{\frac{2\nu}{\theta}} \right)^\nu K_\nu \left(r\sqrt{\frac{2\nu}{\theta}} \right).$$

Above K_ν is a modified Bessel function[27] of the second kind, ν controls smoothness, and θ is a lengthscale as before. As $\nu \to \infty$, i.e., a very smooth parameterization, we get

$$k_\nu(r) \to k_\infty(r) = \exp \left\{ -\frac{r^2}{2\theta} \right\}$$

which can be recognized as (a re-parameterized) Gaussian family, where r is measured on the scale of ordinary (not squared) Euclidean distances.

The code below sets up this "full" Matèrn in R so we can play with it a little.

```
matern <- function(r, nu, theta)
 {
  rat <- r*sqrt(2*nu/theta)
  C <- (2^(1 - nu))/gamma(nu) * rat^nu * besselK(rat, nu)
  C[is.nan(C)] <- 1
  return(C)
 }
```

Sample paths of latent f under a GP with this kernel will be k times differentiable if and

---

[27]https://en.wikipedia.org/wiki/Bessel_function

only if $\nu > k$. Loosely speaking this means that sample paths from smaller ν will be rougher than those from larger ν. This is quite surprising considering how similar the kernel looks when plotted for various ν, as shown in Figure 5.14.

```
r <- seq(eps, 3, length=100)
plot(r, matern(r, nu=1/2, theta=1), type="l", ylab="k(r,nu)")
lines(r, matern(r, nu=2, theta=1), lty=2, col=2)
lines(r, matern(r, nu=10, theta=1), lty=3, col=3)
legend("topright", c("nu=1/2", "n=2", "nu=10"), lty=1:3, col=1:3, bty="n")
```

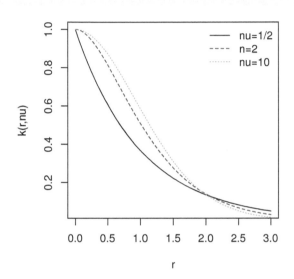

FIGURE 5.14: Matèrn kernels versus Euclidean distance.

In all three cases the lengthscale parameter is taken as $\theta = 1$. R code below calculates covariance matrices for latent field draws F under these three Matèrn specifications. Careful, as in the discussion above, the `matern` function expects its `r` input to be on the scale of ordinary (not squared) distances.

```
X <- seq(0, 10, length=100)
R <- sqrt(distance(X))
K0.5 <- matern(R, nu=1/2, theta=1)
K2 <- matern(R, nu=2, theta=1)
K10 <- matern(R, nu=10, theta=1)
```

Notice that we're not augmenting the diagonal of these matrices with ϵ jitter. One of the great advantages to a Matèrn (for small ν) is that it creates covariance matrices that are better *conditioned*[28], i.e., farther from numerically non-positive definite. Using those covariances, Figure 5.15 plots three sample paths for each case.

```
par(mfrow=c(1,3))
matplot(X, t(rmvnorm(3,sigma=K0.5)), type="l", col=1, lty=1,
```

---

[28]https://en.wikipedia.org/wiki/Condition_number

```
    xlab="x", ylab="y, nu=1/2")
matplot(X, t(rmvnorm(3,sigma=K2)), type="l", col=2, lty=2,
    xlab="x", ylab="y, nu=2")
matplot(X, t(rmvnorm(3,sigma=K10)), type="l", col=3, lty=3, xlab="x",
    ylab="y, nu=10")
```

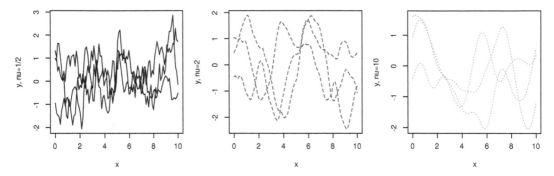

FIGURE 5.15: Sample paths under Matèrn kernels with $\nu = 0.5$ (left), $\nu = 2$ (middle) and $\nu = 10$ (right).

Surfaces under $\nu = 1/2$, shown on the left, are very rough. Since $\nu < 1$ they're nowhere differentiable, yet they still have visible correlation in space. On the right, where $\nu = 10$, surfaces are quite smooth; ones which are nine-times differentiable. In the middle is an intermediate case which is smoother than the first one (being once differentiable) but not as smooth as the last one. To get a sense of what it means to be just once differentiable, as in the middle panel of the figure, consider the following first and second numerical derivatives (via first differences) for a random realization when $\nu = 2$.

```
F <- rmvnorm(1, sigma=K2)
dF <- (F[-1] - F[-length(F)])/(X[2] - X[1])
d2F <- (dF[-1] - dF[-length(dF)])/(X[2] - X[1])
```

Figure 5.16 plots these three sets of values on a common x-axis.

```
plot(X, F, type="l", lwd=2, xlim=c(0,13), ylim=c(-5,5),
    ylab="F and derivatives")
lines(X[-1], dF, col=2, lty=2)
lines(X[-(1:2)], d2F, col=3, lty=3)
legend("topright", c("F", "dF", "d2F"), lty=1:3, col=1:3, bty="n")
```

Whereas the original function realization F seems reasonably smooth, its first differences dF are jagged despite tracing out a clear spatial pattern in x. Since first differences are not smooth, second differences d2F are erratic: this random latent function realization is not twice differentiable. Of course, these are all qualitative statements based on finite realizations, but you get the gist.

We're not usually in the business of generating random (latent) functions. Rather, given observed y-values we wish to estimate the unknown latent function f via a choice of correlation family, and settings for its hyperparameters. What does this mean for ν, governing

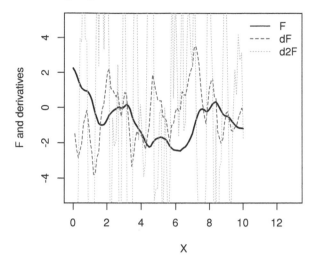

FIGURE 5.16: Numerical derivatives for Matèrn sample paths.

smoothness? Some of the biggest advocates for the Matèrn, e.g., Stein (2012), argue that you should learn smoothness, ν, from your training data. But that's fraught with challenges in practice, the most important being that noisy data provide little guidance for separating noise from roughness, which under the Matèrn is regarded as a form of signal. This results in a likelihood surface which has many dead (essentially flat) spots. More typically, the degree of smoothness is regarded as a modeling choice, ideally chosen with knowledge of underlying physical dynamics.

Another challenge involves the Bessel function K_ν, the evaluation of which demands a cumbersome numerical scheme (say using `besselK` in R) applied repeatedly to $\mathcal{O}(n^2)$ pairs of distances r between training data inputs X_n. For most moderate data sizes, $n < 1000$ say, creating an $n \times n$ covariance matrix under the Matèrn is actually slower than decomposing it, despite the $\mathcal{O}(n^3)$ cost that this latter operation implies.

A useful re-formulation arises when $\nu = p + \frac{1}{2}$ for non-negative integer p, in which case p exactly determines the number of mean-square derivatives.

$$
k_{\nu = p + \frac{1}{2}}(r) = \exp\left\{ r\sqrt{\frac{2\nu}{\theta}} \right\} \frac{\Gamma(p+1)}{\Gamma(2p+1)} \sum_{i=0}^{p} \frac{(p+i)!}{i!(p-i)!} \left(r\sqrt{\frac{8\nu}{\theta}} \right)^{p-1}
$$

Whether or not this is simpler than the previous specification is a matter of taste. From this version it's apparent that the kernel is comprised of a product of an exponential and a polynomial of order $p - 1$. Although tempting to perform model search over discrete p, this is not any easier than over continuous ν. However it is easier to deduce the form from more transparent and intuitive components (i.e., no Bessel functions), at least for small p.

Choosing $p = 0$, yielding only the exponential part which is equivalent to $k(r) = e^{-\frac{r^2}{\theta}}$, is sometimes referred to as the *exponential family*. This choice is also a member of the *power exponential family*, introduced in more detail momentarily. We've already discussed how $p = 0$, implying $\nu = 1/2$, is appropriate for rough surfaces. At the risk of being even more redundant, $p \to \infty$ is a great choice when dynamics are super smooth, although clearly having an infinite sum isn't practical for implementation. From a technical standpoint, both choices are probably too rough or too smooth, respectively, even though they (especially the Gaussian) work fine in many contexts, to the chagrin of purists.

Entertaining a middle ground has its merits. Practically speaking, without explicit knowledge about higher-order derivatives, it'll be hard to distinguish between values $\nu \geq 7/2$, meaning $p \geq 3$, and much larger settings $(p, \nu \to \infty)$ especially with noisy data. That leaves essentially two special cases, which emit relatively tidy expressions:

$$k_{3/2}(r) = \left(1 + r\sqrt{\frac{3}{\theta}}\right) \exp\left(-r\sqrt{\frac{3}{\theta}}\right) \tag{5.18}$$

or
$$k_{5/2}(r) = \left(1 + r\sqrt{\frac{5}{\theta}} + \frac{5r^2}{3\theta}\right) \exp\left(-r\sqrt{\frac{5}{\theta}}\right).$$

The advantage of these cases is that there are no Bessel functions, no sums of factorials nor fractions of gammas. Most folks go for the second one ($p = 2$) without bothering to check against the first, perhaps appealing to physical intuition that many interesting processes are at least twice differentiable. Worked examples with these choices are deferred to Chapter 10 on replication and heteroskedastic modeling. In the meantime, the curious reader is invited to explore them in a homework exercise in §5.5.

The *power exponential family* would, at first blush, seem to have much in common with the Matèrn,

$$k_\alpha(r) = \exp\left\{-\left(\frac{r}{\sqrt{\theta}}\right)^\alpha\right\} \quad \text{for } 0 < \alpha \leq 2,$$

having the same number of hyperparameters and coinciding on two special cases $\alpha = 1 \Leftrightarrow p = 0$ and $\alpha = 2 \Leftrightarrow p \to \infty$. An implementation in R is provided below.

```
powerexp <- function(r, alpha, theta)
  {
  C <- exp(-(r/sqrt(theta))^alpha)
  C[is.nan(C)] <- 1
  return(C)
  }
```

However the process is never mean-square differentiable except in the Gaussian ($\alpha = 2$) special case. Whereas the Matèrn offers control over smoothness, a power exponential provides essentially none. Yet both offer potential for greater numerical stability, owing to better covariance matrix condition numbers[29], than their Gaussian special cases. A common kludge is to choose $\alpha = 1.9$ in lieu of $\alpha = 2$ with the thinking that they're "close" (which is incorrect) but the former is better numerically (which is correct). As with Matèrn ν, or perhaps even more so, it's surprising that fine variation in a single hyperparameter (α) can have such a profound effect on the resulting latent functions, especially when their kernels look so similar as a function of r. See Figure 5.17.

```
plot(r, powerexp(r, alpha=1.5, theta=1), type="l", ylab="k(r,alpha)")
lines(r, powerexp(r, alpha=1.9, theta=1), lty=2, col=2)
lines(r, powerexp(r, alpha=2, theta=1), lty=3, col=3)
legend("topright", c("alpha=1.5", "alpha=1.9", "alpha=2"),
  lty=1:3, col=1:3, bty="n")
```

---

[29]https://en.wikipedia.org/wiki/Condition_number

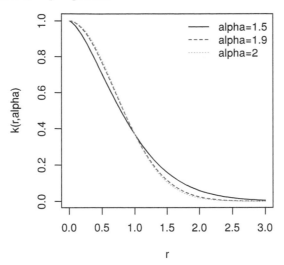

FIGURE 5.17: Power exponential kernels versus Euclidean distance.

Despite the strong similarity for $\alpha \in \{1.5, 1.9, 2\}$ in kernel evaluations, the resulting sample paths exhibit stark contrast, as shown in Figure 5.18.

```
Ka1.5 <- powerexp(R, alpha=1.5, theta=1)
Ka1.9 <- powerexp(R, alpha=1.9, theta=1)
Ka2 <- powerexp(R, alpha=2, theta=1) + diag(eps, nrow(R))
par(mfrow=c(1,3))
ylab <- paste0("y, alpha=", c(1.5, 1.9, 2))
matplot(X, t(rmvnorm(3, sigma=Ka1.5)), type="l", col=1, lty=1,
  xlab="x", ylab=ylab[1])
matplot(X, t(rmvnorm(3, sigma=Ka1.9)), type="l", col=2, lty=2,
  xlab="x", ylab=ylab[2])
matplot(X, t(rmvnorm(3, sigma=Ka2)), type="l", col=3, lty=3,
  xlab="x", ylab=ylab[3])
```

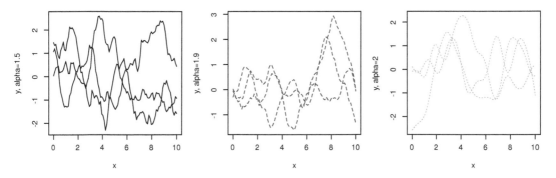

FIGURE 5.18: Sample paths under power exponential kernels with $\alpha = 1.5$ (left), $\alpha = 1.9$ (middle) and $\alpha = 2$ (right).

The $\alpha = 2$ case is a special one: very smooth latent function realizations. Although $\alpha = 1.9$ is not as rough as $\alpha = 1.5$, it still exhibits "sharp turns", albeit fewer. That pair are more similar to one another in that respect than either is to the final one where $\alpha = 2$, being infinitely smooth.

The *rational quadratic* kernel

$$k_{\mathrm{rq}}(r) = \left(1 + \frac{r^2}{2\alpha\theta}\right)^{-\alpha} \quad \text{with } \alpha > 0$$

can be derived as a scale mixture of Gaussian (power exponential with $\alpha = 2$) and exponential (power exponential with $\alpha = 1$) kernels over θ. Since that mixture includes the Gaussian, the process is infinitely mean-square differentiable for all α. An implementation of this kernel in R is provided below.

```
ratquad <- function(r, alpha, theta)
{
  C <- (1 + r^2/(2 * alpha * theta))^(-alpha)
  C[is.nan(C)] <- 1
  return(C)
}
```

To illustrate, panels of Figure 5.19 show the kernel (left) and three MVN realizations (right) for each $\alpha \in \{1/2, 2, 10\}$.

```
par(mfrow=c(1,2))
plot(r, ratquad(r, alpha=1/2, theta=1), type="l",
  ylab="k(r,alpha)", ylim=c(0,1))
lines(r, ratquad(r, alpha=2, theta=1), lty=2, col=2)
lines(r, ratquad(r, alpha=10, theta=1), lty=3, col=3)
legend("topright", c("alpha=1/2", "alpha=2", "alpha=10"), lty=1:3,
  col=1:3, bty="n")
plot(X, rmvnorm(1, sigma=ratquad(R, alpha=1/2, theta=1)), type="l", col=1,
  lty=1, xlab="x", ylab="y", ylim=c(-2.5,2.5))
lines(X, rmvnorm(1, sigma=ratquad(R, alpha=2, theta=1)), col=2, lty=2)
lines(X, rmvnorm(1, sigma=ratquad(R, alpha=10, theta=1)), type="l",
  col=3, lty=3)
```

Although lower α yield rougher characteristics, all are smooth compared to their non-differential analogs introduced earlier.

The idea of hybridizing two kernels, embodied in the particular by the rational quadratic kernel above, nicely generalizes. Given two well-defined kernels, i.e., generating a positive definite covariance structure, there are many ways they can be combined into a single well-defined kernel. The sum of two kernels is a kernel. In fact, if $f(x) = \sum f_k(x_k)$, for random $f_k(x_k)$ with univariate kernels, then the kernel of $f(x)$ arises as the sum of those kernels. Durrande et al. (2012) show how such an additive GP-modeling approach can be attractive in high input dimensions. A product of two kernels is a kernel. The kernel of a convolution is the convolution of the kernel. Anisotropic versions of isotropic covariance functions can be created through products, leading to a separable formulation, or through quadratic forms like $r^2(x, x') - (x - x')^\top A(x - x')$ where A may augment the hyperparameter space. Low rank A are sometimes used to implement linear dimensionality reduction. Rank one choices

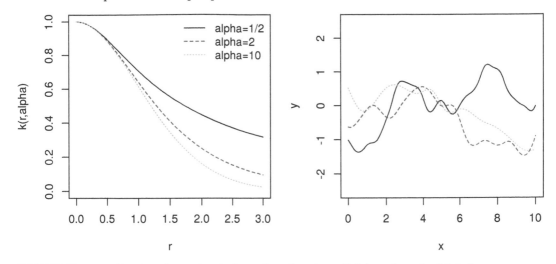

FIGURE 5.19: Rational quadratic kernel evaluations (left) and paths (right).

of A lead to what are known as GP single-index models (GP-SIMs; Gramacy and Lian, 2012).

Although somewhat less popular, there are stationary kernels which are common in specialized situations such as with spherical input coordinates, or for when periodic effects are involved. A good reference for these and several others is an unpublished manuscript by Abrahamsen (1997). Another good one is Wendland (2004), particularly Chapter 9, which details piecewise polynomial kernels with *compact support*, meaning that they go exactly to zero which can be useful for fast computation with sparse linear algebra libraries. The `kergp` package (Deville et al., 2018) on CRAN provides a nice interface for working with GPs under user-customized kernels.

Kernels supporting qualitative factors, i.e., categorical and ordinal inputs, and mixtures thereof with ordinary continuous ones, have appeared in the literature. See work by Qian et al. (2008), Zhou et al. (2011) and a recent addition from Zhang et al. (2018). None of these ideas have made their way into public software, to my knowledge. The `tgp` package supports binarized categorical inputs through treed partitioning. More detail is provided in §9.2.2.

It's even possible to learn all $n(n-1)/2$ entries of the covariance matrix separately, under regularization and a constraint that the resulting Σ_n is positive definite, with semidefinite programming[30]. Lanckriet et al. (2004) describe a transductive learner[31] that makes this tractable implicitly, through the lens of prediction at a small number of testing sites. We'll take a similar approach to thrifty approximation of more conventional GP learning in §9.3.

This section was titled "Stationary kernels". Convenient nonstationary GP modeling is deferred to Chapter 9, however it's worth closing with some remarks connecting back to those made along with linear modeling in §5.3.1 using some simple, yet nonstationary kernel specifications. It turns out that there are choices of kernel which recreate linear and polynomial (mean) modeling.

---

[30]https://en.wikipedia.org/wiki/Semidefinite_programming
[31]https://en.wikipedia.org/wiki/Transduction_(machine_learning)

$$\text{linear} \quad k(x, x') = \sum_{k=1}^{m} \beta_k x_k x'_k$$

$$\text{polynomial} \quad k(x, x') = (x^\top x' + \beta)^p$$

Here β is vectorized akin to linear regression coefficients, although there's no direct correspondence. These kernels cannot be rewritten in terms of displacement $x - x'$ alone. However, their latent function realizations are not nonstationary in a way that many regard in practice as genuine because, as linear and polynomial models, their behavior is rigidly rather than flexibly defined. In other words, the nature of nonstationarity is prescribed rather than emergent in the data-fitting process. The fact that you can do this – choose a kernel to encode mean structure in the covariance – lends real credence to the characterization "it's all in the covariance" when talking about GP models. You can technically have a zero-mean GP with a covariance structure combining linear and spatial structures through sums of kernels, but you'd only do that if you're trying to pull the wool over someone's eyes. (Careful not to do it to yourself.) Most folks would sensibly choose the more conventional model of linear mean and spatial (stationary) covariance. But the very fact that you can do it again begs the question: why get all Bayesian about the function space? If an ordinary linear model – which is not typically thought of as Bayesian without additional prior structure – arises as a special choice of kernel, then why should introduction of a kernel automatically put us in a Bayesian state of mind?

GP regression is just multivariate normal modeling, and you can get as creative as you want with mean and covariance. There are some redundancies, as either can implement linear and polynomial modeling, and much more. Sometimes you know something about prevailing input–output relationships, and for those settings GPs offer potential to tailor as a means of spatial regularization, which is a classical statistician's roundabout way of saying "encode prior beliefs". For example, if $a(x)$ is a known deterministic function and $g(x) = a(x)f(x)$, where $f(x)$ is a random process, then $\mathbb{C}\text{ov}(g(x), g(x')) = a(x)k(x, x')a(x')$. This can be used to normalize kernels by choosing $a(x) = k^{-1/2}(x, x)$ so that

$$\bar{k}(x, x') = \frac{k(x, x')}{\sqrt{k(x, x)}\sqrt{k(x', x')}}.$$

That's a highly stylized example which is not often used in practice, yet it offers a powerful testament to potential for GP customization. GP regression can either be out-of-the-box with simple covariance structures implemented by library subroutines, ready for anything, or can be tailored to the bespoke needs of a particular modeling enterprise and data type. You can get very fancy, or you can simplify, and inference for unknowns need not be too onerous if you stick to likelihood-based criteria paired with mature libraries for optimization with closed-form derivatives.

5.3.4   Signal-to-noise

As with any statistical model, you have to be careful not to get too fancy or it may come back to bite you. The more hyperparameters purporting to offer greater flexibility or a better-tuned fit, the greater the estimation risk. By estimation risk I mean both potential to fit noise as signal, as well as its more conventional meaning (particularly popular in empirical finance) which fosters incorporation of uncertainties, inherent in high variance sampling distributions for optimized parameters, that are often overlooked. Nonparametric models and latent function spaces exacerbate the situation. Awareness of potential sources of such risk is particularly fraught, and assessing its extent even more so. At some point there might be so many knobs that its hard to argue that the "hyperparameter" moniker is apt compared to the more canonical "parameter", giving the practitioner the sense that choosing appropriate settings is key to getting good fits.

To help make these concerns a little more concrete, consider the following simple data generating mechanism.

$$f(x) = \sin(\pi x/5) + 0.2\cos(4\pi x/5), \tag{5.19}$$

originally due to Higdon (2002). Code below observes that function with noise on an $n = 40$ grid.

```
x <- seq(0, 10, length=40)
ytrue <- (sin(pi*x/5) + 0.2*cos(4*pi*x/5))
y <- ytrue + rnorm(length(ytrue), sd=0.2)
```

The response combines a large amplitude periodic signal with another small amplitude one that could easily be confused as noise. Rather than shove these data pairs into a GP MLE subroutine, consider instead a grid of lengthscale and nugget values.

```
g <- seq(0.001, 0.4, length=100)
theta <- seq(0.1, 4, length=100)
grid <- expand.grid(theta, g)
```

Below the MVN log likelihood is calculated for each grid pair, and the corresponding predictive equations evaluated at a second grid of $\mathcal{X} \equiv$ xx inputs, saved for each hyperparameter pair on the first, hyperparameter grid.

```
ll <- rep(NA, nrow(grid))
xx <- seq(0, 10, length=100)
pm <- matrix(NA, nrow=nrow(grid), ncol=length(xx))
psd <- matrix(NA, nrow=nrow(grid), ncol=length(xx))
for(i in 1:nrow(grid)) {
   gpi <- newGP(matrix(x, ncol=1), y, d=grid[i,1], g=grid[i,2])
   p <- predGP(gpi, matrix(xx, ncol=1), lite=TRUE)
   pm[i,] <- p$mean
   psd[i,] <- sqrt(p$s2)
   ll[i] <- llikGP(gpi)
   deleteGP(gpi)
}
l <- exp(ll - max(ll))
```

The code above utilizes isotropic (newGP/predGP) GP functions from laGP. Since the data is in 1d, these are equivalent to separable analogs illustrated earlier in §5.2.6. For now, concentrate on (log) likelihood evaluations; we'll come back to predictions momentarily. Figure 5.20 shows the resulting likelihood surface as an image in the $\theta \times g$ plane. Notice that the final line in the code above exponentiates the log likelihood, so the figure is showing z-values (via color) on the likelihood scale.

```
image(theta, g, matrix(l, ncol=length(theta)), col=cols)
contour(theta, g, matrix(l, ncol=length(g)), add=TRUE)
```

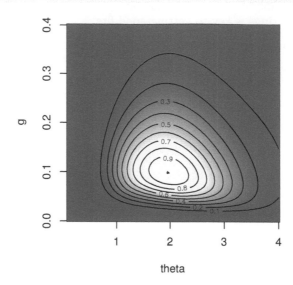

FIGURE 5.20: Log likelihood surface over lengthscale θ and nugget g for mixed sinusoid data (5.19).

Since the data is random, it's hard to anticipate an appropriate range for θ and g axes. It's worth repeating these codes in your own R session to explore variations that arise under new datasets generated under novel random noise, adding another layer to the sense of estimation risk, i.e., beyond that which is illustrated here. What can be seen in Figure 5.20? Maybe it looks like an ordinary log likelihood surface in 2d: pleasantly unimodal, convex, etc., and easy to maximize by eyeball norm. (Who needs fancy numerical optimizers after all?) There's some skew to the surface, perhaps owing to positivity restrictions placed on both hyperparameters.

In fact, that skewness is hiding a multimodal posterior distribution over functions. The modes are "higher signal/lower noise" and "lower signal/higher noise". Some random realizations reveal this feature through likelihood more than others, which is one reason why repeating this in your own session may be helpful. Also keep in mind that there's actually a third hyperparameter, $\hat{\tau}^2$, being optimized implicitly through the concentrated form of the log likelihood (5.8). So there's really a third dimension to this view which is missing, challenging a more precise visualization and thus interpretation. Such signal–noise tension is an ordinary affair, and settling for one MLE tuple in a landscape of high values – even if you're selecting the very highest ones – can grossly underestimate uncertainty. What is apparent in Figure 5.20 is that likelihood contours trace out a rather large area in hyperparameter space. Even the red "outer-reaches" in the viewing area yield non-negligible likelihood, which is consistent across most random realizations. This likelihood surface is relatively flat.

The best view of signal-to-noise tension is through the predictive surface, in particular what that surface would look like for a multitude of most likely hyperparameter settings. To facilitate that, code below pre-calculates quantiles derived from predictive equations obtained for each hyperparameter pair.

```
q1 <- pm + qnorm(0.95, sd=psd)
q2 <- pm + qnorm(0.05, sd=psd)
```

Figure 5.21 shows three sets of lines (mean and quantile-based interval) for every hyperparameter pair, but not all lines are visualized equally. Transparency is used to downweight low likelihood values. Multiple low likelihood settings accumulate shading, when the resulting predictive equations more or less agree, and gain greater opacity.

```
plot(x,y, ylim=c(range(q1, q2)))
matlines(xx, t(pm), col=rgb(0,0,0,alpha=(1/max(l))/2), lty=1)
matlines(xx, t(q1), col=rgb(1,0,0,alpha=(1/max(l))/2), lty=2)
matlines(xx, t(q2), col=rgb(1,0,0,alpha=(1/max(l))/2), lty=2)
```

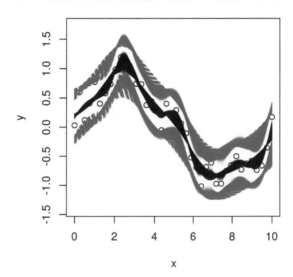

FIGURE 5.21: Posterior predictive equations in terms of means (solid-black) and quantiles (dashed-red).

The hyperparameter grid is 100×100, but clearly there are not 3×10000 distinct lines visible in the figure. Nevertheless it's easy to see two regimes. Some of the black/red lines are more wavy, explaining the low-amplitude periodic structure as signal; others are less wavy, explaining it as noise. Although the likelihood was unimodal, we have a multimodal posterior predictive surface.

For all the emphasis on a Bayesian perspective, marginalizing over latent functions and whatever, it's surprising that Bayesian inference is rarely used where it's needed most. Clearly the MLE/MAP is missing an important element of uncertainty. Only fully Bayesian posterior inference, after specifying priors on hyperparameters and running Markov chain Monte Carlo (MCMC) for posterior sampling, could provide a full assessment of estimation risk and provide posterior predictive quantities with full UQ. Very few libraries offer this functionality, tgp, spBayes and plgp being three important exceptions, yet these rely on rather conventional covariance specifications. The tgp package has some extra, highly non-standard, features which will be discussed in more detail in §9.2.2. As covariance kernels incorporate more hyperparameters – smoothness, separable vectorized lengthscales, rank-one anisotropy, latent noise structures (Chapter 10), whatever Franken-kernel results from adding/convolving/multiplying – and likewise incorporate well-thought-out parametric mean structures, it's obvious that a notionally nonparametric GP framework can become highly and strongly parameterized. In such settings, one must be very careful not to get overconfident about point-estimates so-derived. The only way to do it right, in my opinion, is to be fully Bayesian.

With that in mind, it's a shame to give the (at worst false, at best incomplete) impression of being Bayesian without having to do any of those things. In that light, ML marketing of GPs as Bayesian updating is a double-edged sword. Just because something can be endowed with a Bayesian interpretation, doesn't mean that it automatically inherits all Bayesian merits relative to a more classical approach. A new ML Bayesian perspective on kriging spawned many creative ideas, but it was also a veneer next to the real thing.

5.4   Challenges and remedies

This final section wraps up our GP chapter on somewhat of a lower note. GPs are remarkable, but they're not without limitations, the most limiting being computational. In order to calculate K_n^{-1} and $|K_n|$ an $\mathcal{O}(n^3)$ decomposition is required for dense covariance matrices K_n, as generated by most common kernels. In the case of MLE inference, that limits training data sizes to n in the low thousands, loosely, depending on how many likelihood and gradient evaluations are required to perform numerical maximization. You can do a little better with the right linear algebra libraries installed. See Appendix A.1 for details. (It's easier than you might think.)

Fully Bayesian GP regression, despite many UQ virtues extolled above, can all but be ruled out on computational grounds when n is even modestly large ($n > 2000$ or so), speedups coming with fancy matrix libraries notwithstanding. If it takes dozens or hundreds of likelihood evaluations to maximize a likelihood, it will take several orders of magnitude more to sample from a posterior by MCMC. Even in cases where MCMC is just doable, it's sometimes not clear that posterior inference is the right way to spend valuable computing resources. Surrogate modeling of computer simulation experiments is a perfect example. If you have available compute cycles, and are pondering spending them on expensive MCMC to better quantify uncertainty, why not spend them on more simulations to reduce that uncertainty instead? We'll talk about design and sequential design in the next two chapters.

A full discussion of computational remedies, which mostly boils down to bypassing big matrix inverses, will be delayed until Chapter 9. An exception is GP approximation by convolution which has been periodically revisited, over the years, by geostatistical and computer experiment communities. Spatial and surrogate modeling by convolution can offer flexibility and speed in low input dimension. Modern versions, which have breathed new life into geospatial (i.e., 2d input) endeavours by adding multi-resolution features and parallel computing (Katzfuss, 2017), are better reviewed in another text. With emphasis predominantly being on modestly-larger-dimensional settings common in ML and computer surrogate modeling contexts, the presentation here represents somewhat of a straw man relative to Chapter 9 contributions. More favorably said: it offers another perspective on, and thus potentially insight into, the nature of GP regression.

5.4.1 GP by convolution

In low input dimension it's possible to avoid decomposing a big covariance matrix and obtain an approximate GP regression by taking pages out of a splines/temporal modeling play-book. Higdon (2002) shows that one may construct a GP $f(x)$ over a general region $x \in \mathcal{X}$ by convolving a continuous Gaussian white noise process $\beta(x)$ with smoothing kernel $k(x)$:

$$f(x) = \int_{\mathcal{X}} k(u - x)\beta(u)\, du, \quad \text{for } x \in \mathcal{X}. \tag{5.20}$$

The resulting covariance for $f(x)$ depends only on relative displacement $r = x - x'$:

$$c(r) = \mathbb{C}\text{ov}(f(x), f(x')) = \int_{\mathcal{X}} k(u-x)k(u-x')\, du = \int_{\mathcal{X}} k(u-r)k(u)\, du.$$

In the case of isotropic $k(x)$ there's a 1:1 equivalence between smoothing kernel k and covariance kernel c.

$$\text{e.g.,} \quad k(x) \propto \exp\left\{-\frac{1}{2}\|x\|^2\right\} \rightarrow c(r) \propto \exp\left\{-\frac{1}{2}\left\|\frac{r}{2}\right\|^2\right\}.$$

Note the implicit choice of lengthscale exhibited by this equivalence.

This means that rather than defining $f(x)$ directly through its covariance function, which is what we've been doing in this chapter up until now, it may instead be specified indirectly, yet equivalently, through the latent *a priori* white noise process $\beta(x)$. Sadly, the integrals above are not tractable analytically. However by restricting the latent process $\beta(x)$ to spatial sites $\omega_1, \ldots, \omega_m$, we may instead approximate the requisite integral with a sum. Like knots in splines[32], the ω_j anchor the process at certain input locations. Bigger m means better approximation but greater computational cost.

Now let $\beta_j = \beta(\omega_j)$, for $j = 1, \ldots, \ell$, and the resulting (approximate yet continuous) latent function under GP prior may be constructed as

$$f(x) = \sum_{j=1}^{\ell} \beta_j k(x - \omega_j),$$

where $k(\cdot - \omega_j)$ is a smoothing kernel centered at ω_j. This f is a random function because the β_j are random variables. Choice of kernel is up to the practitioner, with the Gaussian above being a natural default. In spline/MARS regression, a "hockey-stick" kernel $k(\cdot - \omega_j) = (\cdot - \omega_i)_+ \equiv (\cdot - \omega_j) \cdot \mathbb{I}_{\{\cdot - \omega_j > 0\}}$ is a typical first choice.

For a concrete example, here's how one would generate from the prior under this formulation, choosing a normal density with mean zero and variance one as kernel and an evenly spaced grid of $\ell = 10$ locations ω_j, $j = 1, \ldots, \ell$. It's helpful to have the knot grid span a slightly longer range in the input domain (e.g., about 10% bigger) than the desired range for realizations.

---

[32]http://bobby.gramacy.com/surrogates/splines.html

```
ell <- 10
n <- 100
X <- seq(0, 10, length=n)
omega <- seq(-1, 11, length=ell)
K <- matrix(NA, ncol=ell, nrow=n)
for(j in 1:ell) K[,j] <- dnorm(X, omega[j])
```

The last line in the code above is key. Calculating the sum approximating integral (5.20) requires kernel evaluations at every pair of x and ω_j locations. To obtain a finite dimensional realization on an $n = 100$-sized grid, we can store the requisite evaluations in a 100×10 matrix. The final ingredient is random βs – the Gaussian white noise process. For each realization we need ℓ such deviates.

```
beta <- matrix(rnorm(3*ell), ncol=3)
```

To visualize three sample paths from the prior, the code above takes 3ℓ samples for three sets of ℓ deviates in total, stored in an $\ell \times 3$ matrix. The sum is most compactly calculated as a simple matrix–vector product between K and beta values. (Accommodating our three sets of beta vectors, the code below utilizes a matrix–matrix product.)

```
F <- K %*% beta
```

Figure 5.22 plots those three realizations, showing locations of knots ω_j as vertical dashed bars.

```
matplot(X, F, type="l", lwd=2, lty=1, col="gray",
  xlim=c(-1,11), ylab="f(x)")
abline(v=omega, lty=2, lwd=0.5)
```

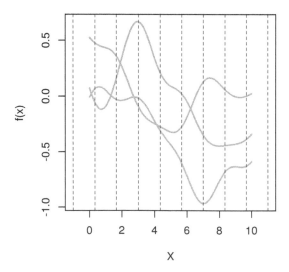

FIGURE 5.22: Three draws from a GP prior by convolution; knots ω_j indicated by vertical dashed bars.

Generating from priors is fun, but learning from data is where real interest lies. When training data come along, possibly observed under noise, the generating mechanism above suggests the following model for the purposes of inference. Let

$$y(x) = f(x) + \varepsilon, \quad \varepsilon \sim \mathcal{N}(0, \sigma^2),$$

which may be fit through an OLS regression, e.g.,

$$Y_n = K_n \beta + \varepsilon, \quad \text{where} \quad K_n^{ij} = k(x_i - \omega_j),$$

and x_i are training input x-values in the data, i.e., with $x_i^\top$ filling out rows of X_n. Whereas previously β was random, in the regression context their role changes to that of unknown parameter. Since that vector can be high dimensional, of length ℓ for ℓ knots, they're usually inferred under some kind of regularization, i.e., ridge, lasso, full Bayes or through random effects. Notice that while K_n is potentially quite large ($n \times \ell$), if ℓ is not too big we don't need to decompose a big matrix. Consequently, such a variation could represent a substantial computational savings relative to canonical GP regression.

So the whole thing boils down to an ordinary linear regression, but instead of using the X_n inputs directly it uses features K_n derived from distances between x_i and ω_j-values. By contrast, canonical GP regression entertains distances between all x_i and x_j in X_n. This swap in distance anchoring set is similar in spirit to inducing point/pseudo input/predictive process approaches, reviewed in greater depth in Chapter 9. To see it in action, let's return to the multi-tiered periodic example (5.19) from §5.3.4, originally from Higdon's (2002) convolution GP paper.

First build training data quantities.

```
n <- length(x)
K <- as.data.frame(matrix(NA, ncol=ell, nrow=n))
for(j in 1:ell) K[,j] <- dnorm(x, omega[j])
names(K) <- paste0("omega", 1:ell)
```

Then fit the regression. For simplicity, OLS is entertained here without regularization. Since n is quite a bit bigger than ℓ in this case, penalization to prevent numerical instabilities or high standard errors isn't essential. Naming the columns of K helps when using `predict` below.

```
fit <- lm(y ~ . -1, data=K)
```

Notice that an intercept is omitted in the regression formula above: we're assuming a zero-mean GP. Also it's worth noting that the $\hat{\sigma}^2$ estimated by `lm` is equivalent to $\hat{\tau}^2 \hat{g}$ in our earlier, conventional GP specification.

Prediction on a grid in the input space at $\mathcal{X} \equiv$ XX involves building out predictive feature space by evaluating the same kernel(s) at those new locations ...

```
xx <- seq(-1, 11, length=100)
KK <- as.data.frame(matrix(NA, ncol=ell, nrow=length(xx)))
for(j in 1:ell) KK[,j] <- dnorm(xx, omega[j])
names(KK) <- paste0("omega", 1:ell)
```

... and then feeding those in as `newdata` into `predict.lm`. It's essential that KK have the same column names as K.

```
p <- predict(fit, newdata=KK, interval="prediction")
```

Figure 5.23 shows the resulting predictive surface summarized as mean and 95% predictive interval(s).

```
plot(x, y)
lines(xx, p[,1])
lines(xx, p[,2], col=2, lty=2)
lines(xx, p[,3], col=2, lty=2)
```

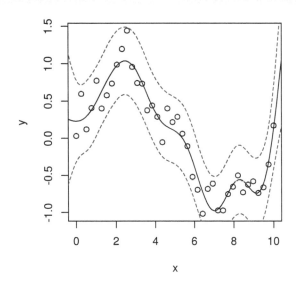

FIGURE 5.23: Posterior predictive under GP convolution.

That surface seems to agree with surfaces provided in Figure 5.21, which synthesized a grid of hyperparameter settings. Entertaining smaller lengthscales is a simple matter of providing a kernel with smaller variance. For example, the kernel below possesses an effective (square-root) lengthscale which is half the original, leading to a wigglier surface. See Figure 5.24.

```
for(j in 1:ell) K[,j] <- dnorm(x, omega[j], sd=0.5)
fit <- lm(y ~ . -1, data=K)
for(j in 1:ell) KK[,j] <- dnorm(xx, omega[j], sd=0.5)
p <- predict(fit, newdata=KK, interval="prediction")
plot(x, y)
lines(xx, p[,1])
lines(xx, p[,2], col=2, lty=2)
lines(xx, p[,3], col=2, lty=2)
```

Fixing the number ℓ and location of kernel centers, the ω_j's, and treating their common scale as unknown, inference can be performed with the usual suspects: likelihood (MLE or Bayes

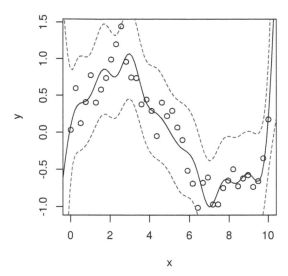

FIGURE 5.24: Predictive surface under a smaller kernel width/effective lengthscale; compare to Figure 5.23.

via least squares), CV, etc. Since this exposition is more of a side note, we'll leave details to the literature. A great starting point is the Ph.D. dissertation of Chris Paciorek (2003), with references and links therein. One feature that this method accommodates rather more gracefully than a canonical GP approach involves relaxations of stationarity, which is the main methodological contribution in Chris' thesis. Allowing kernels and their parameterization to evolve in space represents a computationally cheap and intuitive mechanism for allowing distance-based dynamics to vary geographically. The culmination of these ideas is packaged neatly by Paciorek and Schervish (2006). There has been a recent resurgence in this area with the advent of deep Gaussian processes (Dunlop et al., 2018; Damianou and Lawrence, 2013).

One downside worth mentioning is the interplay between kernel width, determining effective lengthscale, and density of the ω_j's. For fixed kernel, accuracy of approximation improves as that density increases. For fixed ω_j, however, accuracy of approximation diminishes if kernel width becomes narrower (smaller variance in the Gaussian case) because that has the effect of increasing kernel-distance between ω_j's, and thus distances between them and inputs X_n. Try the code immediately above with `sd=0.1`, for example. A kernel width of 0.1 may be otherwise ideal, but not with the coarse grid of ω_j's in place above; an order of magnitude denser grid (much bigger ℓ) would be required. At the other end, larger kernel widths can be problematic numerically, leading to ill-conditioned Gram matrices[33] $K_n^\top K_n$ and thus problematic decompositions when solving for $\hat{\beta}$. This can happen even when the column dimension ℓ is small relative to n.

Schemes allowing scale and density of kernels to be learned simultaneously, and which support larger effective lengthscales (even with fixed kernel density), require regularization and consequently demand greater computation as matrices K and KK become large and numerically unwieldy. Some kind of penalty on complexity, or shrinkage prior on β, is needed to guarantee a well-posed least-squares regression problem and to prevent over-fitting, as can happen in any setting where bases can be expanded to fit noise at the expense of signal. This issue is exacerbated as input dimension increases. Bigger input spaces lead to exponentially

---

[33]https://en.wikipedia.org/wiki/Gramian_matrix

increasing inter-point distances, necessitating many more ω_j's to fill out the void. The result can be exponentially greater computation and potential to waste those valuable resources over-fitting.

Speaking of higher input dimension, how do we do that? As long as you can fill out the input space with ω_j's, and increase the dimension of the kernel, the steps are unchanged. Consider a look back at our 2d data from earlier, which we conveniently saved as X2 and y2, in §5.2.4.

An $\ell = 10 \times 10$ dense grid of ω_j's would be quite big – bigger than the data size $n = 80$, comprised of two replicates of forty, necessitating regularization. We can be more thrifty by taking a page out of the space-filling design literature, using LHSs for knots ω_j's in just the same way we did for design X_n. R code below chooses $\ell = 20$ maximin LHS (§4.3) locations to ensure that ω_j's are as spread out as possible.

```
ell <- 20
omega <- maximinLHS(ell, 2)
omega[,1] <- (omega[,1] - 0.5)*6 + 1
omega[,2] <- (omega[,2] - 0.5)*6 + 1
```

Next build the necessary training data quantities. Rather than bother with a library implementing bivariate Gaussians for the kernel in 2d, code below simply multiplies two univariate Gaussian densities together. Since the two Gaussians have the same parameterization, this treatment is isotropic in the canonical covariance-based GP representation.

```
n <- nrow(X2)
K <- as.data.frame(matrix(NA, ncol=ell, nrow=n))
for(j in 1:ell)
  K[,j] <- dnorm(X2[,1], omega[j,1])*dnorm(X2[,2], omega[j,2])
names(K) <- paste0("omega", 1:ell)
```

Kernel-based features in hand, fitting is identical to our previous 1d example.

```
fit <- lm(y2 ~ . -1, data=K)
```

Now for predictive quantities on testing inputs. Code below re-generates predictive $\mathcal{X} = \mathtt{XX}$ values on a dense grid to ease visualization. Otherwise this development mirrors our build of training data features above.

```
xx <- seq(-2, 4, length=gn)
XX <- expand.grid(xx, xx)
KK <- as.data.frame(matrix(NA, ncol=ell, nrow=nrow(XX)))
for(j in 1:ell)
  KK[,j] <- dnorm(XX[,1], omega[j,1])*dnorm(XX[,2], omega[j,2])
names(KK) <- paste0("omega", 1:ell)
```

Since it's easier to show predictive standard deviation than error-bars in this 2d context, the code below provides se.fit rather than interval="prediction" to predict.lm.

```
p <- predict(fit, newdata=KK, se.fit=TRUE)
```

Figure 5.25 shows mean (left) and standard deviation (right) surfaces side-by-side. Training data inputs are indicated as open circles, and ω_j's as filled circles.

```
par(mfrow=c(1,2))
image(xx, xx, matrix(p$fit, ncol=gn), col=cols, main="mean",
  xlab="x1", ylab="x2")
points(X2)
points(omega, pch=20)
image(xx, xx, matrix(p$se.fit, ncol=gn), col=cols, main="sd",
  xlab="x1", ylab="x2")
points(X2)
points(omega, pch=20)
```

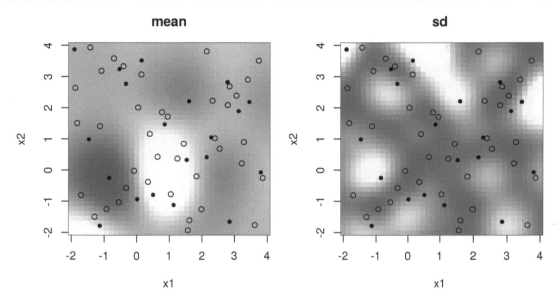

FIGURE 5.25: Convolution GP posterior predictive via mean (left) and standard deviation (right); compare with Figure 5.11. Design is indicated with open circles; knots as filled dots.

For my money, this doesn't look as good as our earlier results in Figure 5.11. The signal isn't as clear in either plot. Several explanations suggest themselves upon reflection. One is differing implicit lengthscale, in particular the one used immediately above is not fit from data. Another has to do with locations of the ω_j, and their multitude: $\ell = 20$. Notice how both mean and sd surfaces exhibit "artifacts" near some of the ω_j. Contrasts are most stark in the sd surface, with uncertainty being much higher nearby filled circles which are far from open ones, rather than resembling sausages as seen earlier. Such behavior diminishes with larger ℓ and when learning kernel widths from data, but at the expense of other computational and fitting challenges.

In two dimensions or higher, there's potential for added flexibility by parameterizing the full covariance structure of kernels: tuning $\mathcal{O}(m^2)$ unknowns in an m-dimensional input space, rather than forcing a diagonal structure with all inputs sharing a common width/effective lengthscale, yielding isotropy. A separable structure is a first natural extension, allowing each coordinate to have its own width. Rotations, and input-dependent scale (i.e., as a function of the ω_j) is possible too, implementing a highly flexible nonstationary capability if a sensible strategy can be devised to infer all unknowns.

The ability to specify a flexible kernel structure that can warp in the input domain (expand, contract, rotate) as inferred by the data is seductive. That was Higdon's motivation in his original 2001 paper, and the main subject of Paciorek's thesis. But details and variations, challenges and potential solutions, are numerous enough in today's literature (e.g., Dunlop et al., 2018), almost twenty years later, to fill out a textbook of their own. Unfortunately, those methods extend poorly to higher dimension because of the big ℓ required to fill out an m-dimensional space, usually $\ell \propto 10^m$ with $\mathcal{O}(\ell^3)$ computation. Why is this a shame? Because it's clearly desirable to have some nonstationary capability, which is perhaps the biggest drawback of the canonical (stationary) GP regression setup, as demonstrated below.

5.4.2   Limits of stationarity

If $\Sigma(x, x') \equiv k(x - x')$, which is what it means for a spatial process to be stationary, then covariance is measured the same everywhere. That means we won't be able to capture dynamics whose nature evolves in the input space, like in our motivating NASA LGBB example (§2.1). Recall how dynamics in lift exhibit an abrupt change across the sound barrier. That boundary separates a "turbulent" lift regime for high angles of attack from a relatively flat relationship at higher speeds. Other responses show tame dynamics away from mach 1, but interesting behavior nearby.

Taking a global view of the three-dimensional input space, LGBB lift exhibits characteristically nonstationary behavior. Locally, however, stationary dynamics reign except perhaps right along the mach 1 boundary, which may harbor discontinuity. How can we handle data of this kind? One approach is to ignore the problem: fit an ordinary stationary GP and hope for the best. As you might guess, ordinary GP prediction doesn't fail spectacularly because good nonparametric methods have a certain robustness about them, as demonstrated in several variations in this chapter. But that doesn't mean there isn't room for improvement.

As a simpler illustration, consider the following variation on the multi-scale periodic process[34] from §5.3.4.

```
X <- seq(0,20,length=100)
y <- (sin(pi*X/5) + 0.2*cos(4*pi*X/5)) * (X <= 9.6)
lin <- X>9.6
y[lin] <- -1 + X[lin]/10
y <- y + rnorm(length(y), sd=0.1)
```

The response is wiggly, identical to (5.19) from Higdon (2002) to the left of $x = 9.6$, and straight (linear) to the right. This example was introduced by Gramacy and Lee (2008a) as a cartoon mimicking LGBB behavior (§2.1) in a toy 1d setting. Our running 2d example from §5.1.2 was conceived as a higher-dimensional variation. To keep the discussion simpler here, we'll stick to 1d and return to the others in Chapter 9.

Consider the following stationary GP fit to these data using methods from laGP.

---

[34]https://www.sfu.ca/~ssurjano/hig02grlee08.html

```
gpi <- newGP(matrix(X, ncol=1), y, d=0.1, g=0.1*var(y), dK=TRUE)
mle <- jmleGP(gpi)
```

Above, isotropic routines are used rather than separable ones. It makes no difference in 1d. As an aside, we remark that the `jmleGP` function is similar to `mleGPsep` with argument `param="both"`; "j" here is for "joint", meaning both lengthscale and nugget. Rather than using a gradient over both parameters, as `mleGPsep` does, `jmleGP` performs a coordinate-wise, or profile-style, maximization iterating until convergence for one hyperparameter conditional on the other, etc. Sometimes this approach leads to more numerically stable behavior; `jmleGPsep` works similarly for separable Gaussian kernels.

Once hyperparameters have been estimated, prediction proceeds as usual.

```
p <- predGP(gpi, matrix(X, ncol=1), lite=TRUE)
deleteGP(gpi)
```

Now we're ready to visualize the fit, as provided by predictive mean and 95% intervals in Figure 5.26.

```
plot(X, y, xlab="x")
lines(X, p$mean)
lines(X, p$mean + 2*sqrt(p$s2), col=2, lty=2)
lines(X, p$mean - 2*sqrt(p$s2), col=2, lty=2)
```

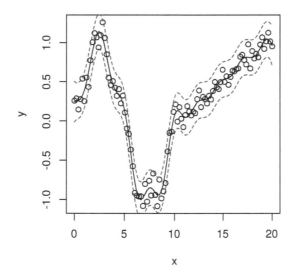

FIGURE 5.26: GP fit to an inherently nonstationary input–output relationship.

Observe how the predictive equations struggle to match disparate behavior in the two regimes. Since only one lengthscale must accommodate the entire input domain, the likelihood is faced with a choice between regimes and in this case it clearly favors the left-hand one. A wiggly fit to the right-hand regime is far better than a straight fit to left. As a result, wiggliness bleeds from the left to right.

Two separate GP fits would have worked much better. Consider ...

```
left <- X < 9.6
gpl <- newGP(matrix(X[left], ncol=1), y[left], d=0.1,
  g=0.1*var(y), dK=TRUE)
mlel <- jmleGP(gpl)
gpr <- newGP(matrix(X[!left], ncol=1), y[!left], d=0.1,
  g=0.1*var(y), dK=TRUE)
mler <- jmleGP(gpr, drange=c(eps, 100))
```

To allow the GP to acknowledge a super "flat" right-hand region, the lengthscale (d) range
has been extended compared to the usual default. Notice how this approximates a (more)
linear fit; alternatively – or perhaps more parsimoniously – a simple lm command could be
used here instead.

Now predicting ...

```
pl <- predGP(gpl, matrix(X[left], ncol=1), lite=TRUE)
deleteGP(gpl)
pr <- predGP(gpr, matrix(X[!left], ncol=1), lite=TRUE)
deleteGP(gpr)
```

... and finally visualization in Figure 5.27.

```
plot(X, y, xlab="x")
lines(X[left], pl$mean)
lines(X[left], pl$mean + 2*sqrt(pl$s2), col=2, lty=2)
lines(X[left], pl$mean - 2*sqrt(pl$s2), col=2, lty=2)
lines(X[!left], pr$mean)
lines(X[!left], pr$mean + 2*sqrt(pr$s2), col=2, lty=2)
lines(X[!left], pr$mean - 2*sqrt(pr$s2), col=2, lty=2)
```

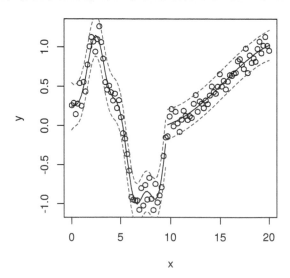

FIGURE 5.27: Partitioned GP fit to a nonstationary input–output relationship. Compare
to Figure 5.26.

Aesthetically this is a much better fit. Partitioning can be a powerful tool for flexible modeling, and not just for GPs. Divide-and-conquer can facilitate nonstationarity, through spatial statistical independence, and yield faster calculations with smaller datasets and much smaller $\mathcal{O}(n^3)$ matrix decompositions. The two fits can even be performed in parallel.

But how to know where to partition without knowing the data generating mechanism? It turns out that there are several clever solutions to that problem. Read on in Chapter 9. In the meantime, we shall see going forward that even stationary GPs have many interesting applications and success stories – as response surfaces, for optimization, calibration and input sensitivity analysis, and more – without worrying (much) about how ideal fits are.

5.4.3   Functional and other outputs

Focus has been on $Y(x) \in \mathbb{R}$; i.e., surrogate modeling for scalar, real-valued outputs. That will remain so throughout the text, but it's worthwhile commenting on what's available in greater generality. Modeling a small handful of real-valued outputs simultaneously is easy and hard at the same time. It's easy because treating each scalar output independently works surprisingly well. Gramacy and Lee (2009) modeled six LGBB outputs (§2.1) independently and without any perceivable ill-effect. It's hard because just about anything else you try can both be unwieldy and underwhelming. Effective, general-purpose multi-output surrogate modeling lies on the methodological frontier, as it were.

If there's a small number p of outputs following the same underlying spatial field, but experiencing correlated random shocks of varying magnitude, then *cokriging* (Ver Hoef and Barry, 1998) could help. The idea is to combine $p \times p$ covariances $\Sigma_p^{(Y)}$ with the usual $n \times n$ inverse distance-based ones $\Sigma_n = \tau^2(K_n + \mathbb{I}_n g)$ in a Kronecker layout[35]. Inference for $\Sigma_n^{(Y)}$ is relatively straightforward. MLE and Bayesian posterior are available in closed form conditional on Σ_n. Trouble is, this isn't a very realistic situation, at least not when data are generated through computer simulation. One exception may be when outputs differ from one another only in resolution or fidelity of simulations. Cokriging has been applied with some success in such *multifidelity* settings (Le Gratiet and Garnier, 2014; Le Gratiet and Cannamela, 2015).

The linear model of coregionalization (LMC; Journel and Huijbregts, 1978; Goovaerts, 1997) is a special case or generalization of cokriging depending upon your perspective. Sometimes cokriging, as described above, is referred to an intrinsic coregionalization model (ICM)[36]. As the name suggests, LMC allows for a more flexible, linear and covarying relationship between outputs. LMC's most prominent success stories in computer surrogate modeling involve simulators providing additional derivative information. Taylor's theorem justifies a linear structure. For examples, see Bompard et al. (2010) and references therein. For a machine learning perspective and Python implementation, see GPy[37].

Functional output is rather more common in our field. Simulators may provide realizations $Y(x, t) \in \mathbb{R}$ across an entire index set $t = 1, \ldots, T$, simultaneously for each x. I've chosen t to represent output indices because functions of time are common. Two-dimensional indexing for image data is also typical. Theoretically, such processes are easy to model with GP surrogates as long as indices are naturally ordered, or otherwise emit a reasonable set of

---

[35]https://en.wikipedia.org/wiki/Kronecker_product
[36]https://en.wikipedia.org/wiki/Kernel_methods_for_vector_output#Intrinsic_coregionalization_model_(ICM)
[37]https://sheffieldml.github.io/GPy/

pairwise distances so that off-the-shelf covariance kernels apply. In other words, treating the output index t as another set of input coordinates is an option. Some have taken to calling this a "left to right mapping": moving output indices from the left side of the (probability) conditioning bar to the right. A similar tack may be taken with small-p outputs (previous paragraph) as long as an appropriate kernel for mixed quantitative and qualitative inputs can be found (Qian et al., 2008; Zhang et al., 2018). The trouble is, this idea is hard to put into practice if $N = n \times T$ is big, as it would be with any nontrivial T. Working with $N \times N$ matrices becomes unwieldy except by approximation (Chapter 9), or when the design in x-space has special structure (e.g., high degrees of replication, as in Chapter 10).[38]

A more parsimonious approach leverages functional bases for outputs and independent surrogate modeling of weights corresponding to a small number of principal components of that basis (Higdon et al., 2008). This idea was originally developed in a calibration setting (§8.1), but has gained wide traction in a number of surrogate modeling situations. MATLAB software is available as part of the GPMSA toolkit (Gattiker et al., 2016). Fadikar et al. (2018) demonstrate use in a quantile regression setting (Plumlee and Tuo, 2014) for an epidemiological inverse problem pairing a disease outbreak simulator to Ebola data from Liberia. Sun et al. (2019b) describe a periodic basis for GP smoothing of hourly simulations of solar irradiance across the continental USA. A cool movie showing surrogate irradiance predictions over the span of a year can be found here[39].

Finally, how about categorical $Y(x)$? This is less common in the computer surrogate modeling literature, but GP classification remains popular in ML. See Chapter 3 of Rasmussen and Williams (2006). Software is widely available in Python (e.g., GPy[40]) and MATLAB/Octave (see gpstuff[41] Vanhatalo et al. (2012)). R implementation is provided in `kernlab` (Karatzoglou et al., 2018) and `plgp` (Gramacy, 2014). Bayesian optimization under constraints (§7.3) sometimes leverages classification surrogates to model binary constraints. GP classifiers work well here (Gramacy and Lee, 2011), but so too do other common nonparametric classifiers like random forests (Breiman, 2001). See §7.3.2 for details.

5.5   Homework exercises

These exercises give the reader an opportunity to explore Gaussian process regression, properties, enhancements and extensions, and related methods.

#1: Bayesian zero-mean GP

Consider the following data-generating mechanism $Y \sim \mathcal{N}_n(0, \tau^2 K_n)$ and place prior $\tau^2 \sim \text{IG}\left(\frac{a}{2}, \frac{b}{2}\right)$ on the scale parameter. Use the following parameterization of inverse gamma $\text{IG}(\theta; \beta, \alpha)$ density, expressed generically for a parameter $\theta > 0$ given shape $\alpha > 0$ and scale $\beta > 0$: $f(\theta) = \frac{\beta^\alpha}{\Gamma(\alpha)} \theta^{-(\alpha+1)} e^{-\beta/\theta}$, where Γ is the gamma function[42].

---

[38]Qian et al. (2008)'s method for categorical inputs exploits a dual relationship with multiple-output modeling. Kronecker structure in the resulting $N \times N$ matrices can make an otherwise unwieldy decomposition manageable.

[39]http://bobby.gramacy.com/solar/

[40]https://sheffieldml.github.io/GPy/

[41]https://research.cs.aalto.fi/pml/software/gpstuff/

[42]https://en.wikipedia.org/wiki/Gamma_function

a. Show that the IG prior for τ^2 is conditionally conjugate by deriving the closed form of the posterior conditional distribution τ^2 given all other hyperparameters, i.e., those involved in K_n.

b. Choosing $a = b = 0$ prescribes a reference prior (Berger et al., 2001, 2009) which is equivalent to $p(\tau^2) \propto 1/\tau^2$. This prior is improper. Nevertheless, derive the closed form of the posterior conditional distribution for τ^2 and argue that it's proper under a condition that you shall specify. Characterize the posterior conditional for τ^2 in terms of $\hat{\tau}^2$.

c. Now consider inference for the hyperparameterization of K_n. Derive the marginal posterior $p(K_n \mid Y_n)$, i.e., as may be obtained by integrating out the scale parameter τ^2 under the IG prior above; however, you may find other means equally viable. Use generic $p(K_n)$ notation for the prior on covariance hyperparameterization, independent of τ^2. How does this (log) posterior density compare to the concentrated log likelihood (5.8) under the reference prior?

d. Deduce the form of the marginal predictive equations $p(Y(x) \mid K_n, Y_n)$ at a new location x, i.e., as may be obtained by integrating out τ^2. Careful, they're not Gaussian but they're a member of a familiar family. How do these equations change in the reference prior setting?

#2: GP with a linear mean

Consider the following data-generating mechanism $Y \sim \mathcal{N}_n(\beta_0 + X_n\beta, \tau^2 K_n)$ where

- $K_n = C_n + g\mathbb{I}_n$,
- C_n is an $n \times n$ correlation matrix defined by a positive definite function $C_\theta(x, x')$ calculated on the n rows of X_n, and which has lengthscale hyperparameters θ,
- and g is a nugget hyperparameter, which must be positive.

There are no restrictions on the coefficients β_0 and β, except that the dimension m of β matches the column dimension of X_n.

a. Argue, at a high level, that this specification is essentially equivalent to the following semiparametric model $y(x) = \beta_0 + x^\top \beta + w(x) + \varepsilon$, and describe what each component means, and/or what distribution it's assumed to have.

b. Conditional on hyperparameters θ and g, obtain closed form expressions for the MLE $\hat{\tau}^2$, $\hat{\beta}_0$ and $\hat{\beta}$. *You might find it convenient to assume, for the purposes of these calculations, that X_n contains a leading column of ones, and that $\beta \equiv [\beta_0, \beta]$.*

c. Provide a concentrated (log) likelihood expression $\ell(\theta, g)$ that plugs-in expressions for $\hat{\tau}^2$, $\hat{\beta}_0$ and $\hat{\beta}$ (or a combined $\hat{\beta}$) which you derived above.

d. Using point estimates for $\hat{\beta}$, $\hat{\tau}^2$ and conditioning on θ and g settings, derive the predictive equations.

#3: Bayesian linear-mean GP

Complete the setup in #2 above with prior $p(\beta, \tau^2) = p(\beta \mid \tau^2)p(\tau^2)$ where $\beta \mid \tau^2 \sim \mathcal{N}_{m+1}(B, \tau^2 V)$ and $\tau^2 \sim \text{IG}\left(\frac{a}{2}, \frac{b}{2}\right)$. Notice that the intercept term β_0 is subsumed into β in this notation.

a. Show that the MVN prior for $\beta \mid \tau^2$ is conditionally conjugate by deriving the closed form of the posterior conditional distribution $\beta \mid \tau^2, Y_n$ and given all other hyperparameters, i.e., those involved in K_n.

b. Show that the IG prior for τ^2 is conditionally conjugate by deriving the closed form of the posterior conditional distribution $\tau^2 \mid \beta, Y_n$ and given all other hyperparameters, i.e., those involved in K_n.

c. Under the reference prior $p(\beta, \tau^2) \propto 1/\tau^2$, which is improper, how do the forms of the posterior conditionals change? Under what condition(s) are these conditionals still proper? (Careful, proper conditionals don't guarantee a proper joint.) Express the β conditional as function of $\hat{\beta}$ from exercise #2.

For the remainder of this question, parts #d–f, use the reference prior to keep the math simple.

d. Derive the marginalized posterior distribution $\tau^2 \mid Y_n$ and given K_n, i.e., as may be obtained by integrating out β, however you may choose to utilize other means. Express your distribution as a function of $\hat{\beta}$ and $\hat{\tau}^2$ from exercise #2.

e. Now consider inference for the hyperparameterization of K_n. Derive the marginal posterior $p(K_n \mid Y_n)$ up to a normalizing constant, i.e., as may be obtained by integrating out both linear mean parameter β and scale τ^2 under their joint reference prior. Use generic $p(K_n)$ notation for the prior on covariance hyperparameterization, independent of τ^2 and β. How does the form of this density compare to the concentrated log likelihood in #2c above? Under what condition(s) is this density proper?

f. Deduce the form of the marginal predictive equations $p(Y(x) \mid K_n, Y_n)$ at a new location x, i.e., as may be obtained by integrating out β and τ^2. Careful, they're not Gaussian but they're a member of a familiar family.

#4: Implementing the Bayesian linear-mean GP

Code up the marginal posterior $p(K_n \mid Y_n)$ from #3e and the marginal predictive equations $p(Y(x) \mid K_n, Y_n)$ from #3f and try them out on the Friedman data. Take the reference prior $p(\beta, \tau^2) \propto 1/\tau^2$ and define $p(K_n)$ as independent gamma priors on isotropic (Gaussian family) lengthscale and nugget as follows:

$$\theta \sim G(3/2, 1) \quad \text{and} \quad g \sim G(3/2, 1/2),$$

providing `shape` and `rate` parameters to `dgamma` in R, respectively.

i. Use the marginal posterior as the basis of a Metropolis–Hastings scheme for sampling from the posterior distribution of lengthscale θ and nugget g hyperparameters. Provide a visual comparison between these marginal posterior densities and the point estimates we obtained in the chapter. How influential was the prior?

ii. Use the marginal posterior predictive equations to augment Table 5.1's RMSEs and scores collecting out-of-sample results from comparators in §5.2.5–5.2.6. (You might consider more random training/testing partitions as in our bakeoff in §5.2.7, extended in #7 below.)

iii. Use boxplots to summarize the marginal posterior distribution of regression coefficients β. Given what you know about the Friedman data generating mechanism, how do these boxplots compare with the "truth"? You will need #3d and #3a for this part.

Suppose you knew, *a priori*, that only the first three inputs contributed nonlinearly to response. How would you change your implementation to reflect this knowledge, and how do the outputs/conclusions (#ii–iii) change?

#5: Matèrn kernel

Revisit noise-free versions of our 1d sinusoidal (§5.1.1 and Figure 5.3) and 2d exponential (§5.1.2 and Figure 5.5) examples with Matèrn $\nu = 3/2$ and $\nu = 5/2$ kernels (5.18). Extend concentrated log likelihood and gradient functions to learn an m-vector of lengthscale hyperparameters $\hat{\theta}$ using nugget $g = 0$ and no ϵ jitter. Define this separable Matèrn in product form as $k_{\nu,\theta}(r) = \prod_{\ell=1}^{m} k_{\nu,\theta_\ell}(r^{(\ell)})$ where $r^{(\ell)}$ is based on (original, not squared) distances calculated only on the ℓth input coordinate. Provide visuals of the resulting surfaces using the predictive grids established along with those examples. Qualitatively (looking at the visuals) and quantitatively (via Mahalanobis distance (5.7) calculated out of sample), how do these surfaces compare to the Gaussian kernel alternative (with jitter and with estimated lengthscale(s))?

For another sensible vectorized lengthscale option, see "ARD Matèrn" here[43]. ARD stands for "automatic relevance determination", which comes from the neural networks/machine learning literature, allowing a hyperparameter to control the relative relevance of each input coordinate. For the Gaussian family the two definitions, product form and ARD, are the same. But for Matèrn they differ ever-so-slightly.

#6: Splines v. GP

Revisit the 2d exponential data (§5.1.2 and Figure 5.5), and make a comparison between spline and GP predictors. For a review of splines, see the supplement linked here[44]. Generate a random uniform design of size $n = 100$ in $[-2, 4]^2$ and observe random responses under additive Gaussian error with a mean of zero and a standard deviation of 0.001. This is your training set. Then generate a dense 100×100 predictive grid in 2d, and obtain (again noisy) responses at those locations, which you will use as a testing set.

Ignoring the testing responses, use the training set to obtain predictions on the testing input grid under

  a. a spline model with a tensor product basis provided in splines2d.R[45];
  b. a zero-mean GP predictor with an isotropic Gaussian correlation function, whose hyperparameters (including nugget, scale, and lengthscale) are inferred by maximum likelihood;
  c. MARS in the `mda` package for R.

You may wish to follow the format in splines2d.R for your GP and MARS comparators. Consider the following benchmarking metrics:

  i. computation time for inference and prediction combined;
  ii. RMSE on the testing set.

Once you're satisfied with your setup using one random training/testing partition, put a "for" loop around everything and do 99 more MC repetitions of the experiment (for 100 total), each with novel random training and testing set as defined above. Make boxplots collecting results for #i–ii above and thereby summarize the distribution of those metrics over the randomized element(s) in the experiment.

---

[43]https://www.mathworks.com/help/stats/kernel-covariance-function-options.html
[44]http://bobby.gramacy.com/surrogates/splines.html
[45]http://bobby.gramacy.com/surrogates/splines2d.R

#7: MARS v. GP redux

Revisit the MARS v. GP bakeoff (§5.2.7) with five additional predictors.

i. MARS via `mda` with `degree=2`.
ii. MARS via `earth` with default arguments.
iii. MARS via `earth` with `degree=2`.
iv. Bayesian MARS via the `bass` function with default arguments from the `BASS` package (Francom, 2017) on CRAN.
v. Bayesian GP with jumps to the limiting linear model (LLM; Gramacy and Lee, 2008b) via `bgpllm` with default arguments in the `tgp` package on CRAN.

Rebuild the RMSE and time boxplots to incorporate these new predictors; ignore proper score unless you'd like to comb `tgp` and `BASS` documentation to figure out how to extract predictive covariance matrices, which are not the default.

`BASS` supports a simulated tempering[46] scheme to avoid Markov chains becoming stuck in local modes of the posterior. Devin Francom recommends the following call for best results on this exercise.

```
fit.bass <- bass(X, y, verbose=FALSE, nmcmc=40000, nburn=30000, thin=10,
   temp.ladder=(1+0.27)^((1:9)-1))
```

A similar importance tempering feature (Gramacy et al., 2010) is implemented in `tgp` and is described in more detail in the second package vignette (Gramacy and Taddy, 2010). Also see `?default.itemps` in the package documentation. The curious reader may wish to incorporate these gold standards for brownie points.

#8: Langley Glide-Back Booster

In this problem, revisit #6 on the "original" LGBB drag response.

```
lgbb <- read.table("lgbb/lgbb_original.txt", header=TRUE)
X <- lgbb[,1:3]
y <- lgbb$drag
```

However note that the scales of LGBB's inputs are heterogeneous, and quite different from #6. In the least, it'd be wise to code your inputs. For best results, you might wish to upgrade the isotropic GP comparator from #6 to a separable version.

A. Consider the subset of the data where the side-slip angle is zero, so that it's a 2d problem. Create a random training and testing partition in the data so that about half are for training and half for testing, and then perform exactly #a–c with #i–ii from #6, above, and report on what you find.
B. Put a "for" loop around everything and do 100 MC repetitions of the above experiment, each with novel random training and testing set as defined above. Then, make boxplots for RMSE results collected for #i–ii above and thereby summarize the distribution of those metrics over randomized element(s) in the experiment.

---

[40]https://en.wikipedia.org/wiki/Parallel_tempering

C. Now return to the full set of data, i.e., for all side-slip angles. Since the number of observations, n, is bigger than 3000, you won't be able to get very far with a random 50:50 split. Instead, do a 20:80 split or whatever you think you can manage. Also, it'll be tough to do a spline approach[47] with a tensor product basis in 3d, so perhaps ignore comparator #6a unless you're feeling particularly brave. Otherwise perform exactly #a–c with #i–ii from #6, above, and report on what you find. (That is, do like in part #A, without the "for" loop in #B.)

D. Repeat #C with a GP scheme using an axis-aligned partition. Fit two GP models in 3d where the first one uses the subset of the training data with `mach < 2` and the second uses `mach >= 2`. Since you're dividing-and-conquering, you can probably afford a 50:50 split for training and testing.

#9: Convolution kernel width

Revisit the 1d multi-tiered periodic example (5.19) as treated by convolution in §5.4.1.

a. Write down a concentrated log likelihood for the kernel width parameter θ, notated as `sd` in the example, and provide an implementation in code.

b. Plot the concentrated log likelihood over the range `sd` $\equiv \theta \in [0.4, 4]$ and note the optimal setting.

c. Verify the result with `optimize` on your concentrated log likelihood.

d. How does the value you inferred compare to the two settings entertained in §5.4.1? How do the three predictive surfaces compare visually?

---

[47]http://bobby.gramacy.com/surrogates/splines.html

6

Model-based Design for GPs

Chapter 4 offered a model-free perspective on design, choosing to spread out X_n where Y_n will be observed, with the aim of fitting a flexible nonlinear, nonparametric, but ultimately generic regression. Gaussian processes (GPs) were on our mind, but only as a notion. Now that we know more about them, we can study how one might design an experiment differently – optimally, in some statistical sense – if committed to a GP and particular choices of its formulation: covariance family, hyperparameters, etc. At first that will seem like a subtle shift, in part because the search algorithms involved are so similar. Ultimately emphasis will be on sequential design – or active learning[1] in machine learning (ML) jargon – motivated by drawbacks inherent in the canonical setting of model-based one-shot (or batch) design, with an eye towards the inherently sequential nature of surrogate-assisted numerical (Bayesian) optimization in Chapter 7.

We'll see how GPs encourage space-filling designs that aren't much different than ones we already know about from Chapter 4, like Latin hypercube sampling (LHS; §4.1) and maximin (§4.2), potentially representing overkill on the one hand (greater computational effort), and unnecessary risk on the other (calculating an "optimally pathological" design that precludes, rather than facilitates, learning). Although a sequential approach may be sub-optimal in theory, we'll see that in practice it helps to avoid such pathologies. Baby steps provide an opportunity to learn about response surface dynamics while simultaneously improving quality of fit.

Before diving in, and at the risk of being redundant, it's worth reiterating the following. Without knowing much about the response surface you intend to model *a priori*, a space filling/Chapter 4 criteria – i.e., LHS, maximin distance, or a hybrid – represents a good choice indeed. Methods herein are not better by default, but they do offer potential for improvements in learning when used correctly, in particular when you do know something about the response surface you intend to model. Plus, our technical development of the subject provides valuable insight into how data influence fitted model and subsequent prediction, which makes the presentation relevant even if most of the designs are ultimately dismissed on practical grounds.

Model-based alternatives all assume something, like a GP kernel hyperparameterization, which is where the risk comes from. Before data are collected, what you think you know is often wrong and, if too highly leveraged, could hurt in unexpectedly bad ways. Seemingly modest assumptions can be amplified into foregone conclusions after data/responses arrive. One way around that risk is to first learn hyperparameters from a small (model-free) space-filling design. Such bootstrapping, in the colloquial rather than statistical sense, is sensible but no free lunch. Spreading points out biases inference towards longer lengthscales. A new sense of space-filling, in terms of pairwise distances, may be more appropriate for most kernel choices.

---

[1] https://en.wikipedia.org/wiki/Active_learning

6.1   Model-based design

A model-based design is one where the model says what X_n it wants according to a criterion targeting some aspect of its fit. Example targets include the quality of estimates for parameters or hyperparameters, or accuracy of predictions at particular inputs out-of-sample, or over the entire input space. With a first-order linear model, you may recall from an elementary design course that maximizing spread in X_n maximizes leverage and therefore minimizes the standard error of $\hat{\beta}$-values. An optimal design for learning regression coefficients places inputs at the corners of the experimental region, for better or worse.

Linear models are highly parametric. Parameter settings are intimately linked to the underlying predictor. With GPs, a nonparametric model whose hyperparameterization is only loosely connected to the predictive distribution, you might think a different strategy is required. But principles are largely the same. Without delving into the subtlety of myriad alternatives, we'll skip to the two most popular approaches. Those are to choose X_n that either

1. "maximize learning" from prior to posterior, thinking about the Bayesian interpretation; or
2. minimize predictive variance averaged over a region of interest (e.g., the entire input space).

Let's begin by looking at those two options in turn. The general program is to specify a criterion $J(X_n)$, and optimize that criterion with respect to X_n. Assuming maximization,

$$X_n^\star = \mathrm{argmax}_{X_n} J(X_n),$$

which is typically solved numerically. Except when essential for clarity, I shall generally drop the $\star$ superscript in $X_n^\star$, indicating an optimally chosen design.

6.1.1   Maximum entropy design

The *entropy*[2] of a density $p(x)$ is defined as

$$H(X) = -\int_{\mathcal{X}} p(x) \log p(x)\, dx.$$

Entropy is larger when $p(x)$ is more uniform. You can think of it as a measure of surprise in random draws from the distribution corresponding to p. A uniform draw always yields a surprising, unpredictable, random value. Entropy is maximized for this choice of p. At the other end of the spectrum, a point-spike p, where all density is concentrated on a single point, offers no surprise. A draw from such a distribution yields the same value every time. Its entropy is zero.

Information is the negative of entropy, $I = -H$. More information is less uniformity, less surprise. To see how information and entropy can relate to design, consider the following. Suppose $\mathcal{X}$ is a fixed, finite set of points in the input space. Place latent functions F at $\mathcal{X}$ under a GP prior $p(F \mid \mathcal{X})$; see §5.3.2. Let $I_{\mathcal{X}}$ denote the information of that prior. Now,

---

[2]https://en.wikipedia.org/wiki/Entropy

denote F restricted to $X \subset \mathcal{X}$ as F_X with implied prior $p(F_X \mid X)$ and information I_X. For a particular choice $X_n \subset \mathcal{X}$, we have that

$$I_{\mathcal{X}} = I_{X_n} + \mathbb{E}_{F_{X_n}}\{I_{\mathcal{X}-X_n|D_n}\}, \tag{6.1}$$

where $D_n = (X_n, Y_n)$ is the completion of X_n with a noise-augmented F_n. Eq. (6.1) partitions information between the amount soaked up, and amount left over after selecting X_n. Choosing X_n to maximize the expected information in prediction of latent function values, i.e., the amount soaked up by second term, is equivalent to minimizing the information I_{X_n} left over (maximizing entropy, H_{X_n}) in the distribution of F_{X_n}. A solution to that optimization problem is a so-called *maximum entropy (or maxent) design*. For more details see Shewry and Wynn (1987), who introduced the concept as a design criterion for spatial models. Maxent was appropriated for computer experiments by Currin et al. (1991) and Mitchell and Scott (1987).

That distribution – either for F_{X_n} or it's noisy analog for $D_n = (X_n, Y_n)$, depending on your preferred interpretation (§5.3.2) – ultimately involves

$$Y_n \sim \mathcal{N}_n(0, \tau^2 K_n) \quad \text{with} \quad K_n^{ij} = C_\theta(x_i, x_j) + g\delta_{ij}. \tag{6.2}$$

MVN conditionals yield the posterior predictive $Y(x) \mid D_n$. One can show that the entropy of that distribution (6.2), for Y_n observed at X_n, is maximized when $|K_n|$ is maximized. For a complete derivation, see Section 6.2.1 of Santner et al. (2018). Recall that K_n depends on design X_n, usually through inverse exponentiated pairwise squared distances.

K_n and thus $|K_n|$ also depend on hyperparameters, exemplified by θ and g in Eq. (6.2), so a maximum entropy design is hyperparameter dependent. This creates a chicken-or-egg problem because we hope to use data, which we've not yet observed since we're still designing the experiment, to learn hyperparameter settings. We'll see shortly that the choice of lengthscale θ can have a substantial impact on maximum entropy designs. The role of the nugget g is more nuanced.

It might help to first perform an initial or seed experiment, perhaps using a small space-filling design from Chapter 4, to help choose sensible values. Rather than discard that data after using it to learn hyperparameter settings, one might instead search for a *sequential maximum entropy* design (§6.2). But we're getting a little ahead of ourselves. Suppose we already had some fortuitously chosen hyperparameter settings to work with. What would a maxent design look like?

Our stochastic exchange `mymaximin` implementation (§4.2.1) may easily be altered to optimize $|K_n|$. Note that the code uses $\log|K_n|$ for greater numerical stability, assumes a separable Gaussian family kernel, but allows the user to tweak default settings of its hyperparameterization.

```
library(plgp)
maxent <- function(n, m, theta=0.1, g=0.01, T=100000)
 {
  if(length(theta) == 1) theta <- rep(theta, m)
  X <- matrix(runif(n*m), ncol=m)
  K <- covar.sep(X, d=theta, g=g)
  ldetK <- determinant(K, logarithm=TRUE)$modulus
```

```
  for(t in 1:T) {
    row <- sample(1:n, 1)
    xold <- X[row,]
    X[row,] <- runif(m)
    Kprime <- covar.sep(X, d=theta, g=g)
    ldetKprime <- determinant(Kprime, logarithm=TRUE)$modulus
    if(ldetKprime > ldetK) { ldetK <- ldetKprime
    } else { X[row,] <- xold }
  }
  return(X)
}
```

As with `mymaximin` this is a simple, yet inefficient implementation from a computational perspective. Many proposed swaps will be rejected either because the outgoing point (`xold`) isn't that bad, or because the incoming `runif(m)` location is chosen too clumsily. Rules of thumb about how to make improvements here are harder to come by, however. One reason is that the effect of hyperparameterization is more difficult to intuit. Heuristics which favor swapping out points with small Euclidean distances can indeed reduce rejections. Derivatives can be helpful for local refinement. A homework exercise (§6.4) asks the curious reader to entertain such enhancements. In spite of its inefficiencies, `maxent` as coded above works well in the illustrative examples below.

First let's try two input dimensions, using the default hyperparameterization.

```
X <- maxent(25, 2)
```

Figure 6.1 offers a visual.

```
plot(X, xlab="x1", ylab="x2")
```

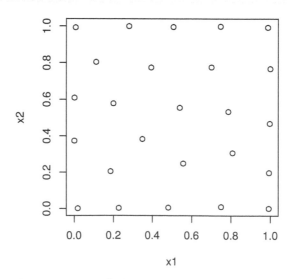

FIGURE 6.1: A maximum entropy design.

What do we see? Actually, it's not much different than a maximin design, at least qualitatively.

Points are, for the most part, all spaced out. Lots of sites have been pushed to the boundary, and marginals are a bit clumpy. The number of unique settings of x_1 and x_2 is low, but no "worse" than with maximin. Worse is in quotes because it can't really be worse. After all, this design is the best for GP learning in a certain sense. Perhaps having a degree of axis alignment is beneficial when it comes to learning – to maximizing information – even if it would seem to be disadvantageous on an intuitive level, or perhaps on more pragmatic terms. For example, if we had doubts about the model, or later decide to study a subset of the inputs (say only x_1), the design in Figure 6.1 is sub-optimal. LHS (§4.1) would fare much better.

One reason for the high degree of similarity between maxent and maximin designs is that the default **maxent** hyperparameterization is isotropic (i.e., radially symmetric), using the same θ_k-values in each input direction, $k \in \{1, 2\}$. So both are using the same Euclidean distances under the hood. On the rare occasion where we prefer a Franken-kernel (§5.3.3), rather than the friendly Gaussian, I suppose we could end up with a very interesting looking maximum entropy design. In a more common separable (Gaussian) setup, one might speculate desire for more spread in some directions rather than others through θ_k. Whether or not that's a good idea depends upon confidence in whatever evidence led to preferring differing lengthscales before collecting any data.

To illustrate, consider a setup where we knew we wanted the lengthscale for x_2 to be five times longer than for x_1, keeping the same value as before.

```
X <- maxent(25, 2, theta=c(0.1, 0.5))
```

Accordingly, Figure 6.2 reveals a design which has more "truly distinct" values in the first coordinate than the second one.

```
plot(X, xlab="x1", ylab="x2")
```

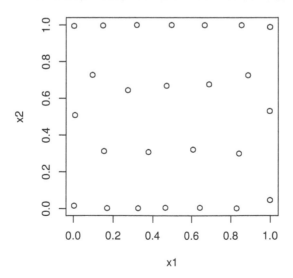

FIGURE 6.2: Maximum entropy design under a separable covariance structure.

This makes sense: more unique values, or denser sampling more generally, are needed in x_1 compared to x_2 because correlation decays more rapidly in the former compared to the latter.

Shorter pairwise distances are needed in x_1 to better "anchor" the predictive distribution, at new x away from X_n-values, in that coordinate. If axis alignment in design benefits GP learning, then the behavior above represents a logical extension to settings where inputs interface unequally with the response.

Surely, you might say, I can measure distance a little differently in a maximin design and accomplish the same effect. Probably. But there's comfort in knowing the design obtained is optimal for the specified kernel and hyperparameter setting. Maximin designs make no promises vis-a-vis the GP being fit, except in a certain asymptotic sense where an equivalence may be drawn (Johnson et al., 1990; Mitchell et al., 1994).

In general, model-based optimality comes at potentially substantial computational cost. Each iteration of `maxent` involves a determinant calculation that requires decomposing an $n \times n$ matrix, at $\mathcal{O}(n^3)$ cost. Maximin, by contrast, is only $\mathcal{O}(n^2)$ via pairwise distances. Maximin can be further improved to $\mathcal{O}(n)$, which you may have discovered in an exercise from §4.4. A similar trick can be performed with the log determinant to get an $\mathcal{O}(n^2)$ implementation when proposing a change in just one coordinate, leveraging the following decomposition.

$$\log |K_n| = \log |K_{n-1}| + \log(1 + g - C_\theta(x_n, X_{n-1}) K_{n-1}^{-1} C_\theta(X_{n-1}, x_n)) \tag{6.3}$$

The origin of this result – see Eq. (6.10) – will be discussed in more detail when we get to sequential updating of GP models later in §6.3. To be useful in the context of maxent search it must be applied twice: first backward (a "downdate") for the input being removed, and then forward for the new one being added.

Before moving on, consider application in 3d.

```
X <- maxent(25, 3)
```

Figure 6.3 shows the position of all pairs of inputs with numbers indexing rows of X_n.

```
Is <- as.list(as.data.frame(combn(ncol(X),2)))
par(mfrow=c(1,length(Is)))
for(i in Is) {
  plot(X[,i], xlim=c(0,1), ylim=c(0,1), type="n",
    xlab=paste0("x", i[1]), ylab=paste0("x", i[2]))
  text(X[,i], labels=1:nrow(X))
}
```

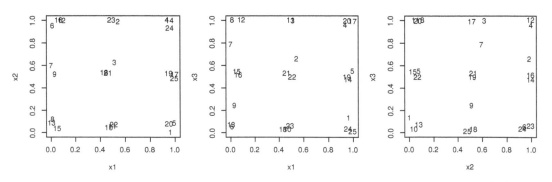

FIGURE 6.3: Projections of pairs of inputs involved in a 3d maximum entropy design.

The picture here is pretty much the same as for maximin design. Projections down into lower dimension don't enjoy any uniformity, so careful inspection across pairs is required to check that points are indeed well-spaced in the 3d study region. Observe that two indices which look close in one projection pair are not close in the others. Still, the outcome is somewhat disappointing on an aesthetic level.

6.1.2 Minimizing predictive uncertainty

As the basis of another criterion, consider predictive uncertainty. At a particular location $x \in \mathcal{X}$, the predictive variance is

$$\sigma_n^2(x) = \tau^2[1 + g - k_n^\top(x)K_n^{-1}k_n(x)] \quad \text{where} \quad k_n(x) \equiv k(X_n, x).$$

This is the same as what some call *mean-squared prediction error (MSPE)*:

$$\text{MSPE}[\hat{y}(x)] = \mathbb{E}\{(\hat{y} - Y(x))^2\} \equiv \sigma_n^2(x).$$

An *integrated MSPE (IMSPE)* criterion is defined as MSPE (divided by τ^2), averaged over the entire input region $\mathcal{X}$, which is expressed below as a function of design X_n.

$$J(X_n) = \int_{\mathcal{X}} \frac{\sigma_n^2(x)}{\tau^2} w(x)\, dx \tag{6.4}$$

Above, function $w(\cdot)$ is an optional, user-specified non-negative weight on inputs satisfying $\int_{\mathcal{X}} w(x)\, dx = 1$. In the simplest setup, $w(x) \propto 1$, bestowing equal importance to predicting well at all locations in $\mathcal{X}$. It's worth noting that the IMSPE criterion is a generalization of A-optimality[3] from classical design.

That's an m-dimensional integral for m-dimensional $\mathcal{X}$. Fortunately if $\mathcal{X}$ is rectangular, and where covariance kernels take on familiar forms such as isotropic or separable Gaussian and Matèrn $\nu \in \{3/2, 5/2, \dots\}$, closed forms are analytically tractable, or at least nearly so (e.g., depending on fast/accurate numerical evaluations of an error function/standard Gaussian distribution function, say.) Details are left to Eq. (10.9) and §10.3.1, on heteroskedastic (i.e., input-dependent noise) GPs, where the substance of such developments is of greater value to the narrative. Here it represents too much of a digression.

We will, however, borrow the implementation in the supporting `hetGP` package (Binois and Gramacy, 2019) on CRAN. The relevant function is called `IMSPE`. Since heteroskedastic GPs involve a few more bells and whistles, the function below strips down `IMSPE`'s interface somewhat so that the setup better matches our ordinary GP setting.

```
library(hetGP)
imspe.criteria <- function(X, theta, g, ...)
 {
  IMSPE(X, theta=theta, Lambda=diag(g, nrow(X)), covtype="Gaussian",
    mult=rep(1, nrow(X)), nu=1)
 }
```

---

[3]https://en.wikipedia.org/wiki/Optimal_design#Minimizing_the_variance_of_estimators

In case you're curious, hetGP involves an n-vector of nuggets (in Lambda), supports the three covariance structures listed above, and nu is our τ^2. The mult argument is a discussion for Chapter 10. Ellipses (. . .) are not used in what immediately follows below, but will come in handy later when we try an alternative approximation.

Now we're ready to use IMSPE in a design search. R code below is ported from maxent (§6.1.1) with modifications to minimize rather than maximize.

```
imspe <- function(n, m, theta=0.1, g=0.01, T=100000, ...)
  {
    if(length(theta) == 1) theta <- rep(theta, m)
    X <- matrix(runif(n*m), ncol=m)
    I <- imspe.criteria(X, theta, g, ...)

    for(t in 1:T) {
      row <- sample(1:n, 1)
      xold <- X[row,]
      X[row,] <- runif(m)
      Iprime <- imspe.criteria(X, theta, g, ...)
      if(Iprime < I) { I <- Iprime
      } else { X[row,] <- xold }
    }
    return(X)
}
```

Like maxent, IMSPE depends on θ and g hyperparameters (and to a lesser extent on $\tau^2 \equiv$ nu), so again calculations require flops in $\mathcal{O}(n^3)$. Similar shortcuts can reduce it to $\mathcal{O}(n^2)$ per iteration, or better; more soon in §6.3, Eq. (6.9). A homework exercise (§6.4) guides the curious reader through suggestions for a more efficient implementation, in particular leveraging other IMSPE-related subroutines from the hetGP package.

Let's illustrate in 2d. The code below provides the search . . .

```
X <- imspe(25, 2)
```

. . . followed by a visualization in Figure 6.4 of the $X_n = $ X that was returned.

```
plot(X, xlab="x1", ylab="x2", xlim=c(0,1), ylim=c(0,1))
```

You can see that the resulting design is quite similar to maxent or maximin analogs, but there's one important distinction: chosen sites avoid the boundary of the input space. There's an intuitive explanation for this. IMSPE considers variance integrated over the entire input space (6.4). Design sites at the boundary don't "cover" that space as efficiently as interior ones. Points on a boundary in 2d cover half as much of the input space as (deep) interior ones do. Points in the corner of a 2d space, i.e., at the intersection of two boundaries, cover one quarter of the space compared to ones in the interior. Thus boundary locations are far less likely to be chosen by IMSPE compared to maxent, say.

This effect is even more pronounced in higher input dimension. Consider 3d.

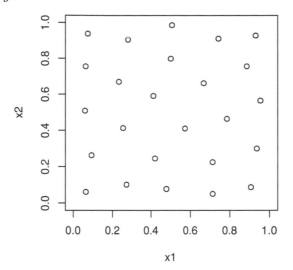

FIGURE 6.4: IMSPE optimal design.

```
X <- imspe(25, 3)
```

Our usual triplet of projections follows in Figure 6.5.

```
Is <- as.list(as.data.frame(combn(ncol(X),2)))
par(mfrow=c(1,length(Is)))
for(i in Is) {
  plot(X[,i], xlim=c(0,1), ylim=c(0,1), type="n",
    xlab=paste0("x", i[1]), ylab=paste0("x", i[2]))
  text(X[,i], labels=1:nrow(X))
}
```

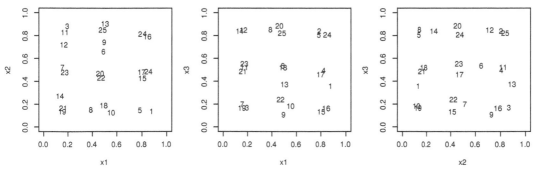

FIGURE 6.5: Two-dimensional projections of a 3d IMSPE design.

Compared to the 2d version, selected sites are even more "off the boundary", but otherwise positioning behavior is quite similar to maxent or maximin. Many practitioners find these designs to be more aesthetically pleasing than the maxent analog. Points off of the boundary make sense, and the criteria itself is easier to intuit. Desire for a design which predicts equally well everywhere is easy to justify to non-experts. One downside to IMSPE, however, is that when good predictions are desired in non-rectangular regions of the input space, or where that space is not weighted equally (i.e., non-constant $w(\cdot)$ in Eq. (6.4)), no known closed form calculation is available.

A simple approximation offers far greater flexibility, but implies a sometimes limiting accuracy-versus-computation trade-off. Our presentation features this approximation in an unfavorable light, however a sequential analog in §6.2.2 and optimization in §7.2.3 fares better, justifying its prominence here. This approach also plays a key role in local GP approximation (§9.3), where the goal is to develop accurate predictors for particular inputs, and sets of inputs lying on submanifolds of the input space.

The idea is to replace the integral in Eq. (6.4) with a (possibly weighted) sum over a reference grid in $\mathcal{X}$, implementing a poor-man's quadrature. Reference grids need not be regular; or they could follow a space-filling construction. So the tail doesn't wag the dog, a cheap LHS may make more sense here than an expensive maximin design. Or the grid can be random, even weighted according to $w(\cdot)$. Many variations supported by this approximation amount to changes in the form or nature of reference grids. Grid density is intimately linked to approximation accuracy.

Concretely, the idea is

$$
J(X_n) = \int_{\mathcal{X}} \frac{\sigma_n^2(x)}{\tau^2} w(x) \, dx \approx
\begin{cases}
\dfrac{1}{T} \displaystyle\sum_{t=1}^{T} \dfrac{\sigma_n^2(x_t)}{\tau^2} w(x_t) & x_t \sim \mathrm{Unif}(\mathcal{X}), \text{ or} \\[3ex]
\dfrac{1}{T} \displaystyle\sum_{t=1}^{T} \dfrac{\sigma_n^2(x_t)}{\tau^2} & x_t \sim \mathcal{W}(\mathcal{X}),
\end{cases}
\tag{6.5}
$$

where $\mathcal{W}(\mathcal{X})$ is the measure of $w(\cdot)$ applied in the input domain $\mathcal{X}$.

To implement and illustrate this scheme, the subroutine below calculates GP predictive variance at reference locations `Xref` for design `X` $= X_n$ and then approximates the integral (6.4) by a mean over `Xref`. The function below re-defines our `imspe.criteria` from above, and ellipses (...) allows re-use of the `imspe` searching function above under the new approximate criterion.

```
imspe.criteria <- function(X, theta, g, Xref)
  {
  K <- covar.sep(X, d=theta, g=g)
  Ki <- solve(K)
  KXref <- covar.sep(X, Xref, d=theta, g=0)
  return(mean(1 + g - diag(t(KXref) %*% Ki %*% KXref)))
  }
```

To apply in 2d, code below creates a regular grid (also in 2d) and passes that into `imspe` as reference locations `Xref`.

```
g <- expand.grid(seq(0,1,length=10), seq(0,1, length=10))
X <- imspe(25, 2, Xref=g)
```

Figure 6.6 shows the resulting design as open circles, with the grid of reference locations indicated as smaller closed gray dots.

```
plot(X, xlab="x1", ylab="x2", xlim=c(0,1), ylim=c(0,1))
points(g, pch=20, cex=0.25, col="gray")
```

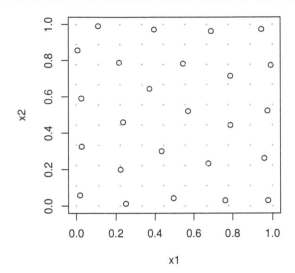

FIGURE 6.6: Approximate IMSPE design (6.5) by averaging over a reference set (gray dots). Compare to Figure 6.4.

This is a good approximation to Figure 6.4's closed-form analog, but not perfect. You might say it's missing a distinct feature of IMSPE-based design: avoiding boundaries. Design sites are not perfectly aligned with the boundary, but nearly so. Trouble is, design locations want to be close to reference set elements, and the reference set has almost as many points on the boundary as in the interior. A continuous analog, by contrast, would put measure zero on the boundary (a 1d manifold) relative to the full volume of the space (a 2d surface).

A different setting of the hyperparameters (say `theta=0.5`) reveals other inefficiencies due to discrete `Xref`. I encourage the curious reader to try this offline. Such drawbacks are an artifact of proposed random swaps having two hurdles to overcome in order to be accepted: 1) be farther from the previous "closest"; 2) but still just as close to a reference location. In the limit of a fully dense reference grid, i.e., an integral over an uncountable space, this second hurdle is always surmounted. Search based on our approximate criteria is more discrete than its closed form cousin, or even than maxent, and thus more difficult to solve. (Discrete and mixed continuous-discrete optimizations are harder than purely continuous ones.) A denser grid offers a partial solution, making the approximate criteria more accurate (at greater computational expense) but the nature of search is no less discrete.

Pivot now to the positive and consider how the approximate method can be applied in greater generality. What if we make the reference set occupy a smaller non-regular space? For example, as a somewhat contrived but certainly illustrative scenario, suppose we're interested in predicting within intersecting ellipsoids centered on $(x_1, x_2) = (0.25, 0.25)$. Code below utilizes a random reference grid from two bivariate Gaussians meeting that description, effectively up-weighting locations closer to $(0.25, 0.25)$; an implicit $w(\cdot)$.

```
Xref <- rmvnorm(100, mean=c(0.25, 0.25),
  sigma=0.005*rbind(c(2, 0.35), c(0.35, 0.1)))
```

```
Xref <- rbind(Xref,
  rmvnorm(100, mean=c(0.25, 0.25),
    sigma=0.005*rbind(c(0.1, -0.35), c(-0.35, 2)))))
X <- imspe(25, 2, Xref=Xref)
```

Figure 6.7 shows the chosen design and reference set.

```
plot(X, xlab="x1", ylab="x2", xlim=c(0,1), ylim=c(0,1))
points(Xref, pch=20, cex=0.25, col="gray")
```

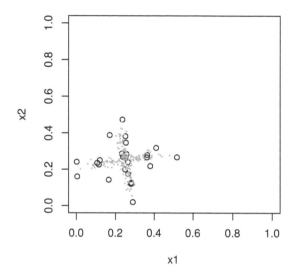

FIGURE 6.7: Approximate IMSPE design (6.5) with a reference set concentrated on a target study area.

Observe how the criteria organically attracts design locations to the reference set, but otherwise encourages some spread among selected sites. Spacing isn't perfect, but that can be attributed to the nature of the approximate criteria, and to imperfect stochastic search. Both are improved with a larger reference set and more stochastic exchange proposals, at the expense of further computation. More efficient variations stem from a sequential design adaptation, presented shortly, and from a derivative-based search, which will have to wait until §9.3.5. Notice that some sites are chosen outside of the outermost Xref locations. It's tempting to attribute this to approximation inefficiency, but in fact that phenomena is a direct consequence of the underlying IMSPE criterion. IMSPE resolves a tension between spreading out sites relative to one another, yet having them close to $\mathcal{X}$. Recall that we have a stationary model, meaning that spread (on multiple scales) in the design is a prime player, even if we're only interested in predicting locally, say in a sub-region of the input space. A more detailed discussion is left to §9.3.

6.2 Sequential design/active learning

A *greedy*[4], sequential approach to design is not just easier, it can also be more practical than the one-shot approach presented above. In many situations, selecting one design point at a time works better than static, single-batch design. For the technical junkie there are connections to non-sequential analogs, framed as approximations, but to settings where "true" hyperparameterization is known. Truth is in quotes since that's never known in practice.

Sequential selection is also more computationally reasonable, being both better behaved numerically and also faster. That's not to say that sequential methods always use fewer flops than batch selection. Often they do not. Yet extra computation versus alternatives (e.g., additional matrix decompositions), in situations where it's warranted, have redeemed value in terms of the guidance that appropriately estimated quantities, tempered over many iterations of design acquisition, provide to searches that target both modeling and design goals. It's easy to forget that design and modeling go hand-in-hand. Design impacts inference and calculations thereupon impact design, and all too often that's not acknowledged when formally describing design goals. In case it's not obvious, I'm uncomfortable with the idea of committing to a design before collecting (a modest amount of) data.

Figure 6.8 summarizes a sequential design setup. Although there are variations, here I shall emphasize the simple setup of augmenting an initial design by one, repeated until a desired size is reached.

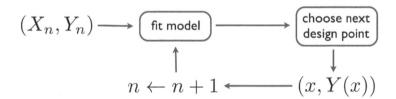

FIGURE 6.8: Diagram of sequential design/active learning/design augmentation.

To supplement that cartoon, consider Algorithm 6.1. The development here is a little backwards, saying what the algorithm is before delving into its key components, in particular Step 3 where a criterion is optimized to choose the next design element. Figure 6.8 is suggestive of indefinite updating, augmenting a design by one in each pass through the circuit. Algorithm 6.1 assumes a fixed total design size N, iterating in $n = n_0, \ldots, N$, starting from a small *seed design* of size n_0.

Some commentary is in order. Consider Step 2. Numerical optimization of the likelihood to infer hyperparameters can require initial values. Random initialization is common when n is small, e.g., $n = n_0$. Initializing with hyperparameters found in earlier iterations of sequential design – a *warm start* – offers computationally favorable results (faster convergence) and negligible deleterious effects when n is large. Earlier on, periodic random reinitialization can confer a certain robustness in the face of multimodal likelihoods (§5.3.4), which are not an uncommon occurrence. Skipping expensive $\mathcal{O}(n^3)$ calculations in Step 2, e.g., using MLE calculations from earlier n, is risky but could represent a huge computational savings: taking flops in $\mathcal{O}(n^3)$ down to $\mathcal{O}(n^2)$ using the updating equations in §6.3.

---

[4]https://en.wikipedia.org/wiki/Greedy_algorithm

Algorithm 6.1 Sequential Design/Active Learning

Assume a flexible surrogate, e.g., a GP model, but with potentially unknown hyper-parameterization.

Require a function $f(\cdot)$ providing outputs $y \sim f(x)$ for inputs x, either deterministic or observed with noise; a choice of initial design size n_0 and final size N; and **criterion** $J(x)$ to search for design augmentation.

Then

1. Run a small seed, or bootstrapping experiment.
   a. Create an initial seed design X_{n_0} with n_0 runs. Typically X_{n_0} is a model-free choice, e.g., derived from a static LHS or maximin design.
   b. Evaluate $y_i \sim f(x_i)$ under each $x_i^\top$ in the ith row of X_{n_0}, for $i = 1, \ldots, n_0$, obtaining $D_{n_0} = (X_{n_0}, Y_{n_0})$.
   c. Set $n \leftarrow n_0$, indexing iterations of sequential design.
2. Fit the surrogate (and hyperparameters) using D_n, e.g., via MLE.
3. Solve **criterion** $J(x)$ based on the fitted model from Step 2, resulting in a choice of $x_{n+1} \mid D_n$: $x_{n+1} = \operatorname{argmax}_{x \in \mathcal{X}} J(x) \mid D_n$.
4. Observe the response at the chosen location by running a new simulation, $y_{n+1} \sim f(x_{n+1})$.
5. Update $D_{n+1} = D_n \cup (x_{n+1}, y_{n+1})$; set $n \leftarrow n + 1$ and **repeat** from Step 2 unless $n = N$.

Return the chosen design and function evaluations D_N, along with surrogate fit (i.e., after a final application of Step 2).

Step 3 in the algorithm presumes maximization, but it may be that minimization is more appropriate for the chosen criterion. Many flavors of optimization can be employed here. For example: evaluating $J(x)$ on a candidate set of inputs $\mathcal{X}$ in a discrete and/or randomized search; numerical optimization with `optim` in the case of compact $\mathcal{X}$, potentially with help from closed-form derivatives. Warm-starting with x_n from previous iterations could be advantageous depending on the criterion, especially under variations considered for optimization in Chapter 7. When targeting global predictive accuracy or maximum information through learning, a more diverse (even random) initialization may be preferred in order to cope with highly multimodal objective surfaces. Note that $J(x)$ evaluations can require forms for the predictive distribution $Y(\mathcal{X}) \mid D_n$, e.g., kriging equations (5.3) in the GP case, or other aspects of fit conditional on hyperparameters from Step 2, and/or derivatives thereof.

In what follows, I introduce several criteria paired with an implementation of Steps 2–3 of Algorithm 6.1. Each is positioned as a pragmatic compromise between myriad alternatives. For example, I always warm start MLE calculations, but optimize criteria with `optim` under both random and warm-started initialization. Derivatives are used for the former but not the latter. Criteria are often easy to differentiate, and pointers are provided for the curious reader, however these are not implemented by our preferred library (`laGP`) under all variations considered below. §9.3.5 and Chapter 10 furnish derivatives accompanied by library routines for two important IMSPE-based sequential criteria in a slightly more ambitious setting.

6.2.1 Whack-a-mole: active learning MacKay

The simplest sequential design scheme for GPs, but it's of course more widely applicable, involves choosing the next point to maximize predictive variance. Given data $D_n = (X_n, Y_n)$, infer unknown hyperparameters (τ^2, θ, g) by maximizing the likelihood, say, and choose x_{n+1} as

$$x_{n+1} = \text{argmax}_{x \in \mathcal{X}} \; \sigma_n^2(x).$$

Obtain $y_{n+1} = Y(x_{n+1}) = f(x_{n+1}) + \varepsilon$; combine to form the new dataset: $D_{n+1} = ([X_n; x_{n+1}^\top], [Y_n; y_{n+1}])$; repeat. In other words, the criterion provided to Algorithm 6.1 is $J(x) \mid D_n \equiv \sigma_n^2(x)$, where $\sigma_n^2(x)$ comes from kriging equations (5.3) applied using $(\hat{\tau}_n^2, \hat{\theta}_n^2, \hat{g}_n)$ trained on data compiled from earlier sequential design iterations: $D_n = (X_n, Y_n)$. The n in the subscript is included to remind readers that the estimator is trained on data D_n.

Simple right? But does it work well? In many cases, yes. MacKay (1992) argues that, in a certain sense, repeated application of such selection approximates a maximum entropy design. MacKay was working with neural networks. Almost a decade later, Seo et al. (2000) ported the idea to GPs, dubbing the method *Active learning MacKay (ALM)*. After deriving GP updating equations in §6.3 we'll be able to make a direct link between $\sigma_n^2(x_{n+1})$ and $|K_{n+1}|$, establishing for GPs the connection MacKay made for neural networks.

Active learning[5] is ML jargon for sequential design. Online data selection is sometimes viewed as an example of reinforcement learning[6], which is itself a special case of optimal control[7]. Criteria for sequential data selection, and their solvers, are known as *acquisition functions* in the active learning literature. In many ways, machine learners have been more "active" in sequential design research of late than statisticians have, so many have adopted their nomenclature. Active learning is where the learning algorithm gets to actively acquire the examples it's trained on, creating a virtuous cycle between information acquisition and inference.

To demonstrate ALM, let's revisit our favorite 2d dataset from §5.1.2, seeding with a small LHS of size $n_0 = 12$. The code chunk below implements Step 1 of Algorithm 6.1 with this choice of $f(\cdot)$.

```
library(lhs)
ninit <- 12
X <- randomLHS(ninit, 2)
f <- function(X, sd=0.01)
  {
    X[,1] <- (X[,1] - 0.5)*6 + 1
    X[,2] <- (X[,2] - 0.5)*6 + 1
    y <- X[,1] * exp(-X[,1]^2 - X[,2]^2) + rnorm(nrow(X), sd=sd)
  }
y <- f(X)
```

Step 2 involves fitting a model. Below, an isotropic GP is fit using `laGP` library routines.

---

[5]https://en.wikipedia.org/wiki/Active_learning
[6]https://en.wikipedia.org/wiki/Reinforcement_learning
[7]https://en.wikipedia.org/wiki/Optimal_control

```
library(laGP)
g <- garg(list(mle=TRUE, max=1), y)
d <- darg(list(mle=TRUE, max=0.25), X)
gpi <- newGP(X, y, d=d$start, g=g$start, dK=TRUE)
mle <- jmleGP(gpi, c(d$min, d$max), c(g$min, g$max), d$ab, g$ab)
```

Lest you spot it on your own and think it's a sleight-of-hand, notice how `darg` specifies a maximum lengthscale of $\theta_{max} = 0.25$, a stark departure from earlier examples with these data. This kludge for `d$max` is crucial to obtaining consistently good results in repeated Rmarkdown builds. Pathological random $n_0 = 12$-sized seed LHS designs X, and associated noisy y-values, sometimes lead to GP fits that characterize a high-noise/low-signal regime via long lengthscales $\hat{\theta}_{n_0}$ in an unconstrained setting. As a consequence, the predictive variance $\sigma^2(\cdot)$ is highest at the edges of the design space, leading to x_{n+1}-values being selected there. Since the true $f(\cdot)$ is nearly zero at all points on the boundary, that acquisition perpetuates a high-noise regime, creating a feedback loop where sites are never (or almost never) selected in the interior. The curious reader may wish to re-run the entirety of the code in this section with `max=0.25` deleted to see what can happen. (It might take a few tries to get a bad result.)

Specifying preference for a relatively low $\hat{\theta}_{n_0}$, effectively providing a strong prior on small lengthscale values and entirely precluding large ones, serves as a workaround. Using a larger n_0 also helps, but would have diminished the value of our illustration. Better workarounds include a more thoughtful initial design, e.g., targeting a diversity of pairwise distances for more reliable $\hat{\theta}_{n_0}$ (Zhang et al., 2020), and/or a more aggregate criteria like ALC (§6.2.2). But let's table that for now, and continue with the illustration.

Before embarking on iterations of sequential design, it'll help to establish some predictive quantities for later comparisons. R code below creates a testing grid and saves true (noiseless) responses at those locations.

```
x1 <- x2 <- seq(0, 1, length=100)
XX <- expand.grid(x1, x2)
yytrue <- f(XX, sd=0)
```

For example, we may use these to calculate RMSE . . .

```
rmse <- sqrt(mean((yytrue - predGP(gpi, XX, lite=TRUE)$mean)^2))
```

. . . in order to benchmark out-of-sample progress over iterations of design acquisition, n, against commensurately-sized designs derived from similar criteria.

Predictive variance $\sigma_n^2(x)$ produces football/sausage-shaped error-bars, so it must have many local maxima. In fact, it's easy to see how the number of local maxima could grow linearly in n. Optimizing globally over that surface presents challenges. Global optimization is a hard sequential design problem in its own right, hence Chapter 7. Here we shall stick to our favorite library-based local solver, `optim` with `method="L-BFGS-B"` (Byrd et al., 1995). Code below establishes the ALM objective as minus predictive variance, with the intention of passing to `optim` whose default is to search for minima.

```
obj.alm <- function(x, gpi)
  - sqrt(predGP(gpi, matrix(x, nrow=1), lite=TRUE)$s2)
```

Our success on a more global front will depend upon how clever we can be about initializing local solvers in a *multi-start scheme*. Suppose we "design" a collection of starting locations placed in parts of the input space known to have high variance, i.e., at the widest part of the sausage – as far as possible from existing design locations. This is basically what our sequential `mymaximin` function (§4.2.1) was created to do, encouraging spread in new (ALM search starting) locations relative to one another and to existing locations (X_n from previous iterations). Rather than paste `mymaximin` in here, it's loaded in the Rmarkdown source behind the scenes. Alternatively, any of the variations entertained in response to §4.4 homework exercises may be preferred for more computationally efficient initialization.

Since there are about as many modes in $\sigma_n^2(x)$ as inputs X_n, it seems sensible to default to n space-filling multi-start locations for ALM searches. The function below facilitates search for x_{n+1}, coded as `xnp1`, implementing of Step 3 of Algorithm 6.1. It's designed to be somewhat generic to searching objective (`obj`), which we provide as `obj.alm` by default but anticipate variations on the horizon. Execution begins by creating a sequential maximin design of multi-start initializers. It then iterates over `optim` calls with `obj`. Notice that this implementation is tailored to 2d inputs, as is everything supporting the rest of this example. However it wouldn't be hard to make things more generic with small edits in a few places.

```
xnp1.search <- function(X, gpi, obj=obj.alm, ...)
  {
  start <- mymaximin(nrow(X), 2, T=100*nrow(X), Xorig=X)
  xnew <- matrix(NA, nrow=nrow(start), ncol=ncol(X) + 1)
  for(i in 1:nrow(start)) {
    out <- optim(start[i,], obj, method="L-BFGS-B", lower=0,
      upper=1, gpi=gpi, ...)
    xnew[i,] <- c(out$par, -out$value)
  }
  solns <- data.frame(cbind(start, xnew))
  names(solns) <- c("s1", "s2", "x1", "x2", "val")
  return(solns)
}
```

On output, a `data.frame` is returned combining starting and ending locations with a final column recording the value of the objective found by `optim`. Such comprehensive output is probably overkill for most situations, because all that matters is `c(x1, x2)` coordinates in the row having the smallest `val`. Other rows are furnished primarily to aid in illustration. For example, Figure 6.9 draws arrows connecting starting `c(s1, s2)` coordinates to locally optimal variance-maximizing solutions, with a red dot at the terminus of the best `val` arrow. (Any arrows with near-zero length are omitted, and occasionally such an arrow "terminates" at the red dot.)

```
solns <- xnp1.search(X, gpi)
plot(X, xlab="x1", ylab="x2", xlim=c(0,1), ylim=c(0,1))
arrows(solns$s1, solns$s2, solns$x1, solns$x2, length=0.1)
m <- which.max(solns$val)
```

```
prog <- solns$val[m]
points(solns$x1[m], solns$x2[m], col=2, pch=20)
```

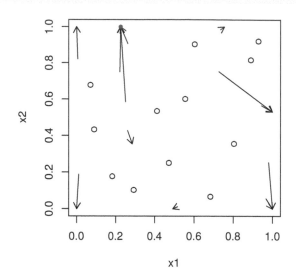

FIGURE 6.9: First iteration of ALM search. Each arrow represents an origin and outcome (terminus) of multi-start exploration of predictive variance. Variance-maximizing location is indicated as a red dot.

Considering the random nature of this Rmarkdown example, it's hard to speculate on details in Figure 6.9. Often several arrows terminate at the same location. At this early stage of sequential design the most likely location for the red dot, and indeed the terminus of many arrows, is somewhere along the boundary of the input space.

Step 4 in Algorithm 6.1 involves running $y \sim f(x_{n+1})$ at the new input location – the red dot. Then add the new pair into the design in Step 5.

```
xnew <- as.matrix(solns[m, 3:4])
X <- rbind(X, xnew)
y <- c(y, f(xnew))
```

Implementing an intermediate calculation, spanning Step 5 of the current iteration and Step 2 of the next, the updateGP function below augments an existing GP object, referenced here by gpi, with new input–output pairs. This is a very fast update, leveraging $\mathcal{O}(n^2)$ calculations detailed in §6.3. However, the penultimate line in the code chunk below updates MLE calculations on hyperparameters at $\mathcal{O}(n^3)$ expense, completing a loop back up to Step 2 of Algorithm 6.1. Often convergence is achieved after very few likelihood evaluations when carrying on from where search terminated in the previous sequential design iteration, i.e., at the MLE setting under just one fewer training data points. The final line below records an RMSE value (under the new design) on the hold-out testing set (which is the same as before). This step is not formally part of Algorithm 6.1.

```
updateGP(gpi, xnew, y[length(y)])
mle <- rbind(mle, jmleGP(gpi, c(d$min, d$max), c(g$min, g$max),
```

```
   d$ab, g$ab))
rmse <- c(rmse, sqrt(mean((yytrue - predGP(gpi, XX, lite=TRUE)$mean)^2)))
```

Updated fit in hand, we're ready for another active learning acquisition. The code below represents a second full pass through the loop in Algorithm 6.1, comprising the sequence of Steps 3 → 4 → 5 → 2, ending with new MLE and RMSE calculations on the augmented design.

```
solns <- xnp1.search(X, gpi)
m <- which.max(solns$val)
prog <- c(prog, solns$val[m])
xnew <- as.matrix(solns[m, 3:4])
X <- rbind(X, xnew)
y <- c(y, f(xnew))
updateGP(gpi, xnew, y[length(y)])
mle <- rbind(mle, jmleGP(gpi, c(d$min, d$max), c(g$min, g$max),
   d$ab, g$ab))
p <- predGP(gpi, XX, lite=TRUE)
rmse <- c(rmse, sqrt(mean((yytrue - p$mean)^2)))
```

The outcome of that search is summarized by Figure 6.10, again with arrows and a red dot. Notice that the red dot from the previous iteration has been promoted to an open circle, now fully incorporated into the design.

```
plot(X, xlab="x1", ylab="x2", xlim=c(0,1), ylim=c(0,1))
arrows(solns$s1, solns$s2, solns$x1, solns$x2, length=0.1)
m <- which.max(solns$val)
points(solns$x1[m], solns$x2[m], col=2, pch=20)
```

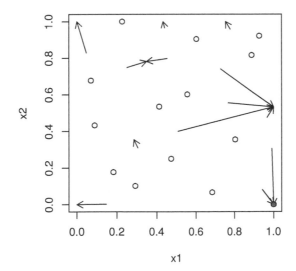

FIGURE 6.10: Second iteration of ALM search after Figure 6.9.

Again, it's hard to speculate about the outcome of this search for the largest variance input location. Usually the result is intuitive considering locations of existing design points (open circles), and the origin and terminus of the arrows. Before turning to a more in-depth investigation into merits of this scheme, let's do several more iterations through the loop, for a total of $N = 25$ runs. Again, this is Steps $3 \rightarrow 4 \rightarrow 5 \rightarrow 2$ wrapped in a `for` loop.

```
for(i in nrow(X):24) {
  solns <- xnp1.search(X, gpi)
  m <- which.max(solns$val)
  prog <- c(prog, solns$val[m])
  xnew <- as.matrix(solns[m, 3:4])
  X <- rbind(X, xnew)
  y <- c(y, f(xnew))
  updateGP(gpi, xnew, y[length(y)])
  mle <- rbind(mle, jmleGP(gpi, c(d$min, d$max), c(g$min, g$max),
    d$ab, g$ab))
  p <- predGP(gpi, XX, lite=TRUE)
  rmse <- c(rmse, sqrt(mean((yytrue - p$mean)^2)))
}
```

The `mle` object, shown in part below (every other iteration to save space), summarizes steps of our search vis-a-vis estimated lengthscale and nugget hyperparameters.

```
mle[seq(1,nrow(mle),by=2),]
```

```
##           d        g tot.its dits gits
## 1   0.05689 0.001575      23    8   15
## 3   0.05241 0.001578       7    4    3
## 5   0.04752 0.001580       7    4    3
## 7   0.04931 0.001578       7    4    3
## 9   0.04947 0.001580       7    4    3
## 11  0.03758 0.001589       9    5    4
## 13  0.04121 0.001587       7    4    3
```

Bigger movements and more MLE iterations, especially for θ (`d` and `dits`), are usually associated with earlier sequential design steps. More often than not, in repeated Rmarkdown builds, later searches are associated with smaller relative MLE moves, converging rapidly in just a few iterations, benefiting from warm starts provided by values stored in the GP object `gpi` from previous iterations. Perhaps full MLE updates, with `jmlegp` above, may not be required for subsequent sequential design iterations. That could represent substantial computational savings as n gets big, making updates take flops in $\mathcal{O}(n^2)$ rather than $\mathcal{O}(n^3)$.

Figure 6.11 summarizes the predictive surface after $N = 25$ design sites have been selected in this way. A mean surface is shown on the left, and standard deviation on the right. Numbers plotted indicate the iteration in which each site was chosen. Recall that the first $n_0 = 12$ came from an LHS.

```
par(mfrow=c(1,2))
cols <- heat.colors(128)
image(x1, x2, matrix(p$mean, ncol=length(x1)), col=cols, main="mean")
text(X, labels=1:nrow(X), cex=0.75)
```

```
image(x1, x2, matrix(sqrt(p$s2), ncol=length(x1)), col=cols, main="sd")
text(X, labels=1:nrow(X), cex=0.75)
```

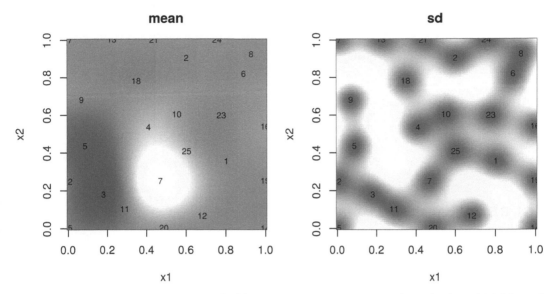

FIGURE 6.11: Predictive mean (left) and standard deviation (right) after ALM-based sequential design.

Those surfaces look reasonable, at least from a purely aesthetic perspective. A fair criticism is that the chosen sites are no better, or could perhaps be even worse, than what would have been obtained with a simple $N = 25$ space-filling design, i.e., calculated in a single batch at the outset. One nice feature, or byproduct, of a sequential model-based alternative is a measure of progress, namely the maximal variance used as the basis of each sequential design decision. We've been saving that as `prog` in the code above, so that these may be plotted like in Figure 6.12. Note that `prog` saves `predGP(...)$s2`, involving a $\hat{\tau}_n^2$ estimate that changes from one acquisition to the next, $n \to n+1$. Within a particular iteration this scale estimate has no impact on selection criteria, but between iterations it can cause `prog` to "jump", up or down, depending on incoming y_{n+1}-values and corresponding hyperparameter updates. At this time, `laGP` doesn't provide direct access to $\hat{\tau}^2$ in order to correct this, if so desired.

```
plot((ninit+1):nrow(X), prog, xlab="n")
```

Whether jumps are present in the figure or not – it depends on the random Rmarkdown build – it's clear that progress has not yet leveled off. Consider 75 more iterations of active learning, stopping at $N = 100$. The code below duplicates the `for` loop above after reinitializing $\hat{\theta} \equiv$ d, and removing the restrictive $\hat{\theta}_{\max} = 0.25$ setting. After $n = 25$ samples are chosen, an aggressive prior is not as important in order to avoid pathological behavior. (A conservative Bayesian approach has greatest value to the practitioner when data are few.)

```
d <- darg(list(mle=TRUE), X)
for(i in nrow(X):99) {
  solns <- xnp1.search(X, gpi)
```

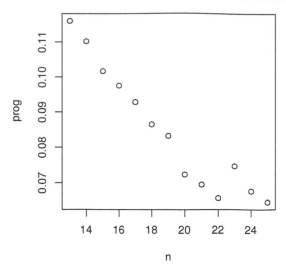

FIGURE 6.12: ALM-progress over sequential design iterations.

```
m <- which.max(solns$val)
prog <- c(prog, solns$val[m])
xnew <- as.matrix(solns[m, 3:4])
X <- rbind(X, xnew)
y <- c(y, f(xnew))
updateGP(gpi, xnew, y[length(y)])
mle <- rbind(mle, jmleGP(gpi, c(d$min, d$max), c(g$min, g$max),
  d$ab, g$ab))
p <- predGP(gpi, XX, lite=TRUE)
rmse <- c(rmse, sqrt(mean((yytrue - p$mean)^2)))
}
```

An inspection of MLEs found during each iteration reveals small changes in $\hat{\theta}_n$ and $\hat{g}_n$ on both relative and absolute terms. The number of iterations required for convergence can vary somewhat more substantially over longer search horizons, but is usually small.

```
mle[seq(14,nrow(mle), by=10),]
```

```
##              d           g tot.its dits gits
## 14 0.03698 0.001588       8    5    3
## 24 0.02567 0.001586       6    3    3
## 34 0.02549 0.001575       7    4    3
## 44 0.03144 0.001513      11    6    5
## 54 0.03413 0.001348      10    5    5
## 64 0.03159 0.001697      11    6    5
## 74 0.03174 0.003289      42   21   21
## 84 0.03102 0.002330      38   18   20
```

As shown in Figure 6.13, progress metrics are starting to level off. The magnitude and frequency of jumps in `prog`, shown on the left, generally decreases over acquisition iterations. (If there were no jumps before, there are almost certainly some in this view.) Accuracy, measured by RMSE in the right panel, shows similar behavior, having leveled off substantially in the latter thirty-five iterations or so. Although an out-of-sample testing set is seldom available for use in this context, it's nice to see a substantial degree of correlation between predictive accuracy and (maximal) predictive uncertainty, which is what was used to guide design.

```
par(mfrow=c(1,2))
plot((ninit+1):nrow(X), prog, xlab="n: design size", ylab="ALM progress")
plot(ninit:nrow(X), rmse, xlab="n: design size", ylab="OOS RMSE")
```

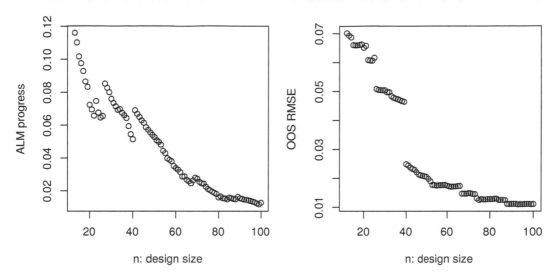

FIGURE 6.13: Maximum variance (left, lower is better) and out-of-sample RMSE (right) over 100 ALM acquisitions.

Figure 6.14 shows the posterior surface and the location of the chosen design sites.

```
par(mfrow=c(1,2))
image(x1, x2, matrix(p$mean, ncol=length(x1)), col=cols, main="mean")
text(X, labels=1:nrow(X), cex=0.75)
image(x1, x2, matrix(sqrt(p$s2), ncol=length(x1)), col=cols, main="sd")
text(X, labels=1:nrow(X), cex=0.75)
```

Observe dense coverage along the boundary, a telltale sign of maxent design. The degree of space-fillingness could be improved, but by proceeding sequentially less faith is required in the quality of an initial hyperparameterization. It can be instructive to repeat selection of these 100 design sites to explore variability in outcomes. ALM can misbehave, especially if the starting design is unlucky to miss strong signal in the data. When that happens, initial fits of hyperparameters favor noise, even when restricted by $\hat{\theta}_{\max} = 0.25$, sparking a feedback loop from which the sequential procedure may fail to recover.

A more graceful solution, besides simply initializing with a larger design, could involve a more aggregate criteria. Maximizing variance is myopic. ALM doesn't recognize that acquisitions impact predictive equations globally. Instead it "wacks down" the highest variance locations, a set of measure zero, potentially ignoring a continuum of fatter regions where uncertainty may cumulatively be much larger. This is why we end up with lots of sites on the boundary.

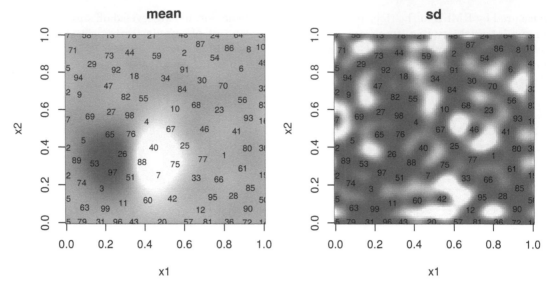

FIGURE 6.14: Predictive mean (left) and standard deviation (right) after ALM-based sequential design.

Variance is high there because there are fewer data points nearby – there are none beyond the boundary, a situation which could not be remedied by any sequential data acquisition scheme. Yet those boundary locations are perhaps least useful when it comes to predicting elsewhere, being as far as possible from almost any notion of elsewhere.

6.2.2 A more aggregate criteria: active learning Cohn

You could say that ALM has somewhat of a scale problem. Just because predictive variance is high doesn't mean that there's value in adding more data. (I'm thinking ahead a little bit to Chapters 9–10 where noise variances may not be uniform in the input space.) One could instead work with epistemic variance (5.16), i.e., the variance of the latent field (§5.3.2), or variance of the mean. Operationally speaking, this amounts to using a zero nugget when calculating out-of-sample covariances among $\mathcal{X} \equiv$ XX. But that only helps to a degree. High variance may well be an accurate inference for the problem at hand, which would mean seriously diminishing returns to further data acquisition. A better metric might be reduction in variance. That is, how helpful is a potential new design site at reducing predictive uncertainty? If a lot, relative to previous reductions and relative to other new design sites, then it could be quite helpful to perform a new run at that location. If not, perhaps another location would be preferred or perhaps we have enough data already.

But this begs the question where; where should the reduction be measured? One option is everywhere, integrating over the entire input space like IMSPE does for global design (§6.1.2). Such an integral is tractable analytically for rectangular input spaces, but I shall defer that discussion to Chapter 10. At the other extreme is measuring reduction in variance at a particular reference location, or at a collection of locations, thereby approximating the integral with a sum like we did with IMSPE (6.5). The first person to suggest such an acquisition heuristic in a nonparametric regression context was Cohn (1994), for neural networks. As with ALM, Seo et al. (2000) adapted Cohn's idea to GPs and called it *active learning Cohn (ALC)*. The result is essentially a sequential analog of IMSPE design, and in fact the sequential version can be shown to approximate a full A-optimal design[8].

How does it all work? Recall that predictive variance follows

$$\sigma_n^2(x) = \hat{\tau}_n^2[1 + \hat{g}_n - k_n^\top(x)K_n^{-1}k_n(x)] \quad \text{where} \quad k_n(x) \equiv C_{\hat{\theta}_n}(X_n, x),$$

written here with an n subscript on all estimated quantities in order to emphasize their dependence on data $D_n = (X_n, Y_n)$ via $\hat{\tau}_n^2$, $\hat{g}_n$, and $\hat{\theta}_n$ hidden inside K_n. Let $\tilde{\sigma}_{n+1}^2(x)$ denote the *deduced variance* based on design X_{n+1} combining X_n and a new input location $x_{n+1}^\top$ residing in its $n + 1^{\text{st}}$ row. Otherwise, $\tilde{\sigma}_{n+1}^2(x)$ conditions on the same estimated quantities as $\sigma_n^2(x)$, i.e., $\hat{\tau}_n^2$, $\hat{g}_n$, and $\hat{\theta}_n$, all estimated from Y_n. Since Y_n doesn't directly appear in $\sigma_n^2(x)$, nor would it directly appear in $\tilde{\sigma}_{n+1}^2(x)$. Therefore, quite simply

$$\tilde{\sigma}_{n+1}^2(x) = \hat{\tau}_n^2[1 + \hat{g}_n - k_{n+1}^\top(x)K_{n+1}^{-1}k_{n+1}(x)] \quad \text{where} \quad k_{n+1}(x) \equiv C_{\hat{\theta}_n}(X_{n+1}, x). \quad (6.6)$$

Using that definition, the ALC criteria is the average (integrated over $\mathcal{X}$ or any other subset of the input space) reduction in variance from $n \to n + 1$ measured through a choice of x_{n+1}, augmenting the design:

$$\Delta\sigma_n^2(x_{n+1}) = \int_{x \in \mathcal{X}} \sigma_n^2(x) - \tilde{\sigma}_{n+1}^2(x) \, dx$$

$$= c - \int_{x \in \mathcal{X}} \tilde{\sigma}_{n+1}^2(x) \, dx.$$

Wishing predictive uncertainty to be reduced as much as possible, that translates into finding an x_{n+1} maximizing $\Delta\sigma_n^2(x_{n+1})$. But that's the same as minimizing the integrated deduced variance. Therefore the criterion that must be solved in each iteration of sequential design, occupying Step 3 of Algorithm 6.1, is

$$x_{n+1} = \text{argmin}_{x \in \mathcal{X}} \int_{x \in \mathcal{X}} \tilde{\sigma}_{n+1}^2(x) \, dx. \quad (6.7)$$

A closed form for rectangular $\mathcal{X}$, complete with derivatives for maximizing, is provided later in Chapter 10. Often in practice the integral is approximated by a sum over a reference set (6.5) as this offers the simplest, most general, implementation. Although a singleton reference set ($\mathcal{X} = x_{\text{ref}}$, dropping the integral or sum entirely) represents a degenerate case where $x_{n+1} = x_{\text{ref}}$, any larger reference set demands x_{n+1} make a trade-off that considers broader impact on variance reduction when new data are added into the design. The laGP package has functions called `alcGP` and `alcGPsep` that automate this objective, providing evaluations of $\Delta\sigma_n^2(x_{n+1})$ to within additive and multiplicative constants. More specifically, the output is scale-free, having "divided out" $\hat{\tau}_n^2$ in a manner similar to IMSPE (§6.1.2). Since we wish to maximize this quantity, the wrapper below negates the output of `alcGP`. It also uses the square root, which conveys a more numerically stable objective for optimization with `optim`. Since the square root is a monotone transformation, the location of stationary points is unchanged.

```
obj.alc <- function(x, gpi, Xref)
  - sqrt(alcGP(gpi, matrix(x, nrow=1), Xref))
```

---

[8]https://en.wikipedia.org/wiki/Optimal_design#Minimizing_the_variance_of_estimators

Our `xnp1.search` from §6.2.1, solving for the next input x_{n+1}, was designed to be generic enough to accommodate other objective functions and arguments, like `obj.alc` and `Xref`, through ellipses (...). Before jumping into an illustration, we must clean up our ALM search and reinitialize a new GP with the same `ninit` starting values. This will help control variability for subsequent comparison between ALC and ALM-based sequential designs.

```
deleteGP(gpi)
X <- X[1:ninit,]
y <- y[1:ninit]
g <- garg(list(mle=TRUE, max=1), y)
d <- darg(list(mle=TRUE, max=0.25), X)
gpi <- newGP(X, y, d=d$start, g=g$start, dK=TRUE)
mle <- jmleGP(gpi, c(d$min, d$max), c(g$min, g$max), d$ab, g$ab)
p <- predGP(gpi, XX, lite=TRUE)
rmse.alc <- sqrt(mean((yytrue - p$mean)^2))
```

The code below invokes the first iteration of ALC search. A 100-element LHS is chosen for reference set `Xref`, but otherwise the process is similar to that for ALM except that `obj.alc` and `Xref` are provided to `xnp1.search`.

```
Xref <- randomLHS(100, 2)
solns <- xnp1.search(X, gpi, obj=obj.alc, Xref=Xref)
m <- which.max(solns$val)
xnew <- as.matrix(solns[m, 3:4])
prog.alc <- solns$val[m]
```

Figure 6.15 offers a visual summary of the search, with arrows and a red dot showing the chosen x_{n+1} location. Smaller filled-gray dots denote `Xref`.

```
plot(X, xlab="x1", ylab="x2", xlim=c(0,1), ylim=c(0,1))
arrows(solns$s1, solns$s2, solns$x1, solns$x2, length=0.1)
points(solns$x1[m], solns$x2[m], col=2, pch=20)
points(Xref, cex=0.25, pch=20, col="gray")
```

Domains of attraction from the ALC criterion are similar to ALM's, however global optima are more likely to be in the interior for ALC. In part, this is because ALC prefers candidates which are far from X_n and close to `Xref`. Since `Xref` is an LHS, very few are found near the boundary. Although it's hard to speculate on a location selected for a particular Rmarkdown build, especially considering the random nature of the reference set used by ALC, the chosen location for x_{n+1} in the version I'm looking at now is indeed in the interior. Code below gathers $y(x_{n+1})$, and updates the GP predictor to get ready for the next acquisition by ALC.

```
X <- rbind(X, xnew)
y <- c(y, f(xnew))
updateGP(gpi, xnew, y[length(y)])
mle <- rbind(mle, jmleGP(gpi, c(d$min, d$max), c(g$min, g$max),
  d$ab, g$ab))
```

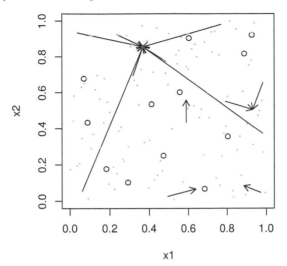

FIGURE 6.15: First iteration of ALC search in the style of Figure 6.9. Gray dots denote reference locations.

```
p <- predGP(gpi, XX, lite=TRUE)
rmse.alc <- c(rmse.alc, sqrt(mean((yytrue - p$mean)^2)))
```

Skipping ahead a bit, the loop below fills out the rest of the design in this way, up to $N = 100$. This **for** loop is identical to the ALM analog, except using ALC with obj.alc and Xref instead. Notice that Xref is refreshed for each acquisition. This helps encourage diversity in search from one iteration to the next. Although any single reference set may lead to a poor sum-based approximation of the integral behind ALC (6.7) in any particular acquisition iteration, a diversity of (even small) reference sets limits the influence of bad behavior on aggregate.

```
d <- darg(list(mle=TRUE), X)
for(i in nrow(X):99) {
  Xref <- randomLHS(100, 2)
  solns <- xnp1.search(X, gpi, obj=obj.alc, Xref=Xref)
  m <- which.max(solns$val)
  prog.alc <- c(prog.alc, solns$val[m])
  xnew <- as.matrix(solns[m, 3:4])
  X <- rbind(X, xnew)
  y <- c(y, f(xnew))
  updateGP(gpi, xnew, y[length(y)])
  mle <- rbind(mle, jmleGP(gpi, c(d$min, d$max), c(g$min, g$max),
    d$ab, g$ab))
  p <- predGP(gpi, XX, lite=TRUE)
  rmse.alc <- c(rmse.alc, sqrt(mean((yytrue - p$mean)^2)))
}
```

Updates to hyperparameter MLEs are saved above, however the values recorded look similar to those from ALM so they're not shown here. Figure 6.16 provides a view into progress over iterations of ALC-based acquisition.

```
par(mfrow=c(1,2))
plot((ninit+1):nrow(X), prog.alc, xlab="n: design size",
  ylab="ALC progress")
plot(ninit:nrow(X), rmse, xlab="n: design size", ylab="OOS RMSE")
points(ninit:nrow(X), rmse.alc, col=2, pch=20)
legend("topright", c("alm", "alc"), pch=c(21,20), col=1:2)
```

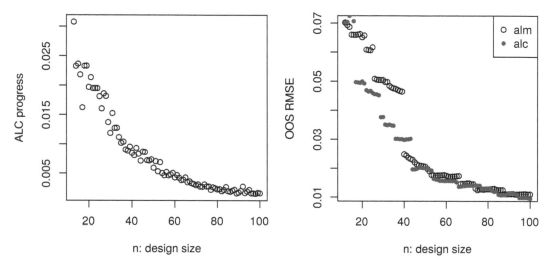

FIGURE 6.16: Progress in ALC sequential design in terms of integrated reduction in variance (left, lower is better) and out-of-sample RMSE (right), with comparison to ALM from Figure 6.12.

The left panel, showing progress on `obj.alc`, is less jumpy than the ALM analog, but also perhaps more noisy. It's less jumpy because `alcGP` and `alcGPsep` factor out scale $\hat{\tau}_n^2$. Noise arises due to novel random `Xref` in each iteration of design; a fixed `Xref` would smooth things out considerably, but bias search. Observe that progress on the ALC criteria is flattening out in later iterations. On the right is out-of-sample RMSE as compared to ALM. Notice how they end at about the same place, but ALC gets there faster. While ALM is busy filling out boundaries in early iterations, ALC is filling in interior regions which are closer to the vast majority of inputs in the testing set. Figure 6.17 shows predictive mean and standard deviation, with locations and order of selected design sites overlaid.

```
par(mfrow=c(1,2))
image(x1, x2, matrix(p$mean, ncol=length(x1)), col=cols, main="mean")
text(X, labels=1:nrow(X), cex=0.75)
image(x1, x2, matrix(sqrt(p$s2), ncol=length(x1)), col=cols, main="sd")
text(X, labels=1:nrow(X), cex=0.75)
```

Coverage of the boundary is sparser, but perhaps not as sparse as what we saw from static IMSPE design in §6.1.2. Predictive uncertainty is higher for ALC than ALM along that boundary, but those regions are not favored by ALC as their remote location makes them

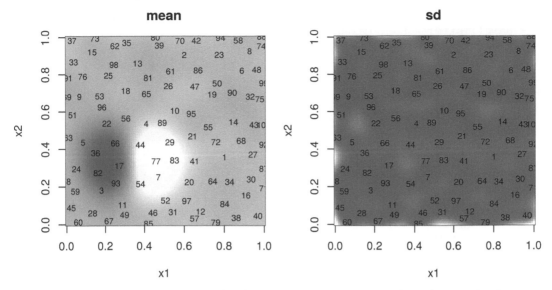

FIGURE 6.17: Predictive mean (left) and standard deviation (right) after ALC-based sequential design.

poor for driving down variance at reference sites; i.e., they're less than ideal for driving down uncertainty aggregated across the entirety of the input space.

6.2.3 Other sequential criteria

Several other heuristics have been proposed in the literature, some of which will be the subject of future chapters – particularly when the target is optimization. One that pops up often involves the Fisher information (FI) of kernel hyperparameters, particularly lengthscale(s) θ. The equations are a little cumbersome, involving second derivatives and expectations as FI calculations often do. Rather than duplicate these here, allow me to refer the interested reader to the appendix of Gramacy and Apley (2015). Search via FI is implemented by fishGP in the laGP package.

```
obj.fish <- function(x, gpi)
  - sqrt(fishGP(gpi, matrix(x, nrow=1)))
```

One attractive aspect of FI, compared to ALC say, is that no reference set is required. The criterion's emphasis on learning parameters means that it's less sensitive to initial choices for those parameters, but since FI is ultimately evaluated under $\hat{\theta}_n$, from the previous iteration n, there's still some degree of sensitivity and so we still have to be careful about initial values. A disadvantage is that FI doesn't directly emphasize predictive accuracy, rather hyperparameter estimation. It usually does not lead to designs with the most accurate predictors. To see why, feed obj.fish into xnp1.search to sequentially build up a design with $N = 100$ input–output pairs.

```
X <- X[1:ninit,]
y <- y[1:ninit]
```

```
d <- darg(list(mle=TRUE, max=0.25), X)
gpi <- newGP(X, y, d=0.1, g=0.1*var(y), dK=TRUE)
mle <- jmleGP(gpi, c(d$min, d$max), c(g$min, g$max), d$ab, g$ab)
rmse.fish <- sqrt(mean((yytrue - predGP(gpi, XX, lite=TRUE)$mean)^2))
prog.fish <- c()
for(i in nrow(X):99) {
  solns <- xnp1.search(X, gpi, obj=obj.fish)
  m <- which.max(solns$val)
  prog.fish <- c(prog.fish, solns$val[m])
  xnew <- as.matrix(solns[m, 3:4])
  X <- rbind(X, xnew)
  y <- c(y, f(xnew))
  updateGP(gpi, xnew, y[length(y)])
  mle <- rbind(mle, jmleGP(gpi, c(d$min, d$max), c(g$min, g$max),
    d$ab, g$ab))
  p <-predGP(gpi, XX, lite=TRUE)
  rmse.fish <- c(rmse.fish, sqrt(mean((yytrue - p$mean)^2)))
}
```

Figure 6.18 shows the design and the resulting fit, which illustrates striking differences compared to earlier analogs.

```
par(mfrow=c(1,2))
image(x1, x2, matrix(p$mean, ncol=length(x1)), col=cols, main="mean")
text(X, labels=1:nrow(X), cex=0.75)
image(x1, x2, matrix(sqrt(p$s2), ncol=length(x1)), col=cols, main="sd")
text(X, labels=1:nrow(X), cex=0.75)
```

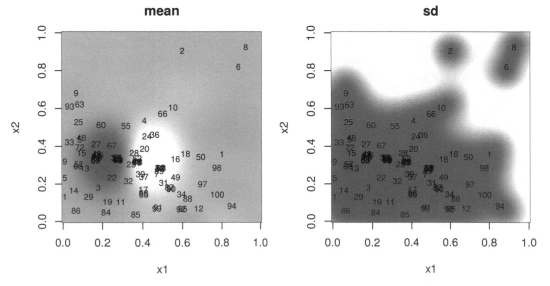

FIGURE 6.18: Predictive mean (left) and standard deviation (right) after an FI-based sequential design.

There's a certain "clumpiness", but also definitely a pattern in selected sites. Whole swaths of the input space are missing locations. This is the result of FI preferring distances between inputs on different scales, which makes sense in order to best learn lengthscale. Whether or not the densely sampled locations coincide with the "interesting part" of the response surface, in this case the "bumps" in the lower-left-hand quadrant, is largely a matter of luck. Repeated builds of this Rmarkdown document lead to dramatically different RMSE progress metrics over time, as shown for one instance in the right panel of Figure 6.19. This happens even though progress on the FI metric, as shown in the left panel, is remarkably consistent. If design acquisitions are lucky to cluster near the bumps, then the fact that the responses are conveniently zero elsewhere – and that our GP is zero-mean reverting – leads to excellent RMSEs compared to ALM and ALC. An unlucky clumping can lead to disastrous RMSE calculations.

```
par(mfrow=c(1,2))
plot((ninit+1):nrow(X), prog.fish, xlab="n: design size",
  ylab="FI progress")
plot(ninit:nrow(X), rmse, xlab="n: design size", ylab="OOS RMSE")
points(ninit:nrow(X), rmse.alc, col=2, pch=20)
points(ninit:nrow(X), rmse.fish, col=3, pch=19)
legend("topright", c("alm", "alc", "fish"), pch=21:19, col=1:3)
```

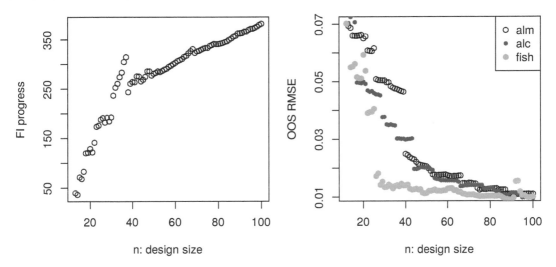

FIGURE 6.19: Progress in terms of FI (left, higher is better) and out-of-sample RMSE as compared to previous heuristics.

Readers are strongly encouraged to repeat the code above to explore that variability. If ALC and ALM are somewhat unsatisfying in that they basically produce space-filling designs with the burden of initializing GP hyperparameters for stable performance over the iterations, FI is unsatisfying in that design targeting good hyperparameters is insufficient for what's usually the primary goal: prediction. A hybrid approach however, first FI to learn parameters then ALC for accurate prediction, may represent an advantageous compromise. Still, FI needs an initial $\hat{\theta}_{n_0}$ from somewhere.

Regarding $\hat{\theta}_n$, sequential design acquisitions above are grossly inefficient from a computational perspective. MLE optimizations in $\mathcal{O}(n^3)$ are carried out in each iteration, from $n = n_0, \ldots, N$, so the overall scheme demands flops in $\mathcal{O}(N^4)$. This is true even when solutions from earlier iterations are used to warm start solvers. If a smartly chosen seed design could yield reliable hyperparameters, perhaps those expensive MLE calculations could be avoided. A *betadist* initializing design (Zhang et al., 2020) represents an attractive alternative in this context, circumventing chicken-or-egg paradoxes in sequential design. A homework exercise in §6.4 takes the diligent student through some of the details, including Monte Carlo (MC) benchmarking exercises. Nevertheless, K_n^{-1} and $|K_n|$ would still be required for subsequent ALC and ALM selection, respectively. Since those are still $\mathcal{O}(n^3)$, the overall scheme would still be quartic. Or are they?

6.3 Fast GP updates

It turns out that they need not be. In fact, updates to all relevant quantities for the model-based sequential design schemes outlined above may be calculated in $\mathcal{O}(n^2)$ time, making the whole scheme $\mathcal{O}(N^3)$. That is, as long as we're content not to repeatedly update hyperparameter estimates. Being able to quickly update a model fit, as data comes in, is often overlooked as an important aspect in model choice. In the context of active learning, and in computer experiments where automation, and loops over inputs to simulations naturally create a sequential inference (and design) environment, fast updating can be crucial to the relevance of nonparametric surrogate models with otherwise large computational demands – especially as datasets get large. Fortunately, GPs offer such a feature.

The key to fast GP updates as new data arrive, as implemented by `updateGP`/`updateGPsep`, is fast decomposition of covariance as $K_n \to K_{n+1}$. In the case of the inverse, the *partitioned inverse equations* are helpful. These are often expressed generically for blocks of arbitrary size, but we only need them for one new (symmetric) row/column.

$$\text{If} \quad K_{n+1}(x_{n+1}) = \begin{bmatrix} K_n & k_n(x_{n+1}) \\ k_n^\top(x_{n+1}) & K(x_{n+1}, x_{n+1}) \end{bmatrix},$$

$$\text{then} \quad K_{n+1}^{-1} = \begin{bmatrix} [K_n^{-1} + g_n(x_{n+1})g_n^\top(x_{n+1})v_n(x_{n+1})] & g_n(x_{n+1}) \\ g_n^\top(x_{n+1}) & v_n^{-1}(x_{n+1}) \end{bmatrix}, \quad (6.8)$$

where $g_n(x_{n+1}) = -v_n^{-1}(x_{n+1})K_n^{-1}k_n(x_{n+1})$ and the scale-free variance follows $v_n(x_{n+1}) = K(x_{n+1}, x_{n+1}) - k_n^\top(x_{n+1})K_n^{-1}k_n(x_{n+1})$. Computational cost for the update is in $\mathcal{O}(n^2)$. Use of Eq. (6.8) here is similar in spirit to a rank one Sherman–Morrison update[9], as may be applied in classical linear regression to update $(X_n^\top X_n)^{-1} \to (X_{n+1}^\top X_{n+1})^{-1}$. Under a linear mean GP specification (e.g., exercise #2 in §5.5), partitioned inverse and Sherman–Morrison fomulae may be combined to efficiently update mean and covariance as new training data arrive.

Observe that $v_n(x)$ in Eq. (6.8) is the same as our scale-free predictive variance. That is, $\sigma_n^2(x) = \hat{\tau}_n^2 v_n(x)$. As a consequence, updating predictive variance at locations x is also fast:

$$v_n(x) - v_{n+1}(x) = k_n^\top(x)G_n(x_{n+1})v_n(x_{n+1})k_n(x) \quad (6.9)$$
$$+ 2k_n^\top(x)g_n(x_{n+1})K(x_{n+1}, x) + K(x_{n+1}, x)^2/v_n(x_{n+1}),$$

where $G_n(x') \equiv g_n(x')g_n^\top(x')$, and $g_n(x') = -K_n^{-1}k_n(x')/v_n(x')$. Taking stock of matrix–vector sizes and calculations thereupon, the computational cost is again in $\mathcal{O}(n^2)$. No quantities bigger than $\mathcal{O}(n^2)$ multiply quantities bigger than $\mathcal{O}(n)$.

As a consequence, ALC (and IMSPE analogs in Chapter 10) may be calculated in $\mathcal{O}(n^2)$ time for each choice of x_{n+1}. We'll use this update again in Chapter 9 when constructing a local approximate GP. An exercise in §6.4 asks the reader to develop a fast update for $\hat{\tau}_n^2 \to \hat{\tau}_{n+1}^2$ as well.

We saw earlier (6.3) how the determinant can be quickly updated, but for completeness I shall restate it here. There's nothing special about this update, as determinants are naturally specified recursively. However, it can be instructive to provide equations in the notation of the partitioned inverse (6.8).

$$\log|K_{n+1}| = \log|K_n| + \log(K(x_{n+1}, x_{n+1}) + g_n^\top(x_{n+1})k_n(x_{n+1})v_n(x_{n+1}))$$
$$= \log|K_n| + \log v_n(x_{n+1}). \tag{6.10}$$

Computational cost is in $\mathcal{O}(n^2)$. Notice that the determinant changes by the variance of the point x_{n+1} added into the old fit. This explains why ALM approximates maximum entropy designs. Choosing x_{n+1} to maximize $v_n(x_{n+1})$ is identical to choosing it to maximize the $(n+1) \times (n+1)$ determinant involved in the maxent criterion.

Although the overall cost, at $\mathcal{O}(N^3)$ for N applications of $\mathcal{O}(n^2)$ operations for $n = 1, \ldots, N$, is the same as for a one-shot calculation, proceeding sequentially represents more work. There must be a bigger constant hidden in the order notation. If you already have a design of the desired size N, just perform inference and build the predictor the old fashioned way, all at once (i.e., with `newGPsep`, `mleGPsep` and `predGPsep`). But if you have decisions to make along the way, based on intermediate designs (such as with ALM or ALC), then it's way better for those to be $\mathcal{O}(n^2)$ rather than $\mathcal{O}(n^3)$ calculations, lest you shall find your scheme is actually in $\mathcal{O}(N^4)$ – unbearably slow for N in the few hundreds.

That said, a tacit goal in sequential design is to limit N – not just because GP calculations get onerous but because running simulations or otherwise collecting data is even more so. If extra resources can be put to getting that right, through careful MLE calculations even after fast updates say, then those may pay dividends in smaller N. Nowhere is the need for striking such a balance more acute than in the context of blackbox, or so-called Bayesian optimization (BO). In BO the goal is to rule out, rather than exhaustively explore, inferior regions of the input space. We want to find the best input configuration in the least number of evaluations N of an expensive to run, complicated, and opaque function.

6.4 Homework exercises

These exercises reinforce themes in batch and sequential model-based design and updates for Gaussian process regression.

#1: Derivative-based maxent design

Consider leveraging chain-rule identities and kernel derivatives (5.9) from §5.2.3 in order to deploy a library-based optimizer (e.g., `optim` with `method="L-BFGS-B"`) toward maximum entropy design. Consider #a and #b below in the context of upgrades to `maxent` with illustrations provided in the $n = 25$ and $m = 2$ case (see §6.1.1) for both $\theta = (0.1, 0.1)$ and $\theta = (0.1, 0.5)$. Throughout use $g = 0.01$.

a. Start by modifying one point in the design at a time in a stochastic exchange format. As in `maxent`, randomly select one point to swap out. Instead of randomly swapping another one back in, consider derivative-based search targeting a locally-optimal location for its coordinates. In other words, if X_{n-1} denotes the $(n-1) \times m$ design comprised of X_n without the randomly chosen "swapped out" row, which you may think of as residing in the nth row without loss of generality, the criteria is $x_n = \arg\max_{x'} |K'_n|$ where K'_n is managed with partition inverse equations. Implement your search in a manner similar to `xnp1.search` in §6.2.1, i.e., with `optim`, except don't bother with a multi-start scheme. Here are a few hints/suggestions which you should address in your solution: i) briefly describe how a similar derivative-based optimization could be applied with `obj.alm`; ii) develop an efficient scheme that avoids calculating a full $n \times n$ determinant, or other decomposition, more than once; iii) why is a multi-start scheme unnecessary?; iv) how should you initialize and what are appropriate bounds for your x'_n searches?; v) is there a variation where you can guarantee progress in each "swap out–in" pair; vi) do you need a stopping rule? Finally, compare progress against `maxent`'s random "swap-in" proposals via compute times over up to `T=10000` search iterations under a common, random initializing design.

b. Now consider how derivatives can be deployed towards optimizing all $n \times m$ coordinates of X_n at once, in a single `optim` call. Provide a mathematical expression for the partial derivative you intend to use, and code functions implementing the corresponding objective and gradient. Feed these to `optim` and try them out. Think about how hint iv), above in #a, on initialization and search bounds might port to this higher-dimensional setting. Considering these bounds, do you need to perform more than one `optim` to converge to a final solution? What stopping criteria could be used to determine how many? *(Hint: It might help to try `optim` without derivatives first.)* Compare timings and progress to your one-at-a-time searches from part #a.

#2: Sequential ISMPE-based design

The `hetGP` package (Binois and Gramacy, 2019) on CRAN offers a bare-bones GP capability similar to `laGP`'s. The two packages differ somewhat in their strategy toward inference, implementation, and support of covariance kernels. But those are largely cosmetic distinctions. Their fancier features, which target computationally tractable mean and variance nonstationarity, are quite distinct and are covered in detail in Chapter 10 and §9.3, respectively. Another difference lies in their support of integrated-variance-based sequential design. As we saw above, `laGP` (Gramacy and Sun, 2018) supports ALC design augmentation (§6.2.2) via sums over reference sets, whereas `hetGP` supports batch IMSPE (§6.1.2) through closed form integration over rectangular domains. The reasons for this disparity are intimately related to the nature of their fancier features. Finally, both `alcGP/alcGPsep` and `IMSPE` support candidate-based search without derivatives, as we have illustrated above.

The `hetGP` package additionally provides a sequential IMSPE analog automating search for x_{n+1} with derivatives in a carefully optimized `Rcpp` implementation. See `IMSPE_optim` in that package. Your task in this exercise is to empirically explore sequential design under IMSPE and compare it to ALC. Note that `alcoptGP` and `alcoptGPsep` support derivatives, so you may optionally/additionally choose to use one of these rather than what's implemented in the chapter.

a. Wrap our 2d ALC sequential design code from §6.2.2 into a function that initializes with a design $X_{n_0} \equiv$ `Xinit`, provided as an argument, and iterates to an $N = 100$-sized design. Use `updateGP` to incorporate each selection, but skip MLE updates in order to better compare apples-to-apples with `hetGP`. Use $(\theta, g) = (0.1, 0.01)$. Keep track of progress over the iterations, as well as total compute time. Calculate RMSE to testing data provided, where both testing and training data come from our 2d test function introduced in §5.1.2 and coded as `f` in §6.2.1. A multi-start scheme with n locations in search of x_{n+1}, which was tailored to ALM's highly multimodal surface, is overkill for ALC. `IMSPE_optim` uses far fewer than that. Modify `xnp1.search` to use just five, specifying `T=100*5` iterations of `mymaximin`.

b. Create a sequential IMPSE analog to #a using `IMSPE_optim` and `hetGP`'s `update` capability. The package doesn't have a `newGP` analog, but calling `mleHomGP` with `known=list(theta=theta, g=g)` and `maxit=0` implements a similar setup – skipping MLE calculations.

c. Compare alternatives from #a and #b (creating two designs initialized with the same $n_0 = 12$-sized seed design $X_{n_0} \equiv$ `X`) qualitatively through views on the design(s), and quantitatively through compute time and progress over sequential design iterations with out-of-sample RMSE.

d. Repeat #c one hundred times, each with a new initializing LHS $X_{n_0} \equiv$ `X`, and report on compute times and the distribution of RMSEs over the iterations.

#3: Distance-distributed design

Zhang et al. (2020) argue that designs based on a diversity of pairwise distances between rows of X_n, i.e., those quantities which appear in evaluating inverse exponentiated distance-based kernels, are better for estimating lengthscales $\hat{\theta}$ than space-filling designs. In fact, random designs are better than maximin and sometimes also better than LHS.

a. Show that maximin designs offer a highly multimodal pairwise (un-squared) Euclidean distance distribution that would bias inference for $\hat{\theta}$ towards certain ranges. Moreover, argue that they may entirely preclude reliable estimation of small $\hat{\theta}$-values. Specifically, use `mymaximin` from §4.2.1, or any of its upgrades from your solution to homework exercises in §4.4, to generate 100 $n = 8$-sized maximin designs in the $m = 2$-dimensional domain $[0, 1]^2$. Calculate all $100 \cdot 8 \cdot 7/2 = 2800$ pairwise distances, plot their histogram or kernel density and comment.

b. One way to obtain a greater diversity of pairwise Euclidean distances would be to deliberately target their *uniform* distribution. Zhang et al. (2020) calculate that, in 2d, pairwise distances distributed as Beta$(2, 5)$ are better than uniform. Develop a scheme in the style of `mymaximin` or `maxent` which searches for a *betadist design* whose distribution of pairwise distances (divided by $\sqrt{2}$ for $[0, 1]^2$) follows Beta(α, β). Uniform and Beta$(2, 5)$ may be calculated as special cases. Use Kolmogorov–Smirnov distance[10] via `ks.test` in R to measure the distance between the empirical distribution of pairwise distances in your `X` designs to the target distribution. Keep track of progress in your search. Visualize

---

[10] https://en.wikipedia.org/wiki/Kolmogorov-Smirnov_test

a couple of $n = 8$-sized Beta$(2, 5)$-dist designs in terms of their locations in 2d, progress in search, and their empirical distribution of distances compared to the target.

c. One of the best uses for betadist designs is in sequential application. Getting reliable $\hat{\theta}_{n_0}$ is crucial to avoiding vicious cycles in learning and sequential selection of augmenting designs. Using the 2d test function introduced in §5.1.2 and coded as f in §6.2.1, along with associated 100×100 out-of-sample testing grid, calculate sequential designs under ALM with the following setup: (i) update MLE calculations for $\hat{\theta}_n \in (0, \sqrt{2}]$ and $\hat{g}_n \in (0, 1)$, as $n \to n + 1$ with searches initialized at $(\theta_0, g_0) = (0.5, 1)$; (ii) seed with $n_0 = 8$ points under maximin, random, LHS, and Beta$(2, 5)$-dist criteria and subsequently perform 56 sequential design iterations for $N = 64$ total; (iii) visualize RMSE progress over the 56 iterations for the four initializing comparators.

d. Repeat #c thirty times in an MC fashion and visualize progress for the four initializing comparators through mean and upper-90% quantile summarizing the distribution of RMSEs.

#4: Sequential GP updates

This exercise fleshes out some of the details behind fast updates of GP hyperparameters in §6.3.

a. Consider the MLE for τ^2 conditional on the hyperparameters (e.g., θ and g) to the covariance structure of a GP. Derive an updating rule for τ^2 as we increase from n to $n + 1$ data points, i.e., $\hat{\tau}_n^2 \to \hat{\tau}_{n+1}^2$ that uses only the previous estimate, i.e., $\hat{\tau}_n^2$, and rows and columns of K_{n+1}^{-1} as prescribed by the partitioned inverse equations (6.8). What is the computational order of your update (i.e., additional to the cost of decomposing K_{n+1}^{-1} using partition inverse equations)?

b. Revisit exercise #3 from §5.5 which augments the GP with a linear mean, via coefficients β. Derive updates for the MLE $\hat{\beta}_n \to \hat{\beta}_{n+1}$, and revise your updates for $\hat{\tau}_n^2 \to \hat{\tau}_{n+1}^2$ accordingly. Note that, in both cases, it's not necessary that you express the update for the $n + 1^{\text{st}}$ estimator (e.g., $\hat{\beta}_{n+1}$) directly in terms of nth one (i.e., $\hat{\beta}_n$). Rather, you may express it in terms of updates to it's relevant sub-calculations (e.g., $X_n^\top K_n^{-1} X_n \to X_{n+1}^\top K_{n+1}^{-1} X_{n+1}$). Say what the computational order is of the updates in each case, i.e., additional to the cost of decomposing K_{n+1}^{-1} with partitioned inverse equations, and other quantities already calculated up to point n.

#5: Sequential design in a nonstationary regime

Consider again the LGBB data (e.g., from exercise #8 in §5.5), but this time with the "fill" version, which is on a much denser grid, and using the side response. Keep things in 2d, where visualization is easiest, and consider only the subset of these data where the side-slip angle beta is two.

```
load("lgbb/lgbb_fill.RData")
X <- lgbb.fill[,1:2]
y <- lgbb.fill$side
btwo <- lgbb.fill$beta == 2
X <- X[btwo,]
y <- y[btwo]
length(y)
```

[1] 4212

Finally, treat these data as deterministic evaluations. Now, we're not going to use these data directly. This will represent our full cache of potential runs. We're going to build up our own active cache by active learning.

a. Appropriate your `mymaximin.cand` function from §4.4 exercise #4, or try `mymaximin.cand` from the `maximin` library (Sun and Gramacy, 2019) on CRAN. Generate a space-filling training design X_n of size $n = 200$ from among the candidates in X. Get the Y_n-values at those inputs (a subset of the full ys) and fit a separable GP to those (X_n, Y_n) pairs. Obtain predictions at the entire set of X locations, including both training and remaining candidates, and report on the RMSE of those predictions by comparing them to the true y values.

b. Write code to build ALM and ALC sequential designs by choosing from the X candidates, and incorporate the y values of the chosen design elements X_n as $n = 31, \ldots, 200$ after initializing with a $n_0 = 30$-sized `maximin.cand` design. Do this for the three different GP specifications as described below. Track and report RMSE on the entire set of X locations, including both training and remaining candidates, for each sequential design iteration $n = 31, \ldots, 200$. Additionally provide image/contour plots of predictive surfaces at the end of the design iterations. *Note that in the case of #2, below, you'll need to think carefully about how you're going to combine across the GPs in the partition(s).*

   1. A separable GP with a parameterization that updates as the designs grow, on a schedule that you determine is sufficient.
   2. Two separable GPs on a partition of the input space: GP_1 governs speeds less than mach 2; GP_2 governs speeds greater than or equal to mach 2. Both GP's hyperparameters should be updated as designs grow, on a schedule that you deem sufficient.

c. Repeat building designs in this way (#b), and tracking progress, as many times as you can (up to thirty, say), in order to report average progress with quantiles.

7

Optimization

In this chapter, the goal is to demonstrate how Gaussian process (GP) surrogate modeling can assist in optimizing a *blackbox* objective function. That is, a function about which one knows little – one opaque to the optimizer – and that can only be probed through expensive evaluation. We view optimization as an example of sequential design or active learning (§6.2}). Many names have been given to this enterprise, and some will be explained alongside their origins below. Recently *Bayesian Optimization (BO)*[1] has caught on, especially in the machine learning (ML) community, and likely that'll stick in part because it's punchier than the alternatives. BO terminology goes back to a paper predating use of GPs toward this end, and refers primarily to decision criteria for sequential selection and model updating. Modern ML vernacular prefers acquisition functions. BO's recent gravitas is primarily due to the prevailing view of GP learning as marginalization over a latent random field (§5.3.2).

The role of modeling in optimization, more generically, has a rich history, and we'll barely scratch the surface here. Models deployed to assist in optimization can be both statistical and non-statistical, however the latter often have strikingly similar statistical analogs. Potential for modern nonparametric statistical surrogate modeling in this context is just recently being recognized by communities for which optimization is bread-and-butter: mathematical programming, statistics, ML, and more. Optimization has played a vital role in stats and ML for decades. All those `"L-BFGS-B"` searches (Byrd et al., 1995) from `optim` in earlier chapters, say to find MLEs or optimal designs, are cases in point. It's intriguing to wonder whether statistical thinking might have something to give back to the optimization world, as it were.

Whereas many communities have long settled for local refinement, statistical methods based on nonparametric surrogate models offer promise for greater scope. True global optimization, whatever that means, may always remain illusive and enterprises motivated by such lofty goals may be folly. But anyone suggesting that wider perspective is undesirable, and that systematic frameworks developed toward that end aren't worth exploring, is ignoring practitioners and clients who are optimistic that the grass is greener just over the horizon.

Statistical decision criteria can leverage globally scoped surrogates to balance exploration (uncertainty reduction; Chapter 6) and exploitation (steepest ascent; Chapter 3) in order to more reliably find global optima. But that's just the tip of the iceberg. Statistical models cope handily with noisy blackbox evaluations – a situation where probabilistic reasoning ought to have a monopoly – and thus lend a sense of robustness to solutions and to a notion of convergence. They offer the means of uncertainty quantification (UQ) about many aspects, including the chance that local or global optima were missed. Extension to related, optimization-like criteria such as level-set/contour finding (e.g., Ranjan et al., 2008) is relatively straightforward, although these topics aren't directly addressed in this chapter. See §10.3.4 for pointers along these lines.

It's important to disclaim that thinking statistically need not preclude application of more

---

[1]https://bayesopt.github.io/

classical approaches. Surrogate modeling naturally lends itself to hybridization. Ideas from mathematical programming not only have merit, but their software is exceedingly well engineered. Let's not throw the baby out with the bath water. It makes sense to borrow strengths from multiple toolkits and leverage solid implementation and decades of stress testing.

Although the main ideas on statistical surrogate modeling for optimization originate in the statistics literature, it's again machine learners who are making the most noise in this area today. This is, in part, because they're more keen to adopt the rich language of mathematical programming, and therefore better able to connect with the wider optimization community. Statisticians tend to get caught up in modeling details and forget that optimization is as much about execution as it's about methodology. Practitioners want to use these tools, but rarely have the in-house expertise required to code them up on their own. General purpose software leveraging surrogates has been slow to come online. An important goal of this chapter is to show how that might work, and to expose substantial inroads along those lines.

We'll see how modern nonparametric surrogate modeling and clever (yet arguably heuristic) criteria, that often can be solved in closed form, may be combined to effectively balance exploration and exploitation. Emphasis here is on GPs, but many methods are agnostic to choices of surrogate. We begin by targeting globally scoped numerical optimization, leveraging only blackbox evaluations of an objective function supplied by the user. Subsequently, we shall embellish that setup with methods for handling constraints, known and unknown (i.e., also blackbox); hybridize with modern methods from mathematical programming; and talk about applications from toy to real data.

7.1   Surrogate-assisted optimization

Statistical methods in optimization, in particular of noisy blackbox functions, probably goes back to Box and Draper (1987), a precursor to a canonical response surface methods text by the same authors (Box and Draper, 2007). Modern Bayesian optimization (BO) is closest in spirit to methods described by Mockus et al. (1978) in a paper entitled "The application of Bayesian methods for seeking the extremum". Yet many strategies suggested therein didn't come to the fore until the late 1990's, perhaps because they emphasized rather crude (linear) modeling. As GPs became established for modeling computer simulations, and subsequently in ML in the 2000s, new life was breathed in.

In the computer experiments literature, folks have been using GPs to optimize functions for some time. One of the best references for the core idea might be Booker et al. (1999), with many ingredients predating that paper. They called it *surrogate-assisted optimization*, and it involved a nice collaboration between optimization and computer modeling researchers. Non-statistical surrogates had been in play in optimization for some time. Nonparametric statistical ones, with more global scope, offered fresh perspective.

The methodology is simple: train a GP on function evaluations obtained so far; minimize the fitted surrogate predictive mean surface of the GP to select the next location for evaluation; repeat. This is an instance of Algorithm 6.1 in §6.2 where the criterion $J(x)$ in Step 3 is based on GP predictive mean $\mu_n(x) = \mathbb{E}\{Y(x) \mid D_n\}$ provided in Eq. (5.2). Although Step 3 deploys its own inner-optimization, minimizing $\mu_n(x)$ is comparatively easy since it doesn't involve evaluating a computationally intensive, and potentially noisy, blackbox. It's something that can easily be solved with conventional methods. As a shorthand, and to connect with other acronyms like EI and PI below, I shall refer to this "mean criterion" as the EY heuristic for surrogate-assisted (Bayesian) optimization (BO).

Before we continue, let's be clear about the problem. Whereas the RSM literature (Chapter 3) orients toward maxima, BO and math programming favor minimization, which I shall adopt for our discussions here. Specifically, we wish to find

$$x^\star = \text{argmin}_{x \in \mathcal{X}} \; f(x)$$

where $\mathcal{X}$ is usually a hyperrectangle, a bounding box, or another simply constrained region. We don't have access to derivative evaluations for $f(x)$, nor do we necessarily want them (or want to approximate them) because that could represent additional substantial computational expense. As such, methods described here fall under the class of *derivative-free optimization*. See, e.g., Conn et al. (2009), for which many innovative algorithms have been proposed, and many solid implementations are widely available. For a somewhat more recent review, including several of the surrogate-assisted/BO methods introduced here, see Larson et al. (2019).

All we get to do is evaluate $f(x)$, which for now is presumed to be deterministic. Generalizations will come after introducing main concepts, including simple extensions for the noisy case. The literature targets scenarios where $f(x)$ is expensive to evaluate (in terms of computing time, say), but otherwise is well-behaved: continuous, relatively smooth, only real-valued inputs, etc. Again, these are relaxable modulo suitable surrogate and/or kernel structure. Several appropriate choices are introduced in later chapters.

Implicit in the computational expense of $f(x)$ evaluations is a tacit "constraint" on the solver, namely that it minimize the number of such evaluations. Although it's not uncommon to study aspects of convergence in these settings, often the goal is simply to find the best input, $x^\star$ minimizing f, in the fewest number of evaluations. So in empirical comparisons we typically track the *best objective value* (BOV) found as a function of the number of blackbox evaluations. In many applied contexts it's more common to have a fixed evaluation budget than it is to enjoy the luxury of running to convergence. Nevertheless, monitoring progress plays a key role when deciding if further expensive evaluations may be required.

7.1.1 A running example

Consider an implementation of Booker et al.'s method on a re-scaled/coded version of the Goldstein–Price[2] function. See §1.4 homework exercises for more on this challenging benchmark problem.

```
f <- function(X)
 {
  if(is.null(nrow(X))) X <- matrix(X, nrow=1)
  m <- 8.6928
  s <- 2.4269
  x1 <- 4*X[,1] - 2
  x2 <- 4*X[,2] - 2
```

---

[2]http://www.sfu.ca/~ssurjano/goldpr.html

```
  a <- 1 + (x1 + x2 + 1)^2 *
    (19 - 14*x1 + 3*x1^2 - 14*x2 + 6*x1*x2 + 3*x2^2)
  b <- 30 + (2*x1 - 3*x2)^2 *
    (18 - 32*x1 + 12*x1^2 + 48*x2 - 36*x1*x2 + 27*x2^2)
  f <- log(a*b)
  f <- (f - m)/s
  return(f)
}
```

Although this $f(x)$ isn't opaque to us, and not expensive to evaluate, we shall treat it as such for purposes of illustration. Not much insight can be gained by looking at its form in any case, except to convince the reader that it furnishes a suitably complicated surface despite residing in modest input dimension ($m = 2$).

Begin with a small space-filling Latin hypercube sample (LHS; §4.1) seed design in 2d.

```
library(lhs)
ninit <- 12
X <- randomLHS(ninit, 2)
y <- f(X)
```

Next fit a separable GP to those data, with a small nugget for jitter. All of the same caveats about initial lengthscales in GP-based active learning – see §6.2.1 – apply here. To help create a prior on θ that's more stable, darg below utilizes a large auxiliary pseudo-design in lieu of X, which at early stages of design/optimization ($n_0 = 12$ runs) may not yet possess a sufficient diversity of pairwise distances.

```
library(laGP)
da <- darg(list(mle=TRUE, max=0.5), randomLHS(1000, 2))
gpi <- newGPsep(X, y, d=da$start, g=1e-6, dK=TRUE)
mleGPsep(gpi, param="d", tmin=da$min, tmax=da$max, ab=da$ab)$msg
```

```
## [1] "CONVERGENCE: REL_REDUCTION_OF_F <= FACTR*EPSMCH"
```

Just like our ALM/C searches from §6.2, consider an objective based on GP predictive equations. This represents an implementation of Step 3 in Algorithm 6.1 for sequential design/active learning, setting EY as sequential design criterion $J(x)$, or defining the acquisition function in ML jargon.

```
obj.mean <- function(x, gpi)
  predGPsep(gpi, matrix(x, nrow=1), lite=TRUE)$mean
```

Now the predictive mean surface (like f, through the evaluations it's trained on) may have many local minima, but let's punt for now on the ideal of global optimization of EY – of the so-called "inner loop" – and see where we get with a search initialized at the current best value. R code below extracts that value: m indexing the best y-value obtained so far, and uses its x coordinates to initialize a "L-BFGS-B" solver on obj.mean.

```
m <- which.min(y)
opt <- optim(X[m,], obj.mean, lower=0, upper=1, method="L-BFGS-B", gpi=gpi)
opt$par
```

```
## [1] 0.4465 0.3256
```

So this is the next point to try. Surrogate optima represent a sensible choice for the next evaluation of the expensive blackbox, or so the thinking goes. Before moving to the next acquisition, Figure 7.1 provides a visualization. Open circles indicate locations of the size-n_0 LHS seed design. The origin of the arrow indicates X[m,]: the location whose y-value is lowest based on those initial evaluations. Its terminus shows the outcome of the optim call: the next point to try.

```
plot(X[1:ninit,], xlab="x1", ylab="x2", xlim=c(0,1), ylim=c(0,1))
arrows(X[m,1], X[m,2], opt$par[1], opt$par[2], length=0.1)
```

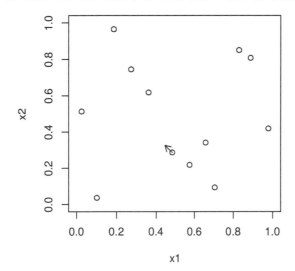

FIGURE 7.1: First iteration of EY-based search initialized from the value of the objective obtained so far.

Ok, now evaluate f at opt$par, update the GP and its hyperparameters ...

```
ynew <- f(opt$par)
updateGPsep(gpi, matrix(opt$par, nrow=1), ynew)
mle <- mleGPsep(gpi, param="d", tmin=da$min, tmax=da$max, ab=da$ab)
X <- rbind(X, opt$par)
y <- c(y, ynew)
```

... and solve for the next point.

```
m <- which.min(y)
opt <- optim(X[m,], obj.mean, lower=0, upper=1, method="L-BFGS-B", gpi=gpi)
opt$par
```

```
## [1] 0.4983 0.2847
```

Figure 7.2 shows what that looks like in the input domain, as an update on Figure 7.1. In particular, the terminus of the arrow in Figure 7.1 has become an open circle in Figure 7.2, as that input–output pair has been promoted into the training data with GP fit revised accordingly.

```
plot(X, xlab="x1", ylab="x2", xlim=c(0,1), ylim=c(0,1))
arrows(X[m,1], X[m,2], opt$par[1], opt$par[2], length=0.1)
```

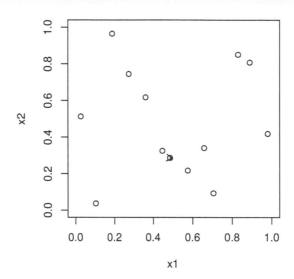

FIGURE 7.2: Second iteration of EY search following Figure 7.1.

If the origin of the new arrow resides at the newly minted open circle, then we have progress: the predictive mean surface was accurate and indeed helpful in finding a new best point, minimizing f. If not, then the origin is back at the same open circle it originated from before.

Now incorporate the new point into our dataset and update the GP predictor.

```
ynew <- f(opt$par)
updateGPsep(gpi, matrix(opt$par, nrow=1), ynew)
mle <- mleGPsep(gpi, param="d", tmin=da$min, tmax=da$max, ab=da$ab)
X <- rbind(X, opt$par)
y <- c(y, ynew)
```

Let's fast-forward a little bit. Code below wraps what we've been doing above into a `while` loop with a simple check on convergence in order to "break out". If two outputs in a row are sufficiently close, within a tolerance `1e-4`, then stop. That's quite crude, but sufficient for illustrative purposes.

```
while(1) {
  m <- which.min(y)
  opt <- optim(X[m,], obj.mean, lower=0, upper=1,
    method="L-BFGS-B", gpi=gpi)
```

```
  ynew <- f(opt$par)
  if(abs(ynew - y[length(y)]) < 1e-4) break;
  updateGPsep(gpi, matrix(opt$par, nrow=1), ynew)
  mle <- mleGPsep(gpi, param="d", tmin=da$min, tmax=da$max, ab=da$ab)
  X <- rbind(X, opt$par)
  y <- c(y, ynew)
}
deleteGPsep(gpi)
```

To help measure progress, code below implements some post-processing to track the best y-value (`bov`: best objective value) over those iterations. The function is written in some generality in order to accommodate application in several distinct settings, coming later.

```
bov <- function(y, end=length(y))
  {
  prog <- rep(min(y), end)
  prog[1:min(end, length(y))] <- y[1:min(end, length(y))]
  for(i in 2:end)
    if(is.na(prog[i]) || prog[i] > prog[i-1]) prog[i] <- prog[i-1]
  return(prog)
  }
```

In our application momentarily, note that we treat the $n_0 = 12$ seed design the same as later acquisitions, even though they were not derived from the surrogate/criterion.

```
prog <- bov(y)
```

That progress meter is shown visually in Figure 7.3. A vertical dashed line indicates n_0, the size of seed LHS design.

```
plot(prog, type="l", col="gray", xlab="n: blackbox evaluations",
  ylab="best objective value")
abline(v=ninit, lty=2)
legend("topright", "seed LHS", lty=2, bty="n")
```

Although it's difficult to comment on particulars due to random initialization, in most variations substantial progress is apparent over the latter 14 iterations of active learning. There may be plateaus where consecutive iterations show no progress, but these are usually interspersed with modest "drops" and even large "plunges" in BOV. When comparing optimization algorithms based on such progress metrics, the goal is to have BOV curves hug the lower-left-hand corner of the plot to the greatest extent possible: to have the best value of the objective in the fewest number of iterations/evaluations of the expensive blackbox.

To better explore diversity in progress over repeated trials with different random seed designs, an R function below encapsulates our code from above. In addition to a tolerance on successive y-values, an **end** argument enforces a maximum number of iterations. The full dataset of inputs X and evaluations y is returned.

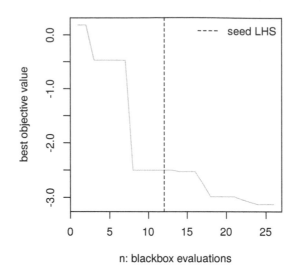

FIGURE 7.3: EY progress in terms of BOV over sequential design iterations.

```
optim.surr <- function(f, m, ninit, end, tol=1e-4)
{
  ## initialization
  X <- randomLHS(ninit, m)
  y <- f(X)
  da <- darg(list(mle=TRUE, max=0.5), randomLHS(1000, m))
  gpi <- newGPsep(X, y, d=da$start, g=1e-6, dK=TRUE)
  mleGPsep(gpi, param="d", tmin=da$min, tmax=da$max, ab=da$ab)

  ## optimization loop
  for(i in (ninit+1):end) {
    m <- which.min(y)
    opt <- optim(X[m,], obj.mean, lower=0, upper=1,
      method="L-BFGS-B", gpi=gpi)
    ynew <- f(opt$par)
    if(abs(ynew - y[length(y)]) < tol) break;
    updateGPsep(gpi, matrix(opt$par, nrow=1), ynew)
    mleGPsep(gpi, param="d", tmin=da$min, tmax=da$max, ab=da$ab)
    X <- rbind(X, opt$par)
    y <- c(y, ynew)
  }

  ## clean up and return
  deleteGPsep(gpi)
  return(list(X=X, y=y))
}
```

Consider re-seeding and re-solving the optimization problem, minimizing f, in this way over 100 Monte Carlo (MC) repetitions. The loop below combines calls to optim.surr with prog post-processing. A maximum number of end=50 iterations is allowed, but often convergence is signaled after many fewer acquisitions.

```
reps <- 100
end <- 50
prog <- matrix(NA, nrow=reps, ncol=end)
for(r in 1:reps) {
  os <- optim.surr(f, 2, ninit, end)
  prog[r,] <- bov(os$y, end)
}
```

It's important to note that these are random initializations, not random searches. Searches (after initialization) are completely deterministic. Surrogate-assisted/Bayesian optimization is not a *stochastic optimization*, like simulated annealing[3], although sometimes deterministic optimizers are randomly initialized as we have done here. Stochastic optimizations, where each sequential decision involves a degree of randomness, don't make good optimizers for expensive blackbox functions because random evaluations of f are considered too wasteful in computational terms.

Figure 7.4 shows the 100 trajectories stored in prog.

```
matplot(t(prog), type="l", col="gray", lty=1,
  xlab="n: blackbox evaluations",  ylab="best objective value")
abline(v=ninit, lty=2)
legend("topright", "seed LHS", lty=2, bty="n")
```

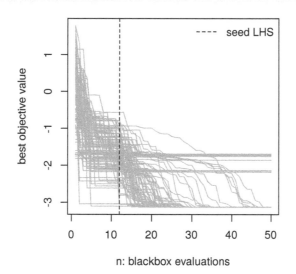

FIGURE 7.4: Multiple BOV progress paths following Figure 7.3 under random reinitialization.

Clearly this is not a global optimization tool. It looks like there are three or four local optima, or at least the optimizer is being pulled toward three or four domains of attraction. Before commenting further, it'll be helpful to have something to compare to.

[3]https://en.wikipedia.org/wiki/Simulated_annealing

7.1.2   A classical comparator

How about our favorite optimization library: `optim` using `"L-BFGS-B"`? Working `optim` in as a comparator requires a slight tweak.

Code below modifies our objective function to help keep track of the full set of y-values gathered over optimization iterations, as these are not saved by `optim` in a way that's useful for backing out a progress meter (e.g., `prog`) for comparison. (The `optim` method works just fine, but it was not designed with my illustrative purpose in mind. It wants to iterate to convergence and give the final result rather than bother you with the details of each evaluation. A `trace` argument prints partial evaluation information to the screen, but doesn't return those values for later use.) So the code below updates a `y` object stored in the calling environment.

```
fprime <- function(x)
 {
  ynew <- f(x)
  y <<- c(y, ynew)
  return(ynew)
 }
```

Below is the same `for` loop we did for EY-based surrogate-assisted optimization, but with a direct `optim` instead. A single random coordinate is used to initialize search, which means that `optim` enjoys `ninit - 1` extra decision-based acquisitions compared to the surrogate method. (No handicap is applied. All expensive function evaluations count equally.) Although `optim` may utilize more than fifty iterations, our extraction of `prog` implements a cap of fifty.

```
prog.optim <- matrix(NA, nrow=reps, ncol=end)
for(r in 1:reps) {
  y <- c()
  os <- optim(runif(2), fprime, lower=0, upper=1, method="L-BFGS-B")
  prog.optim[r,] <- bov(y, end)
}
```

How does `optim` compare to surrogate-assisted optimization with EY? Figure 7.5 shows that surrogates are much better on a fixed budget. New `prog` measures for `optim` are shown in red.

```
matplot(t(prog.optim), type="l", col="red", lty=1,
  xlab="n: blackbox evaluations", ylab="best objective value")
matlines(t(prog), type="l", col="gray", lty=1)
legend("topright", c("EY", "optim"), col=c("gray", "red"), lty=1, bty="n")
```

What makes EY so much better; or `optim` so much worse? Several things. First, `"L-BFGS-B"` spends precious blackbox evaluations on approximating derivatives. That explains the regular plateaus every five or so iterations. Each step spends roughly $2m$ calls to f on calculating a tangent plane, alongside one further call at each newly chosen location, which is often just a short distance away. By comparison, surrogates provide a sense of derivative for "free", not just locally but everywhere. The curious reader may wish to try `method="Nelder-Mead"`

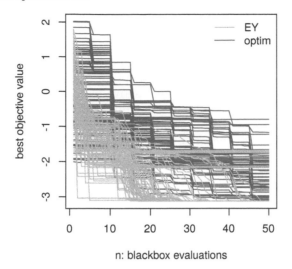

n: blackbox evaluations

FIGURE 7.5: Augmenting Figure 7.4 to include a "L-BFGS-B" comparator under random restarts.

instead (Nelder and Mead, 1965), which is the default in `optim`. Nelder-Mead doesn't support search bounds, so the objective must be modified slightly to prevent search wandering well outside our domain of interest. Nelder-Mead doesn't approximate derivatives, so its progress is a little "smoother" and also a little faster on this 2d problem (when constrained to the box), but is ultimately not as good as the surrogate method. I find that "L-BFGS-B" is more robust, especially in higher dimension, and I like that you can specify a bounding box.

Second, whereas `optim`, using "L-BFGS-B" or "Nelder-Mead", emphasizes local refinements – leveraging some limited memory, which is like using a model of sorts but not a statistical one – surrogates hold potential for large steps because their view of the response surface is more broad. On occasion, that helps EY escape from local minima, which is more likely in early rather than later iterations.

By and large, surrogate-assisted optimization with EY is still a local affair. Its advantages stem primarily from an enhanced sense of perspective. It *exploits*, by moving to the next best spot from where it left off, descending with its own `optim` subroutine on $\mu_n(x)$. As implemented, it doesn't *explore* places that cannot easily be reached from the current best value. It could help to initialize `optim` subroutines elsewhere, but the success of such variations is highly problem dependent. Sometimes it helps a little, sometimes it hurts.

What's missing is some way to balance exploration and exploitation. By the way, notice that we're not actually doing statistics, because at no point is uncertainty being taken into account. Only predictive means are used; predictive variance doesn't factor in. One way to incorporate uncertainty is through the *lower confidence bound (LCB)* heuristic (Srinivas et al., 2009). LCB is a simple linear combination between mean and standard deviation:

$$\alpha_{\mathrm{LCB}}(x) = -\mu_n(x) + \beta_n \sigma_n(x), \quad \text{and} \quad x_{n+1} = \mathrm{argmax}_x \, \alpha_{\mathrm{LCB}}(x). \tag{7.1}$$

LCB introduces a sequence of tuning parameters β_n, targeting a balance between exploration and exploitation as iterations of optimization progress. Larger β_n lead to more conservative searches, until in the limit $\mu_n(x)$ is ignored and acquisitions reduce to ALM (§6.2.1). A family of optimal choices $\hat{\beta}_n$ may be derived by minimizing regret in a multi-armed bandit setting[4], however in practice such theoretically automatic selections remain unwieldy for practitioners.

There's a twist on EY, called *Thompson sampling* (Thompson, 1933), that gracefully incorporates predictive uncertainty. Rather than optimize the predictive mean directly, take a draw from the (full covariance) posterior predictive distribution (5.3), and optimize that instead. Each iteration of search would involve a different, independent surrogate function draw. Therefore, a collection of many optimization steps would feel the effect of a sense of relative uncertainty through a diversity of such draws. Sparsely sampled parts of the input space could see large swings in surrogate optima, which would lead to exploratory behavior. No setting of tuning parameters required.

There are several disadvantages to Thompson sampling however. One is that it's a stochastic optimization. Each iteration of optimization is based on an inherently random process. Another is that you must commit to a predictive grid in advance, in order to draw from posterior predictive equations, which rules out (continuous) library-based optimization in the inner loop, say by `"L-BFGS-B"`. Actually it's possible to get around a predictive grid with a substantially more elaborate implementation involving iterative applications of MVN conditionals (5.1). But there's a better, easier, and more deterministic way to accomplish the same feat by deliberately acknowledging a role for predictive variability in acquisition criteria.

7.2 Expected improvement

In the mid 1990s, Matthias Schonlau (1997) was working on his dissertation, which basically revisited Mockus' Bayesian optimization idea from a GP and computer experiments perspective. He came up with a heuristic called *expected improvement (EI)*, which is the basis of the so-called *efficient global optimization (EGO)* algorithm. This distinction is subtle: one is the sequential design criterion (EI), and the other is its repeated application toward minimizing a blackbox function (EGO). In the literature, you'll see the overall method referred to by both names/acronyms.

Schonlau's key insight was that predictive uncertainty is underutilized by surrogate frameworks for optimization, which is especially a shame when GPs are involved because they provide such a beautiful predictive variance function. The basic tenets, however, are not limited to GP surrogates. The key is to link prediction to potential for optimization via a measure of *improvement*. Let $f_{\min}^n = \min\{y_1, \ldots, y_n\}$ be the smallest, best blackbox objective evaluation obtained so far, i.e., the BOV quantity we tracked with `prog` in our earlier illustrations. Schonlau defined potential for improvement over $f_{\min}^n$ at an input location x as

$$ I(x) = \max\{0, f_n^{\min} - Y(x)\}. $$

$I(x)$ is a random variable. It measures the amount by which a response $Y(x)$ could be below the BOV obtained so far. Here $Y(x)$ is shorthand for $Y(x) \mid D_n$, the predictive distribution obtained from a fitted model. If $Y(x) \mid D_n$ has non-zero probability of taking on any value on the real line, as it does under a Gaussian predictive distribution, then $I(x)$ has nonzero probability of being positive for any x.

---

[4]https://en.wikipedia.org/wiki/Multi-armed_bandit

Now there are lots of things you could imagine doing with $I(x)$. (Something must be done because in its raw form, as a random variable, it's of little practical value.) One option is to convert it into a probability. The *probability of improvement (PI)* criterion is $\mathrm{PI}(x) = P(I(x) > 0 \mid D_n)$, which is equivalent to $P(Y(x) < f^n_{\min} \mid D_n)$. Maximizing PI is sensible, but could result in very small steps. The most probable input $x^\star = \max_x \mathrm{PI}(x)$ may not hold the greatest potential for large improvement, which is important when considering the tacit goal of minimizing the number of expensive blackbox evaluations. Instead, maximizing *expected improvement (EI)*, $\mathrm{EI}(x) = \mathbb{E}\{I(x) \mid D_n\}$, more squarely targets potential for large improvement.

The easiest way to calculate PI or EI, where "easy" means agnostic to the form of $Y(x) \mid D_n$, is through MC approximation. Draw $y^{(t)} \sim Y(x) \mid D_n$, for $t = 1, \ldots, T$, from their posterior predictive distribution, and average

$$\mathrm{PI}(x) \approx \frac{1}{T} \sum_{t=1}^{T} \mathbb{I}_{\{y^{(t)} > 0\}} \quad \text{or} \quad \mathrm{EI}(x) \approx \frac{1}{T} \sum_{t=1}^{T} \max\{0, f^n_{\min} - y^{(t)}\}.$$

In the limit as $T \to \infty$ these approximations become exact. This approach works no matter what the distribution of $Y(x)$ is, so long as you can simulate from it. With fully Bayesian response surface methods leveraging Markov chain Monte Carlo (MCMC) posterior sampling, say, such approximation may represent the only viable option.

However if $Y(x) \mid D_n$ is Gaussian, as it's under the predictive equations of a GP surrogate conditional on a particular set of hyperparameters, both have a convenient closed form. PI involves a standard Gaussian CDF (Φ) evaluation, as readily calculated with built-in functions in R.

$$\mathrm{PI}(x) = \Phi\left(\frac{f^n_{\min} - \mu_n(x)}{\sigma_n(x)}\right) \tag{7.2}$$

Transparent in the formula above is that both predictive mean and uncertainty factor into the calculation.

Deriving EI takes a little more work, but nothing an A+ student of calculus couldn't do using substitution and integration by parts. Details are in an appendix of Schonlau's thesis. I shall simply quote the final result here.

$$\mathrm{EI}(x) = (f^n_{\min} - \mu_n(x)) \, \Phi\left(\frac{f^n_{\min} - \mu_n(x)}{\sigma_n(x)}\right) + \sigma_n(x) \, \phi\left(\frac{f^n_{\min} - \mu_n(x)}{\sigma_n(x)}\right), \tag{7.3}$$

where ϕ is the standard Gaussian PDF. Notice how EI contains PI as a component in a larger expression. "One half" of EI is PI multiplied (or weighted) by the amount by which the predictive mean is below $f^n_{\min}$. The other "half" is predictive variance weighted by a Gaussian density evaluation. In this way, maximizing EI organically and transparently balances competing goals of exploitation ($\mu_n(x)$ below $f^n_{\min}$) and exploration (large predictive uncertainty $\sigma_n(x)$).

In what follows the discussion drops PI and focuses exclusively on EI. An exercise in §7.4 encourages the curious reader to rework examples below by swapping in a simple PI implementation, in addition to other alternatives.

7.2.1   Classic EI illustration

As a first illustration of EI, code chunks below recreate an example and visuals presented in the first published/journal manuscript describing EI/EGO (Jones et al., 1998). I like to introduce Schonlau's thesis first to give him proper credit, whereas the Journal of Global Optimization article linked in the previous sentence is often cited as Jones, et al. (1998). Consider data $D_n = (X_n, Y_n)$ hand-coded in R below, which was taken by eyeball from Figure 10 in that paper, and re-centered to have a mean near zero. Observe the deliberate gap in the input space between fourth and fifth inputs, scanning from the left.

```
x <- c(1, 2, 3, 4, 12)
y <- c(0, -1.75, -2, -0.5, 5)
```

Code below initializes a GP fit on these data with hyperparameterization chosen to match that figure, again by eyeball. (It's such a great example from a pedagogical perspective, I didn't want to blow it. We'll do a more dynamic and novel illustration shortly.) Predictions may then be obtained on a dense grid in the input space.

```
gpi <- newGP(matrix(x, ncol=1), y, d=10, g=1e-8)
xx <- seq(0, 13, length=1000)
p <- predGP(gpi, matrix(xx, ncol=1), lite=TRUE)
```

Ok, we have everything we need to calculate EI (7.3). R code below evaluates that equation using predictive quantities stored in p after calculating $f_{\min}^n$ from the small initial set of y values.

```
m <- which.min(y)
fmin <- y[m]
d <- fmin - p$mean
s <- sqrt(p$s2)
dn <- d/s
ei <- d*pnorm(dn) + s*dnorm(dn)
```

The left panel in Figure 7.6 shows the predictive surface in terms of mean and approximate 95% error-bars. The predictive mean clearly indicates potential for a solution in the left half of the input space. Our simple surrogate-assisted EY optimizer (§7.1.2) would exploit that low mean and acquire the next point there. However, error-bars suggest great potential for minima in the right half of the input space. Although means are high there, error-bars fall well below the current best value $f_{\min}^n$, suggesting exploration might be warranted. EI, shown in the right panel of the figure, synthesizes this information to strike a balance between exploration and exploitation.

```
par(mfrow=c(1,2))
plot(x, y, pch=19, xlim=c(0,13), ylim=c(-4,9), main="predictive surface")
lines(xx, p$mean)
lines(xx, p$mean + 2*sqrt(p$s2), col=2, lty=2)
lines(xx, p$mean - 2*sqrt(p$s2), col=2, lty=2)
abline(h=fmin, col=3, lty=3)
```

```
legend("topleft", c("mean", "95% PI", "fmin"), lty=1:3,
   col=1:3, bty="n")
plot(xx, ei, type="l", col="blue", main="EI", xlab="x", ylim=c(0,0.15))
```

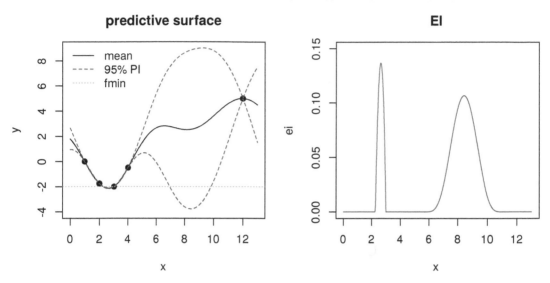

FIGURE 7.6: Classic EI illustration showing predictive surface (left) and corresponding EI surface (right).

That synthesis creates a multimodal EI surface. It's a close call between the left and right options in the input space. Outside of those regions EI is essentially zero, although theoretically positive. The left mode is higher but narrower. Maximizing EI results in choosing the input for the next blackbox evaluation, x_{n+1}, somewhere between 2 and 3. Options other than maximization, towards perhaps a better accounting for breadth of expected improvement, are discussed in §7.2.3. R code below makes a more precise selection and incorporates the new pair (x_{n+1}, y_{n+1}).

```
mm <- which.max(ei)
x <- c(x, xx[mm])
y <- c(y, p$mean[mm])
```

Notice that the predictive mean is being used for y_{n+1} in lieu of a "real" evaluation. For a clean illustration we're supposing that our predictive mean was accurate, and that therefore we have a new global optima. A more realistic example follows soon. Code below updates the GP fit based on the new acquisition and re-evaluates predictive equations on our grid.

```
updateGP(gpi, matrix(xx[mm], ncol=1), p$mean[mm])
p <- predGP(gpi, matrix(xx, ncol=1), lite=TRUE)
deleteGP(gpi)
```

Cutting-and-pasting from above, next convert those predictions into EIs based on new $f_{\min}^n$.

```
m <- which.min(y)
fmin <- y[m]
d <- fmin - p$mean
s <- sqrt(p$s2)
dn <- d/s
ei <- d*pnorm(dn) + s*dnorm(dn)
```

Updated predictive surface (left) and EI criterion (right) are provided in Figure 7.7.

```
par(mfrow=c(1,2))
plot(x, y, pch=19, xlim=c(0,13), ylim=c(-4,9), main="predictive surface")
lines(xx, p$mean)
lines(xx, p$mean + 2*sqrt(p$s2), col=2, lty=2)
lines(xx, p$mean - 2*sqrt(p$s2), col=2, lty=2)
abline(h=fmin, col=3, lty=3)
legend("topleft", c("mean", "95% PI", "fmin"), lty=1:3,
  col=1:3, bty="n")
plot(xx, ei, type="l", col="blue", main="EI", xlab="x", ylim=c(0,0.15))
```

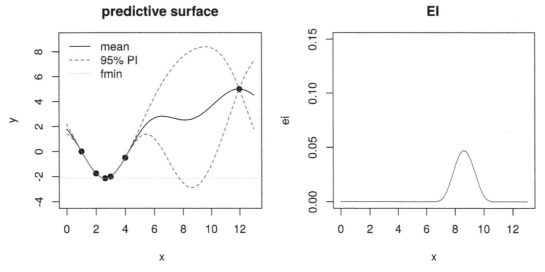

FIGURE 7.7: Predictive surface (left) and EI (right) after the first acquisition; see Figure 7.6.

There are several things to notice from the plots in the figure. Perhaps the most striking feature lies in the updated EI surface, which now contains just one bump. Elsewhere, EI is essentially zero. Choosing max EI will result in exploring the high variance region. The plot is careful to keep the same y-axis compared to Figure 7.6 in order to make transparent that EI tends to decrease as data are added. It need not always decrease, especially when hyperparameters are re-estimated when new data arrive. Since $\hat{\tau}^2$ is analytic, laGP-based subroutines always keep this hyperparameter up to date. The implementation above doesn't update lengthscale and nugget hyperparameters, but it could. Comparing left panels between Figures 7.6 and 7.7, notice that predictive variability has been reduced globally even though a point was added in an already densely-sampled area. Data affect GP fits globally under this stationary, default specification, although influence does diminish exponentially as gaps between training and testing locations grow in the input space.

We're at the limits of this illustration because it doesn't really involve a function f, but rather some data on evaluations gleaned by eyeball from a figure in a research paper. A more interactive demo is provided as gp_ei_sin.R[5] in supplementary material on the book web page. The demo extends our basic sinusoid example from §5.1.1, but is quite similar in flavor to the illustration above. Specifically, it combines an EI function from the `plgp` package (Gramacy, 2014) with GP fitting from `laGP` (Gramacy and Sun, 2018) – somewhat of a hodgepodge but it keeps the code short and sweet. For enhanced transparency on our running Goldstein–Price example (§7.1.1), we'll code up our own EI function here, basically cut-and-paste from above, for use with `laGP` below.

```
EI <- function(gpi, x, fmin, pred=predGPsep)
  {
    if(is.null(nrow(x))) x <- matrix(x, nrow=1)
    p <- pred(gpi, x, lite=TRUE)
    d <- fmin - p$mean
    sigma <- sqrt(p$s2)
    dn <- d/sigma
    ei <- d*pnorm(dn) + sigma*dnorm(dn)
    return(ei)
  }
```

Observe how this implementation combines prediction with EI calculation. That's a little inefficient in situations where we wish to record both predictive and EI quantities, as we did in the example above, especially when evaluating on a dense grid. Rather, it's designed with implementation in Step 3 of Algorithm 6.1 in mind. The final argument, **pred**, enhances flexibility in specification of fitted model and associated prediction routine.

7.2.2 EI on our running example

To set up EI as an objective for minimization, the function below acts as a wrapper around – EI.

```
obj.EI <- function(x, fmin, gpi, pred=predGPsep)
  - EI(gpi, x, fmin, pred)
```

We've seen that EI can be multimodal, but it's not pathologically so like ALM and, to a lesser extent, ALC. ALM/C inherit multimodality from the sausage-shaped GP predictive distribution. Although EI is composed of a measure of predictive uncertainty, its hybrid nature with the predictive mean has a calming effect. Consequently, EI is only high when both mean is low and variance is high, each contributing to potential for low function realization. The number of modes in EI may fluctuate throughout acquisition iterations, but in the long run should resemble the number of troughs in f, assuming minimization. By contrast, the number of ALM/C modes would increase with n. Like with ALM/C, a multi-start scheme for EI searches is sensible, but need not include $\mathcal{O}(n)$ locations and these

---

[5]http://bobby.gramacy.com/surrogates/gp_ei_sin.R

need not be as carefully placed, e.g., at the widest parts of the sausages. A fixed number, perhaps composed of the input location of the BOV (that corresponding to $f_{\min}^n$, where we know the function is low) and a few other points spread around the input space, is often sufficient for decent performance. If not for numerical issues that sometimes arise when EI evaluations are numerically zero for large swaths of inputs, having just two multi-start locations (one at $f_{\min}^n$ and one elsewhere) often suffices. Working with $\log \text{EI}(x)$ sometimes helps in such situations, but I won't bother in our implementation here.

With those considerations in mind, the function below completes our solution to Algorithm 6.1, specifying $J(x)$ in Step 3 for EI-based optimization. EI.search is similar in flavor to xnp1.search from §6.2.1, but tailored to EI and the multi-start scheme described above.

```
eps <- sqrt(.Machine$double.eps) ## used lots below

EI.search <- function(X, y, gpi, pred=predGPsep, multi.start=5, tol=eps)
  {
    m <- which.min(y)
    fmin <- y[m]
    start <- matrix(X[m,], nrow=1)
    if(multi.start > 1)
      start <- rbind(start, randomLHS(multi.start - 1, ncol(X)))
    xnew <- matrix(NA, nrow=nrow(start), ncol=ncol(X)+1)
    for(i in 1:nrow(start)) {
      if(EI(gpi, start[i,], fmin) <= tol) { out <- list(value=-Inf); next }
      out <- optim(start[i,], obj.EI, method="L-BFGS-B",
        lower=0, upper=1, gpi=gpi, pred=pred, fmin=fmin)
      xnew[i,] <- c(out$par, -out$value)
    }
    solns <- data.frame(cbind(start, xnew))
    names(solns) <- c("s1", "s2", "x1", "x2", "val")
    solns <- solns[solns$val > tol,]
    return(solns)
}
```

Although a degree of stochasticity is being injected through the multi-start scheme, this is simply to help the inner loop (maximizing EI), where evaluations are cheap. From the perspective of the outer loop – iterations of sequential design, with details coming momentarily – the intention is still to perform a deliberate and deterministic search. Actually more multi-start locations translate into more careful acquisitions x_{n+1} and subsequent expensive blackbox evaluation. The data.frame returned on output has one row for each multi-start location, although rows yielding effectively zero EI are culled. Multi-start locations whose first EI evaluation is effectively zero are immediately aborted. There are many further ways to "robustify" EI.search, but it's surprising how well things work without much fuss. For example, Jones et al. (1998) suggested branch and bound[6] search rather than multi-start derivative-based numerical optimization (optim in our implementation above). Providing gradients to optim can help too. But the improvements that those enhancements offer are at best slight in my experience.

All right, let's initialize an EI-based optimization – same as for the two earlier comparators.

---

[6]https://en.wikipedia.org/wiki/Branch_and_bound

```
X <- randomLHS(ninit, 2)
y <- f(X)
gpi <- newGPsep(X, y, d=0.1, g=1e-6, dK=TRUE)
da <- darg(list(mle=TRUE, max=0.5), randomLHS(1000, 2))
```

Code below solves for the next input to try, extracting the best row from the output data.frame.

```
solns <- EI.search(X, y, gpi)
m <- which.max(solns$val)
maxei <- solns$val[m]
```

Before acting on that solution, Figure 7.8 summarizes the outcome of search with arrows indicating starting and ending locations of each multi-start EI optimization. Open circles mark the original/existing design. The red dot corresponds to the best row.

```
plot(X, xlab="x1", ylab="x2", xlim=c(0,1), ylim=c(0,1))
arrows(solns$s1, solns$s2, solns$x1, solns$x2, length=0.1)
points(solns$x1[m], solns$x2[m], col=2, pch=20)
```

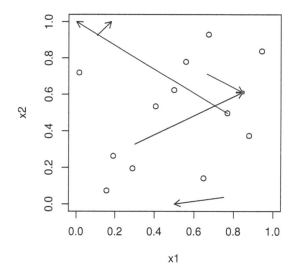

FIGURE 7.8: First iteration of EI search on Goldstein–Price objective in the style of Figure 7.1.

One of the arrows originates from an open circle. This is the multi-start location corresponding to $f_{\min}^n$. Up to four other arrows come from an LHS. If the total number of arrows is fewer than 5, the default multi.start in EI.search, that's because some were initialized in numerically-zero EI locations, and consequently that search was voided at the outset. Usually two or more arrows appear, sometimes with distinct terminus indicating a multi-modal criterion.

Moving on now, code below incorporates the new data at the chosen input location (red dot) and updates the GP fit.

```
xnew <- as.matrix(solns[m,3:4])
X <- rbind(X, xnew)
y <- c(y, f(xnew))
updateGPsep(gpi, xnew, y[length(y)])
mle <- mleGPsep(gpi, param="d", tmin=da$min, tmax=da$max, ab=da$ab)
```

We're ready for the next acquisition. Below the search is combined with a second GP update, after incorporating the newly selected data pair.

```
solns <- EI.search(X, y, gpi)
m <- which.max(solns$val)
maxei <- c(maxei, solns$val[m])
xnew <- as.matrix(solns[m,3:4])
X <- rbind(X, xnew)
y <- c(y, f(xnew))
updateGPsep(gpi, xnew, y[length(y)])
mle <- mleGPsep(gpi, param="d", tmin=da$min, tmax=da$max, ab=da$ab)
```

Figure 7.9 offers a visual, accompanied by a story that's pretty similar to Figure 7.8.

```
plot(X, xlab="x1", ylab="x2", xlim=c(0,1), ylim=c(0,1))
arrows(solns$s1, solns$s2, solns$x1, solns$x2, length=0.1)
points(solns$x1[m], solns$x2[m], col=2, pch=20)
```

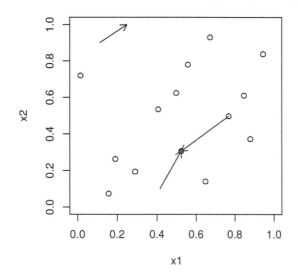

FIGURE 7.9: Second iteration of EI search after Figure 7.8.

You get the idea. Rather than continue with pedantic visuals, a `for` loop below repeats in this way until fifty samples of f have been collected. Hopefully one of them will offer a good solution to the optimization problem, an $x^\star$ with a minimal objective value $f(x^\star)$.

```
for(i in nrow(X):end) {
  solns <- EI.search(X, y, gpi)
  m <- which.max(solns$val)
  maxei <- c(maxei, solns$val[m])
  xnew <- as.matrix(solns[m,3:4])
  ynew <- f(xnew)
  X <- rbind(X, xnew)
  y <- c(y, ynew)
  updateGPsep(gpi, xnew, y[length(y)])
  mle <- mleGPsep(gpi, param="d", tmin=da$min, tmax=da$max, ab=da$ab)
}
deleteGPsep(gpi)
```

Notice that no stopping criteria are implemented in the loop. Our previous y-value-based criterion would not be appropriate here because progress isn't measured by the value of the objective, but instead by potential for (expected) improvement. A relevant such quantity is recorded by `maxei` above. Still, y-progress is essential to drawing comparison to earlier results.

```
prog.ei <- bov(y)
```

Figure 7.10 presents these two measures of progress side-by-side, with `y` on the left and `maxei` on the right.

```
par(mfrow=c(1,2))
plot(prog.ei, type="l", xlab="n: blackbox evaluations",
  ylab="EI best observed value")
abline(v=ninit, lty=2)
legend("topright", "seed LHS", lty=2)
plot(ninit:end, maxei, type="l",  xlim=c(1,end),
  xlab="n: blackbox evaluations", ylab="max EI")
abline(v=ninit, lty=2)
```

First, notice how BOV (on the left) is eventually in the vicinity of -3, which is as good or better than anything we obtained in previous optimizations. Even in this random Rmarkdown build I can be pretty confident about that outcome for reasons that'll become more apparent after we complete a full MC study, shortly. The right panel, showing EI progress, is usually spiky. As the GP learns about the response surface, globally, it revises which areas it believes are high variance, and which give low response values, culminating in dramatic shifts in potential for improvement. Eventually maximal EI settles down, but there's no guarantee that it won't "pop back up" if a GP update is surprised by an acquisition. Sometimes those "pops" are big, sometimes small. Therefore `maxei` is more useful as a visual confirmation of convergence than it is an operational one, e.g., one that can be engineered into a library function. Hence it's common to run out an EI search to exhaust a budget of evaluations. The function below, designed to encapsulate code above for repeated calls in an MC setting, demands a `stop` argument in lieu of more automatic convergence criteria.

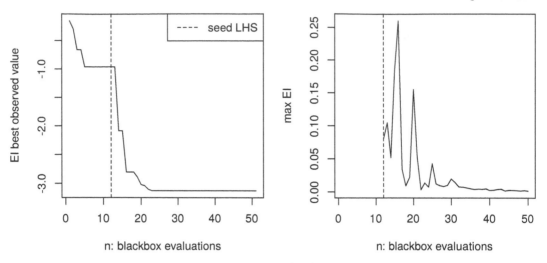

FIGURE 7.10: EI progress in terms of BOV (left) and maximal EI used for acquisition (right).

```
optim.EI <- function(f, ninit, end)
{
  ## initialization
  X <- randomLHS(ninit, 2)
  y <- f(X)
  gpi <- newGPsep(X, y, d=0.1, g=1e-6, dK=TRUE)
  da <- darg(list(mle=TRUE, max=0.5), randomLHS(1000, 2))
  mleGPsep(gpi, param="d", tmin=da$min, tmax=da$max, ab=da$ab)

  ## optimization loop of sequential acquisitions
  maxei <- c()
  for(i in (ninit+1):end) {
    solns <- EI.search(X, y, gpi)
    m <- which.max(solns$val)
    maxei <- c(maxei, solns$val[m])
    xnew <- as.matrix(solns[m,3:4])
    ynew <- f(xnew)
    updateGPsep(gpi, xnew, ynew)
    mleGPsep(gpi, param="d", tmin=da$min, tmax=da$max, ab=da$ab)
    X <- rbind(X, xnew)
    y <- c(y, ynew)
  }

  ## clean up and return
  deleteGPsep(gpi)
  return(list(X=X, y=y, maxei=maxei))
}
```

This optim.EI function hard-codes a separable laGP-based GP formulation, but is easily modified for other settings or an isotropic analog. Using that function, let's repeatedly solve the problem in this way (and track progress) with 100 random initializations, duplicating work similar to our EY and "L-BFGS-B" optimizations from §7.1.1–7.1.2.

```
reps <- 100
prog.ei <- matrix(NA, nrow=reps, ncol=end)
for(r in 1:reps) {
  os <- optim.EI(f, ninit, end)
  prog.ei[r,] <- bov(os$y)
}
```

Because showing three sets of 100 paths (300 total) would be a hot spaghetti mess, Figure 7.11 shows averages of those sets of 100 for our three comparators. Variability in those paths, which is mostly of interest for latter iterations/near the budget limit, is shown in a separate figure momentarily.

```
plot(colMeans(prog.ei), col=1, lwd=2, type="l",
  xlab="n: blackbox evaluations", ylab="average best objective value")
lines(colMeans(prog), col="gray", lwd=2)
lines(colMeans(prog.optim, na.rm=TRUE), col=2, lwd=2)
abline(v=ninit, lty=2)
legend("topright", c("optim", "EY", "EI", "seed LHS"),
  col=c(2, "gray", 1, 1), lwd=c(2,2,2,1), lty=c(1,1,1,2),
  bty="n")
```

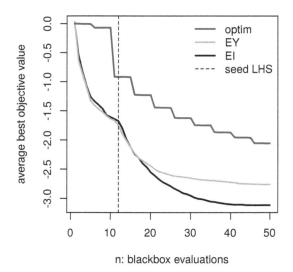

FIGURE 7.11: Average BOV progress for the three comparators entertained so far.

Although EI and EY perform similarly for the first few acquisitions after initialization, EI systematically outperforms in subsequent iterations. Whereas the classical surrogate-assisted EY heuristic gets stuck in inferior local optima, EI is better able to pop out and explore other alternatives. Both are clearly better than derivative-based methods like "L-BFGS-B", labeled as `optim` in Figure 7.11.

Figure 7.12 shows the diversity of solutions in the final, fiftieth iteration. Only in a small handful of 100 repeats does EI not find the global min after 50 iterations. Classical surrogate-assisted EY optimization fails to find the global optima about half of the time. Numerical "L-BFGS-B" optimization fails more than 75% of the time.

```
boxplot(prog.ei[,end], prog[,end], prog.optim[,end],
  names=c("EI", "EY", "optim"), border=c("black", "gray", "red"),
  xlab="comparator", ylab="best objective value")
```

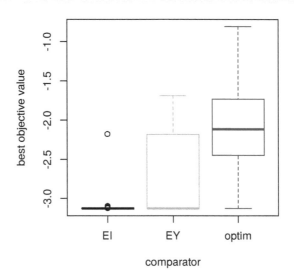

FIGURE 7.12: Boxplots summarizing the distribution of progress after the final acquisition.

All of the methods would benefit from more iterations, but only marginally unless random restarts are implemented for EY and `optim` comparators. Once those become stuck in a local minima they lack the ability to pop out on their own. EI, on the other hand, is in a certain sense guaranteed to become unstuck. Under certain regularity conditions, like that hyperparameters are fixed and known and the objective function isn't completely at odds with typical GP modeling assumptions, the EGO algorithm (i.e., repeated EI searches) will eventually converge to a global optimum. As usual with such theorems, the devil is in the details of the assumptions hidden in those regularity conditions. When are those satisfied, or easily verified, in practice? The real sleight-of-hand here is in the notion of convergence, not in the assumptions. EGO really only converges in the sense that eventually it'll explore everywhere. Since completely random search also has that property, the accomplishment isn't all that extraordinary. But the fact that its searches do not look random, but work as well as random in the limit, offers some comfort.

Perhaps a more important, or more relevant, theoretical result is that you can show that each EI-based acquisition is best in a certain sense: selection of x_{n+1} by EI is optimal if $n + 1 = N$, where N is the total budget available for evaluations of f. That is, EI is best if the next sample is the last one you plan to take, and your intention is to predict that $\min_x \mu_{n+1}(x)$ is the global minimum. For a sample of technical details relating to convergence of EI-like methods, see Bull (2011). Another perspective views EI as a *greedy*[7] heuristic approximating a fuller "look-ahead" scheme. In the burgeoning BO literature there are several alternative *acquisition functions* (Snoek et al., 2012), i.e., sequential design heuristics,

---

[7]https://en.wikipedia.org/wiki/Greedy_algorithm

with similar behavior and theory. Bect et al. (2016) provide a compelling framework for studying properties and variations of EGO/EI-like methods.

7.2.3 Conditional improvement

As an example of a scheme intimately related to EI, but which is in principle forward thinking and targets even broader scope, consider *integrated expected conditional improvement* (IECI; Gramacy and Lee, 2011). IECI was designed for optimization under constraints, our next topic, and that transition motivates its discussion here amid myriad similarly motivated extensions (see, e.g., Snoek et al., 2012; Bect et al., 2016) and methods based on the *knowledge gradient* (KG, as described by Wu et al., 2017, and references therein). I make no claims that IECI is superior to these alternatives, or superior to EI or KG. Rather it's a well-motivated choice I know well, adding some breadth to a discussion of acquisition functions for optimization, paired with convenient library support.

In essence, IECI is to EI what ALC is to ALM. Instead of optimizing directly, first integrate then optimize. Like with ALM to ALC, developing intermediate conditional criteria represents a crucial first step. Recall that ALC (§6.2.2) entails measuring variance at a reference location x after a new location x_{n+1} is added into the design. Then that gets integrated, or more approximately summed, over a wider set of $x \in \mathcal{X}$.

Similarly, *conditional improvement* measures improvement at reference location x, after another location x_{n+1} is added into the design.

$$I(x \mid x_{n+1}) = \max\{0, f_{\min}^n - Y(x \mid x_{n+1})\}$$

"Deduce" moments of the distribution of $Y(x \mid x_{n+1}) \mid D_n$ as follows:

- $\mathbb{E}\{Y(x \mid x_{n+1}) \mid D_n\} = \mu_n(x)$ since y_{n+1} has not come yet.
- $\mathbb{V}\mathrm{ar}[Y(x \mid x_{n+1}) \mid D_n] = \sigma_{n+1}^2(x)$ follows the ordinary GP predictive equations (5.2) with $X_{n+1} = (X_n^\top; x_{n+1}^\top)^\top$ and hyperparameters learned at iteration n. See Eq. (6.6) in §6.2.2. In practice, this quantity is most efficiently calculated with partitioned inverse equations (6.8), just like with ALC.

Integrating $I(x \mid x_{n+1})$ with respect to $Y(x \mid x_{n+1})$ yields the *expected conditional improvement (ECI)*, i.e., the analog of EI for conditional improvement.

$$\mathbb{E}\{I(x \mid x_{n+1}) \mid D_n\} = (f_{\min}^n - \mu_n(x)) \, \Phi\left(\frac{f_{\min}^n - \mu_n(x)}{\sigma_{n+1}(x)}\right) + \sigma_{n+1}(x) \, \phi\left(\frac{f_{\min}^n - \mu_n(x)}{\sigma_{n+1}(x)}\right)$$

To obtain a function of x_{n+1} only, i.e., a criterion for sequential design, integrate over the reference set $x \in \mathcal{X}$

$$\mathrm{IECI}(x_{n+1}) = -\int_{x \in \mathcal{X}} \mathbb{E}\{I(x \mid x_{n+1}) \mid D_n\} w(x) \, dx. \tag{7.4}$$

Such is the *integrated expected conditional improvement (IECI)*, for some (possibly uniform) weights $w(x)$. As with ALC, in practice that integral is approximated with a sum.

$$\mathrm{IECI}(x_{n+1}) \approx -\frac{1}{T} \sum_{t=1}^{T} \mathbb{E}\{I(x^{(t)} \mid x_{n+1})\} w(x^{(t)}) \quad \text{where } x^{(t)} \sim p(\mathcal{X}), \text{ for } t = 1, \dots, T,$$

and $p(\mathcal{X})$ is a measure on the input space $\mathcal{X}$ which may be uniform. Alternatively, one may combine weights and measures on $\mathcal{X}$ into a single measure. For now, take both to be uniform in $[0, 1]^m$.

The minus in IECI may look peculiar at first glance. Since $I(x \mid x_{x+1})$ is, in some sense, a two-step measure, small values are preferred over larger ones. ALC similarly prefers smaller future variances. Instead of measuring improvement directly at x_{n+1}, as EI does, IECI measures improvement in a roundabout way, assessing it at a reference point x under the hypothetical scenario that x_{n+1} is added into the design. If x still has high improvement after x_{n+1} has been added in, then x_{n+1} must not have had much influence on potential for improvement at x. If x_{n+1} is influential at x, then improvement at x should be small after x_{n+1} is added in, not large. Observe that the above argument makes tacit use of the assumption that $\mathbb{E}\{I(x \mid x_{n+1})\} \leq \mathbb{E}\{I(x)\}$, for all $x \in \mathcal{X}$, a kind of *monotonicity condition*.

Alternately, consider instead expected *reduction in improvement (RI)*, analogous to reduction in variance from ALC.

$$\mathrm{RI}(x_{n+1}) = \int_{x \in \mathcal{X}} (\mathbb{E}\{I(x) \mid D_n\} - \mathbb{E}\{I(x \mid x_{n+1}) \mid D_n\})w(x)\,dx$$

Clearly we wish to maximize RI, to reduce the potential for future improvement as much as possible. And observe that $\mathbb{E}\{I(x)\}$ doesn't depend on x_{n+1}, so it doesn't contribute substantively to the criterion. What's left is exactly IECI. In order for RI to be positive, the very same monotonicity condition must be satisfied. For ALC we don't need to worry about monotonicity since, conditional on hyperparameters, future $(n+1)$ variance is always lower than past (n) variance. It turns out that the definition of $f_{\min}$ is crucial to determining whether or not monotonicity holds.

A drawing in Figure 7.13 illustrates how $f_{\min}$ can influence ECI. Two choices of $f_{\min}$ are entertained, drawn as horizontal lines. One uses only observed y-values, following exactly our definition above for $f_{\min}$. The other takes $f_{\min}$ from the extremum of the GP predictive mean, drawn as a solid parabolic curve: $f_{\min} = \min_x \mu_n(x)$. EI is related to the area of the predictive density drawn as a solid line, plotted vertically and centered at $\mu_n(x)$, which lies underneath the horizontal $f_{\min}$ line(s). ECI is likewise derived from the area of the predictive density drawn as a dashed line lying below the horizontal $f_{\min}$ line(s). This dashed density has the same mean/mode as the solid one, but is more sharply peaked by influence from x_{n+1}.

If we suppose that these densities, drawn as bell-shaped curves in the figure, are symmetric (as they are for GPs), then it's clear that the relationship between ECI and EI depends upon $f_{\min}$. As the dashed line is more peaked, the left-tail cumulative distributions have the property that $F_n(f_{\min} \mid x_{n+1}) \geq F_n(f_{\min})$ for all $f_{\min} \geq \mathbb{E}\{Y(x \mid x_{n+1})\} = \mathbb{E}\{Y(x)\}$. Since $f_{\min} = \min\{y_1, \ldots, y_n\}$ is one such example, we could observe $\mathbb{E}\{I(x \mid x_{n+1})\} \geq \mathbb{E}\{I(x)\}$, violating the monotonicity condition. Only $f_{\min} = \min_x \mu_n(x)$ guarantees that ECI represents a reduction compared to EI. That said, in practice the choice of $f_{\min}$ matters little. But while we're on the topic of what constitutes improvement – i.e., improvement upon what? – let's take a short segue and talk about noisy blackbox objectives.

7.2.4 Noisy objectives

The BO literature over-accentuates discord between surrogate-assisted (EI or EY) optimization algorithms for deterministic and noisy blackboxes. The first paper on BO of stochastic

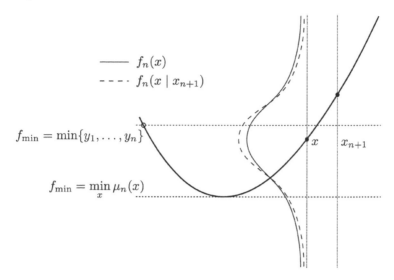

FIGURE 7.13: Cartoon illustrating how choice of $f_{\min}$ affects ECI. Adapted from Gramacy and Lee (2011).

simulations is by Huang et al. (2006). Since then the gulf between noisy and deterministic methods for BO has seemed to widen despite a surrogate modeling literature which has, if anything, downplayed the state(s) of affairs: Need a GP for noisy simulations? Just estimate a nugget. (With `laGP`, use `jmleGP` for isotropic formulations, or `mleGPsep` with `param="both"` for separable ones.) EI relies on the same underlying surrogates, so no additional changes are required there either, at least not to modeling aspects. Whenever noise is present, a modest degree of replication – especially in seed designs – can be helpful as a means of separating signal from noise. More details on that front are deferred to Chapter 10.

The notion of improvement requires a subtle change. Actually, nothing is wrong with the form of $I(x)$, but how its components $f_{\min}^n$ and $Y(x) \mid D_n$ are defined. If there's noise, then $Y(x_i) \neq y_i$. Responses $Y_1, \ldots, Y_n$ at X_n are random variables. So $f_{\min}^n$ is also a random variable. At face value, this substantially complicates matters. MC evaluations of EI, extending sampling of $Y(x)$ to $\min\{Y_1, \ldots, Y_n\}$, may be the only faithful means of taking expectation over all random quantities. Fully analytic EI in such cases seems out of reach. Unfortunately, MC thwarts library-based optimization of EI. For example, simple `optim` methods wouldn't work because the EI approximation would lack smoothness, except in the case of prohibitively large MC sampling efforts.

A common alternative is to use $f_{\min} = \min_x \mu_n(x)$, exactly as suggested above to ensure the monotonicity condition, and proceed as usual. Picheny et al. (2013) call this the "plug-in" method. Plugging in deterministic $\min_x \mu_n(x)$ for random $f_{\min}$ works well despite under-accounting for a degree of uncertainty. It's a refreshing coincidence that this choice of $f_{\min}$ addresses both issues (monotonicity and noise), and that this choice makes sense intuitively. The quantity $\min_x \mu_n(x)$ is our model's best guess about the global function minimum, and $\min_x \mu_{n+1}(x)$ is the one we intend to report when measuring progress. It stands to reason, then, that that value is sensible as a means of assessing potential to improve upon said predictions in subsequent acquisition iterations.

A second consideration for noisy cases involves the distribution of $Y(x) \mid D_n$, via $\mu_n(x)$ and $\sigma_n^2(x)$ from the GP predictive equations (5.2). In the deterministic case, and when GP modeling without a nugget, $\sigma_n^2(x) \to 0$ for all x as $n \to \infty$. Here we're assuming that X_n

becomes dense in the input space as $n \to \infty$, as improvement-based heuristics essentially guarantee. With enough data there's no predictive uncertainty, so there's eventually no value to any location by improvement $I(x)$, either through EI or IECI. This makes sense: if you sample everywhere you'll know all of the optima in the study region, both global and local. But in the stochastic case, and when GP modeling with a non-zero nugget, $\sigma_n^2(x) \nrightarrow 0$ as $n \to \infty$. No matter how much data is collected, there will always be predictive uncertainty in $Y(x) \mid D_n$, so there will always be nonzero improvement $I(x)$, through EI, IECI or otherwise. Our algorithm will never converge.

A simple fix is to redefine improvement on the latent random field $f(x)$, rather than directly on $Y(x)$. See §5.3.2 for details. Eventually, with enough data, there's no uncertainty about the function – no epistemic uncertainty[8] – even though there would be some *aleatoric uncertainty* about its noisy measurements. What that means, operationally speaking, is that when calculating EI one should use a predictive standard deviation without a nugget, as opposed to the full version (5.2). Specifically, and at slight risk of redundancy (5.16), use

$$\breve{\sigma}_n^2(x) = \hat{\tau}^2(1 - C(x, X_n)K_n^{-1}C(x, X_n)^\top) \tag{7.5}$$

rather than the usual $\sigma_n^2(x) = \hat{\tau}^2(1 + \hat{g}_n - C(x, X_n)K_n^{-1}C(x, X_n)^\top)$ in EI acquisitions. IECI is similar via $\breve{\sigma}_{n+1}^2(x)$. Recall that $\hat{g}_n$ is still involved in K_n^{-1}, so the nugget is still being "felt" by predictions. Crucially, $\breve{\sigma}_n^2(x) \to 0$ as $n \to \infty$ as long as the design eventually fills the space. Library functions `predGP` and `predGPsep` provide $\breve{\sigma}_n^2(x)$, or its multivariate analog, when supplied with argument `nonug=TRUE`.

Finally, Thompson sampling, which was dismissed earlier in §7.1.2 as a stochastic optimization, is worth reconsidering in noisy contexts. Noisy evaluation of the blackbox introduces a degree of stochasticity which can't be avoided. A little extra randomness in acquisition criteria doesn't hurt much and can sometimes be advantageous, especially when ambiguity between signal and noise regimes is present. See exercises in §7.4 for a specific example. Hybrids of LCB with EI have been successful in noisy optimization contexts. For example, quantile EI (QEI; Picheny et al., 2013) works well when noise level can be linked to simulation fidelity; more on this and similar methods when we get to optimizing heterskedastic processes in Chapter 10.3.4.

7.2.5 Illustrating conditional improvement and noise

Let's illustrate both IECI and optimization of a noisy blackbox at the same time. Consider the following data, which is in 1d to ease visualization.

```
fsindn <- function(x)
 sin(x) - 2.55*dnorm(x,1.6,0.45)
X <- matrix(c(0, 0.3, 0.6, 0.8, 1.0, 1.2, 1.4, 1.6, 1.8, 2.0, 2.2, 2.5,
   2.8, 3.1, 3.4, 3.7, 4.4, 5.3, 5.7, 6.1, 6.5, 7), ncol=1)
y <- fsindn(X) + rnorm(length(X), sd=0.15)
```

This seed design deliberately omits one of the two local minima of `fsindn`, although it's otherwise uniformly spaced in the domain of interest, $\mathcal{X} = [0, 7]$. The code below initializes a GP fit, and performs inference for scale, lengthscale and nugget hyperparameters.

---

[8]https://en.wikipedia.org/wiki/Uncertainty_quantification#Aleatoric_and_epistemic_uncertainty

```
gpi <- newGP(X, y, d=0.1, g=0.1*var(y), dK=TRUE)
mle <- jmleGP(gpi)
```

Before calculating EI and IECI, let's peek at the predictive distribution. Code below extracts predictive quantities for both $Y(x)$ and latent $f(x)$, i.e., using $\sigma_n^2(x)$ or $\breve{\sigma}_n^2(x)$ from Eq. (7.5), respectively.

```
XX <- matrix(seq(0, 7, length=201), ncol=1)
pY <- predGP(gpi, XX, lite=TRUE)
pf <- predGP(gpi, XX, lite=TRUE, nonug=TRUE)
```

Figure 7.14 shows the mean surface, which is the same under both predictors, and two sets of error-bars. Red-dashed lines correspond to the usual error-bars, based on the full distribution of $Y(x) \mid D_n$, using $\sigma_n^2(x)$ above. Green-dotted ones use $\breve{\sigma}_n^2(x)$ instead.

```
plot(X, y, xlab="x", ylab="y", ylim=c(-1.6,0.6), xlim=c(0,7.5))
lines(XX, pY$mean)
lines(XX, pY$mean + 1.96*sqrt(pY$s2), col=2, lty=2)
lines(XX, pY$mean - 1.96*sqrt(pY$s2), col=2, lty=2)
lines(XX, pf$mean + 1.96*sqrt(pf$s2), col=3, lty=3)
lines(XX, pf$mean - 1.96*sqrt(pf$s2), col=3, lty=3)
legend("bottomright", c("Y-bars", "f-bars"), col=2:4, lty=2:3, bty="n")
```

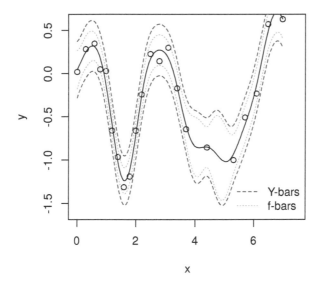

FIGURE 7.14: Comparing predictive quantiles under $Y(x)$ and $f(x)$.

Observe how "f-bars" quantiles are uniformly narrower than their "Y-bars" counterpart. With enough data D_n, as $n \to \infty$ and X_n filling the space, "f-bars" would collapse in on the mean surface, shown as the solid black line.

Now we're ready to calculate EI and IECI under those two predictive distributions. `EI` with `nonug=TRUE` can be achieved by passing in a predictor `pred` with that option pre-set, e.g., by copying the function and modifying its default arguments (its `formals`). The `ieciGP` function provided by laGP has its own nonug argument. By default, ieciGP takes candidate locations, XX in the code below, identical to reference locations Xref, similar to alcGP.

```
fmin <- min(predGP(gpi, X, lite=TRUE)$mean)
ei <- EI(gpi, XX, fmin, pred=predGP)
ieci <- ieciGP(gpi, XX, fmin)
predGPnonug <- predGP
formals(predGPnonug)$nonug <- TRUE
ei.f <- EI(gpi, XX, fmin, pred=predGPnonug)
ieci.f <- ieciGP(gpi, XX, fmin, nonug=TRUE)
```

To ease visualization, it helps to normalize acquisition function evaluations so that they span the same range.

```
ei <- scale(ei, min(ei), max(ei) - min(ei))
ei.f <- scale(ei.f, min(ei.f), max(ei.f) - min(ei.f))
ieci <- scale(ieci, min(ieci), max(ieci) - min(ieci))
ieci.f <- scale(ieci.f, min(ieci.f), max(ieci.f) - min(ieci.f))
```

Figure 7.15 shows these four acquisition comparators. Solid lines are for EI, and dashed for IECI; the color scheme matches up with predictive surfaces above, where "-f" is the nonug=TRUE version, i.e., based on the latent random field f. So that all are maximizable criteria, negative IECIs are plotted.

```
plot(XX, ei, type="l", ylim=c(0, 1), xlim=c(0,7.5), col=2, lty=1,
  xlab="x", ylab="improvements")
lines(XX, ei.f, col=3, lty=1)
points(X, rep(0, nrow(X)))
lines(XX, 1-ieci, col=2, lty=2)
lines(XX, 1-ieci.f, col=3, lty=2)
legend("topright", c("EI", "EI-f", "IECI", "IECI-f"), lty=c(1,1,2,2),
  col=c(2,3,2,3), bty="n")
```

Since training data were randomly generated for this Rmarkdown build, it's hard to pinpoint exactly the state of affairs illustrated in Figure 7.15. Usually EI prefers (in both variants) to choose x_{n+1} from the left half of the space, whereas IECI (both variants) weighs both minima more equally. Occasionally, however, EI prefers the right mode instead, and occasionally both prefer the left, all depending on the random data. Both represent local minima, but the one on the right experiences higher aggregate uncertainty due both to lower sampling and to a wider domain of attraction. The left minima has a narrow trough. As a more aggregate measure, IECI usually up-weights x_{n+1} from the right half, pooling together large ECI in a bigger geographical region, and consequently putting greater value on their potential to offer improvement globally in the input space.

Both EI and IECI cope with noise just fine. Variations with and without nugget are subtle, at least as exemplified by Figure 7.15. A careful inspection of the code behind that figure reveals that we didn't actually minimize $\mu_n(x)$ to choose $f^n_{\min}$. Rather it was sufficient to use $\min_i \mu_n(x_i)$, which is less work computationally because no auxiliary numerical optimization is required.

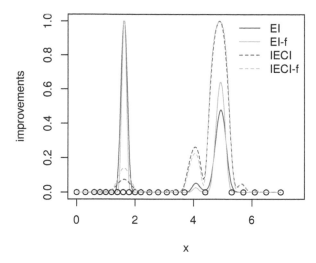

FIGURE 7.15: EI versus IECI in $Y(x)$ and $f(x)$ alternatives.

Rather than exhaust the reader with more `for` loops, say to incorporate IECI and noise into a bakeoff or to augment our running example (§7.1.1) on the (possibly noise-augmented) Goldstein–Price[9] function, these are left to exercises in §7.4. IECI, while enjoying a certain conservative edge by "looking ahead", rarely outperforms the simpler but more myopic ordinary EI in practice. This is perhaps in part because it's challenging to engineer a test problem, like the one in our illustration above, exploiting just the scenario IECI was designed to hedge against. Both EI and IECI can be recast into the *stepwise uncertainty reduction (SUR)* framework (Bect et al., 2016) where one can show that they enjoy a supermartingale property[10], similar to a submodularity property[11] common to many active learning techniques. But that doesn't address the extent to which IECI's limited degree of lookahead may or may not be of benefit, by comparison to EI say, along an arc of future iterations in the steps of a sequential design. Advantages or otherwise are problem dependent. In a context where blackbox objective functions $f(x)$ almost never satisfy technical assumptions imposed by theory – the Goldstein–Price function isn't a stationary GP – intuition may be the best guide to relative merits in practice.

Where IECI really shines is in a constrained optimization context, as this is what it was originally designed for – making for a nice segue into our next topic.

7.3 Optimization under constraints

To start off, we'll keep it super simple and assume constraints are known, but that they take on non-trivial form. That is, not box/bound constraints (too easy), but something tracing out non-convex or even unconnected regions. Then we'll move on to unknown, or *blackbox constraints* which require expensive evaluations to probe. First we'll treat those as binary,

---

[9]http://www.sfu.ca/~ssurjano/goldpr.html
[10]https://en.wikipedia.org/wiki/Martingale_(probability_theory)#Submartingales,
_supermartingales,_and_relationship_to_harmonic_functions
[11]https://en.wikipedia.org/wiki/Submodular_set_function

valid or invalid, and then as real-valued. That distinction, binary or real-valued, impacts how constraints are modeled and how they fold into an effective sequential design criteria for optimization. Throughout, the goal is to find the *best valid value (BVV)* of the objective in the fewest number of blackbox evaluations. More details on the problem formulation(s) are provided below.

7.3.1   Known constraints

For now, assume constraints are known. Specifically, presume that we have access to a function, $c(x) : \mathcal{X} \rightarrow \{0, 1\}$, returning zero (or equivalently a negative number) if the constraint is satisfied, or one (or a positive number) if the constraint is violated, and that we can evaluate it for "free", i.e., as much as we want. Evaluation expense accrues on blackbox objective $f(x)$ only.

The problem is given formally by the mathematical program[12]

$$x^\star = \mathrm{argmin}_{x \in \mathcal{X}} \; f(x) \quad \text{subject to} \quad c(x) \leq 0. \tag{7.6}$$

One simple surrogate-assisted solver entails extending EI to what's called *expected feasible improvement (EFI)*, which was described in a companion paper to the EI one by the same three authors, but with names of the first two authors swapped (Schonlau et al., 1998).

$$\mathrm{EFI}(x) = \mathbb{E}\{I(x)\}\mathbb{I}(c(x) \leq 0),$$

with $I(x)$ using an $f_{\min}^n$ defined over the valid region only. In deterministic settings, that means $f_{\min}^n = \min_{i=1,\dots,n} \{y_i : c(x_i) \leq 0\}$. When noise is present $f_{\min}^n = \min_{x \in \mathcal{X}} \{\mu_n(x) : c(x) \leq 0\}$. The former may be the empty set, in which case the latter is a good backup except in the pathological case that none of the study region $\mathcal{X}$ is valid.

A verbal description of EFI might be: do EI but don't bother evaluating, nor modeling, outside the valid region. Seems sensible: invalid evaluations can't be solutions. We know in advance which inputs are in which set, valid or invalid, so don't bother wasting precious blackbox evaluations that can't improve upon the best solution so far.

Yet that may be an overly simplistic view. Mathematical programmers long ago found that approaching local solutions from outside of valid sets can sometimes be more effective than the other way around, especially if the valid region is comprised of disjoint or highly non-convex sets. One example is the augmented Lagrangian method[13], which we shall review in more detail in §7.3.4.

Surrogate modeling considerations also play a role here. Data acquisitions under GPs have a global effect on the updated predictive surface. Information from observations of the blackbox objective outside of the valid region might provide more insight into potential for improvement (inside the region) than any point inside the region could. Think of an invalid region sandwiched between two valid ones (see Figures 7.16–7.17) and how predictive uncertainty might look, and in particular its curvature, at the boundaries. One evaluation splitting the difference between the two boundaries may be nearly as effective as two right on each boundary. EFI would rule out that potential economy. Or, in situations where one really doesn't want, or can't perform an "invalid run", but is still faced with a choice between high

---

[12]https://en.wikipedia.org/wiki/Mathematical_optimization
[13]https://en.wikipedia.org/wiki/Augmented_Lagrangian_method

EI on either side of the invalid-region boundary, it might make sense to weigh potential for improvement by the amount of the valid region a potential acquisition would cover. That's exactly what IECI was designed to do.

Adapting IECI to respect a constraint, but to still allow selection of evaluations "wherever they're most effective", is a matter of choosing indicator $\mathbb{I}(c(x) \leq 0)$ as weight $w(x)$:

$$\mathrm{IECI}(x_{n+1}) = -\int_{x \in \mathcal{X}} \mathbb{E}\{I(x \mid x_{n+1})\} \mathbb{I}(c(x) \leq 0) \, dx$$

This downweights reference x-values not satisfying the constraint, and thus also x_{n+1}-values similarly, however it doesn't preclude x_{n+1}-values from being chosen in the invalid region. Rather, the value of x_{n+1} is judged by its ability to impact improvement within the valid region. An alternative implementation which may be computationally more advantageous, especially when approximating the integral with a sum over a potentially dense collection $\mathtt{Xref} \subseteq \mathcal{X}$, is to exclude from $\mathtt{Xref}$ any input locations which don't satisfy the constraint. More precisely,

$$\mathrm{IECI}(x_{n+1}) \approx -\frac{1}{T} \sum_{t=1}^{T} \mathbb{E}\{I(x^{(t)} \mid x_{n+1})\} \quad \text{where } x^{(t)} \sim p_c(\mathcal{X}), \text{ for } t = 1, \dots, T,$$

and $p_c(\mathcal{X})$ is uniform on the valid set $\{x \in \mathcal{X} : c(x) \leq 0\}$.

To illustrate, consider the same 1d data-generating mechanism as in §7.2.5 except with an invalid region $[2, 4]$ sandwiched between two valid regions occupying the first and last third of the input space, respectively.

```
X <- matrix(c(0, 0.3, 0.6, 0.8, 1.0, 1.2, 1.4, 1.6, 1.8, 4.4, 5.3, 5.7,
    6.1, 6.5, 7), ncol=1)
y <- fsindn(X) + rnorm(length(X), sd=0.15)
```

No observations lie in the invalid region, yet. For dramatic effect in this illustration, it makes sense to hard-code a longer lengthscale when fitting the GP.

```
gpi <- newGP(X, y, d=5, g=0.1*var(y), dK=TRUE)
```

Readers are strongly advised to tinker with this (also with `jmleGP`) to see how it effects the results. Next, establish a dense grid in the input space $\mathcal{X} \equiv \mathtt{XX}$ and take evaluations of GP predictive equations thereupon.

```
XX <- matrix(seq(0, 7, length=201), ncol=1)
p <- predGP(gpi, XX, lite=TRUE)
```

Figure 7.16 shows the resulting predictive surface.

```
plot(X, y, xlab="x", ylab="y", ylim=c(-3.25, 0.7))
lines(XX, p$mean)
lines(XX, p$mean + 1.96*sqrt(p$s2), col=2, lty=2)
lines(XX, p$mean - 1.96*sqrt(p$s2), col=2, lty=2)
```

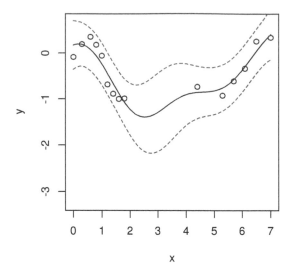

FIGURE 7.16: Predictive surface for a 1d problem with invalid region $[2, 4]$ sandwiched between two valid ones.

Because there are no evaluations in the middle third of the space, predictive uncertainty is very high there. Since the lengthscale is long, the posterior mean on functions favors a trough in that region, which translates into a promising solution to the constrained optimization problem – a valid global minimum. It seems quite likely that one of two local valid minima reside on either side of the invalid region.

First consider calculating EI for each **XX** location on the predictive grid, then normalizing to ease later visualization. In this illustration, only the **nonug=TRUE** option is shown, primarily to simplify visuals.

```
fmin <- min(predGP(gpi, X, lite=TRUE)$mean)
ei <- EI(gpi, XX, fmin, pred=predGPnonug)
ei <- scale(ei, min(ei), max(ei) - min(ei))
```

Ordinarily here, we'd have to be careful to calculate f_{min}^n based only on valid input locations. However in this simple example all seed design **X** locations are valid. (And we know EFI won't choose any invalid ones.) The code below extracts EIs for the valid region, effectively implementing EFI but in a way that's handy for our visualization below.

```
lc <- 2
rc <- 4
eiref <- c(ei[XX < lc], ei[XX > rc])
```

To facilitate IECI calculation, the next code chunk establishes a set of reference locations **Xref** dense in $\mathcal{X} \equiv$ **XX**, but similarly omitting invalid region $[2, 4]$.

```
Xref <- matrix(c(XX[XX < lc,], XX[XX > rc,]), ncol=1)
```

Using that pre-selected **Xref** in lieu of weights, **laGP**'s IECI subroutine may be evaluated for all of **XX** as follows.

```
ieci <- ieciGP(gpi, XX, fmin, Xref=Xref, nonug=TRUE)
ieci <- scale(ieci, min(ieci), max(ieci) - min(ieci))
```

Figure 7.17 compares EI/EFI and IECI in this known constraints setting. A solid-black line corresponds to EFI extracted from EI which is dashed within the invalid region.

```
plot(XX, ei, type="l", ylim=c(0, max(ei)), lty=2, xlab="x",
  ylab="normalized improvements")
lines(Xref[Xref < lc], eiref[Xref < lc])
lines(Xref[Xref > rc], eiref[Xref > rc])
points(X, rep(0, nrow(X)))
lines(XX, 1-ieci, col=2, lty=2)
legend("topright", c("EI", "IECI"), lty=1:2, col=1:2, bty="n")
abline(v=c(lc,rc), col="gray", lty=2)
text(lc,0,"]")
text(rc,0,"[")
text(seq(lc+.1, rc-.1, length=20), rep(0, 20), rep("/", 20))
```

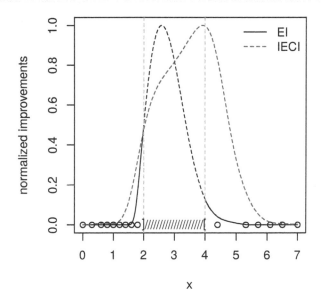

FIGURE 7.17: EFI and IECI surfaces for a simple 1d constrained optimization with invalid region sandwiched between two valid regions.

Notice how EI is maximized within the invalid region, but we won't choose x_{n+1} there because its EFI is zero. Instead, EFI is maximized at the boundary of the invalid region, with nearly identical height on either side. IECI, by contrast, is maximized outside the valid region – at least in this random Rmarkdown instance – offering clear adjudication between the two local EFI modes. IECI prefers the right-hand mode, and not on the boundary with the invalid region but instead splitting the difference between boundary and design locations X_n already in hand. For my money, IECI's is a better choice for x_{n+1}.

In sequential application, over repeated acquisitions, admittedly EFI and IECI perform similarly. Which is better is highly problem specific, depending in particular on how complicated the valid region is, and how pathological interaction is between constraint boundary and blackbox objective. For many problems, IECI is likely overkill since it's more computationally demanding, like ALC.

On the other hand, known constraint settings are specialized – you might even say highly peculiar. How often do you encounter a known constraint region that's complicated in practice?

7.3.2 Blackbox binary constraints

Having a blackbox that simultaneously evaluates both objective and constraint at great computational expense is far more common. The mathematical program (7.6) is unchanged, but now constraints $c(x)$ cannot be evaluated at will. Often blackbox constraint functions are vector-valued. In that case, the simple program (7.6) still applies, but $c(x) \leq 0$ must be interpreted component-wise and valid re-defined to mean that all components satisfy the inequality simultaneously. Like for objective $f(x)$, we'll need a surrogate model for constraint $c(x)$, and an appropriate model will depend upon the nature of the function(s). When binary, either $c(x) \in \{0, 1\}$ or $c(x) \in \{0, 1\}^m$ for multiple constraints, a classification model may be appropriate. When real-valued $c(x) \in \mathbb{R}$ or $c(x) \in \mathbb{R}^m$, a regression model is needed.

We can get as fancy or simple as we want with constraint models and fitting schemes. GPs are an obvious choice for real-valued cases, as we illustrate in §7.3.4. For now stick with binary constraints. GPs are an option for binary outputs too, e.g., through a logistic link (see, e.g., Chapter 3 of Rasmussen and Williams, 2006), but there are simpler off-the-shelf classifiers that often work as well or better in practice. Details are forthcoming in our empirical work below. For now, the presentation is agnostic to choice of classifier. Let $p_n^{(j)}(x)$ stand in for the predicted probability, from fitted surrogate model classifier(s), that input x satisfies the jth constraint, for $j = 1, \ldots, m$.

Extending EFI to blackbox constraints is trivial. Simply replace the indicator (from the known constraint) with the surrogate's predicted probability of satisfaction:

$$\text{EFI}(x) = \mathbb{E}\{I(x)\} \prod_{j=1}^{m} p_n^{(j)}(x).$$

IECI is no different, except the product moves inside the integral.

$$\text{IECI}(x_{n+1}) = -\int_{x \in \mathcal{X}} \mathbb{E}\{I(x \mid x_{n+1})\} \prod_{j=1}^{m} p_n^{(j)}(x) \, dx$$

In both cases EI is being weighted by the joint probability of constraint satisfaction, assuming mutually independent constraint surrogates. Explicitly for IECI as in Eq. (7.4), take $w(x) = \prod_{j=1}^{m} p_n^{(j)}(x)$.

To illustrate, revisit our earlier 1d known constraint problem fsindn from §7.2.5. This time imagine the constraint, which was invalid in $[2, 4]$, as residing inside the blackbox. Expensive evaluations provide noisy $Y(x) = f(x) + \varepsilon$ and deterministic $c(x) \in \{0, 1\}$, simultaneously. Model f with a GP, as before. In order to learn the constraint function from data, choose a simple yet flexible model from off the shelf: a random forest (RF; Breiman, 2001) via randomForest (Breiman et al., 2018) on CRAN.

Consider EFI first; returning to IECI shortly. A setup like this (GP/RF/EFI) was first entertained by Lee et al. (2011). The code below generates a small quasi-space-filling design to start out with, and evaluates the blackbox at those locations.

```
ninit <- 6
X <- matrix(c(1, 3, 4, 5, 6, 7))
y <- fsindn(X) + rnorm(length(X), sd=0.15)
const <- as.numeric(X > lc & X < rc)
```

Now fit surrogates. The block of code below targets GP fitting of the objective. With such a small amount of noisy data, it's sensible to constrain inference a little to help the GP separate signal from noise. Recall our discussion in §6.2.1 on care in initialization with sequential design. Below the lengthscale is fixed, and the nugget estimated conditionally under a sensible prior. As more data are gathered, such precautions become less necessary.

```
gpi <- newGP(X, y, d=1, g=0.1*var(y), dK=TRUE)
ga <- garg(list(mle=TRUE, max=var(y)), y)
mle <- mleGP(gpi, param="g", tmin=eps, tmax=var(y), ab=ga$ab)
```

Next load **randomForest** and fit constraint evaluations, being careful to coerce them into **factor** form so that the library knows to fit a classification model rather than the default regression option for real-valued responses.

```
library(randomForest)
cfit <- randomForest(X, as.factor(const))
```

Below, f_{min} is calculated by a smoothing over outputs predicted at valid input locations in X_n. The number and location of samples in the seed design was chosen (based on oracle knowledge of the constraint) so that the set of valid locations is nonempty in order to simplify, somewhat, the exposition here. (In fact, a single invalid input is guaranteed.)

```
Xv <- X[const <= 0,,drop=FALSE]
fmin <- min(predGP(gpi, Xv, lite=TRUE)$mean)
```

We now have all ingredients needed to evaluate EI and the probability of constraint satisfaction, globally in the input space over predictive grid **XX**.

```
pc <- predict(cfit, XX, type="prob")[,1]
ei <- EI(gpi, XX, fmin, pred=predGPnonug)
ei <- scale(ei, min(ei), max(ei) - min(ei))
```

The EI surface, probability of satisfaction, and their product yielding EFI are shown together in Figure 7.18. Constraint evaluations are indicated as open circles. So that the parity of constraint evaluations matches probability of satisfaction, easing visualization, $1 - c_i$, $i = 1, \ldots, n$ are plotted instead of raw c_i.

```
plot(XX, ei, type="l", xlim=c(0,8), ylim=c(0,1), xlab="x", ylab="ei & pc")
lines(XX, pc, col=2, lty=2)
```

```
points(X, 1 - const, col=2)
lines(XX, ei*pc, lty=2)
legend("right", c("EI", "p(c<0)", "EFI"), col=c(1,2,1),
  lty=c(1,2,2), bty="n")
```

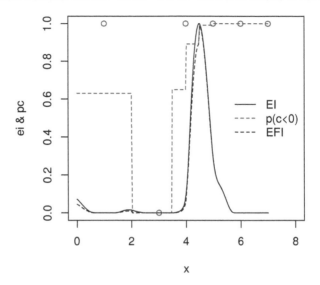

FIGURE 7.18: EI(x) and $p(c(x) < 0)$ combining to form EFI(x). Open circles indicate $1 - c_i$ to match parity with RF's $p(c_i < 0)$. EI and EFI are normalized.

Several interesting observations in the figure are worth remarking upon. Notice that $p_n(x) \equiv$ p(c<0) is zero nearby the single input having violated the constraint (plotted as $c_i = 0$, but actually measured as $c_i = 1$). This will cause EFI to evaluate to zero regardless of what EI is, but actually it's essentially zero there anyways. Likewise, throughout the rest of the input space, EFI is a version of EI modulated slightly downwards, more so approaching the middle of the input space where the RF classifier is less sure about the transition from valid to invalid region. EFI inherits discontinuities from RF's stepwise regime changes. The extent to which such discontinuities are detectable visually depends upon the Rmarkdown build.

After numerically maximizing EFI on the grid, code below acquires new x_{n+1}, updating $n \leftarrow n + 1$ with a new random objective and deterministic constraint evaluation.

```
m <- which.max(ei*pc)
X <- rbind(X, XX[m,])
y <- c(y, fsindn(XX[m,]) + rnorm(1, sd=0.15))
const <- c(const, as.numeric(XX[m,] > lc && XX[m,] < rc))
```

Next, update GP and RF fits to ready those surrogates for the next acquisition.

```
updateGP(gpi, X[nrow(X),,drop=FALSE], y[length(y)])
mle <- mleGP(gpi, param="g", tmin=eps, tmax=var(y), ab=ga$ab)
cfit <- randomForest(X, as.factor(const))
```

In that subsequent iteration, f^n_{min} is recalculated based on the corpus of valid inputs obtained so far. Predictions under both models are evaluated at candidates $\mathcal{X} \equiv$ XX, and EI's are calculated, collecting all ingredients for EFI.

```
Xv <- X[const <= 0,,drop=FALSE]
fmin <- min(predGP(gpi, Xv, lite=TRUE)$mean)
pc <- predict(cfit, XX, type="prob")[,1]
ei <- EI(gpi, XX, fmin, pred=predGPnonug)
ei <- scale(ei, min(ei), max(ei) - min(ei))
```

Updated EFI surface, and its requisite components are shown visually in Figure 7.19.

```
plot(XX, ei, type="l", xlim=c(0,8), ylim=c(0,1), xlab="x", ylab="ei & pc")
lines(XX, pc, col=2, lty=2)
points(X, 1 - const, col=2)
lines(XX, ei*pc, lty=2)
legend("right", c("EI", "p(c<0)", "EFI"), col=c(1,2,1),
  lty=c(1,2,2), bty="n")
```

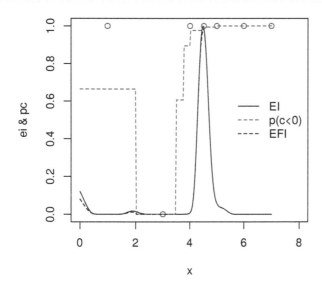

FIGURE 7.19: Second iteration of EFI following Figure 7.18.

In the version I'm viewing as I write this, updated surfaces strongly resemble those from the previous iteration (Figure 7.18) except perhaps with EI/EFI relatively higher on the valid side, i.e., that side not chosen for acquisition in the previous iteration. To fast-forward a little, R code below wraps our data augmenting, model updating and EFI-calculating from above into a `for` loop, selecting ten more evaluations in this fashion. Notice that `jmleGP` is used in this loop to tune lengthscales as well as nugget.

```
for(i in 1:10) {
  m <- which.max(ei*pc)
  X <- rbind(X, XX[m,])
  y <- c(y, fsindn(XX[m,]) + rnorm(1, sd=0.15))
  const <- c(const, as.numeric(XX[m,] > lc && XX[m,] < rc))
```

```
    updateGP(gpi, X[nrow(X),,drop=FALSE], y[length(y)])
    mle <- jmleGP(gpi, drange=c(eps, 2), grange=c(eps, var(y)), gab=ga$ab)
    cfit <- randomForest(X, as.factor(const))
    Xv <- X[const <= 0,,drop=FALSE]
    fmin <- min(predGP(gpi, Xv, lite=TRUE)$mean)
    pc <- predict(cfit, XX, type="prob")[,1]
    ei <- EI(gpi, XX, fmin, pred=predGPnonug)
    ei <- scale(ei, min(ei), max(ei) - min(ei))
}
```

After choosing a total of twelve points beyond the seed space-filling design, Figure 7.20 shows the nature of $p_n(x)$, EI and EFI from model fits based on a total of $n = 17$ runs.

```
plot(XX, ei, type="l", xlim=c(0,8), ylim=c(0,1), xlab="x", ylab="ei & pc")
lines(XX, pc, col=2, lty=2)
points(X, 1 - const, col=2)
lines(XX, ei*pc, lty=2)
legend("right", c("EI", "p(c<0)", "EFI"), col=c(1,2,1),
    lty=c(1,2,2), bty="n")
```

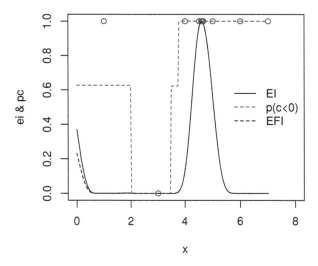

FIGURE 7.20: EFI view, ten more iterations after Figure 7.19.

That view reveals little to no exploration in the left half of the space, a behavior quite consistent across Rmarkdown builds. Although EI and EFI on the other side of the invalid region may increase in relative terms, it would take many right-side acquisitions before the criterion is maximized on the left. That's a shame because, as Figure 7.21 reveals, the other side is definitely worth exploring.

```
p <- predGP(gpi, XX, lite=TRUE)
plot(XX, fsindn(XX), col="gray", type="l", lty=2)
points(X, y)
```

```
lines(XX, p$mean)
legend("top", c("truth", "mean"), col=c("gray", 1), lty=c(2,1), bty="n")
```

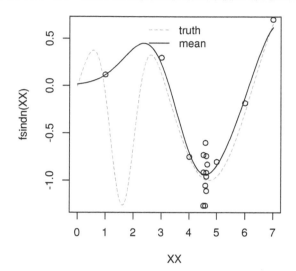

FIGURE 7.21: Final predictive mean (solid) fit to data (open circles), versus true surface (gray-dashed).

In the figure, black-open circles indicate sampled evaluations in the design, whereas the dashed-gray line is the truth. EFI's exploration of the valid region was quite myopic in this instance.

Does IECI fare any better? Suppose we reinitialize the fit back at the same starting position (for a fair comparison) and go through the first couple of steps carefully with IECI, just as for EFI above. But first, don't forget to free up the old GP fit.

```
deleteGP(gpi)
X <- X[1:ninit,,drop=FALSE]
y <- y[1:ninit]
const <- const[1:ninit]
gpi <- newGP(X, y, d=1, g=0.1*var(y), dK=TRUE)
mle <- mleGP(gpi, param="g", tmin=eps, tmax=var(y), ab=ga$ab)
cfit <- randomForest(X, as.factor(const))
pc <- predict(cfit, XX, type="prob")[,1]
```

Predictions from RF, obtained on the last line above, can be passed directly into `ieciGP` through its weight argument, `w`. The original IECI paper (Gramacy and Lee, 2011) used a classification GP for $p_n(x)$, coupled with an ordinary regression GP for IECI calculations in the `plgp` package. Sequential updating of both GPs is described in detail by Gramacy and Polson (2011). The `ieciGP` and `ieciGPsep` features in `laGP` are rather more generic, with a `w` argument allowing predicted probabilities to come from any classifier. Besides simplifying the narrative, we shall retain an RF classification surrogate here for a fairer benchmark against EFI. Readers curious about the original version are encouraged to inspect demos provided with `plgp`, in particular `demo("plconstgp_1d_ieci")`.

```
Xv <- X[const <= 0,,drop=FALSE]
fmin <- min(predGP(gpi, Xv, lite=TRUE)$mean)
ieci <- ieciGP(gpi, XX, fmin, w=pc, nonug=TRUE)
ieci <- scale(ieci, min(ieci), max(ieci) - min(ieci))
```

Figure 7.22 shows the resulting surfaces. Since the seed design and responses are the same here as for EFI, the RF $p_{n=6}(x)$ surface is identical to Figure 7.18. For consistency with EI-based figures above, one minus IECI and c_i are shown.

```
plot(XX, 1 - ieci, type="l", xlim=c(0,8), ylim=c(0,1),
  xlab="x", ylab="ieci & pc")
lines(XX, pc, col=2, lty=2)
points(X, 1 - const, col=2)
legend("right", c("IECI", "p(c<0)"), col=c(1,2,2), lty=c(1,2,1), bty="n")
```

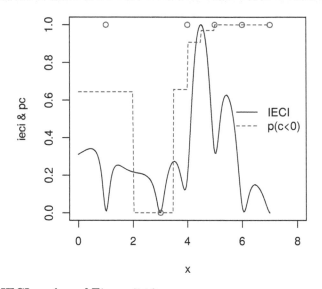

FIGURE 7.22: IECI analog of Figure 7.18.

By contrast with Figure 7.18, IECI is high on both sides of the invalid region. Moreover, IECI is smooth everywhere even though $p_n(x)$ is piecewise constant. It's never zero, although nearly so at the location of the solitary invalid input. IECI owes all three of those features, which are aesthetically more pleasing than the EFI analog, to its aggregate nature. A continuum of $p_n(x)$-values, as well as a continuum of conditional improvements, weigh in to determine which potential x_{n+1} offers the greatest promise for improvement in the valid region.

Code below selects the next point to minimize IECI, collects objective and constraint responses at that location, and augments the dataset.

```
m <- which.min(ieci)
X <- rbind(X, XX[m,])
y <- c(y, fsihdn(XX[m,]) + rnorm(1, sd=0.15))
const <- c(const, as.numeric(XX[m,] > lc && XX[m,] < rc))
```

Next, the GP is updated and RF refit.

```
updateGP(gpi, X[ncol(X),,drop=FALSE], y[length(y)])
mle <- mleGP(gpi, param="g", tmin=eps, tmax=var(y), ab=ga$ab)
cfit <- randomForest(X, as.factor(const))
```

New IECI calculations are then derived from updated $f^n_{\min}$ and $p_n(x)$ evaluations.

```
Xv <- X[const <= 0,,drop=FALSE]
fmin <- min(predGP(gpi, Xv, lite=TRUE)$mean)
pc <- predict(cfit, XX, type="prob")[,1]
ieci <- ieciGP(gpi, XX, fmin, w=pc, nonug=TRUE)
ieci <- scale(ieci, min(ieci), max(ieci) - min(ieci))
```

Figure 7.23 shows the updated surfaces. Since both sides of the invalid region had high IECI in the previous iteration, it's perhaps not surprising to see that IECI is now much higher on the side opposite to where the most recent acquisition was made.

```
plot(XX, 1 - ieci, type="l", xlim=c(0,8), ylim=c(0,1),
  xlab="x", ylab="ieci & pc")
lines(XX, pc, col=2, lty=2)
points(X, 1 - const, col=2)
legend("right", c("IECI", "p(c<0)"), col=c(1,2,2), lty=c(1,2,1), bty="n")
```

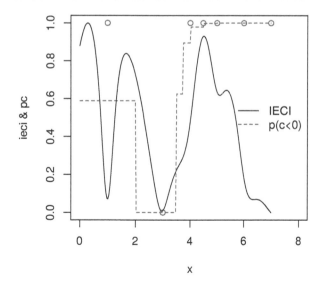

FIGURE 7.23: IECI updates after first IECI acquisition from Figure 7.22.

All right, lets see what happens when we do ten more steps like this, just as we did for EFI.

```
for(i in 1:10) {
  m <- which.min(ieci)
  X <- rbind(X, XX[m,])
```

```
y <- c(y, fsindn(XX[m,]) + rnorm(1, sd=0.15))
const <- c(const, as.numeric(XX[m,] > lc && XX[m,] < rc))
updateGP(gpi, X[nrow(X),,drop=FALSE], y[length(y)])
mle <- jmleGP(gpi, drange=c(eps, 2), grange=c(eps, var(y)), gab=ga$ab)
cfit <- randomForest(X, as.factor(const))
Xv <- X[const <= 0,,drop=FALSE]
fmin <- min(predGP(gpi, Xv, lite=TRUE)$mean)
pc <- predict(cfit, XX, type="prob")[,1]
ieci <- ieciGP(gpi, XX, fmin, w=pc, nonug=TRUE)
ieci <- scale(ieci, min(ieci), max(ieci) - min(ieci))
}
```

As shown in Figure 7.24, multiple evaluations have been taken on both sides of the invalid
region. Sometimes, depending on the Rmarkdown build, runs are even taken within the
invalid region. Compared to EFI, IECI has done a better job of exploring potential for
improvement. Even after seventeen iterations, IECI is still high in appropriate places, on
both sides of the invalid region. Surely it won't be too many more iterations until the
problem is nearly solved.

```
plot(XX, 1 - ieci, type="l", xlim=c(0,8), ylim=c(0,1),
  xlab="x", ylab="ieci & pc")
lines(XX, pc, col=2, lty=2)
points(X, 1 - const, col=2)
legend("right", c("IECI", "p(c<0)"), col=c(1,2,2), lty=c(1,2,1), bty="n")
```

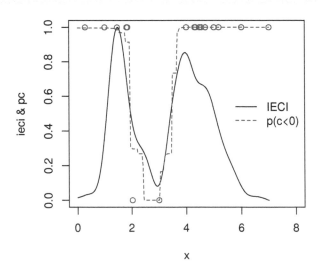

FIGURE 7.24: Predictive surface after ten more IECI iterations; compare with Figure
7.20 for EFI.

The curious reader is encouraged to perform a few more iterations under both EFI and IECI
acquisition functions. About 30 runs in total is sufficient for IECI to learn with confidence
that the global (valid) minimum is on the left side of the input space. EFI takes rather more
iterations, although differences between the two are by no means stark. Getting a good feel

for how these two methods compare on this problem requires repeated experiments, in an MC fashion, to average over the random nature of the objective evaluations.

```
deleteGP(gpi)
```

7.3.3 Real-valued constraints

Most often constraint functions are real-valued. Usually there are multiple such constraints. Paradoxically, this situation simplifies matters somewhat, because there are generally more modeling choices for real-valued simulations (e.g., GPs). Tractable nonlinear classifiers are somewhat harder to come by. Evaluations of $c_n^{(j)}(x) \in \mathbb{R}$ provide extra information (compared to $\{0, 1\}$), namely distance to feasibility. This can help accelerate convergence if used appropriately. But multiple constraints obviously make the overall problem more challenging: more things to model, making navigation toward valid optima a more complex enterprise.

EFI and IECI remain unchanged as general strategies as long as probability of constraint satisfaction $p_n^{(j)}(x) = \mathbb{P}(c_n^{(j)}(x) \leq 0)$ can be backed out of fitted surfaces $c_n^{(j)}(x)$ for each constraint $j = 1, \ldots, m$.[14] Under GPs, or any fitted response surface emitting Gaussian predictive equations, probabilities are readily available from the standard Gaussian CDF Φ, i.e., pnorm given $\mu_n^{(j)}(x)$ and $\sigma_n^{2(j)}(x)$

$$p_n^{(j)}(x) = \Phi\left(-\frac{\mu_n^{(j)}(x)}{\sigma_n^{(j)}(x)}\right).$$

Consequently EFI and IECI may utilize the magnitude of observed constraint values at best indirectly through Φ, and through the product $\prod_{j=1}^{m} p_n^{(j)}(x)$ in the case of multiple constraints.

Sometimes constraints are where all the action is. So far in this chapter, focus has concentrated on the objective, with constraints being a nuisance. In many optimization problems constraints steal the show. Recall our motivating Lockwood groundwater remediation application from §2.4. The Lockwood objective is simple/known (e.g., linear), but simulation-based constraints require heavy computation, tracing out highly non-linear and non-convex valid regions. These create many deceptive local minima in the input domain. That can make for a hard search indeed.

Here's a toy problem to fix ideas: a linear objective in two variables

$$\min_{x} \left\{ x_1 + x_2 : c_1(x) \leq 0, \, c_2(x) \leq 0, \, x \in [0, 1]^2 \right\}, \tag{7.7}$$

where two non-linear constraints are given by

---

[14]Here I'm re-purposing m, previously notating dimension of the input space where x lives, for the number of constraints. None of the discussion in this chapter demands clarity on input dimension, relieving potential for ambiguity.

$$c_1(x) = \frac{3}{2} - x_1 - 2x_2 - \frac{1}{2}\sin\left(2\pi(x_1^2 - 2x_2)\right) \tag{7.8}$$

$$\text{and} \quad c_2(x) = x_1^2 + x_2^2 - \frac{3}{2}.$$

The function `aimprob` below, named for the American Institute of Mathematics (AIM)[15] where it was cooked up (Gramacy et al., 2016), implements this toy problem as a blackbox. It was designed to synthesize many of the elements in play in problems such as Lockwood, yet in lower dimension and with a transparent and easy to evaluate form.

```
aimprob <- function(X, known.only=FALSE)
 {
  if(is.null(nrow(X))) X <- matrix(X, nrow=1)
  f <- rowSums(X)
  if(known.only) return(list(obj=f))
  c1 <- 1.5 - X[,1] - 2*X[,2] - 0.5*sin(2*pi*(X[,1]^2 - 2*X[,2]))
  c2 <- X[,1]^2 + X[,2]^2 - 1.5
  return(list(obj=f, c=drop(cbind(c1, c2))))
 }
```

A `known.only` argument allows "free" objective evaluations to be returned, if desired. This is required by one of our library implementations below. For now it may be ignored. While on the subject, however, even with known $f(x) = x_1 + x_2$ this is a hard problem when $c(x)$ is treated as an expensive blackbox. Figure 7.25 shows why. (The code chunk below first builds a plotting macro to quickly draw surfaces of this kind for use in later examples.) Colored contours show the objective. Being linear, the unconstrained global optimum is at the origin, however that location is deeply invalid. Red-dashed contours represent invalid regions; green-solid ones satisfy the constraint.

```
## establishing the macro
plotprob <- function(blackbox, nl=c(10,20), gn=200)
 {
  x <- seq(0,1, length=gn)
  X <- expand.grid(x, x)
  out <- blackbox(as.matrix(X))
  fv <- out$obj
  fv[out$c[,1] > 0 | out$c[,2] > 0] <- NA
  fi <- out$obj
  fi[!(out$c[,1] > 0 | out$c[,2] > 0)] <- NA
  plot(0, 0, type="n", xlim=c(0,1), ylim=c(0,1), xlab="x1", ylab="x2")
  C1 <- matrix(out$c[,1], ncol=gn)
  contour(x, x, C1, nlevels=1, levels=0, drawlabels=FALSE, add=TRUE, lwd=2)
  C2 <-  matrix(out$c[,2], ncol=gn)
  contour(x, x, C2, nlevels=1, levels=0, drawlabels=FALSE, add=TRUE, lwd=2)
  contour(x, x, matrix(fv, ncol=gn), nlevels=nl[1],
    add=TRUE, col="forestgreen")
  contour(x, x, matrix(fi, ncol=gn), nlevels=nl[2], add=TRUE, col=2, lty=2)
 }
```

```
## visualizing the aimprob surface(s)
plotprob(aimprob)
text(rbind(c(0.1954, 0.4044), c(0.7191, 0.1411), c(0, 0.75)),
  c("A", "B", "C"), pos=1)
```

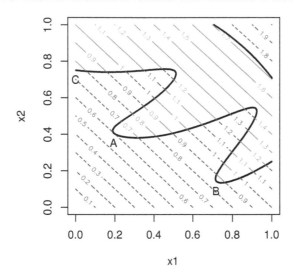

FIGURE 7.25: Linear objective (7.7) via contours colored by two nonlinear constraints (7.8). Locations of valid local minima are indicated by A, B and C.

Observe that there are three local minima, notated by points A, B, and C, respectively. Coordinates of these locations are provided below.

$$x^A \approx [0.1954, 0.4044] \qquad x^B \approx [0.7191, 0.1411] \qquad x^C = [0, 0.75],$$
$$f(x^A) \approx 0.5998 \qquad\qquad f(x^B) \approx 0.8609 \qquad\qquad f(x^C) = 0.75$$

Local optimum A is the global solution. A highly nonlinear $c_1(x)$ makes for a challenging surface to optimize in search of minima. The second constraint, $c_2(x)$ may seem uninteresting, but it reminds us that solutions may not exist on every boundary. In math programming jargon, c_2 is a *non-binding constraint*. Search algorithms which place undue emphasis on boundary exploration could be fooled in the face of a plethora of non-binding constraints.

For problems like `aimprob`, techniques from the mathematical programming literature could prove quite valuable. Math programming has efficient algorithms for non-linear blackbox optimization under constraints with provable local convergence properties (see, e.g., Nocedal and Wright, 2006), paired with lots of polished open source software[16]. Whereas statistical approaches enjoy global convergence properties, excel when simulation is expensive, noisy, and non-convex, they offer limited support for constraints. Very few well-engineered libraries exist. There are almost none that handle blackbox constraints. Some kind of hybrid, leveraging math programming for constraints and statistical surrogates for global scope, could offer powerful synergy.

As somewhat of an aside, it's perhaps telling of the current state of affairs in statistical (Bayesian) optimization that a special issue of the Journal of Statistical Software on *Numerical*

---

[16]http://plato.asu.edu/guide.html

Optimization in R: Beyond `optim`[17] overlooks BO methods such as EI. Despite its many attractive properties, the word is not yet out in mainstream computational statistics circles. A partial explanation may be that statisticians can be cagey about releasing code, and implementation is essential to effective use of optimization methodology. Machine learners have released open Python libraries featuring EI with contributions from dozens of academic and industrial authors. For example, spearmint[18] and MOE[19] both offer limited support for constraints. A very nice exception in R is `DiceOptim` (Picheny et al., 2016b). Not coincidentally, `DiceOptim` implements some of the hybrids we're about to discuss momentarily. As does `laGP`. Perhaps a JSS article featuring these packages is just around the corner.

7.3.4 Augmented Lagrangian

A framework ripe for hybridization with statistical surrogate-assisted optimization involves an apparatus called the *augmented Lagrangian* (AL; Bertsekas, 2014). Also see Chapter 17 of Nocedal and Wright (2006). The AL is a composite of objective $f(x)$ and vectorized constraint $c(x)$:

$$L_A(x; \lambda, \rho) = f(x) + \lambda^\top c(x) + \frac{1}{2\rho} \sum_{j=1}^{m} \max\left(0, c_j(x)\right)^2 \qquad (7.9)$$

where $\rho > 0$ is a penalty parameter, and $\lambda \in \mathbb{R}_+^m$ serves as Lagrange multiplier[20]. The formulation above is purely mathematical. Later when we introduce statistical surrogates we'll bring back n subscripts, etc.

An AL optimization utilizes the AL composite (7.9) to transform a constrained optimization problem into a sequence of simply constrained ones. Solutions $x^\star = \mathrm{argmin}_x L_A(x; \lambda, \rho)$ to the so-called "AL subproblem", can guide an optimizer toward solutions to the original problem (7.6) through a dynamically determined sequence of λ and ρ-values. Basically, a schedule of delicate increases to λ and ρ when $x^\star$ is invalid, as detailed in Algorithm 7.1, coaxes subproblems toward valid optima. Omitting the Lagrange multiplier term $\lambda^\top c(x)$ leads to (an example of) a so-called *additive penalty method (APM)* composite.

$$\mathrm{APM}(x; \rho) = f(x) + \frac{1}{2\rho} \sum_{j=1}^{m} \max\left(0, c_j(x)\right)^2$$

Without considerable care in choosing the form and scale of penalization (ρ), APMs can introduce ill-conditioning in the resulting subproblems. By introducing Lagrange multiplier λ, working together with penalty ρ to define the subproblem, with automatic updates as the method iterates, local convergence can be guaranteed under relatively mild conditions.

Many of the details are provided by Algorithm 7.1. There are basically two steps: 1) optimize the subproblem; and 2) update parameters (λ, ρ). While most of the work is in Step 1, off-loading the subproblem to an ordinary optimizer (`"L-BFGS-B"`), the action is all in Step 2. Updates of λ and ρ serve to nudge subproblems toward good and valid local evaluations of the objective function. When x^k satisfies all constraints, i.e., $c(x^k) \leq 0$ meaning that all components of that vectorized logical statement are true, the penalty parameter ρ is

---

[17]https://www.jstatsoft.org/issue/view/v060
[18]https://github.com/JasperSnoek/spearmint
[19]https://github.com/Yelp/MOE
[20]https://en.wikipedia.org/wiki/Lagrange_multiplier

Algorithm 7.1 Augmented Lagrangian Constrained Optimization

Assume the search region is $\mathcal{X}$, perhaps a unit hypercube.

Require a blackbox evaluating $f(x)$ and vectorized $c(x)$ jointly; initial values x^0, λ^0 and ρ^0, and maximum number of "outer loop" iterations $k_{\max}$.

For $k = 1, \ldots, k_{\max}$ comprising the "outer loop":

1. "Inner loop": approximately solve the *subproblem*:

$$x^k = \arg\min_{x \in \mathcal{X}} \left\{ L_A(x; \lambda^{k-1}, \rho^{k-1}) : x \in \mathcal{X} \right\}.$$

2. Update:
   - $\lambda_j^k = \max\left(0, \lambda_j^{k-1} + \frac{1}{\rho^{k-1}} c_j(x^k)\right)$, for $j = 1, \ldots, m$;
   - if $c(x^k) \leq 0$, set $\rho^k = \rho^{k-1}$; otherwise, set $\rho^k = \frac{1}{2}\rho^{k-1}$.

End For

Return $x^{k_{\max}}$, the solution to the most recently solved subproblem.

---

unchanged and λ moves closer to zero, if not identically so. This situation is akin to the overall scheme being "on the right track". On the other hand, if any of the constraints are not satisfied, i.e., some $c_j(x^k) > 0$, then the corresponding components of the Lagrange multiplier λ_j are increased, and the penalty $1/\rho$ doubles. This is like a course correction from Chapter 3.

Although updates of (λ, ρ) from one iteration to the next are key, the range specified for the "outer (for) loop" in the algorithm isn't much more than window dressing. A presumption that there's a budget $k_{\max}$ is somewhat unrealistic, as the real expense lies in the evaluations which happen in the "inner loop" of Step 1, solving the subproblem. Determining convergence within the inner loop [Step 2], is highly dependent on the choice of inner loop solver. Thankfully, theory for global convergence of the overall AL scheme is forgiving about criteria used to end each inner loop search. As long as some progress is made on the subproblem, perhaps to economize on the number of blackbox evaluations, convergence is eventually guaranteed – although not necessarily before exhausting a run budget. The AL framework for constrained optimization is robust to inner loop dynamics in this sense.

That state of affairs is rather akin to expectation maximization (EM)[21], to choose an example more familiar to a statistical audience. EM will converge even if the M-step is inefficient. As long as progress can be made, increasing the (E)xpected log likelihood rather than fully (M)aximizing it, outer iterations between E- and M-steps will converge to a local optima of the observed data log likelihood.

When reducing evaluations of expensive blackbox functions is less of a concern, outer loop convergence may be called when all constraints are satisfied and the (approximated) gradient of the Lagrangian is sufficiently small; for example, given thresholds $\eta_1, \eta_2 \geq 0$, one could stop when

$$\left\| \max\left\{c(x^k), 0\right\} \right\| \leq \eta_1 \quad \text{and} \quad \left\| \nabla f(x^k) + \sum_{j=1}^{m} \lambda_i^k \nabla c_j(x^k) \right\| \leq \eta_2.$$

---

[21]https://en.wikipedia.org/wiki/Expectation-maximization_algorithm

Presuming that blackbox run budgets preclude iterating exhaustively until convergence, a more pragmatic return value (compared to what's prescribed in Algorithm 7.1) may be warranted. The final subproblem solution, $x^{k_{max}}$, could well be invalid if a budget short-circuits search. Instead it's safer to report $x^\star$ yielding the BVV $y^\star = f(x^\star)$ of the objective recorded so far. Within the inner loop, where blackbox evaluations of f and c are made in pursuit of a subproblem solution, simply update $(x^\star, y^\star) \leftarrow (x, y)$ whenever $y = f(x) \leq y^\star$ and $c(x) \leq 0$ are encountered.

To illustrate, consider the following application on our toy `aimprob` (7.7)–(7.8). Setting up the subproblem solved in Step 2 of Algorithm 7.1, the code below wraps an arbitrary `blackbox` (e.g., `aimprob`) in AL clothes. A variable in the global environment, `evals`, is updated to keep track of the number of blackbox evaluations within the inner loop.

```
ALwrap <- function(x, blackbox, B, lambda, rho)
 {
  if(any(x < B[,1]) | any(x > B[,2])) return(Inf)
  fc <- blackbox(x)
  al <- fc$obj + lambda %*% fc$c +
    rep(1/(2*rho), length(fc$c)) %*% pmax(0, fc$c)^2
  evals <<- evals + 1
  return(al)
 }
```

The beauty of constrained optimization by AL is how easy it is to code the outer loop, as long as that a good unconstrained optimization library is on hand to handle the inner loop. For our illustration, we'll keep it simple and use the default `method="Nelder-Mead"` from `optim`, acting directly on `blackbox` (i.e., `aimprob`) through `ALwrap`. Since that `method` doesn't support bound constraints, bounding box checking against argument `B` (the analog of our search region/input domain $\mathcal{X}$ from earlier in the chapter) is essential as a first step in the wrapper above. Combining those elements and the updates to (λ, ρ), the function below implements the entirety of Algorithm 7.1.

```
ALoptim <- function(blackbox, B, start=runif(ncol(B)),
  lambda=rep(0, ncol(B)), rho=1/2, kmax=10, maxit=15)
 {
  ## initialize AL wrapper
  evals <- 0
  formals(ALwrap)$blackbox <- blackbox

  ## initialize outer loop progress
  prog <- matrix(NA, nrow=kmax + 1, ncol=nrow(B) + 1)
  prog[1,] <- c(start, NA)

  ## "outer loop" iterations
  for(k in 1:kmax) {

    ## solve subproblem ("inner loop")
    out <- optim(start, ALwrap, control=list(maxit=maxit),
      B=B, lambda=lambda, rho=rho)
```

```
## extract the x^* that was found, and keep track of progress
start <- out$par
fc <- blackbox(start)
prog[k+1,1:ncol(B)] <- start
if(all(fc$c <= 0)) prog[k+1,ncol(B)+1] <- fc$obj

## update augmented Lagrangian parameters
lambda <- pmax(0, lambda + (1/rho)*fc$c)
if(any(fc$c > 0)) rho = rho/2
}

## collect for returning
colnames(prog) <- c(paste0("x", 1:ncol(B)), "obj")
return(prog)
}
```

The only required arguments are `blackbox`, written in a form amenable to wrapping in `ALwrap` as exemplified by `aimprob`, and study region $\mathcal{X} \equiv$ B. Sensible defaults are provided for the other tunable parameters, at least as a jumping-off point. Notice that `maxit=15` limits the inner loop to fifteen iterations, however the number of `blackbox` evaluations may be much greater than that depending on how the inner loop attacks the subproblem. Also notice that an extra `blackbox` call is made outside of `optim` to extract the constraint evaluation at the solution to the subproblem. A more clever implementation, working with a custom inner loop optimizer that saves function evaluations, might be able to avoid this redundant, potentially expensive, evaluation. Likewise, the implementation tracks the best valid subproblem solution, rather than the best of all valid evaluations (BVV).

Let's see how `ALoptim` works on `aimprob` with a somewhat pathological initialization: about as far as you can get from any of the valid local minima.

```
evals <- 0
B <- matrix(c(rep(0,2), rep(1,2)), ncol=2)
prog <- ALoptim(aimprob, B, start=c(0.9, 0.9))
```

Figure 7.26 illustrates progress by plotting outer loop iteration number (plus one for the starting location), with "cross-hairs" at the final value to ease visualization.

```
plotprob(aimprob)
text(prog[,1], prog[,2], 1:nrow(prog))
m <- which.min(prog[,3])
abline(v=prog[m,1], lty=3, col="gray")
abline(h=prog[m,2], lty=3, col="gray")
```

It would appear that after just ten outer loop iterations the global solution has been found, at least to within a decent tolerance. However, this summary hides a huge computational expense in terms of the number of blackbox evaluations.

```
evals
```

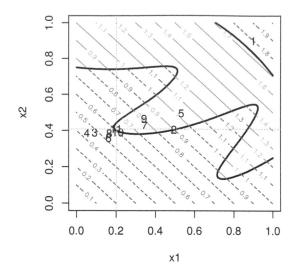

FIGURE 7.26: One application of optimization by augmented Lagrangian. The numbers indicate k at location x^k solving the AL subproblem. Cross-hairs highlight the final solution.

```
## [1] 168
```

This situation could potentially have been much worse with a higher value of `maxit`. To what extent might one improve upon this state of affairs? Before heading down that road, consider first what happens when we randomly reinitialize at a few more places.

```
reps <- 30
all.evals <- rep(NA, reps)
start <- end <- matrix(NA, nrow=reps, ncol=2)
for(i in 1:reps) {
  evals <- 0
  prog <- ALoptim(aimprob, B, kmax=20)
  start[i,] <- prog[1,-3]
  m <- which.min(prog[,3])
  end[i,] <- prog[m,-3]
  all.evals[i] <- evals
}
```

To ensure that a good answer is obtained in each repetition, a higher number of outer loop iterations, `kmax=20`, is used. Figure 7.27 shows the (random) starting and best valid ending value of the objective from each repetition as arrows.

```
plotprob(aimprob)
arrows(start[,1], start[,2], end[,1], end[,2], length=0.1, col="gray")
```

It's quite clear from this view that the AL is adequately finding local solutions, but sometimes it's surprising what local solution it finds. In at least one occasion out of thirty, the ending value lands in a trough that's both farther away from its starting location and traverses shallower gradients in order to do so. So while local AL convergence is fairly robust to subproblem progress in inner loop iterations, it can be hard to predict which of several

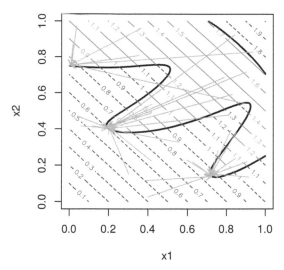

FIGURE 7.27: Outcome of repeated applications of AL-optimization. The origin of each arrow indicates a randomly chosen initialization; terminus indicates the outcome of search.

local (valid) optima it'll converge to. And all that while spending frivolously on blackbox evaluations, at least compared to what I'm about to propose momentarily.

```
summary(all.evals)
```

```
##     Min. 1st Qu.  Median    Mean 3rd Qu.    Max.
##      244     332     337     322     339     340
```

For a more interactive variation on the above illustration, see aimprob.R[22]. This demo allows the user to step through outer loop iterations, under random initialization, to get a better feel for how an AL guides search. An attentive reader will notice from the demo that the AL method often approaches valid solutions from the invalid side of constraint boundaries, with at least as many evaluations in invalid as valid regions – a behavior well-studied in the literature. As a result, many software libraries implementing AL composites (7.9) consider as solutions $x^\star$ those inputs corresponding to the BOV whose constraints are within an ϵ tolerance of valid (e.g., $c(x^\star) \leq 1e^{-4}$). Others feel that it's safer to simply report the BVV of the objective.

The calculation below shows that solutions to the subproblem, the x^ks, are about equally likely to be valid as invalid on this problem, which can be interpreted as a symptom of that behavior.

```
mean(is.na(prog[,3]))
```

```
## [1] 0.4286
```

Without a cache of the entirety of **blackbox** evaluations, and without a much larger **maxit** and **kmax**, it's difficult to illustrate this phenomenon in greater detail. Of course, we don't want to dwell too much here on review of math programming methods. The takeaway

---

[22]http://bobby.gramacy.com/surrogates/aimprob.R

message is that the AL method is a potentially expensive, local affair, that otherwise has many nice empirical and theoretical properties. And it's quite easy to implement.

7.3.5 Augmented Lagrangian Bayesian optimization (ALBO)

Since the inner loop solver can be anything, as long as it makes progress on the subproblem, why not try a surrogate-assisted/BO solver? That might be the best of both worlds. Surrogate methods enjoy global scope, heuristics like EI are good at navigating exploration–exploitation trade-offs, and setups adapted to the constrained setting are few. AL methods converge to valid solutions, but are profligate with blackbox evaluations despite their distinctly local scope. This idea, of combining the AL with Bayesian optimization (BO) was originally proposed by Gramacy et al. (2016), and later dubbed ALBO by Picheny et al. (2016c) who extended the methodology to equality and mixed (inequality and equality) constraints.

The crux of the approach involves training a surrogate, which will ultimately guide the inner loop, on the entire corpus of blackbox evaluations obtained so far, i.e., over all inner and outer iterations of the loops in Algorithm 7.1. Whereas in a more conventional AL approach, each subproblem would be solved without any memory of solvers run in earlier outer loop iterations. Not only that, but typically inner loop optimizers (like those from `optim`) are themselves also memoryless, or at best have limited memory like `"L-BFGS-B"`. Therefore the proposed application of surrogates here represents a regime shift in the *modus operandi* compared to canonical AL methodology. Surrogate-based inner loops will have not only global scope on the specific subproblems they're solving, but also knowledge of the surface(s) gathered from previous subproblems.

Suppose that n such evaluations of the blackbox have been made so far, across all inner and outer loops of Algorithm 7.1. Denote by

$$D_n = (x_1, f(x_1), c(x_1)), \ldots, (x_n, f(x_n), c(x_n))$$

the data available for training surrogates. We've been largely able to avoid double-indexing so far in this chapter, over constraint coordinates $j = 1, \ldots, m$ and observation indices $i = 1, \ldots, n$. (In some places I was downright sloppy, but context – usually a 1d constraint – helped.) Going forward, I shall invest in more precise notation. Let $c_j^{(i)}$ be the jth coordinate of $c(x_i)$, the vector of constraint evaluations from the ith run of the blackbox. To have notation match for objective evaluations, even though they're not vectorized, I shall write $f^{(i)}$ as a shorthand for $f(x_i)$. So our training data can now be more compactly notated as $D_n = \{(f^{(i)}, c^{(i)}\}_{i=1}^n$ for m-vectors $c^{(i)}$. As usual, focus is on GP surrogates for D_n, although much of the discussion is agnostic about that choice with the caveat that a few of the closed-form results leverage Gaussian predictive equations.

There are several options for how exactly to proceed, as regards modeling data D_n in a means effective for solving subproblems in Step 1 at each iteration k. One is easy to rule out. Perhaps it's strange to begin by focusing attention on a flawed notion. But there's good reason for doing so. It represents a rather straightforward approach; the first thing one might think to try. Ultimately it serves to illustrate pitfalls in use of GP surrogates in practice, whether for optimization or otherwise. You might call it the gestalt approach: model the whole rather than the sum of its parts. Let $y_i = L_A(x_i; \lambda^{k-1}, \rho^{k-1})$ via $f^{(i)}$ and $c^{(i)}$. That is,

$$y_i = f^{(i)} + (\lambda^{k-1})^\top c^{(i)} + \frac{1}{2\rho^{k-1}} \sum_{j=1}^m \max\left(0, c_j^{(i)}\right)^2.$$

Then fit a surrogate to $(x_1, y_1), \ldots, (x_n, y_n)$ in the usual way, and guide the inner loop search using your preferred acquisition function: predictive mean (EY), EI, IECI, etc.

Benefits to this scheme are immediately self-evident. Foremost, it's highly modular. Plug in any (objective-only) surrogate-assisted optimizer trained on y_i's, whether of your own design, borrowing one of the functions provided earlier, or from an existing library. Such a scheme ought to have many desirable features. It'll enjoy global scope, being trained on all data encountered so far. Yet it'll also act locally: the AL method focuses search on promising parts of the input space from the perspective of satisfying constraints. This hybrid scheme would seem to embody an attractive means of balancing exploration, exploitation, constraint satisfaction, and (perhaps above all) simple implementation.

So what's the problem? The biggest issue is that the response surface created by mapping $D_n \to \{(x_i, y_i)\}_{i=1}^n$ possesses two pathologies known to thwart effective spatial modeling, all wrapped into a tidy package. The squared term creates inherent nonstationarity by amplifying dynamics away from valid regions. The max creates kinks in the surface, breaking smoothness. Although it's definitely possible to engineer a covariance kernel, for GP modeling say, that anticipates response surface features such as these, I know of none readily available for such purposes.

A somewhat smaller issue is one of efficiency. The AL composite (7.9) takes on a quadratic form of sorts, yet that information isn't leveraged in the surrogate model specification. Again, at least not if a canonical, library-based GP modeling is deployed. In situations where f is known, for example linear in our toy **aimprob** and Lockwood problems, the apparatus would needlessly model a known quantity. Many challenging problems have known objectives, so that special case ought to be explicitly acknowledged by the methodological development. In fairness, classical AL doesn't leverage a known quadratic form or a potentially known objective either. Yet more specialized setups have been proposed in the literature, paying dividends in practice (Kannan and Wild, 2012). Since we're drawing up schematics for something new, why not lay out the full wish-list at the very start?

All of these shortcomings are addressed simultaneously by separately/independently surrogate modeling each component of the AL. That is, fit a surrogate to the objective data $\{(x_i, f^{(i)})\}_{i=1}^n$, if needed, and fit m separate surrogates to $\{x_i, c_j^{(i)}\}_{i=1}^n$ for each of m constraints. As shorthand, write f^n for fitted objective surrogate, and $c^n = (c_1^n, \ldots, c_m^n)$ for mutually independent fitted constraint surrogates. Let $Y_f(x) \equiv Y_{f^n}(x) \sim f^n$ denote a random variable characterizing predictive uncertainty at novel input x under the objective surrogate, which may simply be $y_f(x) = f(x)$ if the objective is treated as known. Similarly, let $c^n = (c_1^n, \ldots, c_m^n)$ emit predictive random variables $Y_c(x) \equiv Y_c^n(x) = (Y_{c_1}^n(x), \ldots, Y_{c_m}^n(x))$ following the distribution of novel predictions under the m constraint surrogates. Then, the distribution of the composite random variable

$$Y(x) = Y_f(x) + \lambda^\top Y_c(x) + \frac{1}{2\rho} \sum_{j=1}^m \max(0, Y_{c_j}(x))^2 \qquad (7.10)$$

can serve as a surrogate for $L_A(x; \lambda, \rho)$.

What can you do with this composite random variable? If surrogate predictive equations are Gaussian, as they are from fitted GPs, then the AL composite posterior mean (EY) is available in closed form.

$$\mathbb{E}\{Y(x)\} = \mu_f^n(x) + \lambda^\top \mu_c^n(x) + \frac{1}{2\rho} \sum_{j=1}^m \mathbb{E}\{\max(0, Y_{c_j}(x))^2\}$$

A result from generalized EI (Schonlau et al., 1998) furnishes a closed form for the expectation inside that sum above. Specifically, an expression for $\mathbb{E}\{\max(0, Y_{c_j}(x))^2\}$ follows by recognizing its argument as a powered improvement for $-Y_{c_j}(x)$ over zero, that is, $I^{(0)}_{-Y_{c_j}}(x) = \max\{0, 0 + Y_{c_j}(x)\}$. Since the power is 2, an expectation-variance relationship can be exploited to obtain

$$\mathbb{E}\{\max(0, Y_{c_j}(x))^2\} = \mathbb{E}\{I^{(0)}_{-Y_{c_j}}(x)\}^2 + \mathbb{V}\mathrm{ar}[I^{(0)}_{-Y_{c_j}}(x)]$$

$$= \sigma^{2n}_{c_j}(x)\left[\left(1 + \left(\frac{\mu^n_{c_j}(x)}{\sigma^n_{c_j}(x)}\right)^2\right)\Phi\left(\frac{\mu^n_{c_j}(x)}{\sigma^n_{c_j}(x)}\right) + \frac{\mu^n_{c_j}(x)}{\sigma^n_{c_j}(x)}\phi\left(\frac{\mu^n_{c_j}(x)}{\sigma^n_{c_j}(x)}\right)\right].$$

It's hard to imagine many other quantities relevant for optimization taking on analytic closed forms. The max in the AL composite random variable (7.10) is hard to work around. We got lucky with the mean above. Variances may similarly be available, enabling LCB-based acquisition (7.1), however theoretical guidance for choosing appropriate weights β_n in this AL context isn't readily available. The simplest way to evaluate EI under the composite AL is through MC. Sample T deviates $y_f^{(t)}(x)$ and $y_c^{(t)}(x)$, form $y^{(t)}(x)$ through Eq. (7.9), and average:

$$\mathrm{EI}(x) \approx \frac{1}{T}\sum_{t=1}^{T}\max\{0, y^n_{\min} - y^{(t)}(x)\}. \tag{7.11}$$

Surprisingly, this works quite well in practice and even applies when surrogates emit non-Gaussian predictors. T as small as 100 is often sufficient to achieve stable relative EI comparisons in $\mathcal{X}$-space. Such crude numerical approximation can actually be better, in terms of cost-benefit trade-off, than other fancier alternatives. This is borne out in our empirical work. But hold that thought for a moment.

A big downside to MC-based acquisition is lack of determinism. Stochastic optimization dulls the deliberate weighing of trade-offs that an expensive blackbox deserves. Perhaps more practically, an MC EI is hard to meta-optimize, say through an `optim` subsubroutine. After all, `optim`-like inability to cope with noisy evaluations was one of the big motivators for BO alternatives.

What alternatives? Well if the story ended there, this whole ALBO thing would be somewhat underwhelming. Although the max thwarts analytics, there's a well-known technique from math programming for softening such hard thresholds. One can equivalently reformulate AL subproblems without the max by introducing so-called *slack variables*. The setup is as follows. Introduce s_j, for $j = 1, \ldots, m$, i.e., one for each $c_j(x)$; convert inequality into equality constraints: $c_j(x) - s_j = 0$; and augment the program with additional bound constraints $s_j \geq 0$, for $j = 1, \ldots, m$. In practice these latter, simply defined constraints are subsumed into the box $\mathcal{B} = \mathcal{X} \times \mathcal{S}$ where $\mathcal{S} = [0, \infty]^m$. In this way, the problem is mapped from p-dimensional $\mathcal{X}$ space to $(p + m)$-dimensional $\mathcal{X} \times \mathcal{S}$ space. Although our focus here is on inequality constrained problems, it's worth commenting that equality (and thus mixed constraints) are implemented by fixing $s_j = 0$ for relevant coordinates, j, of the constraint vector.

Slack-equality-based AL is sometimes presented as the canonical form in textbooks (Nocedal and Wright, 2006, Chapter 17) because that framework can accommodate both equality and inequality constraints, even simultaneously. Introducing slacks into ALBO, as described above, facilitates the only known EI-based method for handling mixed (equality and inequality) constraints (Picheny et al., 2016c) in surrogate-assisted/BO literatures. But I'll try not to unduly complicate the narrative here by over-emphasizing the mixed constraints setting. Instead, let's focus on using slacks to reformulate the ALBO composite random variable (7.10), with an eye toward a more analytical EI calculation.

Picheny et al. (2016c) showed that, for the slack-inequality constrained problem, the AL composite becomes

$$L_A(x, s; \lambda, \rho) = f(x) + \lambda^\top (c(x) + s) + \frac{1}{2\rho} \sum_{j=1}^m (c_j(x) + s_j)^2, \quad \text{so that}$$

$$Y(x, s) = Y_f(x) + \sum_{j=1}^m \lambda_j s_j + \frac{1}{2\rho} \sum_{j=1}^m s_j^2 + \frac{1}{2\rho} \sum_{j=1}^m \left[(\alpha_j + Y_{c_j}(x))^2 - \alpha_j^2 \right], \quad (7.12)$$

where $\alpha_j = \lambda_j \rho + s_j$. Observe that if $s_j = 0$, encoding an equality constraint, the quadratic penalty eventually forces $c_j(x) = 0$ with small enough ρ.

It's helpful to re-arrange the expression for $Y(x, s)$ in Eq. (7.12) in order to focus on interplay between slacks s and ordinary inputs x. Let $g(s) = \sum_{j=1}^m \lambda_j s_j + \frac{1}{2\rho} \sum_{j=1}^m s_j^2 - \alpha_j^2$ capture the part of $Y(x, s)$ which is a function of slacks s only, and let $W(x, s) = \sum_{j=1}^m [(\alpha_j + Y_{c_j}(x))^2]$ capture that which depends on both types of inputs. With those definitions, we may more compactly write

$$Y(x, s) = Y_f(x) + g(s) + \frac{1}{2\rho} W(x, s).$$

Using $Y_{c_j} \sim \mathcal{N} \left(\mu_{c_j}(x), \sigma_{c_j}^2(x) \right)$, W can be written as

$$W(x, s) = \sum_{j=1}^m Z_j^2, \quad \text{with} \quad Z_j \sim \mathcal{N} \left(\mu_{c_j}(x) + \alpha_j, \sigma_{c_j}^2(x) \right)$$

$$= \sum_{j=1}^m \sigma_{c_j}^2(x) \bar{Z}_j^2, \quad \text{with} \quad Z_j \sim \mathcal{N} \left(\frac{\mu_{c_j}(x) + \alpha_j}{\sigma_{c_j}(x)}, 1 \right)$$

$$= \sum_{j=1}^m \sigma_{c_j}^2(x) X_j, \quad \text{with} \quad X_j \sim \chi_{\nu=1}^2 \left(\left(\frac{\mu_{c_j}(x) + \alpha_j}{\sigma_{c_j}(x)} \right)^2 \right),$$

where that final random variable is a *weighted sum of non-central chi-square (WSNC)* variates (Duchesne and de Micheaux, 2010). WSNC density, distribution and quantile functions are provided by R packages `CompQuadForm` (de Micheaux, 2017) and `sadists` (Pav, 2017).

Those library functions – for CDF evaluation in particular – enable EI approximation to a high degree of accuracy with straightforward numerics. Under the AL-composite, that involves working with $\text{EI}(x, s) = \mathbb{E} \left[(y_{\min}^n - Y(x, s)) \, \mathbb{I}_{\{Y(x,s) \leq y_{\min}^n\}} \right]$, given the current minimum $y_{\min}^n$ of the AL over all n runs. Consider the somewhat simpler case where $f(x)$ is treated as known. Let $w_{\min}^n = 2\rho \left(y_{\min}^n - f(x) - g(s) \right)$, and D_W denote the WSNC (cumulative) distribution of $W(x, s)$. It can be shown that

$$\text{EI}(x, s) = \frac{1}{2\rho} \mathbb{E} \left[(w_{\min}^n - W(x, s)) \, \mathbb{I}_{W(x,s) \leq w_{\min}^n} \right]$$

$$= \frac{1}{2\rho} \int_{-\infty}^{w_{\min}^n} D_W(t) \, dt = \frac{1}{2\rho} \int_0^{w_{\min}^n} D_W(t) \, dt.$$

In other words, EI calculation involves a one-dimensional definite integral whose integrand may be evaluated with a library subroutine. For example, if `wts` is a vector holding the m components of $\sigma^2_{c_j}(x)$, and `ncp` holds arguments of $\chi^2_{\nu=1}(\cdot)$ above, i.e.,

$$\text{ncp}_j = \left(\frac{\mu_{c_j}(x) + \alpha_j}{\sigma_{c_j}(x)} \right)^2,$$

then the following commands yield EI under the AL composite.

```
R ex> library(sadists)
R ex> obj.EI <- function(t)
+          psumchisqpow(q=t, wts, rep(1, m), ncp, rep(1, m))
R ex> EI <- integrate(obj.EI, 0, wmin)$value / (2*rho)
```

Note these lines are just an example; that R code is not included in the Rmarkdown build in order to produce any output shown here or below. For more details, see Appendix C of the arXiv version[23] of Picheny et al. (2016c), which illustrates both `sadists` and `CompQuadForm` application. A worked example involving those calculations under the hood will be provided momentarily. Adjustments are trivial when f is treated as unknown; see Picheny et al. (2016c) for details. An example for that case is coming soon too. It's worth acknowledging that the scheme outlined above is still numeric. However, unlike the MC version (7.11), which collected random deviates in $m+1$ dimensions, the numerics here are in 1d. Simple quadrature suffices, like that implemented by `integrate` in base R, which provides approximations to tolerances low enough to meta-optimize, with `optim` say.

7.3.6 ALBO implementation details

Before jumping headlong into illustrations, several details are worthy of lip service. The attentive reader may have noticed some logical gaps in the passages above, and may have justifiably been worried they were being swept under a rug. Two are engineering details: a choice of initializing (λ^0, ρ^0); and how to determine inner loop convergence of the surrogate-assisted sub-solver. I shall summarize default solutions to these which work well, but could potentially be sub-optimal in particular settings. Two others involve the slack re-formulation and are more technical, in the sense that there's a right way from a certain point of view. The first involves re-describing Algorithm 7.1 for slacks. Changes here are largely semantic, however the crucial step of updating (λ^k, ρ^k) requires explicit adjustment. Finally, although slack variables expand the search space by up to m dimensions, one additional per constraint beyond the dimension of x, an optimal setting of s^k_j can be expressed as a function of x, mapping the search space back down again to the dimension of x only.

A sensible initialization strategy for (λ^0, ρ^0) balances the scales of objective and constraint in the AL on the basis of outputs obtained from a seed space-filling design of size n_0. The description here, and in the following passages, favors generality with respect to a mixed (inequality and equality) constraint setting. Let $v(x)$ be a logical vector of length $m = m_\leq + m_=$ recording the validity of x in a zero-slack setting. Let $v_j(x) = 1$ if the jth inequality constraint is satisfied, $c_j(x) \leq 0$ for $j = 1, \ldots, m_\leq$; and let $v_\ell(x) = 1$ if the $(j - m_\leq)^{\text{th}}$ equality constraint is satisfied, $|c_j(x)| \leq \epsilon$ for $j = m_\leq + 1, \ldots, m$; otherwise let $v_j(x) = 0$. Then take

---

[23]https://arxiv.org/abs/1605.09466

$$\rho^0 = \frac{\min_{i=1,\ldots,n_0}\{\sum_{j=1}^{m} c_j^2(x_i) : \exists j, v_j(x_i) = 0\}}{2\min_{i=1,\ldots,n_0}\{f(x_i) : \forall j, v_j(x_i) = 1\}}, \tag{7.13}$$

and $\lambda^0 = 0$. The denominator above isn't defined if the initial design has no valid values (i.e., if there's no x_i with $v_j(x_i) = 1$ for all j). When that happens, use the median of $f(x_i)$ in the denominator instead. On the other hand, if the initial design has no invalid values and hence the numerator isn't defined, take $\rho^0 = 1$.

Initial implementations of ALBO (Gramacy et al., 2016) ran the solver, optimizing either $\mathbb{E}\{Y(x)\}$ or EI(x), until progress had been made on the composite objective; i.e., until an x^k was found such that $L_A(x^k, \lambda^{k-1}, \rho^{k-1}) < L_A(x^{k-1}, \lambda^{k-1}, \rho^{k-1})$. The idea was to economize on the number of blackbox evaluations from the outer loop while satisfying assumptions underpinning theory providing for convergence of the overall AL scheme. In a discussion of that paper, Picheny et al. (2016a) suggested that one might get away with even fewer evaluations – potentially as few as one – when EI is used. Their reasoning was that such "single-acquisition (inner loop) termination", taking x^k from EI even if it produces a worse AL than the current value, matches the spirit of EI-based search as optimal, in a certain sense, if it's the final one (see Bull (2011) and the end of §7.2.2). Original/prototype ALBO, which continued the inner loop by taking steps of EI until AL improved, represented overkill in terms of empirical outer-loop progress.

Single-acquisition termination also meshes well with an updating scheme analogous to Step 2 in Algorithm 7.1: updating only when *no* actual improvement (in terms of constraint violation) is realized by that choice. Technically, that updating scheme must be re-written in order to cope with a slack variable formulation, incorporating a slight twist for situations where some constraints specify equality (i.e., $s_j = 0$). For completeness at the expense of slight redundancy, reworked lines of Algorithm 7.1 are provided in Algorithm 7.2. Rather than single-indexing $m_\leq$ inequality constraints first, followed $m_=$ equality ones, Algorithm 7.2 uses an equality set $\mathcal{E}$ as a shorthand.

Algorithm 7.2 Slack Variable and Mixed Constraint Adjustments

---

Require the same as Algorithm 7.1, with (potentially empty) set $\mathcal{E} \subseteq \{1, \ldots, m\}$ indicating equality constraints, and a tolerance $\epsilon > 0$ deeming such constraints satisfied.

Assume a search region augmented to $\mathcal{B} = \mathcal{X} \times \mathcal{S}$.

Let (x^k, s^k) approximately solve $\min_{x,s} \{L_A(x, s; \lambda^{k-1}, \rho^{k-1}) : (x, s) \in \mathcal{B}\}$ in iteration k of the outer loop, taking $s_j^k = 0$ for all $j \in \mathcal{E}$.

2. Update:
   - $\lambda_j^k = \lambda_j^{k-1} + \frac{1}{\rho^{k-1}}(c_j(x^k) + s_j^k)$, for $j = 1, \ldots, m$;
   - if $c_{j \notin \mathcal{E}}(x^k) \leq 0$ and $|c_{j \in \mathcal{E}}(x^k)| \leq \epsilon$, set $\rho^k = \rho^{k-1}$;
   - else $\rho^k = \frac{1}{2}\rho^{k-1}$

Otherwise proceed as usual.

Above, the first part of Step 2 is the same as in the non-slack AL in Algorithm 7.1 without the "max", and with slacks augmenting constraint values. The following "if" checks for validity at x^k, deploying a threshold $\epsilon \geq 0$ on equality constraints. Such softening is essential since a real-valued blackbox would never (or exceedingly rarely) evaluate to exactly zero (or any other particular number). If validity holds at (x^k, s^k), the current AL iteration is deemed to have made progress and the penalty remains unchanged; otherwise it's doubled. An alternate formulation may entertain $|c(x^k)+s^k| \le \epsilon$ for all coordinates $j = 1, \ldots, m$ without separating out inequality and equality constraints. While the latter is cleaner diagrammatically, the version presented in the algorithm limits exposure to choices of threshold ϵ. When there are no equality constraints, no softening is required.

In Algorithm 7.2, as well as in the preceding discussion, an optimization over a larger dimensional $\mathcal{X} \times \mathcal{S}$ space is implied by $\min_{x,s} \left\{ L_A(x, s; \lambda^{k-1}, \rho^{k-1}) : (x, s) \in \mathcal{B} \right\}$. With some of the slacks $s_j = 0$ for $j \in \mathcal{E}$, indexing the set of equality constraints, that search dimension is somewhat reduced. Still, that's more space to search than in the original AL version outlined in Algorithm 7.1. It turns out that the remaining elements of $\mathcal{S}$, i.e., $j \notin \mathcal{E}$, can also be dispensed with, although perhaps not as trivially.

For observed $c_j(x)$, associated slack variables minimizing the composite $L_A(x, s; \lambda^{k-1}, \rho^{k-1})$ can be obtained analytically, and thereby be concentrated out. Using the form of $Y(x, s)$ from Eq. (7.12), note that $\min_{s \in \mathbb{R}^m} y(x, s)$ is equivalent to $\min_{s \in \mathbb{R}^m} \sum_{j=1}^{m} (2\lambda_j \rho s_j + s_j^2 + 2s_j c_j(x))$. For fixed x, this is strictly convex in s. Therefore, its unconstrained minimum can only be it's stationary point, which satisfies $0 = 2\lambda_j \rho + 2s_j^\star(x) + 2c_j(x)$, for $j = 1, \ldots, m$. Accounting for the nonnegativity constraint, we obtain the following optimal slacks:

$$s_j^\star(x) = \max \left\{ 0, -\lambda_j \rho - c_j(x) \right\}, \qquad j \notin \mathcal{E}. \tag{7.14}$$

Above $s^\star$ is expressed as a function of x to convey that x remains a free quantity in $y(x, s^\star(x))$.

In the blackbox $c(x)$ setting, $y(x, s^\star(x))$ is only directly accessible at data locations x_i. At other x-values, however, surrogates provide a useful approximation. When $Y_c(x)$ is (approximately) Gaussian it's straightforward to show that the optimal setting of the slack variables, solving $\min_{s \in \mathbb{R}^m} \mathbb{E}[Y(x, s)]$, are $s_j^\star(x) = \max\{0, -\lambda_j \rho - \mu_{c_j}(x)\}$, i.e., like in Eq. (7.14) with a prediction $\mu_{c_j}(x)$ for $Y_{c_j}(x)$ replacing the unknown $c_j(x)$ value.

Other criteria may be used to choose slacks. Instead of minimizing the mean of the composite, one could maximize EI. Appendix A of Picheny et al. (2016c) explains how this is of dubious practical value. Compared to the EY settings described above, setting optimal slacks by searching over EI is both more computationally intensive and provides near identical results in practice. Even when acquisitions are ultimately made by EI on the AL composite, analytically concentrating out slacks via EY is sufficient.

As a final remark here, before getting on to illustrations, it's worth noting that there's a downside to EI calculations on the AL compared to simpler alternatives like $\mathbb{E}\{Y(x)\}$, whether in original or slack formulations. $\mathrm{EI}(x)$ can be exactly zero for some x-values – and not just a few, but on an uncountably infinite subset of the study region $\mathcal{X}$. This is true even when Gaussian predictive equations are in play. The AL composite random variable, $Y(x)$ or $Y(x, s)$, is a quadratic function of surrogate random variables $Y_f(x)$ and $Y_c(x)$. The image of a quadratic need not span the entire real line. The same is not true for EI in unconstrained settings (i.e., not with the AL), where the Gaussian form of $Y(x)$ guarantees that improvement $I(x)$ has positive probability of being non-zero. As a result, it can be hard to find regions of positive EI, especially in latter outer loop iterations where the penalty $1/\rho^{k-1}$ is high. For a more detailed discussion of when such situations may arise, see Section 3.3 of Gramacy et al. (2016). If an inner loop solver can't find any non-zero EI regions, a failsafe switch to EY-based search can kick-in. A method successful in deploying EI in early iterations, navigating exploration and exploitation trade-offs, is quite likely to finish with purely exploitative EY-search – not necessarily a bad thing.

7.3.7 ALBO in action

We've gone longer than usual without concrete examples. The simplicity of ALBO masks myriad alternatives and implementation details: choosing formulations (original or slack), integrating with libraries for surrogate modeling (i.e., GPs), initialization and inner loop convergence, choices for acquisition (EY or EI), etc. That's too much code to wield all at once in a real-time Rmarkdown setting. So we'll borrow the `optim.auglag` implementation in `laGP`.

Known objective

Our `aimprob` function, implementing Eqs. (7.7)–(7.8), is provided in exactly the format required for `optim.auglag`, complete with a `known.only` argument allowing objective to be probed without cost, if desired. The only other essential argument is search region $\mathcal{X}$, which we already set to be the unit cube `B` in our earlier `ALoptim` example(s). By default, `optim.auglag` utilizes MC approximated EI acquisition on a sequence of increasingly dense random space-filling candidate search grids. Unless otherwise specified, a budget of `end=100` blackbox runs will be performed, seeding with an initial design of size `start=10`. Far fewer runs are required to find good valid solutions for `aimprob`, so we'll take that down to `end=50` for this example, and switch off progress printing.

```
ei.mc <- optim.auglag(aimprob, B, end=50, verb=0)
```

Before inspecting the output, what shall we compare it to? Using $\mathbb{E}\{Y(x)\}$ for acquisitions is an option, and may be invoked as follows.

```
ey <- optim.auglag(aimprob, B, end=50, ey.tol=1, verb=0)
```

Analog EFI acquisition, but otherwise using the same GP surrogates and sharing all other implementation details, is provided by `optim.efi`.

```
efi <- optim.efi(aimprob, B, end=50, verb=0)
```

A slack-variable-based formulation can be invoked in two different ways. The first, with `slack=TRUE`, offers a setup identical to the MC version except that exact EI evaluations are calculated by numerically integrating WSNC distribution functions at random candidates. The second, with `slack=2`, is similar but finishes off with a meta-`optim`-based maximization initialized from the highest EI candidate, leveraging the deterministic nature and high accuracy of slack EI calculations. A similar multi-start EI is provided by `DiceOptim`, specifically for AL-based acquisitions as well as many others implemented therein.

```
ei.sl <- optim.auglag(aimprob, B, end=50, slack=TRUE, verb=0)
ei.slopt <- optim.auglag(aimprob, B, end=50, slack=2, verb=0)
```

Although no timings are reported here, slack-based methods are more computationally intensive. Non-slack/MC versions can perform all fifty iterations in under a second. Using `slack=TRUE` takes about five seconds; `slack=2` can take upwards of thirty seconds on modern workstations. Figure 7.28 shows BVVs for each method over fifty blackbox evaluations.

```
plot(efi$prog, type="l", ylim=c(0.6, 1.6), ylab="best valid value",
  xlab="n: blackbox evaluations")
lines(ey$prog, col=2, lty=2)
lines(ei.mc$prog, col=3, lty=3)
lines(ei.sl$prog, col=4, lty=4)
lines(ei.slopt$prog, col=5, lty=5)
legend("topright", c("EFI", "EY", "EI.mc", "EI.sl", "EI.slopt"),
  col=1:5, lty=1:5)
```

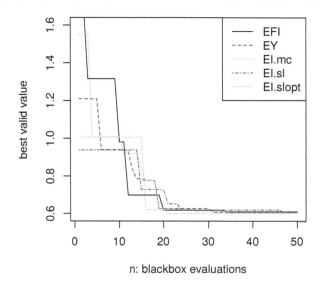

FIGURE 7.28: Comparing EFI and AL-based comparators on the toy problem (7.7)–(7.8).

Although they start at different places, because they are seeded with different (random) space-filling designs, they all end up at about the same place after just 25 blackbox evaluations – at what we know from Figure 7.25 to be the global (valid) optimum. That's pretty impressive considering that ALoptim (§7.3.4) required 332 evaluations, but offered only local convergence. Despite starting from different spots, the curve labeled "EI.slopt" (corresponding to slack=2) is consistently first to get down to a BVV of about 0.6, the global minimum.

The curious reader may wish to wrap those calls in for loops to explore average case behavior. Panels of Figure 7.29, which are based on a similar exercise from Picheny et al. (2016c), used 100 random restarts and included several other methods for comparison.

The left panel shows BVV. All four methods entertained are pretty speedy compared to ordinary, non surrogate-assisted AL, consistently finding the global minimum after 25 blackbox evaluations. The blue comparator, PESC for predictive entropy search (Hernández-Lobato et al., 2015) leverages an implementation in spearmint[24], a Python library. Comparisons are drawn to methods from Gramacy et al. (2016) in gray: ALBO but without slack variables, automatic initialization (7.13) or single-acquisition (inner loop) termination. Taken together, those engineering details offer a dramatic improvement over the initial prototype.

The right panel shows *log utility gap*, $\log(\text{BVV}_n - y^\star)$ over acquisitions $n = n_0, \ldots, N$, tracking log differences between the theoretical BVV of the objective, $y^\star$, and those found by

---

[24]https://github.com/HIPS/Spearmint/tree/PESC

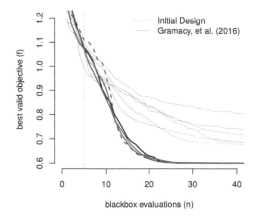 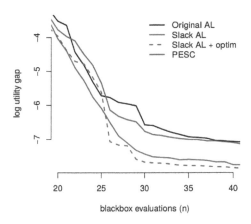

FIGURE 7.29: Cribbed summary of comparison from Picheny et al. (2016c) of raw BVV (left) progress, and log utility gap (right) in order to zoom in on the best methods in later iterations.

search (empirical BVV). Under this progress metric, separation between the best comparators is accentuated. Perhaps more notably, all methods continue to make progress well beyond the twenty-fifth evaluation, albeit only incrementally. PESC and slack AL with meta-optimization of EI are best; pairwise t-tests reveal that these differences are indeed statistically significant. See Picheny et al. (2016c) for more details.

Unknown objective

How about if the objective function is more complicated and, along with constraints, expensive to evaluate jointly in the blackbox? Sure. To illustrate, how about augmenting our toy `aimprob` with the objective below?

```
herbtooth <- function(X)
  {
    if(!is.matrix(X)) X <- matrix(X, ncol=2)
    g <- function(z)
      return(exp(-(z - 1)^2) + exp(-0.8*(z + 1)^2) - 0.05*sin(8*(z + 0.1)))
    return(-g(X[,1])*g(X[,2]))
  }
```

Some have been calling this function "Herbie's tooth" because it was cooked up by Herbie Lee[25], my PhD advisor, as a challenging surface to model and optimize (Lee et al., 2011; Gramacy and Lee, 2011). Herbie's tooth has featured in several recent papers, including the local approximate Gaussian process (LAGP) paper (Gramacy and Apley, 2015). When visualized as a perspective plot, as we shall do in Chapter 9's Figure 9.25 when we get to LAGP, it looks like a molar. A mathematical depiction is provided by Eq. (9.3). Without dense sampling in the input space it's hard to accurately emulate this test function. More relevant in our current context is that molars have lots of nooks and crannies, i.e., local minima, which makes for a challenging optimization problem.

---

[25]https://users.soe.ucsc.edu/~herbie/

For visuals, and ultimately to optimize under `aimprob` constraints, an updated `aimprob2` function is defined below, using `herbtooth` in place of the known linear objective. Herbie's tooth is usually evaluated in $[-2, 2]^2$, whereas observe that `aimprob2` assumes coded inputs.

```
aimprob2 <- function(X, known.only=FALSE)
 {
   if(is.null(nrow(X))) X <- matrix(X, nrow=1)
   if(known.only) stop("no outputs are treated as known")
   f <- herbtooth(4*(X - 0.5))
   c1 <- 1.5 - X[,1] - 2*X[,2] - 0.5*sin(2*pi*(X[,1]^2 - 2*X[,2]))
   c2 <- rowSums(X^2) - 1.5
   return(list(obj=f, c=drop(cbind(c1, c2))))
}
```

Leveraging our plotting macro from earlier, Figure 7.30 shows the surfaces in play in this updated toy problem. Plainly, there's a multitude of local minima, in both valid and invalid parts of the input space.

```
plotprob(aimprob2, nl=c(10,13))
```

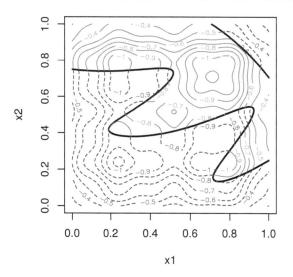

FIGURE 7.30: Herbie's tooth objective paired with constraints from the original toy problem (7.8).

By inspection, it would appear that there are three local minima which are more-or-less equally good as global optima in three of the four quadrants of the space. A fourth one, located in the remaining quadrant, would be similarly attractive if not invalid. These are, roughly speaking, at combinations of coordinates 0.25 and 0.75 up from zero and down from one. To see how well surrogate AL and MC EI work on this problem, code below performs searches in thirty replicates, averaging over randomized seed designs. Specifying `fhat=TRUE` causes a GP surrogate to be fit to objective function evaluations returned by `blackbox`.

```
prog <- matrix(NA, nrow=30, ncol=100)
```

```
xbest <- matrix(NA, nrow=30, ncol=2)
for(r in 1:30) {
  out2 <- optim.auglag(aimprob2, B, fhat=TRUE, start=20, end=100, verb=0)
  prog[r,] <- out2$prog
  v <- apply(out2$C, 1, function(x) { all(x <= 0) })
  X <- out2$X[v,]
  obj <- out2$obj[v]
  xbest[r,] <- X[which.min(obj),]
}
```

Full BVV progress is recorded over acquisition iterations, as are x-coordinates corresponding to the best solution found in each replicate. Figure 7.31 shows our thirty progress trajectories over 100 blackbox evaluations.

```
matplot(t(prog), type="l", ylab="best valid value",
  xlab="blackbox evaluations", col="gray", lty=1)
```

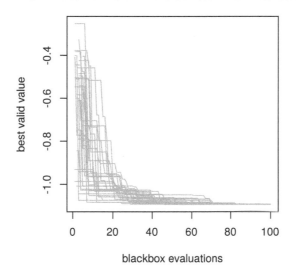

FIGURE 7.31: Best valid value for ALBO search of Herbie's tooth with toy constraints (7.8) in thirty random restarts.

All searches have converged on an agreed-upon objective evaluation of about -1.0929 for the BVV at the final iteration. Figure 7.32 shows the geographical locations of the valid argmin_x found in each restart.

```
plotprob(aimprob2, nl=c(10,13))
points(xbest[,1], xbest[,2], pch=18, col="blue")
```

Apparently, only two of the three local valid minima are reasonable as global solutions. There's some jitter in solutions found owing to MC approximation of EI on space-filling candidate grids. A more precise solution can be calculated, at substantially greater computational expense, with `slack=2`. A thriftier option might be to initialize a classical, local AL-based

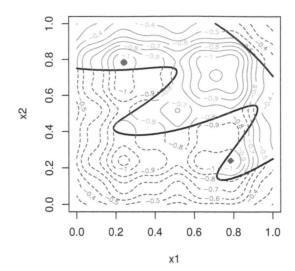

FIGURE 7.32: Locations of solutions $x^\star$ from each of thirty repetitions.

search staring off where the surrogate/EI one ended. For a more compelling illustration, let's back off the number of blackbox iterations to half of what was used above.

```
out2 <- optim.auglag(aimprob2, B, fhat=TRUE, start=20, end=40, verb=0)
v <- apply(out2$C, 1, function(x) { all(x <= 0) })
X <- out2$X[v,]
obj <- out2$obj[v]
xbest <- X[which.min(obj),]
```

Then feed the input corresponding to the BVV, calculated above, into **ALoptim**. Finishing a surrogate-assisted optimization with a classical one, to drill down a local trough in order to increase precision of the final solution, is a common tactic (e.g., Taddy et al., 2009).

```
end <- length(out2$rho)
evals <- 0
drill <- ALoptim(aimprob2, B, start=xbest, lambda=out2$lambda[end,],
  rho=out2$rho[end], kmax=1, maxit=100)
evals
```

```
## [1] 49
```

Figure 7.33 shows how **xbest**, based on just 40 evaluations, compares to our earlier results obtained after 100 runs. Although sub-optimal, just 49 further **ALoptim** evaluations – depending on the Rmarkdown build – yields an answer (**drill**) that's at least as good.

```
hist(prog[,100], main="", xlab="best valid value",
  xlim=range(c(prog[,100], min(obj))))
abline(v=min(obj), col=2, lty=2)
abline(v=drill[2,3], col=2)
legend("top", c("xbest", "drill"), col=2, lty=2:1)
```

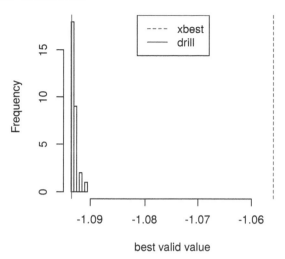

FIGURE 7.33: Histogram of BVVs in thirty repetitions augmented with $f(x^\star)$ from a limited AL-based search (vertical red-dashed line) followed by ordinary AL-search $f(x^{\text{drill}})$ initialized at $x^\star$ (red-solid).

Surrogate-based AL search, via MC EI, finds the local domain of attraction of the global solution in relatively few evaluations. Embarking on a local search from there usually yields an answer that's indistinguishable from one obtained from a more laborious search enjoying greater global scope, and a more judicious balance between exploration and exploitation.

7.3.8 Equality constraints and more

Several other examples, with comparison to EFI and classic AL, are illustrated by demos provided with the `laGP` package. For example, `demo("ALfhat")` duplicates the example immediately above, offering comparison to a classical AL (`ALoptim`-like) implementation. Two other demos in the package entertain a mixed constraints setting. EFI isn't directly applicable in the presence of equality constraints; ALBO methods represent the only surrogate-assisted option expressly targeting equality (and mixed) constraints – at least that I'm aware of at the time of writing.

It's tempting to try transforming an equality constraint into two inequalities (one $\leq$ and one $\geq$, or whose negation is $\leq$). To the extent that the effect of such a reformulation is understood, the outlook is bleak. It puts double-weight on equalities and violates certain regularity conditions. Numerical issues have been reported in empirical evaluations (Sasena, 2002). This EFI-enabling hack is included in empirical comparisons automated by the two demos.

See `demo("GSB")` for a 2d problem involving a Goldstein–Price objective (§7.1.1), the toy sinusoidal inequality constraint c_1 from `aimprob`, and two equality constraints that together trace out four ribbons of valid region. EFI under the dual-inequality transformation is competitive on this problem, but AL methods, especially with `slack=2`, work better. See `demo("LAH")` for a 4d problem with known linear objective, an inequality constraint derived from the Ackley function[26], and an equality constraint derived from the Hartmann

---

[26]https://www.sfu.ca/~ssurjano/ackley.html

4 function[27]. EFI is pretty bad on this one because the dual inequality transformation over-emphasizes the equality constraint relative to the inequality one.

It would be remiss not to mention that `optim.auglag` isn't equipped to handle noisy black-boxes at this time, whether in the objective, constraint or both. An adaptation estimating nuggets would be relatively straightforward. Still, such a setup is untested at this time, and thus potential pitfalls are unknown. A recent paper by Letham et al. (2019) targets noisy constrained optimization under an updated EFI. Calculations therein utilize quasi MC numerics to fully quantify, and thus balance, relevant uncertainties in acquisition of new runs.

Whenever there's noise, and especially when it's substantial, balance between exploration and exploitation is nuanced. Understanding what's signal and what's noise is essential to knowing whether there's potential for improvement. Authors in recent literature have been quick to criticize EI-like methods for acquiring replicates, or nearly so, prompting a search for alternative heuristics. Replicating may seem wasteful when the blackbox is expensive, but repeated sampling represents the only fool-proof mechanism for separating signal from noise. It's similarly wasteful to sample heavily in a region under the belief that the signal shows potential for improvement when a small handful of replicates could summarily dismiss that potential as noise.

I think this setting represents one of the big open problems in BO, especially when noise levels may change over the input region. New sequential design heuristics, surrogate models, and methods for coping with constraints (like the AL) will be needed in this setting. See, for example, Jalali et al. (2017). Chapter 10 introduces heteroskedastic GPs, illustrating how replication in design is key to effective learning, prediction, and quantification of uncertainty. Learning for optimization is different than learning for prediction, but similar themes are often in play, albeit to varying degrees.

Math programming has owned optimization for a century, or perhaps longer. And they may yet, at least in some generality, for yet a century more. That said, statistical methods have a near monopoly when observations are noisy. Therefore, the landscape of research into methods for BO – as opposed to the more classical math programming option – will undoubtedly be a space to watch for developments targeting optimization under uncertainty. In many such contexts, such as in e-commerce, those methods (particularly the most exciting developments from the ML/BO literature) already make up the vanguard.

As researchers in applied science become more comfortable with simulation as a means of exploring complex relationships in biology, epidemiology, economics, sociology, physics, chemistry, engineering, and more, they will eventually turn to statisticians for help optimizing and calibrating those systems as a means of affecting policy, implementation, design of new systems and modernization of old. Descriptive power, scope for synthesis, and potential for automation inherent in modern nonparametric and hierarchical models (and their experimental design) is unprecedented in modern times. BO is but one fine example of the confluence of these mathematical and technological advances. Sensitivity analysis and calibration are another.

---

[27]https://www.sfu.ca/~ssurjano/hart4.html

7.4 Homework exercises

These problems explore Bayesian optimization in objective-only and constrained settings.

#1: Thompson sampling and probability of improvement

Augment the running Goldstein–Price example (§7.1.1) to include `optim` with `method="Nelder-Mead"`, Thompson sampling (TS; §7.1.2) and PI (7.2) comparators. For TS, base acquisitions on a predictive draw obtained on candidates formed randomly in two variations:

i. A novel size-100 LHS in the full $[0,1]^2$ input space for each iteration of search.
ii. A novel size-90 LHS in $[0,1]^2$ augmented with ten candidates selected at random from the smallest rectangle containing the best five inputs found so far.

Considering the similarity between these four new methods and ones presented in the chapter (`"L-BFGS-B"`, EY, EI), you may wish to develop a more modular implementation where code and initializing points can be shared. That investment will pay dividends in #2 and #3 below.

Report average progress and provide boxplots of the best objective value (BOV) for all seven comparators over one hundred repetitions.

#2: Six-dimensional problem

Consider the Hartmann 6 function[28] as implemented by `hartman6` in the `DiceOptim` package (Picheny et al., 2016b) on CRAN. Re-tool the running Goldstein–Price example (§7.1.1) and EY-v-`optim`-v-EI comparison for this setting. As in #1, perform up to fifty blackbox evaluations where surrogate-assisted variations are seeded with a random design of size $n_0 = 12$. Report average progress over optimization acquisitions and provide a boxplot of the BOV for each at the end. If you worked on #1, perhaps include those comparators as well.

#3: Adding noise

Revisit #1 and #2, above, with blackbox evaluations observed with additive noise: $Y(x) = f(x) + \varepsilon$. Take $\varepsilon \sim \mathcal{N}(0, \sigma^2 = 0.1^2)$ for both problems. Noise makes the problem harder, and it can help to seed with replicates in order to separate signal from noise. So take $N = 75$ with $n_0 = 20$ composed of two replicates on an LHS of size ten. Since `"L-BFGS-B"` isn't a reasonable comparator in this context, replace with an `optim`-based IECI measured on novel size-100 LHS reference sets in each iteration. Report average progress in terms of the true, no-noise output over optimization acquisitions and provide a boxplot of the true BOV for each at the end. If you worked on #1, perhaps include those comparators as well.

#4: EFI versus IECI for optimization under constraints

Treat the function `htc` below, which returns objective and constraint, as a blackbox.

---

[28] https://www.sfu.ca/~ssurjano/hart6.html

```
library(splancs)
htc <- function(X, epoly)
 {
  if(!is.matrix(X)) X <- matrix(X, ncol=2)
  z <- herbtooth(X)
  cl <- rep(1, length(z))
  if(!is.null(epoly)) {
    xy <- as.points(X[,1], X[,2])
    io <- inout(xy, epoly)
    cl[!io] <- 0
  }
  return(list(obj=z, c=as.numeric(!cl)))
 }
```

Valid region is defined by the polygon `epoly`, which we shall take to approximate an ellipse.

```
library(ellipse)
epoly <- ellipse(rbind(c(1,-0.5), c(-0.5,1)), scale=c(0.75, 0.75))
formals(htc)$epoly <- epoly
```

Work with a pre-defined 200×200 grid in $[-2, 2]^2$. Starting with $n_0 = 25$ random grid-points, find the best valid value (BVV) of the objective (on that grid) in a budget of 50 further evaluations, for $N = 75$ total, in the following variations. *Hint: you may wish to plot objective and constraint surfaces on a grid first to see what you're dealing with.*

i. Use a GP surrogate for the objective, and a random forest (RF) surrogate (via **randomForest**) for the constraint. Base acquisitions on EFI over remaining candidates from the grid.
ii. Similarly with GP for the objective and RF for the constraint, base acquisitions on IECI over remaining candidates from the grid, taking as reference locations 100 novel random candidates. (A reference set of 1000 would be even better, if you can afford the computation.)

While you're gathering those acquisitions, keep track of the truly best valid value. That is, among the grid elements which are actually valid, ignoring the RF classifier, save the objective value of the grid location which your model predicts is lowest. Also save the entire cache of x, and (y, c)-evaluations collected over the fifty acquisitions. Repeat the experiment fifty times and report visually, and comment verbally, on patterns in these summaries of performance.

Finally, explore `demo("plconstgp_2d_ieci.R")` in `plgp` on CRAN which couples a regression GP for the objective with a classification GP for the constraints in a fully Bayesian setup (Gramacy and Lee, 2011). Compare and contrast performance against your implementation. (Note there's no grid in the demo.)

#5: An old friend

Revisit exercise #3 from §1.4, which combined the Goldstein–Price[29] function as an objective, with the sinusoidal constraint from `aimprob` (7.8), via EFI (`optim.efi`) and augmented

---

[29]http://www.sfu.ca/~ssurjano/goldpr.html

Lagrangian (`optim.auglag`) support provided by laGP. For the AL, try both EY (`ey.tol=1`), and EI, the default, with and without slack variables. Use arguments `start=10` and `end=60` for a total of fifty acquisitions after initializing with a small LHS of size ten. How do your new results compare to the mathematical programming approach you took earlier? Considering the random nature of initialization(s), you may wish to average over thirty or so runs in order to explore variability. Track BVV of the objective over the iterations, reporting its average over the repetitions and full distribution after the last acquisition. *Hint: this exercise is a simplified version of what you would find as* `demo("GSBP")` *in laGP.*

#6: Remediation by optimization

Revisit the Lockwood solvent groundwater plume site case study from §2.4. The program described therein is reproduced below. With x_j denoting the pumping rate for well j, solve

$$\min_{x} \left\{ f(x) = \sum_{j=1}^{6} x_j : c_1(x) \le 0,\ c_2(x) \le 0,\ x \in [0, 2 \cdot 10^4]^6 \right\}.$$

If you've not done so already, see exercise #4a from §2.6 in order to compile and test the `runlock` back-end.

Your task is to perform up to 500 iterations of optimization under each of EFI, IECI and the augmented Lagrangian (AL) with EY and EI alternatives.

- For AL variations use `optim.auglag` from laGP with the following recommended settings.
  - Work with coded inputs by providing `Bscale=10000` and use bounding box `B=matrix(c(rep(0,6), rep(2,6)), ncol=2)`.
  - Use a separable covariance specification with `sep=TRUE`.
  - You may also find it helpful, but slower, to use `ncandf=function(t) { 1000 }` rather than the default search candidate setting.
  - For EY rather than EI, specify `ey.tol=1`.
- For EFI use `optim.efi` in laGP.
  - Same suggestions as above for `optim.auglag` apply here.
- For for IECI you'll have to do it yourself, but you may use the GP fitting capability built into laGP.
  - Note that `ieciGPsep` models the objective f, whereas our objective, above, is a known linear function.
  - Describe how it could be modified to accommodate the known objective. A description is sufficient; you don't need to actually do it (because the setup will still work, but perhaps not as efficiently).
- On all of the above, you might find it helpful to try an easier, less expensive blackbox where you know the answer, like in #4 or #5, first.

Compare your BVV progress to results from the Matott et al. (2011) study, which you can find in `runlock/pato_results.csv`.

- In each case, initialize your search with 30 space-filling candidates (e.g., `start=30` in `optim.*`).
- Optionally, use `Xstart` to seed those comparators with identical initial designs, which could help facilitate a lower variance comparison.
- Put a `for` loop around your optimizations and report average BVV progress over as many random restarts as you have time for (or you think are sufficient to stamp out MC error).

8

Calibration and Sensitivity

Many scientific phenomena are studied with mathematical (i.e., computer) models and field experiments simultaneously. Real experiments are expensive, and for this and other reasons (ethics, lack of materials/infrastructure, etc.) limited configurations can be entertained. Computer simulations are lots cheaper, but usually not so cheap as to allow infinite exploration of configuration space(s). Plus simulations usually idealize reality, contributing bias, and engage more "knobs", or tuning parameters, than can be controlled or even known in the field.

So the goal in the first part of this chapter is to build an apparatus that can harmonize two data types, computer simulated and field observation, for the purpose of learning about/predicting the real underlying process, or possibly optimizing some aspect of it. We want to learn about any discrepancies, or bias, between computer model and field data; learn best settings of the computer model's knobs; meta-model/emulate computer model runs as a surrogate for new predictions, while compensating for its bias relative to reality as measured in the field. Ideally those predictions will offer a full accounting of uncertainty, for all things being estimated at all levels.

What's meant by full, and what's reasonable pragmatically, is always a matter of perspective. Uncertainty quantification (UQ)[1] is a loaded term from the applied math/numerical analysis community. A lot of UQ focuses on understanding distributions of outputs, or observables from a process, as a function of uncertain or random inputs. Most applications amount to *uncertainty propagation*. A key component of that is understanding how inputs affect outputs when layers of fitted models are used, like GPs for surrogates and additionally as models of discrepancy in the calibration context. So the second half of the chapter takes a diversion to detail estimating *main effects* and *sensitivity indices* for GPs and related nonparametric predictors.

These two topics, calibration and sensitivity, could easily stand alone in their own chapter(s). See Chapters 7–8 of Santner et al. (2018). Sensitivity analysis has filled entire textbooks, although the context of those presentations is different. My aim in combining them here is to frame them as two important applications of GP surrogates where UQ, i.e., faithful propagation of uncertainty, is key.

8.1 Calibration

Computer model *calibration* juggles three processes. Real process R represents an ideal, depicting unknown dynamics of phenomena under study. The goal is to learn as much as

---

[1]https://en.wikipedia.org/wiki/Uncertainty_quantification

possible about R through mathematical modeling, computer and physical experimentation, which leads us to the other two processes. Field F is where a physical experiment observing R takes place. Computer model M implements/solves a mathematical model that idealizes R.

Let $Y^F(x)$ denote a field observation under m_x-dimensional conditions x, and $y^R(x)$ denote the real output under condition x. Assume R and F are related as follows.

$$Y^F(x) = y^R(x) + \varepsilon, \quad \text{where} \quad \varepsilon \stackrel{\text{iid}}{\sim} \mathcal{N}(0, \sigma_\varepsilon^2)$$

This isn't much different from typical modeling apparatuses where observations are corrupted by independent and identically distributed idiosyncratic Gaussian noise. Considering the expense of setting up a physical experiment in the field, we presume that only a small number n_F of field observations Y_{n_F} are available at x locations X_{n_F}. Sometimes it's easier to obtain repeated observations under a single setting x, rather than changing x which may involve manually re-configuring a complex system, so n_F may embed a nontrivial degree of replication. Replicates can be helpful for separating signal from noise, especially when σ_ε^2 is large. That is, the number of unique settings in X_{n_F} may be many fewer than n_F. Chapter 10 considers modeling and design under replication in more detail. For now let me simply remark that replication is common in field experiments, and computer model calibration settings are no different. However this detail is largely ignored for the remainder of the chapter.

Let $y^M(x, u)$ denote output from a computer model run under conditions x and tuning or calibration parameters u. We shall presume that $y^M(\cdot, \cdot)$ is deterministic to simplify the following discussion. There's no reason why stochastic simulation must be precluded by the framework, however such setups are far less well investigated in the literature. Inputs x to computer model $y^M(x, u)$ coincide with x's from the field experiment(s). Inputs u, in dimension m_u, represent any aspect of M which can't be controlled in F and/or are unknown in R. It's quite typical for a mathematical model, or its computer implementation, to have more knobs than can be controlled in the field. Example u coordinates may arise from an artificial aspect of computer implementation, like mesh size. Or they might have real physical meaning, like acceleration due to gravity, which is not known (precisely enough) to be recorded in the field. Some practitioners make a distinction between the two, calling the former a *tuning* parameter (omitting from probabilistic modeling enterprises), and treating only the latter as a *calibration* parameter u. I'll be lazy by using those two terms interchangeably and modeling in a unified fashion.

The goal is to study the relationship between the computer model $y^M(x, u)$, its fine-tuning through u, and the field $Y^F(x)$ as a means of learning about real phenomena $y^R(x)$. In this way, calibration is an example or generalization of a statistical inverse problem[2]. Which it is – example or generalization – depends on your perspective. Inverse problems emphasize learning u, attempting to attribute causal links between unknown factors in a simulation and empirical, physical observation. Calibration is more ambitious in its attempt to synthesize multiple information sources and to assimilate functional relationships through an acknowledgment of bias between computer simulation and field observation. Often such assimilation is at odds with the establishment of causal links, however, and can suffer from confounding and identification issues.

Although there are many ways you could imagine undertaking such an analysis, one has percolated to the top as canonical: the Kennedy and O'Hagan framework. Another approach,

---

[2]https://en.wikipedia.org/wiki/Inverse_problem

called *history matching*[3], is a popular alternative (Craig et al., 1996; Vernon et al., 2010; Williamson et al., 2013). History matching is a more hands-on process, as is perhaps exemplified by the flow diagram found at that link. Although there's much to recommend a more careful approach to marrying disparate sources of information, the presentation below emphasizes a more easily automated Kennedy and O'Hagan alternative.

8.1.1 Kennedy and O'Hagan framework

Kennedy and O'Hagan (2001) proposed a Bayesian framework for coupling M and F. KOH, hereafter, represent a real process R as the computer model output at the best setting of calibration parameters, $u^\star$, plus a discrepancy term acknowledging that there can be systematic disagreement between model and truth.

$$y^R(x) = y^M(x, u^\star) + b(x)$$

$$\text{so that} \quad Y^F(x) = y^M(x, u^\star) + b(x) + \varepsilon$$

The quantity $b(\cdot)$ is a functional discrepancy, or *bias* correction. Although I may shorten and casually refer to $b(\cdot)$ as "bias", the actual bias (which is a property of M not R) would actually work out to

$$-b(x) = y^M(x, u^\star) - y^R(x).$$

The point here is that a computer model has systematic imperfections, even under its best tuning $u^\star$, but KOH specify an *a priori* belief that reasonable correction can be learned through $b(\cdot)$. Errors ε are independent zero-mean Gaussian with variance σ_ε^2.

Altogether, unknowns are $u^\star$, σ_ε^2, and discrepancy $b(\cdot)$. KOH emphasized Bayesian inference, particularly averaging over trade-offs between calibration values u and discrepancies $b(\cdot)$ under a GP prior. Known information or restrictions on u-values can be specified through prior $p(u)$. Otherwise a uniform prior (over a finite domain) can be used. Often, and especially when little prior information is available on u, a regularizing prior with mass somewhat more concentrated on a default or midway value can prevent over-concentration of posterior density on boundary settings. Reference priors for σ_ε^2 are typical (Berger et al., 2001). KOH utilized a GP specification with linear mean for $b(\cdot)$, but the presentation here considers a zero-mean for simplicity and for consistency with GP treatments elsewhere in this monograph. Results analogous to those from a homework exercise in §5.5 offer ready extension.

If evaluating the computer model is fast, then inference (Bayesian or otherwise) is made rather straightforward via residuals between computer model outputs and field observations at n_F field locations X_{n_F}

$$Y_{n_F}^{b|u} \equiv y^b(X_{n_F}, u) \equiv Y_{n_F} - Y_{n_F}^{M|u} \equiv Y_{n_F} - y^M(X_{n_F}, u) \tag{8.1}$$

which can be computed at will for any u (Higdon et al., 2004). An "r" superscript may have been more appropriate for residuals. Besides avoiding clash with "R" for "real", superscript "b" was chosen instead to emphasize the role of residuals in training $b(\cdot)$. Eq. (8.1) is characterizing a new n_F-dimensional response vector $Y_{n_F}^{b|u}$ at inputs X_{n_F}. Each u-setting gives a different such vector measuring noise and bias between field data and computer

---

[3]https://www.streamsim.com/technology/history-matching

model. With a GP prior for $b(\cdot)$, $Y_{n_F}^{b|u}$ is n_F-variate MVN with covariance derived through inverse exponentiated squared Euclidean distances between rows of X_{n_F}. This implies a likelihood on parameters (u, θ_b), where θ_b may collect scale, m_x lengthscales and nugget hyperparameters. Let $\Sigma_{n_F}^b$ denote the $n_F \times n_F$ covariance matrix built from X_{n_F} and θ_b. Note that by including both scale and nugget in θ_b, $\Sigma_{n_F}^b$ captures field data variance σ_ε^2 implicitly through their product. The likelihood is thus proportional to

$$|\Sigma_{n_F}^b|^{-1/2} \exp\left\{ -\frac{1}{2}(Y_{n_F}^{b|u})^\top (\Sigma_{n_F}^b)^{-1} Y_{n_F}^{b|u} \right\}. \tag{8.2}$$

That likelihood can be maximized over all unknown coordinates, or fully Bayesian inference may be used to sample from the joint posterior.

If evaluating the computer model is expensive or otherwise indirectly available, a surrogate $\hat{y}^M(\cdot, \cdot)$ can be fit to n_M simulations of M run over a design $[X_{n_M}; U_{N_N}]$ in (x, u)-space. KOH recommend a GP prior for y^M, i.e., a coupled pair of GPs including $b(\cdot)$. Rather than performing inference for y^M separately, using just n_M runs as typical of computer experiments in isolation, KOH recommend joint posterior inference for all unknowns $\Theta = (y^M, b(\cdot), u^\star, \sigma_\varepsilon^2)$ using the full corpus of data from computer model and field experiment $[Y_{n_M}, Y_{n_F}]$. From a Bayesian perspective, this is the coherent thing to do: infer all unknowns jointly given all data with $p(\Theta \mid Y_{n_F}, Y_{n_M}) \propto p([Y_{n_F}, Y_{n_M}] \mid \Theta) \times p(\Theta)$. When the computer model M is very slow, limiting n_M, joint inference facilitates efficient use of observational quantities as both field data and computer model runs can inform about $\hat{y}^M$ in addition to $\hat{b}$. As in the "Higdon free-M" setting (8.2), the likelihood involves evaluating a mean-zero MVN density, but this time it's $n_M + n_F$ variate for stacked computer model and field data.

$$\begin{bmatrix} Y_{n_M} \\ Y_{n_F} \end{bmatrix} \sim \mathcal{N}_{n_M + n_F}\left(\begin{bmatrix} 0 \\ 0 \end{bmatrix}, \begin{bmatrix} \Sigma_{n_M} & \Sigma_{n_M}(X_{n_F}, u) \\ \Sigma_{n_M}(X_{n_F}, u)^\top & \Sigma_{n_F}(u) + \Sigma_{n_F}^b \end{bmatrix} \right)$$

Above, $\Sigma_{n_M} \equiv \Sigma([X_{n_M}, U_{n_M}])$ is the usual $n_M \times n_M$ covariance matrix defining a GP surrogate for simulations Y_{n_M}, tacitly conditioned on $m_x + m_u$ hyperparameters θ scaling pairwise distances between inputs in (x, u)-space. The nugget may be omitted in this deterministic setting. Note that all Σ's lacking a superscript reference the surrogate $y^{\hat{M}}(\cdot)$, not the covariance structure from the bias-correcting GP. $\Sigma_{n_F}^b$ belongs to the bias GP (8.2), an $n_F \times n_F$ matrix based on distances in x-space and hyperparameters θ_b. $\Sigma_{n_M}(X_{n_F}, u)$ is an $n_M \times n_F$ matrix based on θ-scaled pairwise distances between computer model design $[X_{n_M}, U_{n_M}]$ and field data design X_{n_F} augmented by columns $u^\top$, concatenated to all rows identically. Finally, $\Sigma_{n_F}(u)$ is similar to $\Sigma_{n_M}(X_{n_F}, u)$ except to itself rather than to computer simulation data. Specifically, $\Sigma_{n_F}(u)$ is an $n_F \times n_F$ matrix containing θ-hyperparameterized pairwise inverse distances between rows of X_{n_F} augmented by columns $u^\top$.

Choices of u and sets of hyperparameters θ and θ_b may be entertained through MVN density evaluations, either to maximize or sample from the posterior (after completing with appropriate priors, of course). It's quite common to maximize the likelihood first to find $(\hat{u}, \hat{\theta}, \hat{\theta}_b)$, then fix the hyperparameters at $(\hat{\theta}, \hat{\theta}_b)$ and subsequently sample from the posterior for u only, say with Metropolis–Hastings (MH)[4] style Markov chain Monte Carlo (MCMC)[5]. The degree to which sampling in an m_u-dimensional space is more manageable than an $(m_x + m_u)$-dimensional one depends, of course, on the size of the coordinate systems involved. For specifics on Bayesian inference by MCMC, see, e.g., Hoff's excellent (2009) text. For

---

[4]https://en.wikipedia.org/wiki/Metropolis–Hastings_algorithm
[5]https://en.wikipedia.org/wiki/Markov_chain_Monte_Carlo

particular implementation in the KOH calibration setting, see Kennedy and O'Hagan (2001). Although pseudocode is provided in Algorithm 8.1, with a worked example later in §8.1.5, the nuances of Bayesian inference/MCMC implementation are largely beyond the scope of this text.

Algorithm 8.1 KOH Metropolis–Hastings (MH) Sampler

Assume known hyperparameterization(s) $(\hat{\theta}, \hat{\theta}_b)$ for coupled GPs as priors for the surrogate for computer model $y^M(\cdot, \cdot)$ and bias $b(\cdot)$. Let $\ell(u)$ represent the joint MVN log (marginal) likelihood for $[Y_{n_M}, Y_{n_F}]$:

$$\ell(u) = c - \frac{1}{2}\log|\mathbb{V}(u)| - \frac{1}{2}\begin{bmatrix} Y_{n_M} \\ Y_{n_F} \end{bmatrix}^{\top} \mathbb{V}(u)^{-1} \begin{bmatrix} Y_{n_M} \\ Y_{n_F} \end{bmatrix}, \qquad (8.3)$$

$$\text{where} \quad \mathbb{V}(u) = \begin{bmatrix} \Sigma_{n_M} & \Sigma_{n_M}(X_{n_F}, u) \\ \Sigma_{n_M}(X_{n_F}, u)^{\top} & \Sigma_{n_F}(u) + \Sigma_{n_F}^{b} \end{bmatrix}.$$

Require prior density $p(u)$ and conveniently sampled (possibly random walk) proposal density $q(u, u')$, computer model observations Y_{n_M} at inputs (X_{n_M}, U_{n_M}) used to define Σ_{n_M} above, field data observations Y_{n_F} at locations X_{n_F}, and an initial value $u^{(0)}$.

For $t = 1, \ldots, T$ desired samples from the Markov chain with stationary distribution $p(u \mid [Y_{n_M}, Y_{n_F}], \hat{\theta}, \hat{\theta}_b)$, do ...

1. Propose a new $u' \sim q(u^{(t-1)}, \cdot)$.
2. Calculate the ratio of (marginal) likelihoods in log space as

$$\Delta\ell(u', u^{(t-1)}) = \ell(u') - \ell(u^{(t-1)}).$$

3. Complete the MH acceptance ratio in log space as

$$\log\alpha = \Delta\ell(u', u^{(t-1)}) + \log p(u') - \log p(u^{(t-1)}) + \log q(u', u^{(t-1)}) - \log q(u^{(t-1)}, u').$$

4. Draw $v \sim \text{Unif}[0, 1]$.
   - If $v < \alpha$, accept u' and take $u^{(t)} \leftarrow u'$;
   - else reject u' and take $u^{(t)} \leftarrow u^{(t-1)}$.

End For

Return the collection $\{u^{(t)}\}_{t=B}^{T}$ of samples from the posterior of calibration parameter u, possibly after discarding some number of samples $B \in \{0, 1, 2, \ldots,\}$ as *burn-in*.

It's worth remarking that Algorithm 8.1 emphasizes posterior inference for calibration parameter u, but actually implicitly samples from the joint posterior for u, $b(\cdot)$, and σ_ε^2 since latent quantities from those processes (i.e., bias corrections and their residuals to field observations Y_{n_F}) are analytically marginalized out (§5.3.2) through log likelihood evaluations $\ell(\cdot)$. Only samples from the marginal posterior for u are returned, however. Samples $b^{(t)}(\mathcal{X}) \mid u^{(t)}$, and thereby $Y^F(\mathcal{X})^{(t)}$ from the marginal posterior predictive distribution, could be gathered at a later time provided predictive locations of interest $\mathcal{X}$. A homework exercise in §8.3 guides the reader through a derivation of those equations by augmenting the MVN in $[Y_{n_M}, Y_{n_F}]$ to $[Y_{n_M}, Y_{n_F}, Y^F(\mathcal{X})]$, and deducing the conditional Gaussian $Y^F(\mathcal{X}) \mid [Y_{n_M}, Y_{n_F}], u^{(t)}$ using identities similar to those used for prediction with ordinary GPs (5.3).

8.1.2   Modularization

Before jumping headlong into an example, allow me a substantial digression. There are many things going on in the KOH apparatus, and appreciating them can be obscured by the complexities of Bayesian inference and MCMC. It's natural, if mostly for historical trends in pedagogy, to think first about optimizing before integrating and that's what I'd like to do here. This is not common in the computer model calibration literature, but I hope that simplifying first, in several directions, will set a stronger foundation and offer some perspective.

KOH has tremendous flexibility – perhaps even too much! Coupled $b(\cdot)$ and $y^M(\cdot, \cdot)$, with u acting as weak adhesive binding them together, might lead to parameter/process identification, confounding and MCMC mixing issues. Imagine poor $\hat{y}^M$ being compensated for by $\hat{b}(\cdot)$ and a "far-away" u-setting, obscuring our view of the best approximating computer model and its calibration $u^\star$. Moreover, the approach is fraught with computational challenges. If n_M and n_F are of any moderate size, testing the limits of cubic covariance decomposition for their respective MVNs, that problem is severely exacerbated when $(n_M + n_F) \times (n_M + n_F)$ matrices are involved. Plus, why should $\hat{y}^M$ worry about anything other than y^M? One could argue that the surrogate's purview should comprise computer model runs only. Coupling with field data may be advantageous from an information theoretic perspective, leading to the most efficient posterior learning, but at the expense of both computation and interpretation. As an unabashed pragmatist, I think those two latter facets must be squared before statistical efficiency concerns are raised.

Liu et al. (2009) proposed going "back to basics" by fitting the surrogate $\hat{y}^M(\cdot, \cdot)$ independent of field data, using only the n_M simulations. They gave this approach a fancy name: *modularization*. Perhaps this is what anyone would have done instinctively, were it not for KOH's suggestion otherwise. Compartmentalization, a synonym of modularization, is good engineering practice. Components should perform robustly in isolation, irrespective of their anticipated role in a larger system. Liu et al. were careful to clarify that modularized KOH is no less Bayesian, and no less joint *a posteriori*. Unknowns u and $b(\cdot)$ are still inferred conditional on both computer model surrogate $\hat{y}^M$ and field data Y_{n_F}. However the setup does imply an independence structure in the prior – one which was, in ordinary KOH, deliberately *not* imposed.

Reasons for enforcing independence stem from statistical analogs of engineering principles. In short, the ordinary/original KOH is perhaps unnecessarily complicated, both technically and intuitively. It also sometimes leads to bizarre inferences. Using a simple example first presented by Joseph (2006), originally from Santner et al. (2003), Liu et al. showed that fully Bayesian KOH calibration yields surrogate model fits that can be unfaithful to computer model simulations, being biased by field data. Separately, the authors went through several other examples where Bayesians had gone off the deep end. A more modular approach, which they describe, helps protect against pathologies.

Running example: acceleration due to gravity

You might wonder: why be Bayesian at all? For one thing, regularization through priors has its merits as an inferential tool. Prior distributions promote stability in estimates in a means that's intuitive, at least on statistical grounds. Big MCMC, as fully Bayesian settings often demand, comes with a big otiokor prioo, copccially when each likelihood evaluation incurs cubic costs in n_M, n_F or their sum. Can the regularizing effects of a well-designed prior be appropriated without the expense of Monte Carlo (MC) inference?

Liu et al. (2009) modular KOH is suggestive of a much simpler alternative through a maximum *a posteriori* (MAP) estimator, calculated as follows: 1) fit a surrogate $\hat{y}^M$; 2) for each setting of u, build residuals $Y_{n_F} - \hat{y}^M(X_{n_F})$ and use those to train a GP bias correction; 3) optimize u via the marginal likelihood of that bias, modulo a choice of prior $p(u)$; 4) optionally, use bootstrap[6] or jackknife[7] resampling to quantify uncertainty. Details are left to Algorithm 8.2 shortly. For now, it's easier to illustrate by example. Plus, we've gone too far into the chapter without concrete illustration.

An excellent class of examples involves free-falling objects. Simulating the time it takes for an object to fall from a certain height is either an elementary or potentially intricate, if well-understood, enterprise. It all depends on how complicated you want to get with modeling. Acceleration due to gravity might be known, but possibly not precisely. Coefficients of drag may be completely unknown. A model incorporating both factors, but not others such as ambient air disturbance or rotational velocity, could be biased or inconsistent in unpredictable ways.

Consider the amount of time it takes for a wiffle ball to hit the ground when dropped from certain heights. Thankfully, performing the field experiment is rather trivial, if cumbersome. Just drop wiffle balls from different heights and measure how long it takes them to hit the ground. Saving us the tedium of performing the experiment ourselves, Derek Bingham[8] and Jason Loeppky[9] have graciously provided their own measurements, collecting $n_F = 63$ field observations at 21 heights, with three replicates at each height, measured in meters. See ball.csv[10].

A visualization of these data is provided by Figure 8.1. Time, on the y-axis, is measured in seconds.

```
ball <- read.csv("ball.csv")
plot(ball, xlab="height", ylab="time")
```

Apparently, the ladder or stairs they were using prevented them from recording measurements at heights between two and 2.5 meters. Suppose we're interested in accurately predicting the time it takes for a ball to drop from certain heights, particularly in this under-sampled region. One option is, of course, to fit a GP directly to the field data. (Never mind the extra effort of choosing a mathematical model, implementing it in code, performing simulations, calibrating, and correcting for bias, etc.) Code below provides one such potential fit $\hat{y}^F(\cdot)$.

```
library(laGP)
field.fit <- newGP(as.matrix(ball$height), ball$time, d=0.1,
  g=var(ball$time)/10, dK=TRUE)
eps <- sqrt(.Machine$double.eps)
mle <- jmleGP(field.fit, drange=c(eps, 10), grange=c(eps, var(ball$time)),
  dab=c(3/2, 8))
```

Next consider predictions $\hat{y}^F(\mathcal{X})$ on a testing grid $\mathcal{X}$. Code below utilizes a grid hs of heights in terms of coded inputs, mapping them back to the scale on which these data were recorded. This will help streamline some of our later analyses. More details soon.

---

[6]https://en.wikipedia.org/wiki/Bootstrapping_(statistics)
[7]https://en.wikipedia.org/wiki/Jackknife_resampling
[8]http://people.stat.sfu.ca/~dbingham/
[9]https://stat.ok.ubc.ca/faculty/loeppky.html
[10]http://bobby.gramacy.com/surrogates/ball.csv

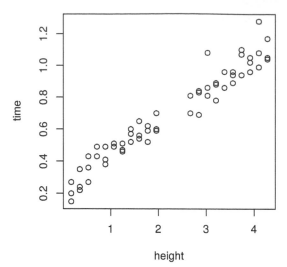

FIGURE 8.1: Bingham and Loeppky's ball drop data.

```
hr <- range(ball$height)
hs <- seq(0, 1, length=100)
heights <- hs*diff(hr) + hr[1]
p <- predGP(field.fit, as.matrix(heights), lite=TRUE)
deleteGP(field.fit)
```

Figure 8.2 provides a summary of that predictive distribution in terms of means and central 90% quantiles. Along the *x*-axis, as a red-dashed line, a summary of the predictive standard deviation is provided to aid visualization.

```
plot(ball, xlab="height", ylab="time")
lines(heights, p$mean, col=4)
lines(heights, qnorm(0.05, p$mean, sqrt(p$s2)), lty=2, col=4)
lines(heights, qnorm(0.95, p$mean, sqrt(p$s2)), lty=2, col=4)
lines(heights, 10*sqrt(p$s2)-0.6, col=2, lty=3, lwd=2)
legend("topleft", c("Fhat summary", "Fhat sd"), lty=c(1,3),
  col=c(4,2), lwd=1:2)
```

For my taste, this predictive surface is too wiggly. Surely these data ought to follow a monotonic, if noisy relationship. Uncertainty is too high in the gap, compared against what I would expect intuitively. Maybe some extra modeling could be useful after all. Perhaps coupling with known physics can mitigate those unsightly effects.

What does "Physics 101" say? Time t to drop a distance h for gravity g follows

$$t = \sqrt{2h/g}.$$

Somewhat realistically, we don't know the value of g for the location where the balls were dropped. So gravity is our calibration parameter, our u. Of course there are other unknowns, like air resistance – which will interact deferentially with height/terminal velocity. In other words that mathematical model is biased, and thus there's scope to improve upon it through

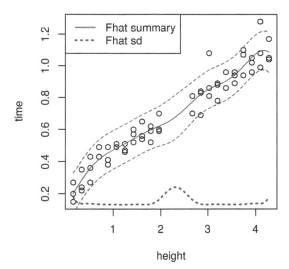

FIGURE 8.2: GP fit to field data; predictive standard deviation is along the bottom in dashed-red.

hybridization, by coupling with field data. So field data hold the potential to help the computer model as much as the other way around.

At the same time, g could not just be determined by acceleration due to gravity. At least not with the field data in hand. With more things slowing the ball down than speeding it up, g will almost certainly be forced into a role of compromise. To account for air resistance, say, estimated g will probably be shifted downward from its true value. This setup lacks a degree of identifiability no matter how we perform statistical inference. But that doesn't mean the enterprise isn't worthwhile.

Consider the following computer implementation of our mathematical model, simultaneously mapping natural inputs (h, g) to coded ones (x, u) in $[0, 1]^2$.

```
timedrop <- function(x, u, hr, gr)
 {
  g <- diff(gr)*u + gr[1]
  h <- diff(hr)*x + hr[1]
  return(sqrt(2*h/g))
 }
```

Two-vector `hr` is derived from the field data range, and was defined above for the purpose of generating a predictive grid of heights. The range for gravity specified below restricts our study to $[6, 14]$, equivalently defining a (uniform) prior $p(u)$ in what follows.

```
gr <- c(6, 14)
```

Suppose we're prepared to run `timedrop` at $n_M = 21$ input locations, commensurate in size to the number of unique inputs in the field data experiment, but in two dimensions. R code below constructs a maximin LHS (§4.3) in 2d and performs computer model simulations at those locations.

```
library(lhs)
XU <- maximinLHS(21, 2)
yM <- timedrop(XU[,1], XU[,2], hr, gr)
```

Now let's train a GP on those realizations, fixing the nugget to a small jitter value to acknowledge the deterministic nature of `timedrop` simulations.

```
yMhat <- newGPsep(XU, yM, d=0.1, g=1e-7, dK=TRUE)
mle <- mleGPsep(yMhat, tmin=eps, tmax=10)
```

Recall that `mleGPsep` modifies `yMhat` with updated `mle` values as a side effect. Next, extract surrogate predictive mean evaluations over a grid of heights, for a span of six equally-spaced potential u-values.

```
us <- seq(0, 1, length=6)
XX <- expand.grid(hs, us)
pm <- predGPsep(yMhat, XX, lite=TRUE)
```

Figure 8.3 offers a visual of those surfaces, with separate curves for each u-value.

```
plot(ball)
matlines(heights, matrix(pm$m, ncol=length(us)))
```

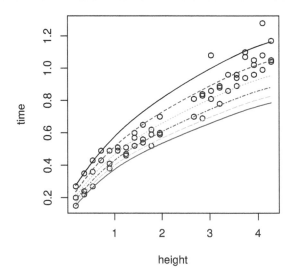

FIGURE 8.3: Computer model surrogates for several settings of calibration parameter u.

Aesthetically, the curves and data in that plot are largely in agreement. Some u-values generate curves that are better fits than others, yet none are perfect. All exhibit bias. To my eye, the best one is the green-dotted curve in the middle, but it's clearly biased low for balls dropped from greater heights. When calibrating, it makes sense to account for that bias. When predicting with the computer model, it makes sense to correct for it.

A modularized apparatus calibrates u via fits for bias. Settings of u which make residuals between surrogate predictions and field data observations "easier" to model are preferred. There are many options for assessing goodness of fit, and throughout this text we've preferred maximizing the likelihood, so that's where we shall start here. The function `bhat.fit` coded below takes field data (`X`, `yF`) and computer model surrogate predictions (e.g., predictive means from a GP) `Ym`. It also takes prior/initializing specifications for lengthscales (`da`) and nugget (`ga`), about which further discussion is provided below.

```
bhat.fit <- function(X, yF, Ym, da, ga, clean=TRUE)
 {
   bhat <- newGPsep(X, yF - Ym, d=da$start, g=ga$start, dK=TRUE)
   if(ga$mle) cmle <- mleGPsep(bhat, param="both", tmin=c(da$min, ga$min),
     tmax=c(da$max, ga$max), ab=c(da$ab, ga$ab))
   else cmle <- mleGPsep(bhat, tmin=da$min, tmax=da$max, ab=da$ab)
   cmle$nll <- - llikGPsep(bhat, dab=da$ab, gab=ga$ab)
   if(clean) deleteGPsep(bhat)
   else { cmle$gp <- bhat; cmle$gptype <- "sep" }
   return(cmle)
 }
```

As you can see, the function initializes and fits a GP to the discrepancy between field data and surrogate (`yF` - `Ym` at `X`) and returns the value of the minimizing negative log likelihood so obtained. As a mathematical abstraction, $\hat{b}$ is measuring goodness-of-fit for bias no matter how `Ym` are obtained. Below we shall take $Ym \equiv \hat{y}^M(X_{n_F}, u)$, and treat $\hat{b}$ as a merit function for choices of u. Optionally, `bhat.fit` returns a reference to the fitted GP, although by default this step is skipped (`clean=TRUE`), causing the object to be freed instead. When searching for $\hat{u}$ through $\hat{b}$, we don't need to save every GP fit en-route, but we will need the last one at the end in order to tap fitted models for prediction. Finally, observe that `bhat.fit` combines $\hat{b}$ and $\hat{\sigma}_\varepsilon^2$ fits via inference for a nugget.

Next create an objective to optimize, over coded gravity u-values, to find the best setting $\hat{u}$ estimating unknown $u^\star$. The `calib` function below takes argument `u` in the first position, which is helpful when optimizing with `optim`, and `yMhat` in the fourth position. The idea is to provide `fit=bhat.fit` from above, so that GPs are fit to residuals, where arguments `da` and `ga` have been assigned as defaults in advance, as illustrated momentarily. Setting things up in this way, rather than passing `da` and `ga` and then calling `bhat.fit` allows `bhat.fit` to be swapped out later for another model/fit, if desired, without altering `calib`. Later in §8.1.4 I shall utilize this feature when presenting a "nobias" alternative.

```
calib <- function(u, XF, yF, yMhat, fit, clean=TRUE)
 {
   XFu <- cbind(XF, matrix(rep(u, nrow(XF)), ncol=length(u), byrow=TRUE))
   Ym <- predGPsep(yMhat, XFu, lite=TRUE)$mean
   cmle <- fit(XF, yF, Ym, clean=clean)
   return(cmle)
 }
```

Argument `yMhat` should be a fitted GP surrogate for computer model runs (X_{n_M}, Y_{n_M}). Field data locations `X` are combined with an extra column of u values and fed into `predGPsep` to get surrogate means `Ym`, which are then fed in to `fit=bhat.fit`.

Rather than shoving `calib` right into `optim`, which is exactly what we shall do momentarily, consider first evaluating on a u-grid to aid visualization. After all, optimizing a deterministic function in 1d is easy with the eyeball norm. The R chunk below sets up such a grid and codes height inputs into $[0, 1]$,

```
u <- seq(0, 1, length=100)
XF <- as.matrix((ball$height - hr[1])/diff(hr))
```

Before setting this running, arguments `da` and `ga` must be specified. Default priors for lengthscales and nuggets may be calculated through `darg` and `garg` functions provided with `laGP`. Observe how these are set up to occupy default values of `bhat.fit` arguments (i.e., `formals`).

```
formals(bhat.fit)$da <- darg(d=list(mle=TRUE), X=XF)
formals(bhat.fit)$ga <- garg(g=list(mle=TRUE), y=ball$time)
```

Although `darg` and `garg` have been used previously to set search ranges and starting values, we've not yet discussed their full prior-generating capacity. This is as good a time as any. Respectively, `darg` and `garg` offer light regularization through the distribution of pairwise distances in `X`, and marginal variances in `y`. In detail, `darg` calculates default lengthscale search ranges and initializing values from the empirical range of (non-zero) squared Euclidean distances between rows of `X`, and their 10% quantile, respectively. Gamma prior hyperparameters are chosen to have shape $a = 3/2$ and rate b derived by the incomplete Gamma inverse function (DiDonato and Morris Jr, 1986) to put 95% of the cumulative Gamma distribution below the maximum such distance observed. The `garg` routine is similar except that it works with `(y - mean(y))^2` instead of pairwise `X` distances. Another difference is that the starting value is chosen as the 2.5% quantile. Keen readers will note that `garg` is more squarely targeting priors on $\sigma_\varepsilon^2 \equiv \tau^2 g$. If knowledge of τ^2 is available *a priori* of fitting $\hat{\tau}^2 \mid g$, then some minor adjustments could help fine-tune priors for g.

Ok, now evaluating on the grid . . .

```
unll <- rep(NA, length(u))
for(i in 1:length(u))
  unll[i] <- calib(u[i], XF, ball$time, yMhat, bhat.fit)$nll
```

Before plotting that surface, the code below implements the more hands-off `optim` solution, which we can add to the visualization.

```
obj <- function(x, XF, yF, yMhat, fit) calib(x, XF, yF, yMhat, fit)$nll
soln <- optimize(obj, lower=0, upper=1, XF=XF, yF=ball$time,
  yMhat=yMhat, fit=bhat.fit)
uhat <- soln$minimum
```

Figure 8.4 shows that surface and its numerical optima, our calibrated setting $\hat{u} \equiv$ `uhat`.

```
plot(u, unll, type="l", xlab="u", ylab="negative log likelihood")
abline(v=uhat, col=2, lty=2)
legend("top", "uhat", lty=2, col=2)
```

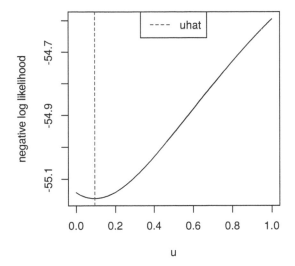

FIGURE 8.4: Negative log likelihood surface for u and its calibrated value $\hat{u}$.

Converting back to natural units for the calibration parameter, we can see that our estimate is too low given what we know about acceleration due to gravity on Earth.

```
ghat <- uhat*diff(gr) + gr[1]
ghat
```

[1] 6.75

In fact it's about 1/3 lower than it should be. Recall that we're asking gravity to do double-duty. Wiffle balls are exceptionally buoyant compared to other balls, being riddled with holes that trap air. Our mathematical model acknowledges no air resistance. Ideally, such unknowns would be swept entirely into an estimate of bias $\hat{b}$, however no aspect of the calibration apparatus (whether KOH or modularized) precludes compensation by $\hat{u}$ instead. Consequently our $\hat{g}$ above, via $\hat{u}$, loses some of its physical interpretation. Typically $\hat{b}$ and $\hat{u}$ work together to compensate for an imperfect mathematical model and surrogate, challenging identifiability. Further discussion shall have to wait until we get a chance to inspect $\hat{b}$ in our example below. First, it makes sense to pause that example momentarily to codify the procedure mathematically and algorithmically.

8.1.3 Calibration as optimization

We optimized something (that's what `optim` was doing); it ought to be possible, and possibly helpful, to back out a formal criterion and discuss its properties. It all flows from the surrogate $\hat{y}^M$ and a twist on some notation introduced in §8.1.1, specifically Eqs. (8.1)–(8.2). Let $\hat{Y}_{n_F}^{M|u} = \hat{y}^M(X_{n_F}, u)$ denote a vector of n_F emulated output y-values at inputs X_{n_F} obtained under a setting u of the calibration parameter(s). Then, let the computer model surrogate residual $Y_{n_F}^{b|u} = Y_{n_F} - \hat{Y}_{n_F}^{M|u}$ denote the n_F-vector of fitted discrepancies. Given these quantities, the quality of a particular u may be measured by the implied joint probability of observing Y_{n_F} at inputs X_{n_F}, under our model $b(\cdot)$ for discrepancies $Y_{n_F}^{b|u}$.

A GP prior on $b(\cdot)$ implies that $\hat{Y}_{n_F}^{b|u} \sim \mathcal{N}(0, \Sigma_n)$, where Σ_n is specified through scaled inverse

exponentiated squared Euclidean distances between inputs X_{n_F}, and thus fully specifies that joint probability. The best-fitting GP regression $\hat{b}(\cdot)$ trained on data $D_{n_F}^b(u) = (X_{n_F}, \hat{Y}_{n_F}^{b|u})$, viewed as a function of u, defines a likelihood for u. Values of u which lead to higher such MVN likelihood, i.e., higher probability of observing Y_{n_F} through those discrepancies, are preferred. So in symbols we have the following mathematical program:

$$\hat{u} = \arg\max_u \left\{ p(u) \left[\max_{\theta_b} p_b(\theta_b \mid D_{n_F}^b(u)) \right] \right\}. \tag{8.4}$$

Recall that $p(u)$ is a (possibly uniform) prior for u, and $p_b(\theta_b \mid \cdots)$ denotes a marginal likelihood implied by a GP prior for $b(\cdot)$, having hyperparameters θ_b including scale, lengthscales and nugget. Algorithm 8.2 provides some of the details in pseudocode, with calculations commencing in log space as usual.

Algorithm 8.2 Modularized KOH Calibration by Optimization

Assume computer model simulations $y^M(\cdot, \cdot)$ are deterministic, and field data are noisy as $Y^F(\cdot) = y^M(\cdot, u^\star) + b(\cdot) + \varepsilon$, where $\varepsilon \sim \mathcal{N}(0, \sigma_\varepsilon^2)$. GP priors are not assumed, however examples are given for that canonical choice.

Require prior density $p(u)$, computer model observations Y_{n_M} at inputs (X_{n_M}, U_{n_M}), field data observations Y_{n_F} at configurations X_{n_F}, and an initial value $u^{(0)}$.

Then

1. Fit $\hat{y}^M(\cdot, \cdot)$ to data (X_{n_M}, Y_{n_M}),
   - e.g., by GP and estimated hyperparameters $\hat{\theta}$ including scale and lengthscales, with `mleGPsep(..., param="d")`.

2. Build an objective to put into an optimizer for calibration parameter u, `obj(u)`, defined as follows:
   a. Obtain surrogate predictive mean values $\hat{Y}_{n_F}^{M|u} = \mu(X_{n_F}, u)$ from $\hat{y}^M(\cdot, u)$, e.g., using `predGPsep(...)$mean`.
   b. Calculate residuals between surrogate predictions and field data locations: $Y_{n_F}^{b|u} = Y_{n_F} - \hat{Y}_{n_F}^{M|u}$.
   c. Fit $\hat{b}(\cdot)$ to residual data $(X_{n_F}, Y_{n_F}^{b|u})$,
      - e.g., by GP and estimated hyperparameters $\hat{\theta}_b$ including scale, lengthscale(s) and nugget with `mleGPsep(..., param="both")`.
   d. Provide as scalar output `obj(u)` the sum of log prior $\log p(u)$ plus the maximizing log likelihood value under hyperparameters $\hat{\theta}_b$,
      - e.g., with `llikGPsep(...)`.

3. Solve $\hat{u} = \text{argmin}_u - \texttt{obj(u)}$, represented mathematically by Eq. (8.4), with library-based numerical methods,
   - e.g., using `optim` with `method="NelderMead"`, or `method="L-BFGS-B"` if $p(u)$ has support in a hyperrectangle.

4. Rebuild $\hat{b}$ like in Step 2c above, using $\hat{u}$.

Return $\hat{u}$, $\hat{y}^M(\cdot, \cdot)$, and $\hat{b}(\cdot)$ so that predictive calculations may be made at new field data locations $\mathcal{X}$ as $\hat{y}^M(\mathcal{X}, \hat{u}) + \hat{b}(\mathcal{X})$.

In contrast to "full KOH" in Algorithm 8.1, notice that estimating hyperparameters $\hat{\theta}$ and

$\hat{\theta}_b$ plays a fundamental role in the inferential process for $\hat{u}$ and $\hat{b}$. The algorithm does not presume these to be known at the outset. Accordingly, the outcome of this modularized maximization could be used to set KOH hyperparameters, if desired. Observe in Step 2c that scale ($\hat{\tau}^2$) and nugget ($\hat{g}$) coordinates of $\hat{\theta}_b$ are being used implicitly to estimate the field data noise variance as $\hat{\sigma}_\varepsilon^2 = \hat{\tau}^2 \hat{g}$. In the case where returned predictors $\hat{y}^M(\mathcal{X}, \hat{u})$ and $\hat{b}(\mathcal{X})$ are both (approximately) MVN, as they would be under a GP prior, the distribution of their sum would also be MVN because they're modeled as conditionally independent given $\hat{u}$. Finally, Step 3 calls for library-based numerical optimization.

One caveat here is that nested `optim` calls with identical `method` specifications, as might happen when choosing `method="L-BFGS-B"` for optimizing over u and `laGP`-based methods for calculating MLE hyperparameters for $\hat{b}(\cdot)$, cause R to crash. A robust but antique BFGS implementation using C static variables confuses the two optimizations and leads to invalid memory access, bringing the whole session down. A simple fix is to use the `optim` default of `method="NelderMead"` for u-optimization instead, perhaps after suitably modifying the objective to check any bound constraints required by $p(u)$.

It's worth emphasizing that calibration parameter $\hat{u}$ is not chosen to minimize bias, but rather is chosen jointly with $\hat{b}$ to obtain the best correction for that bias, e.g., under a GP prior. The optimization/algorithm above makes this transparent, whereas in Algorithm 8.1 such nuances – which also apply – are somewhat obscured by Metropolis–Hastings details. In fact, it may be that estimates $(\hat{b}, \hat{u})$ impart large amplitudes on the bias correction, preferring $\hat{u}$ that push $\hat{y}^M(\cdot, \hat{u})$ away from the real process $y^R(\cdot)$, rather than toward it. For details and further discussion, see Brynjarsdóttir and O'Hagan (2014) and Tuo and Wu (2016).

That's all to say that we have to be satisfied with u, or gravity g in our example, being a tuning parameter rather than a primary quantity of interest. If minimal bias is really what we want, then adjustments are needed.[11] Plumlee (2017) proposes forcing $b(\cdot)$ to be orthogonal to $\hat{y}^M$ as a means of obtaining a bias correction that accounts for effects that are missing from computer model simulations, as opposed to those just being "off". Tuo and Wu (2015) suggest least squares for b rather than a full GP.[12] You might ask why we didn't do this from the very start? Some people do. Both suggestions sacrifice prediction for enhanced interpretation, but unfortunately don't guarantee identifiability of $\hat{u}$ except under regularity conditions that are hard to verify/justify in practice. Although there are very good reasons to diverge from the ordinary KOH, in particular to entertain more restrictive models for bias correction, the original GP formulation will be hard to dethrone from it's canonical position because it provides accurate predictions for $Y^F(\cdot)$ out of sample.

The KOH calibration apparatus, and its modularized variation, couples two highly flexible yet well-regularized nonparametric GP models, linked by calibration parameter u. That flexibility is most potent when coping with a data-generating mechanism that may not be faithful to modeling assumptions. (It's easy to forget that all real data are met with misspecified models in practice.) Authors looking for higher fidelity GP modeling have deliberately deployed similar tactics outside of the calibration setting. Ba and Joseph (2012) coupled two GPs to deal with heteroskedasticity, a form of variance nonstationarity. Bornn et al. (2012) introduced a latent input dimension, just like u, to gain mean-field nonstationarity; Johnson et al. (2018) used a similar trick to select between a small number of mean–variance regimes in a real-time disease forecasting framework. Surprisingly, KOH nests both "tricks", yet precedes them by more than a decade. None cite KOH as inspiration.

---

[11]Some go so far as to suggest that one should not fit a KOH-type model without having informative prior information on u, $b(\cdot)$, or both.
[12]Tuo and Wu (2016) prefer a so-called *native norm* instead.

Back to the example: bias-corrected prediction

It's time to return to our ball drop example. Fitted $\hat{u}$ in hand, we must revisit some of the calculations involved in optimization to back out $\hat{b}$. Repeated calls to `calib` created `gpi` references which, if not immediately destroyed, could have represented a massive memory leak. Providing `clean=FALSE` to `bhat` prevents that memory from being recycled, and augments the output object to contain a reference thereto.

```
bhat <- calib(uhat, XF, ball$time, yMhat, bhat.fit, clean=FALSE)
```

To visualize that discrepancy, code below gathers predictions over our height grid. Since the fitted GP captures both bias correction $\hat{b}$ and field data noise $\hat{\sigma}_\varepsilon^2$, `nonug=TRUE` is provided to get uncertainty in $\hat{b}$ only.

```
pb <- predGPsep(bhat$gp, as.matrix(hs), nonug=TRUE)
sb <- sqrt(diag(pb$Sigma))
q1b <- qnorm(0.95, pb$mean, sb)
q2b <- qnorm(0.05, pb$mean, sb)
```

Figure 8.5 shows that $\hat{b}(\cdot)$ surface with means and quantiles extracted above.

```
plot(heights, pb$mean, type="l", xlab="height", ylab="time bias",
  ylim=range(c(q1b, q2b)), col=3)
lines(heights, q1b, col=3, lty=2)
lines(heights, q2b, col=3, lty=2)
```

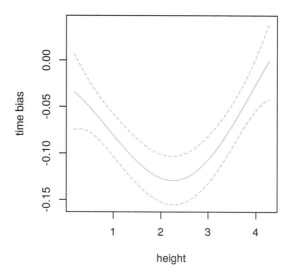

FIGURE 8.5: Estimated bias correction for the ball drop example.

Observe how bias correction is predominantly negative, or downward, being most extreme for middle heights. Lowest and highest heights require almost no correction, statistically speaking. To combine bias with computer model surrogate, code below obtains predictive equations for $\hat{y}^M(\cdot)$ on the same height grid, paired with $\hat{u}$. Quantiles are saved to display the usual three-line summary.

```
pm <- predGPsep(yMhat, cbind(hs, uhat))
q1m <- qnorm(0.95, pm$mean, sqrt(diag(pm$Sigma)))
q2m <- qnorm(0.05, pm$mean, sqrt(diag(pm$Sigma)))
```

Means and covariances from $\hat{y}^M$ and $\hat{b}$ may be combined additively, leveraging conditional independence.

```
m <- pm$mean + pb$mean
Sigma.Rhat <- pm$Sigma + pb$Sigma - diag(eps, length(m))
```

Both GP predictors, representing $\hat{y}^M$ and $\hat{b}$ respectively, have covariances augmented with eps jitter along the diagonal in lieu of a fitted nugget to ensure numerical positive definiteness. Adding two of them together results in 2*eps along the diagonal, which is large enough to impart visual jitter on draws from that distribution, calculated momentarily in R below. To compensate, one of those eps augmentations is taken back off. Sample paths so obtained represent approximations to the real process R; they're realizations from a $\hat{y}^R(\cdot)$.

```
library(mvtnorm)
yR <- rmvnorm(30, m, Sigma.Rhat)
```

In order to accommodate two views into the uncertainty in the reconstructed real process, a second set of predictive equations is calculated for $\hat{b} + \sigma_\varepsilon^2$ as well, yielding $\hat{Y}^F$. This is essentially the same predict command on bhat$gp as above, but without nonug=TRUE. Quantiles are saved for the three-line predictive summary.

```
pbs2 <- predGPsep(bhat$gp, as.matrix(hs))
s <- sqrt(diag(pm$Sigma + pbs2$Sigma))
q1 <- qnorm(0.95, m, s)
q2 <- qnorm(0.05, m, s)
deleteGPsep(bhat$gp)
```

Figure 8.6 provides these two views, with computer model fit $\hat{y}^M$ and samples from $\hat{y}^R$ on the left, and a summary of $\hat{Y}^F$ on the right.

```
par(mfrow=c(1,2))
plot(ball)
lines(heights, pm$mean)
lines(heights, q1m, lty=2)
lines(heights, q2m, lty=2)
matlines(heights, t(yR), col="gray", lty=1)
legend("bottomright", c("Mhat summary", "Mhat + bhat draws"),
  lty=c(1,1), col=c("black", "gray"))
plot(ball)
lines(heights, pm$mean + pb$mean, col=4)
lines(heights, q1, col=4, lty=2)
lines(heights, q2, col=4, lty=2)
legend("topleft", "yMhat + bhat + s2", bty="n")
```

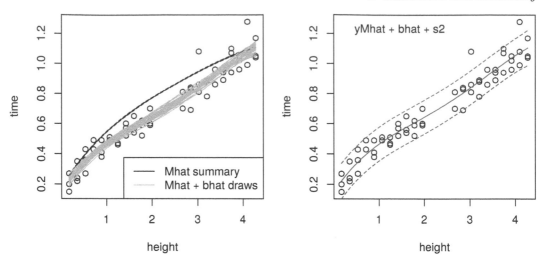

FIGURE 8.6: Modularized KOH predictive distribution via draws (left) and summaries (right).

In the left panel, notice how the surrogate with our estimated calibration parameter, $\hat{y}^M(\cdot, \hat{u})$, way over-predicts. Recall that KOH considers likelihood of the residual process under a GP; it doesn't target minimal bias. Also observe how uncertainty on $\hat{y}^M(\cdot, \hat{u})$ is very low across the heights entertained. Black-dashed quantile lines nearly cover the solid mean line. Samples from the joint predictive distribution of $\hat{y}^R$ in gray are tight. That distribution could similarly have been represented as a three-line summary, with means and quantiles. However, I chose to show sample paths instead in order to remind readers of the value of full covariance structures. Some lines are more "bendy" than others. But all are much smoother than the mean line from Figure 8.2, back at the very start of this example, where field data were fit directly, without the aid of a computer model. The resulting predictive surface for field measurements, shown in the right panel, is much smoother than the surface from that first fit. Error-bars remain narrow even across the gap in training data. KOH predictors borrow strength from computer model surrogates in regions absent of field data.

That the computer model surrogate is monotonically increasing inside the range of heights under study, but predictive distributions for real and field processes are not, is interesting to note. We may be observing an inflection point in the process where dominant dynamics change. If I were to guess, I'd say that wiffle balls don't reach their terminal velocity until they're dropped from heights above 2.5 meters or so, at which point turbulent air and other factors dominate dynamics. Perhaps it takes a few meters for air to freely circulate within the ball, through its Swiss cheese-like holes, ultimately causing the ball to slow down a bit. This aspect may be interesting to investigate further through a more elaborate mathematical model and computer code.

8.1.4 Removing bias

An alternative explanation is that we're doing a bad job of estimating $\hat{u}$ and $\hat{b}$. Perhaps we'd be better off with a simpler apparatus: one without bias correction, say. To entertain that notion, riffing on themes first described by Cox et al. (2001), the code below implements an alternative to `bhat.fit`.

```
se2.fit <- function(X, yF, Ym, clean=TRUE)
 {
  gp <- newGP(X, yF - Ym, d=0, g=0)
  cmle <- list(nll=-llikGP(gp))
  if(clean) deleteGP(gp)
  else { cmle$gp <- gp; cmle$gptype <- "iso" }
  return(cmle)
 }
```

Providing d=0 and g=0 tells newGP not to fit a covariance structure, but instead calculate quantities depicting a zero-mean, iid noise, process ignoring inputs X. No mleGP commands are required since scale $\hat{\tau}^2$ is estimated automatically, in closed form, within newGP. I thought ahead and built calib to accept any discrepancy-fitting function as an argument, even bias-free se2.fit. R code below evaluates calib with fit=se2.fit on our u-grid from earlier, to help visualize, and then creates an objective for optimization in order to more precisely estimate $\hat{u}$.

```
unll.se2 <- rep(NA, length(u))
for(i in 1:length(u))
  unll.se2[i] <- calib(u[i], XF, ball$time, yMhat, fit=se2.fit)$nll
obj.nobias <- function(x, XF, yF, yMhat, fit)
  calib(x, XF, yF, yMhat, fit)$nll
soln <- optimize(obj.nobias, lower=0, upper=1, XF=XF, yF=ball$time,
  yMhat=yMhat, fit=se2.fit)
uhat.nobias <- soln$minimum
```

Figure 8.7 shows the resulting surface and estimate of $\hat{u}$, with old $\hat{u}$ (under GP bias correction) added on for reference.

```
plot(u, unll.se2, type="l", xlab="u", ylab="negative log likelihood")
abline(v=uhat, col=2, lty=2)
abline(v=uhat.nobias, col=3, lty=3)
legend("bottomright", c("uhat", "uhat nobias"), lty=2:3, col=2:3)
```

Our new estimate of $\hat{u}$ is higher than before, but after converting back to natural units (of gravity) we see that it's still probably too low; perhaps estimates are still compensating for air resistance.

```
ghat.nobias <- uhat.nobias*diff(gr) + gr[1]
ghat.nobias
```

```
## [1] 8.166
```

Visualizing the resulting predictive surface for $\hat{Y}^F$ requires running back through calib with clean=FALSE, and then combining computer model predictions $\hat{y}^M(\cdot, \hat{u})$ with noise.

```
cmle.nobias <- calib(uhat.nobias, XF, ball$time, yMhat,
  se2.fit, clean=FALSE)
```

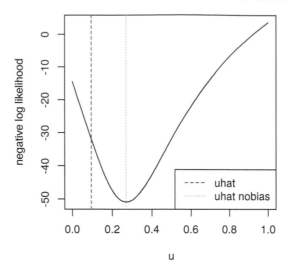

FIGURE 8.7: Likelihood surface for u and $\hat{u}$ under the nobias alternative.

```
se2.p <- predGP(cmle.nobias$gp, as.matrix(hs), lite=TRUE)
pm.nobias <- predGPsep(yMhat, cbind(hs, uhat.nobias), lite=TRUE)
q1nob <- qnorm(0.05, pm.nobias$mean, sqrt(pm.nobias$s2 + se2.p$s2))
q2nob <- qnorm(0.95, pm.nobias$mean, sqrt(pm.nobias$s2 + se2.p$s2))
deleteGP(cmle.nobias$gp)
```

Figure 8.8 provides the usual three-line summary.

```
plot(ball)
lines(heights, pm.nobias$mean, col=4)
lines(heights, q1nob, col=4, lty=2)
lines(heights, q2nob, col=4, lty=2)
legend("topleft", c("yMhat + se2"), col=4, lty=1, lwd=2)
```

Compared to our initial bias corrected version, the curves in the figure are more monotonic
but they also perhaps systematically under-predict for all but the lowest drops. So we
have two competing fits. How, besides aesthetically, can one choose between them? Answer:
out-of-sample validation, e.g., cross validation (CV). The code chunk below collects fitting
and prediction code from above into a stand-alone function that can be called repeatedly, in
a leave-one-out fashion. The first argument takes a set of predictive locations, **XX**, whereas
the rest accept field data, computer model fit (i.e., pre-fit `gpi` reference), and discrepancy
fitting method. Like `calib`, implementation here is designed to be somewhat modular to
this final choice, say using `fit=bhat.fit` or `fit=se2.fit`.

```
calib.pred <- function(XX, XF, yF, yMhat, fit)
  {
  soln <- optimize(obj, lower=0, upper=1, XF=XF, yF=yF,
    yMhat=yMhat, fit=fit)
  bhat <- calib(soln$minimum, XF, yF, yMhat, fit, clean=FALSE)
```

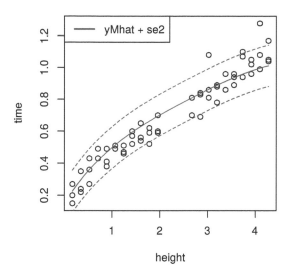

FIGURE 8.8: Nobias predictive surface; compare with Figure 8.6.

```
if(bhat$gptype == "sep") pb <- predGPsep(bhat$gp, XX, lite=TRUE)
else pb <- predGP(bhat$gp, XX, lite=TRUE)
pm <- predGPsep(yMhat, cbind(XX, soln$minimum), lite=TRUE)
m <- pm$mean + pb$mean
s2 <- pm$s2 + pb$s2
q1 <- qnorm(0.95, m, sqrt(s2))
q2 <- qnorm(0.05, m, sqrt(s2))
if(bhat$gptype == "sep") deleteGPsep(bhat$gp)
else deleteGP(bhat$gp)
return(list(mean=m, s2=s2, q1=q1, q2=q2, uhat=soln$minimum))
}
```

Next comes a leave-one-out CV (LOO-CV) loop over $n_F = 63$ field data points, alternately holding out each as a testing set, training on the others, and then predicting. Note that throughout we are conditioning on the same computer model surrogate yMhat, fit to the full computer experiment. CV is over field data only. Each iteration first considers the usual bias correcting modularized KOH setup, and then a simpler nobias alternative.

```
uhats <- q1 <- q2 <- m <- s2 <- rep(NA, nrow(XF))
uhatsnb <- q1nb <- q2nb <- mnb <- s2nb <- uhats
for(i in 1:nrow(XF)) {
  train <- XF[i,,drop=FALSE]
  test <- XF[-i,,drop=FALSE]
  cp <- calib.pred(train, test, ball$time[-i], yMhat, bhat.fit)
  m[i] <- cp$mean
  s2[i] <- cp$s2
  q1[i] <- cp$q1
  q2[i] <- cp$q2
  uhats[i] <- cp$uhat
  cpnb <- calib.pred(train, test, ball$time[-i], yMhat, se2.fit)
```

TABLE 8.1: Jackknife distribution for $\hat{u}$.

	Min.	1st Qu.	Median	Mean	3rd Qu.	Max.
u	0.0559	0.0907	0.0938	0.0946	0.0972	0.1551
g	6.4476	6.7252	6.7502	6.7568	6.7776	7.2411

TABLE 8.2: Jackknife distribution for nobias $\hat{u}$.

	Min.	1st Qu.	Median	Mean	3rd Qu.	Max.
u	0.2633	0.2683	0.2705	0.2708	0.2716	0.2876
g	8.1068	8.1462	8.1637	8.1664	8.1726	8.3012

```
  mnb[i] <- cpnb$mean
  s2nb[i] <- cpnb$s2
  q1nb[i] <- cpnb$q1
  q2nb[i] <- cpnb$q2
  uhatsnb[i] <- cpnb$uhat
}
```

Summary statistics, such as field data predictive quantities at x_i along with $\hat{u}_i$, for $i = 1, \ldots, n_F$, have been saved for subsequent inspection. For example, Table 8.2 shows what $\hat{u}$-values we get in the bias correcting case, including values mapped back to natural units of gravity.

```
kable(rbind(u=summary(uhats), g=summary(uhats)*diff(gr) + gr[1]),
  caption="Jackknife distribution for $\\hat{u}$.")
```

Quite a big range actually. Some are very small indeed. Others are substantially larger, but none nearly as large as the nominal value of $9.8m/s^2$ on Earth. That summary is of a so-called jackknife[13] sampling distribution, a precursor to the bootstrap[14] which would perhaps represent a more standard alternative to studying the sampling distribution of $\hat{u}$ in modern times. (That is, beyond the Bayesian option we started the chapter with, and to which we shall return in §8.1.5.) Resampling methods like the bootstrap are simple and sufficient when field data are plentiful.

Table 8.2 summarizes the jackknife distribution for $\hat{u}$ from the nobias alternative.

```
kable(rbind(u=summary(uhatsnb), g=summary(uhatsnb)*diff(gr) + gr[1]),
  caption="Jackknife distribution for nobias $\\hat{u}$.")
```

Although substantially higher than their bias-correcting analog, this distribution is still shifted substantially lower than nominal. In both cases the computer model surrogate is being asked to compensate for dynamics not accounted for by the underlying mathematical model. Since both come up short on that metric – although we never expected either to

---

[13]https://en.wikipedia.org/wiki/Jackknife_resampling
[14]https://en.wikipedia.org/wiki/Bootstrapping_(statistics)

do particularly well – a comparison on out-of-sample predictive accuracy grounds seems more practical. Figure 8.9 offers visual inspection on those terms. Each prediction is shown as a filled circle, bias-correcting in red (left panel) and nobias in green (right panel), with similarly colored vertical line segments indicating 90% prediction interval(s).

```
par(mfrow=c(1,2))
plot(ball, main="bias correcting")
points(ball$height, m, col=2, pch=20)
segments(ball$height, q1, ball$height, q2, col=2)
plot(ball, main="nobias")
points(ball$height, mnb, col=3, pch=20)
segments(ball$height, q1nb, ball$height, q2nb, col=3)
```

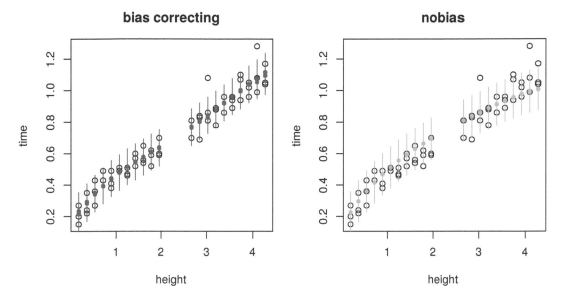

FIGURE 8.9: Leave-one-out CV results in ordinary (left) and nobias (right) alternatives. Filled dots show predictive means; vertical lines indicate 90% intervals.

For my money, the panel on the left looks better: only two dots are left uncovered. On the right I count six dots without a green line going through them at least part way. It's easy to make such a comparison quantitative with proper scores (Gneiting and Raftery, 2007). Eq. (27) from that paper covers pointwise cases: means and variances, without full covariance; i.e., Eq. (5.6) from §5.2.1 forcing diagonal $\Sigma(\mathcal{X})$.

```
b <- mean(- (ball$time - m)^2/s2 - log(s2))
nb <- mean(- (ball$time - mnb)^2/s2nb - log(s2nb))
scores <- c(bias=b, nobias=nb)
scores
```

```
##    bias nobias
## 4.205  3.948
```

Higher scores are better, so correcting for bias wins.

Modularized KOH calibration isn't perfect, but it's relatively simple to solve with library methods for GP fitting and for optimization to find $\hat{u}$. Resulting predictions are accurate owing to flexibility as much as to information quality, provided down the chain of mathematical model, simulation data and field experiment. Although there are many variations, and we briefly discussed a few, the presentation above mainly covers highlights targeting robust implementation. Opportunities for stress-testing are provided by homework exercises both here in §8.3 and (with minor modification to handle big simulation data) on our motivating radiative shock hydrodynamics example (§2.2) in §9.4.

8.1.5   Back to Bayes

As a capstone, and because I said we would, let's return back to full KOH and think about posterior sampling conditional on GP hyperparameters estimated with the modularized variation above. We shall follow Algorithm 8.1, whose most important calculation is covariance $\mathbb{V}(u)$. The top-left block of $\mathbb{V}(u)$ is Σ_{n_M}, capturing computer model surrogate covariance on design $[X_{n_M}, U_{n_M}]$. Observe that Σ_{n_M} doesn't depend upon u. Using $\hat{\theta}$ stored in `mle` (and scale $\hat{\tau}^2$ derived thereupon, in closed form), Σ_{n_M} may be calculated as follows.

```
library(plgp)
KM <- covar.sep(XU, d=mle$d, g=1e-7)
tau2M <- drop(t(yM) %*% solve(KM) %*% yM / length(yM))
SigmaM <- tau2M*KM
```

A portion of bottom-right block of $\mathbb{V}(u)$, namely $\Sigma_{n_F}^b$, also doesn't depend on u. This matrix measures bias covariance between field data locations, conditioned on parameters $\hat{\theta}_b$, including noise level σ_ε^2 via an estimated nugget. Calculations below borrow $\hat{\theta}_b$ from `bhat`. Completing the hyperparameter specification with $\hat{\tau}_b^2$ requires predictions from the computer model, which may be obtained by applying `predGPsep` on `yMhat` with $\hat{u}$ calculated above.

```
KB <- covar.sep(XF, d=bhat$theta[1], g=bhat$theta[2])
XFuhat <- cbind(XF,
  matrix(rep(uhat, nrow(XF)), ncol=length(uhat), byrow=TRUE))
Ym <- predGPsep(yMhat, XFuhat, lite=TRUE)$mean
YmYm <- ball$time - Ym
tau2B <- drop(t(YmYm) %*% solve(KB) %*% YmYm / length(YmYm))
SigmaB <- tau2B * KB
deleteGPsep(yMhat)
```

Remaining components of $\mathbb{V}(u)$ depend upon u and must be rebuilt on demand for each newly proposed value of u as MCMC iterations progress. $\Sigma_{n_M}(X_{n_F}, u)$ measures covariance between $[X_{n_M}, U_{n_M}]$ and $[X_{n_F}, u^\top]$ under the surrogate. Notation $[X_{n_F}, u^\top]$ is shorthand for a design derived by concatenating u to each row of field data inputs X_{n_F}. Similarly $\Sigma_{n_F}(u)$ captures surrogate covariance between $[X_{n_F}, u^\top]$ and itself. Both are calculated by the function coded below, which subsequently completes $\mathbb{V}(u)$ by combining with Σ_{n_M} and $\Sigma_{n_F}^b$, e.g., as saved from earlier calculations like those immediately above.

```
ViVldet <- function(u, XU, XF, SigmaM, tau2M, mle, SigmaB)
  {
```

```
## build blocks
XFu <- cbind(XF, u)
SMXFu <- tau2M * covar.sep(XU, XFu, d=mle$d, g=0)
SMu <- tau2M * covar.sep(XFu, XFu, d=mle$d, g=1e-7)

## build V from blocks
V <- cbind(SigmaM, SMXFu)
V <- rbind(V, cbind(t(SMXFu), SMu + SigmaB))

## return inverse and determinant
## (improvements possible with partitioned inverse equations)
return(list(inv=solve(V),
  ldet=as.numeric(determinant(V, log=TRUE)$modulus)))
}
```

Observe that `ViVldet` doesn't return $\mathbb{V}(u)$ but rather its inverse and determinant as required by the log likelihood (8.3). Code below implements a function calculating that quantity up to an additive constant.

```
llik <- function(u, XU, yM, XF, yF, SigmaM, tau2M, mle, SigmaB)
{
  V <- ViVldet(u, XU, XF, SigmaM, tau2M, mle, SigmaB)
  ll <- - 0.5*V$ldet
  Y <- c(yM, yF)
  ll <- ll - 0.5*drop(t(Y) %*% V$inv %*% Y)
  return(ll)
}
```

Although `llik` takes a multitude of arguments, repeated calls in an MCMC loop would only vary its first argument, u. In order to simplify calls to `llik` in the Metropolis–Hastings (MH) scheme coming shortly, the code below sets the latter eight arguments as defaults using quantities from/derived for our ball drop experiment. That leaves u as the only unspecified argument in `llik`, establishing a convenient shorthand.

```
formals(llik)[2:9] <- list(XU, yM, XF, ball$time, SigmaM, tau2M,
  mle, SigmaB)
```

The final ingredient is the prior. One option is uniform over the study area $g \in [6, 14]$, mapping to $u \in [0, 1]$ in coded units. Beta priors, generalizing the uniform, are also popular, often with shape parameters (> 1) that mildly discourage concentration of posterior on boundaries of the study region. For our example below, I chose a beta prior of this kind but over a somewhat wider space, $u \in [-0.75, 2]$ which maps to $g \in [0, 22]$. The result is a relatively flat prior over the study region $[6, 14]$, emphasizing values nearby $9.8m/s^2$ and discouraging pathological/extreme values such as $g = 0$ on the low end, and g bigger than two-times nominal on the high end.

```
lprior <- function(u, shape1=1.1, shape2=1.1, lwr=-0.75, upr=2)
{
```

```
  u <- (u - lwr)/(upr - lwr)
  dbeta(u, shape1, shape2, log=TRUE)
}
```

Ok, we're ready for MCMC. Below, space is allocated for 10^4 samples from the posterior. Initial values are specified for the chain using $\hat{u}$ calculated above.

```
T <- 10000
lpost <- u <- rep(NA, T)
u[1] <- uhat
lpost[1] <- llik(u[1]) + lprior(u[1])
```

A `for` loop implements a random walk Metropolis sampler (see, e.g., Sherlock et al., 2010) using Gaussian proposals with variance 0.3^2 (`sd=0.3`), tuned to ensure good mixing of the Markov chain. Symmetry in that proposal choice simplifies expressions involved in MH accept–reject calculations since the ratio of proposal densities (q), or equivalently the difference of their logs, cancel. Since such a scheme can generate innovations outside the study area, a check on the prior is required in order to short circuit handling of proposals which violate support constraints.

```
for(t in 2:T) {

  ## random walk Gaussian proposal
  u[t] <- rnorm(1, mean=u[t - 1], sd=0.3)
  lpu <- lprior(u[t])
  if(is.infinite(lpu)) { ## prior reject
    u[t] <- u[t - 1]
    lpost[t] <- lpost[t - 1]
    next
  }

  ## calculate log posterior
  lpost[t] <- llik(u[t]) + lpu

  ## Metropolis accept-reject calculation
  if(runif(1) > exp(lpost[t] - lpost[t - 1])) { ## MH reject
    u[t] <- u[t - 1]
    lpost[t] <- lpost[t - 1]
  }
}
```

Figure 8.10 shows a trace plot of samples obtained from the Markov chain targeting the posterior distribution for u under KOH. Mixing is visually very good, and by initializing with the MLE the chain reaches stationarity instantaneously.

```
plot(u, type="l", xlab="MCMC iteration 1:T", ylab="u")
```

One way to assess MCMC quality, and thereby efficiency of the sampler, is through effective

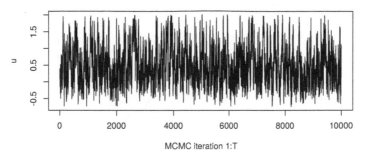

FIGURE 8.10: MCMC trace plot for KOH calibration.

sample size (ESS; Kass et al., 1998). Out of a total of $T = 10^4$ sequentially correlated MCMC samples, ESS measures about how many independent samples that's worth through a measurement of autocorrelation in the chain.

```
library(coda)
ess <- effectiveSize(u)
ess
```

```
##   var1
## 413.7
```

In this case, about one in each of 24 samples can be regarded as independent, having "forgotten" the past from which it came. One way to improve upon that may be to adjust proposal variance. For our purposes 414 samples is good enough to summarize the empirical distribution, say with a kernel density.

```
d <- density(u*diff(gr) + gr[1], from=0, to=22)
```

Before plotting that density in Figure 8.11, R code below evaluates the prior over a grid in u-space for comparison. To ease interpretation, the x-axis in the plot is provided in natural units of gravity.

```
ugrid <- seq(-0.75, 2, length=1000)
ggrid <- ugrid*diff(gr) + gr[1]
lp <- lprior(ugrid)
plot(d, xlab="g", lwd=2, main="")
lines(ggrid, exp(lp)/30, col="gray")
abline(v=ghat, col=2, lty=2)
legend("topright", c("uhat", "prior", "posterior"), lty=c(2,1,1),
  col=c("red", "gray", "black"), lwd=c(1,1,1,2), bty="n")
```

A takeaway from the figure is that there is, at best, modest information in the likelihood relative to the prior. We have a posterior density whose mass is shifted slightly to the right, away from the likelihood (or posterior under a uniform prior) and toward the prior. As a result, the most probable setting for the unknown gravitational acceleration parameter is substantially lower than the nominal value.

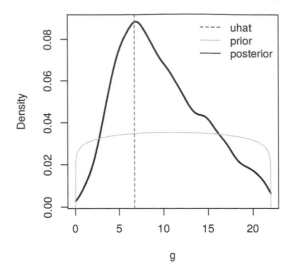

FIGURE 8.11: Comparing posterior, prior and $\hat{u}$ for KOH calibration.

```
w <- which.max(lpost)
u[w]*diff(gr) + gr[1]
```

```
## [1] 6.923
```

Another way to view this state of affairs is through the lens of bias correction. It would appear that our GP fit to discrepancies between computer model and field data is able to cope with almost any reasonable value of u. Some cause less trouble than others, but on the whole u-coupled GPs enjoy more than enough flexibility to explain dynamics exhibited by field and simulation data in a variety of ways. Some of those ways entail seemingly contradictory hypotheses through extremely low and high degrees of gravitational force.

Samples from the posterior distribution for u in hand, and implicitly also for $b(\cdot)$, the next step is to convert those into samples from the posterior predictive distribution for $Y^F(x)$, potentially for many $x \in \mathcal{X}$. Each sample from the posterior, $u^{(t)}$, can be used to define a conditional predictive distribution for $Y^F(\mathcal{X}) \mid Y_{n_F}, Y_{n_M}, u^{(t)}$. A homework exercise in §8.3, which we've alluded to once before but it bears repeating, asks the curious reader to derive that distribution by first expanding MVN covariance structure $\mathbb{V}(u^{(t)})$ to include new rows/columns representing the distribution of all three sets of Y-values jointly, and subsequently applying MVN conditioning identities (5.1). Then, averaging over all $t = 1, \ldots, T$ yields an empirical predictive density that marginalizes over uncertainty in the calibration parameter, approximating

$$p(Y^F(x) \mid Y_{n_F}, Y_{n_M}) = \int_{\mathcal{U}} p(Y^F(x) \mid u, Y_{n_F}, Y_{n_M}) \cdot p(u \mid Y_{n_F}, Y_{n_M}) \, du.$$

Predictive distributions which integrate out all unknowns are a hallmark of Bayesian analysis. Recall that we've conditioned upon hyperparameters from the earlier modularized analysis. A fully Bayesian calibration, extending MCMC to hyperparameters for both GPs, can dramatically expand the complexity of the overall scheme. This usually represents overkill, however there are settings where a full accounting of all uncertainties is essential.

Theoretically minded but practically aware

There are many schools of thought on the right way to calibrate and simultaneously quantify uncertainty, but few recipes which are readily deployable out of the box. MATLAB® software called GPMSA is perhaps the first suite of its kind, providing surrogate modeling and computer model calibration capabilities (Gattiker et al., 2016). Two packages on CRAN called `CaliCo` (Carmassi, 2018) and `BACCO` (Hankin, 2013) offer a Bayesian approach similar to the one implemented above, with some of the extensions left here to exercises in §8.3.

One of the great advantages of the Bayesian paradigm is that it exposes model inadequacies by giving ready access to estimation uncertainties. In the KOH case "inadequacy" manifests as extreme flexibility, which is a paradox. Modularization helps because it limits flexibility somewhat through a more constrained prior, allowing only computer model runs to influence surrogate fits. Nevertheless confounding and identifiability are ever-present concerns (Gu and Wang, 2018; Gu, 2019).

There are many reasons to calibrate, with KOH or otherwise. One is simply predictive; another is to get a sense of how the apparatus could be tuned, or to quantify how much information is in the data (and prior) about promising u settings. Both are very doable, and worth doing, even in the face of confounding. When causal interpretation is essential, further constraints such as limiting forms of bias correction (Plumlee, 2017; Tuo and Wu, 2015) can help with posterior identifiability of u, but often at the expense of predictive accuracy.

Fully Bayesian prediction at the field level, $Y^F(\mathcal{X})$ via ordinary KOH as above and in the homework, is hard to beat. Other, more complicated but also more satisfying examples are offered up as further homework exercises. More ambitious "big simulation" analogs in §9.3.6, as motivated by the radiative shock hydrodynamics example of §2.2, benefit from the thriftier, more modular approach.

8.2 Sensitivity analysis

In any nonparametric regression setting, but especially when two nonparametric regressions are coupled together as in §8.1, it's important to understand the role inputs play in predicted outputs. When inputs change, how do outputs change? In simple linear regression, estimated slope coefficients $\hat{\beta}_j$ and their standard errors speak volumes, resulting in t-tests or F-tests to ascertain relevance. Or, we may inspect leverage or Cook's distance[15] to focus on particular input–output pairs. By contrast, the effect of fitted GP hyperparameters and input settings on predictive surfaces is subtle and sometimes counter-intuitive. In a way, that's what nonparametric means: parameters don't unilaterally dictate what's going on. Model fits and predictive equations gain flexibility from data, sometimes with the help of – but equally often in spite of – any estimated tuning or hyperparameters.

Because of the complicated nature by which data affect fit and predictor in nonparametric regression, approaches to decomposing the effects of inputs x on outputs Y in that setting are varied. Many methods focus in particular on GP regression, but with no less diversity despite (model) specificity. Oakley and O'Hagan (2004) offer what is perhaps the first complete

---

[15]https://en.wikipedia.org/wiki/Cook's_distance

treatment, although efforts date back to Welch et al. (1992) with revisions from Morris et al. (2008).

The approach presented here has many aspects in common with these ideas, and with Marrel et al. (2009), whose analysis tailors *first-order Sobol indices* (Sobol, 1993, 2001) to surrogates derived from GP predictive equations. Yet our presentation will lean more generic. By not focusing expressly on GP equations we can, among other things, accommodate a higher-order analysis under a more unified umbrella. Our approach follows the Saltelli (2002) school of thought in numerical integration: what happens to predictive means and variances when fixing some input coordinates and integrating out others? While illustrations will focus on GP regression, as our preferred surrogate, I won't leverage GPs in our methodological development. The edited volume by Saltelli et al. (2000) provides an overview of the field. Valuable recent work on smoothing methods (Storlie and Helton, 2008; Da Veiga et al., 2009; Storlie et al., 2009) provide a nice overview of nonparametric regression coupled with sensitivity analysis.

8.2.1   Uncertainty distribution

If we're going to say how sensitive outputs are to changes in inputs, it makes sense to first say what inputs we expect/care about, and how much they themselves may change/vary. Underlying the Saltelli method is a reference distribution for x, sometimes called an *uncertainty distribution* $U(x)$. U can represent uncertainty about future values of x, or the relative amount of research interest in various areas of the input space. In many applications, the uncertainty distribution is simply uniform over a bounded region.

In Bayesian optimization (Chapter 7), U can be used to express prior information from experimentalists or modelers on where to look for solutions. For example, when there's a large number of input variables over which an objective function is to be optimized, typically only a small subset will be influential within the confines of their uncertainty distribution. Sensitivity analysis can be used to reduce the volume of the search space of such optimizations (Taddy et al., 2009). Finally, in the case of observational systems such as air-quality or smog levels (§8.2.4), $U(x)$ may derive from an estimate of the density governing natural occurrence of x factors, e.g., air pressure, temperature, wind and cloud cover. In such scenarios, sensitivity analysis attempts to resolve natural variability in responses $Y(x)$.

Although one can adapt the type of sampling described shortly to account for correlated inputs in U (Saltelli and Tarantola, 2002), we treat here the standard and computationally convenient independent specification,

$$U(x) = \prod_{k=1}^{m} u_k(x_k),$$

where u_k, for $k = 1, \ldots, m$, represent densities assigned to the margins of x. With U being specified probabilistically, readers may not be surprised to see sampling feature as a principal numerical device for averaging over uncertainties, i.e., over variability in U. Such averages approximate expectations, which are integrals. Latin hypercube sampling (LHS; §4.1) was conceived to reduce variability in exactly that sort of Monte Carlo (MC) approximation to integrals. Accordingly, LHSs with margins u_k feature heavily in our Saltelli-style calculation of Sobol sensitivity indices.

8.2.2   Main effects

The simplest sensitivity indices are *main effects*, which deterministically vary one input variable, j, while averaging others over U_{-j}:

$$\text{me}(x_j) \equiv \mathbb{E}_{U_{-j}}\{y \mid x_j\} = \iint_{\mathcal{X}_{-j}} y p(y \mid x_1, \ldots, x_m) \cdot u_{-j}(x_1, x_{j-1}, x_{j+1}, x_m)\, dx_{-j} dy. \quad (8.5)$$

Above, $u_{-j} = \prod_{k \neq j} u_k(x_k)$ represents density derived from the joint distribution U without coordinate j, i.e. U_{-j} with $\mathcal{X}_{-j}$ and x_{-j} defined similarly, and $p(y \mid x_1, \ldots, x_m) \equiv p(Y \mid x) \equiv p(Y(x) = y)$ comes from the surrogate, e.g., a GP predictor.

Algorithmically, calculating main effects proceeds as follows: grid out $\mathcal{X}$-space in each coordinate with values x_{ji}, for $i = 1, \ldots, G$ say; gather $\text{me}(x_{ji})$-values holding the jth coordinate fixed at each x_{ji}, in turn, and average over the rest (and y) in an MC fashion; finally plot all $\text{me}(x_{ji})$ on a common x-axis. A pseudocode in Algorithm 8.3 formalizes that sequence of steps with some added numerical detail. The algorithm utilizes samples of size N in its MC approximations. Although Eq. (8.5) and Algorithm 8.3 showcase mean main effects $\mu(x) = \int y p(y \mid x)\, dy$, any aspect of $p(y \mid x)$ which may be expressed as an integral can be averaged with respect to U in this manner. Quantiles are popular, for example, since they may be plotted on the same axes as means.

Algorithm 8.3 Main Effects

Assume m-dimensional coded inputs $x \in [0, 1]^m$ and interest in mean main effects $\mu(x) = \int y p(y \mid x)\, dy$, although $\mu(x)$ may stand in below for other quantities available from predictive equations.

Require uncertainty distribution U which may readily be sampled, say with LHS via margins $u_1, \ldots, u_m$; a desired sample size N controlling accuracy of the MC calculation; surrogate or predictive equations $p(y \mid x)$, say from a GP, generating $\mu(x)$ or another quantity of interest; grid $0 \leq g_1, \ldots, g_G \leq 1$ which may be applied separately but identically to each margin of x.

For each coordinate of x_j, as $j = 1, \ldots, m$, and each grid element of g_i, as $i = 1, \ldots, G$, i.e., a double-"for" loop ...

1. Draw N samples from U_{-j} and combine with g_i to create $N \times m$ predictive matrix $\mathcal{X}_j$ whose j^{th} column has N copies of g_i and remaining columns hold the samples.
2. Evaluate the surrogate at $\mathcal{X}_j$, saving predictive means $\mu(\mathcal{X}_j)$ in N-vector $\hat{y}$,
   - or similarly for another predictive quantity of interest.
3. Average over the N samples to save approximate $\text{me}(x_{ji})$ as $\text{me}_{ji} = \frac{1}{N} \sum_{s=1}^{N} \hat{y}_s$.

End For

Return $m \times G$ matrix "me" or, more commonly, plot each of m row-vectors me_j on a common x-axis with coordinates $g_1, \ldots, g_G$.

In situations where sampling from U is computationally expensive, one may prefer instead to pre-sample a single $N \times m$ matrix $\mathcal{U}$ of deviates in lieu of the $m \times G$ separate $N \times (m-1)$ samples implied by each iterate of Step 1 within the "for" loop. In that setup, a revised Step 1 would instead de-select the jth column to create $N \times (m-1)$ matrix $\mathcal{U}_{-j}$, combining with grid entries to create $\mathcal{X}_j$ in Step 2. Reusing deviates from U in this manner, where each iterate of the double-"for" loop involves similar $\mathcal{U}_{-j}$, induces correlation in otherwise independent calculations. That increases MC error which can be mitigated, to a certain extent, with larger N. Whether by pre-sampling, or with the original Step 1, LHS from U is common as a means of reducing variability for fixed MC sample size N.

Before jumping into an illustration, it's worth commenting on a common alternative "main effect" that avoids integration over "$-j$" input coordinates, instead replacing them with their mean value.

$$\overline{\text{me}}(x_j) \equiv \mathbb{E}\{y \mid x_j, \bar{x}_{-j}\}$$

This is not the same thing, and it's not even really a poor man's approximation. The behavior of responses as a function of typical (average) input values is not the same as average behavior over all input values. But that doesn't make it wrong or mean it's not an interesting thing to look at. However most would argue that it provides, at best, a limited view into the effect of inputs on outputs. I tend to agree. Situations where this alternative is, for the most part, adequate correspond to settings where far simpler surrogates (e.g., linear/polynomial) may have been sufficient to begin with. In those situations, the methods of Chapter 3 provide more precise and satisfying results.

To kick the tires a bit, return to the Friedman data (5.12) from §5.2.5 which combines a nice span of nonlinear, linear, and useless effects. The function is pasted below.

```
fried <- function (n, m=6)
 {
  if(m < 5) stop("must have at least 5 cols")
  X <- randomLHS(n, m)
  Ytrue <- 10*sin(pi*X[,1]*X[,2]) + 20*(X[,3] - 0.5)^2 + 10*X[,4] + 5*X[,5]
  Y <- Ytrue + rnorm(n, 0, 1)
  return(data.frame(X, Y, Ytrue))
 }
```

To create a dataset for surrogate learning, let's evaluate `fried` on a random design of size 250. Then fit a GP surrogate.

```
data <- fried(250)
gpi <- newGPsep(as.matrix(data[,1:6]), data$Y, d=0.1,
  g=var(data$Y)/10, dK=TRUE)
mle <- mleGPsep(gpi, param="both", tmin=rep(eps, 2),
  tmax=c(10, var(data$Y)))
```

Begin by keeping it simple with uniform $U(x)$ in $[0,1]^6$. The code chunk below allocates space necessary to calculate MC averages of size $N = 10K$ on a uniform grid of $G = 30$ values in each x_j.

```
N <- 10000
G <- 30
m <- q1 <- q2 <- matrix(NA, ncol=6, nrow=G)
grid <- seq(0, 1, length=G)
XX <- matrix(NA, ncol=6, nrow=N)
```

Then, the double-"for" loop may commence. Below main effects are collected for predictive means, and central 90% predictive quantiles so that we may also inspect error-bars on main effects. Since emphasis here is on the mean, nonug=TRUE is provided when gathering predictive summaries. Although these data are inherently noisy, epistemic uncertainty in main effects should decrease to zero as the number of training data points grows toward infinity (uniformly within a finite study region).

```
for(j in 1:6) {
  for(i in 1:G) {
    XX[,j] <- grid[i]
    XX[,-j] <- randomLHS(N, 5)
    p <- predGPsep(gpi, XX, lite=TRUE, nonug=TRUE)
    m[i,j] <- mean(p$mean)
    q1[i,j] <- mean(qnorm(0.05, p$mean, sqrt(p$s2)))
    q2[i,j] <- mean(qnorm(0.95, p$mean, sqrt(p$s2)))
  }
}
```

Figure 8.12 provides a visualization of these mean main effects. Error-bars will be added momentarily, but it's easier to focus on one thing at a time. There are already six lines in the figure; adding error-bars will make eighteen.

```
plot(0, xlab="grid", ylab="main effect", xlim=c(0,1),
  ylim=range(c(q1,q2)), type="n")
for(j in 1:6) lines(grid, m[,j], col=j, lwd=2)
legend("bottomright", paste0("x", 1:6), fill=1:6, horiz=TRUE, cex=0.75)
```

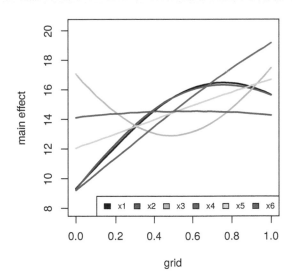

FIGURE 8.12: Mean main effects for Friedman data observed with irrelevant input x_6.

First, note that the scale of the y-axis isn't directly meaningful. Apparently, mean responses vary from about ten to twenty. Considering what we know about the Friedman function, the lines in the figure make sense. For example, x_1 and x_2 contribute in a similar, non-linear (perhaps sinusoidal) manner. Since they only interact in a product, their effects are indistinguishable from one another when holding one fixed and averaging over the other. Input x_3 augments with a face-up parabola; x_4 and x_5 factor in linearly. Finally, x_6 is flat. It has no effect on the response when averaging over other inputs.

Figure 8.13 updates with 90% central error-bars. Observe that there's substantial variability in these main effects. Still, changes in mean main effects across the x-axis far exceeds the width of the interval(s) for all except input x_6. We have our first indication that x_6 is a useless input.

```
plot(0, xlab="grid", ylab="main effect", xlim=c(0,1),
  ylim=range(c(q1,q2)), type="n")
for(j in 1:6) {
  lines(grid, m[,j], col=j, lwd=2)
  lines(grid, q1[,j], col=j, lty=2)
  lines(grid, q2[,j], col=j, lty=2)
}
legend("bottomright", paste0("x", 1:6), fill=1:6, horiz=TRUE, cex=0.75)
```

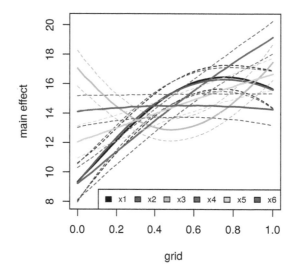

FIGURE 8.13: Main effects from Figure 8.12 augmented with quantiles.

A larger training set would drive down predictive variability and make these error-bars even narrower. Two-hundred fifty data points isn't much in six input dimensions. Were it a gridded design, that would be equivalent to a mere three unique settings for each input. That we're able to view a bit of nuance in mean main effects with a sample of this size is a testament to LHS design, offering more diversity in the margins than a regular grid.

8.2.3   First-order and total sensitivity

Main effects make pretty pictures but they promote a qualitative impression of variable influence. By marginalizing out other inputs, they ignore how variables work together to affect changes in the response. The most common notion of sensitivity is tied to the relationship between conditional and marginal variance for Y. Variance-based methods decompose uncertainties in objective function evaluations and their surrogates into variances of conditional expectations, all with respect to $U(x)$. These are a natural measure of output association with specific sets of variables and provide a basis upon which the importance of individual inputs may be judged.

The variance-based approach presented here is originally due to Sobol, wherein a deterministic objective function is decomposed into summands of functions on lower dimensional subsets of the input space. Consider the functional decomposition

$$f(x_1, \ldots, x_m) = f_0 + \sum_{j=1}^{m} f_j(x_j) + \sum_{1 \leq i < j \leq m} f_{ij}(x_j, x_i) + \cdots + f_{1,\ldots,m}(x_1, \ldots, x_m).$$

When f is modeled as a stochastic process $Y(x)$ conditional on inputs x, we can develop a similar decomposition of response distributions which arise when $Y(x)$ has been integrated over one subset of covariates $x_J = \{x_j : j \in J\}$, where $J \subseteq \{1, \ldots, m\}$, and where the complement of this subset, $x_{-J} = \{x_j : j \notin J\}$ is allowed to vary according to a marginalized uncertainty distribution. For example, we may study the marginal conditional expectation

$$\mathbb{E}_{U_{-J}}\{\mu(x) \mid x_J\} = \int_{\mathcal{X}_{-J}} \mathbb{E}\{y \mid x\} u_{-J}(x_{-J}) \, dx_{-J},$$

where the subsequent marginal uncertainty density is given by $u_J(x_J) = \int_{\mathcal{X}_{-J}} u(x) \, dx_{-J}$. Observe that this generalizes the expectation used to define main effects (8.5).

Sobol-based sensitivity analysis attempts to decompose, and quantify, variability in $\mathbb{E}\{y \mid x_J\}$ with respect to changes in x_J according to $U_J(x_J)$. If U is such that inputs are uncorrelated, the variance decomposition is available as

$$\mathbb{V}\mathrm{ar}(\mathbb{E}_{U_{-J}}\{y \mid x\}) = \sum_{j=1}^{m} V_j + \sum_{1 \leq i < j \leq m} V_{ij} + \cdots + V_{1,\ldots,m},$$

where $V_j = \mathbb{V}\mathrm{ar}_{U_j}(\mathbb{E}_{U_{-j}}\{y \mid x_j\})$, $V_{ij} = \mathbb{V}\mathrm{ar}_{U_{ij}}(\mathbb{E}_{U_{-ij}}\{y \mid x_i, x_j\}) - V_i - V_j$, and so on. When inputs are correlated this identity no longer holds, although a "less-than-or-equal-to" inequality is always true. But nevertheless it's still useful to retain an intuitive interpretation of the V_J's as a portion of overall marginal variance.

First-order sensitivity

With that motivation in mind, define *first-order sensitivity indices* as

$$S_j = \frac{\mathbb{V}\mathrm{ar}_{U_j}(\mathbb{E}_{U_{-j}}\{y \mid x_j\})}{\mathbb{V}\mathrm{ar}_U(y)}, \quad j = 1, \ldots, m. \tag{8.6}$$

In words, S_j is the proportion of variability in the (mean surrogate) response attributable to the jth input, i.e., response sensitivity to variable main effects. As you can see, S_j is scalar, so a first-order analysis reports m numbers, which is of lower dimension than main effects, providing m functions.

Perhaps main effects could be categorized as "zeroth-order" sensitivity indices, but that sells them short because of the high dimensional, functional view they provide. Yet upon conditioning there's no notion of variability. By replacing that grid-wise evaluation with a variance calculation – i.e., another integral – first-order sensitivities offer more global perspective wrapped in a neat package. As a proportion of variability, S_j also provides a relative notion of variable importance.

Observe that lower-case y's are used above, even though the posterior predictive quantity $Y(x) \mid Y_n$, as in our intended surrogate GP application, is clearly a random variable. This is to emphasize that, in calculating S_j, all expectations and variances are taken with respect to uncertainty distribution U. Integrating over y to extract posterior predictive means is implicit for GPs, or can be ignored in the case where $y = f(x)$ can be probed directly and without noise. Throughout our sensitivity index presentation, inner expectations are taken over $X_{-j} \sim U_{-j}$ and outer variances over the only remaining random quantity, $X_j \sim U_j$. When variances/expectations are not nested, as in the variance in the denominator (8.6), integration over all of $X \sim U$ is implied. Below we shall drop distributional subscripts Var_{U_j} and $\mathbb{E}_{U_{-j}}$ to streamline the notation, but the same principles are in play.

MC approximation of first-order indices S_j benefits from the following development. Using the definition for variance, we have

$$S_j = \frac{\mathbb{E}\{\mathbb{E}^2\{y \mid x_j\}\} - \mathbb{E}^2\{y\}}{\mathrm{Var}(y)} \tag{8.7}$$

since $\mathbb{E}_{U_j}^2\{\mathbb{E}_{U_{-j}}\{y \mid x_j\}\} = \mathbb{E}_U^2\{y\} \equiv \mathbb{E}^2\{y\}$. Eq. (8.7) follows (8.6) by taking inner expectation over X_{-j}, and outer one over X_j, combining to integrate over all of $X \sim U$. To calculate those expectations, let M and M' be samples of size N from U, e.g., from LHSs respecting U, being comprised of m-length row vectors s_k and s'_k, for $k = 1, \ldots, N$ respectively. Approximate unconditional quantities as

$$\widehat{\mathbb{E}\{y\}} = \frac{1}{N}\sum_{k=1}^{N}\mathbb{E}\{y \mid s_k\} \quad \text{and} \quad \widehat{\mathrm{Var}(y)} = \frac{1}{N}\mathbb{E}\{y \mid M\}^\top\mathbb{E}\{y \mid M\} - \widehat{\mathbb{E}^2\{y\}},$$

where $\mathbb{E}\{y \mid M\}$ is the column vector $[\mathbb{E}\{y \mid s_1\}, \cdots, \mathbb{E}\{y \mid s_N\}]^\top$ and $\widehat{\mathbb{E}^2\{y\}} = \widehat{\mathbb{E}\{y\}}\widehat{\mathbb{E}\{y\}}$. Approximating the rest of S_j, through conditional expectations, requires mixing columns of M with a similarly built matrix M' of s'_k conditioned columns. Independence in coordinates of U is crucial here. Let M'_j be M' with j^{th} column replaced by the jth column of M, and likewise let M_j be M with the jth column of M'. Then, the conditional second moment required for S_j may be approximated as

$$\mathbb{E}\{\widehat{\mathbb{E}^2\{y \mid x_j\}}\} = \frac{1}{N-1}\mathbb{E}\{y \mid M\}^\top\mathbb{E}\{y \mid M'_j\}.$$

A formal algorithm is coming shortly, but to save a little space we'll wait until after providing details on total sensitivity indices as the two share many subroutines. First, an illustration: let's run through the math above in code below on the Friedman data. Unconditional quantities may be approximated as follows, based on a single m-dimensional LHS M of size N.

```
M <- randomLHS(N, 6)
pM <- predGPsep(gpi, M, lite=TRUE, nonug=TRUE)
Ey <- mean(pM$mean)
Vary <- (t(pM$mean) %*% pM$mean)/N - Ey^2
```

Then, using mixtures of columns of a second LHS with those of the first, approximations to conditional quantities may be combined with the unconditional ones above to estimate S_j, for $j = 1, \ldots, m$.

```
Mprime <- randomLHS(N, 6)
S <- EE2j <- rep(NA, 6)
for(j in 1:6) {
  Mjprime <- Mprime
  Mjprime[,j] <- M[,j]
  pMprime <- predGPsep(gpi, Mjprime, lite=TRUE, nonug=TRUE)
  EE2j[j] <- (t(pM$mean) %*% pMprime$mean)/(N - 1)
  S[j] <- (EE2j[j] - Ey^2)/Vary
}
```

Estimated first-order indices are quoted below.

```
S
```

```
## [1] 0.22156 0.19778 0.09057 0.38071 0.09384 0.01413
```

According to those numbers, our surrogate is most sensitive to input four. This agrees with Figures 8.12–8.13 providing main effects plot(s) which show x_4's marginal response varying over the largest range of the y-axis. Given what we know about the true data generating mechanism, this summary is correct. Although input x_3 (being quadratic) has the potential for bigger effect (x_4's is but linear), that could only happen for $|x_3 - 0.5| > 1$, lying outside the study area. Inputs x_1 and x_2 have the second highest first-order sensitives, being similar to one another. This too makes sense: they only work together in interaction and, as products of numbers less than one in absolute value, span a smaller range of outputs despite sharing the same amplitude as x_4, namely 10.

Total sensitivity

Total sensitivity indices are the mirror image of first-order indices:

$$T_j = \frac{\mathbb{E}\{\mathrm{Var}(y \mid x_{-j})\}}{\mathrm{Var}(y)}.$$

Observe that $\mathbb{E}\{\mathrm{Var}(y \mid x_{-j})\} = \mathrm{Var}(y) - \mathrm{Var}(\mathbb{E}\{y \mid x_{-j}\})$, so T_j measures residual variance in conditional expectation and thus represents all influence connected to a given variable. Consequently the difference between first-order and total sensitivities, $T_j - S_j$, measures variability in y due to the interaction between input j and the other inputs. A large difference $T_j - S_j$ can trigger additional local analysis to determine its functional form.

Again, expanding out the definition for variance as a difference in squared expectations, we have

$$T_j = 1 - \frac{\mathbb{E}\{\mathbb{E}^2\{y \mid x_{-j}\}\} - \mathbb{E}^2\{y\}}{\mathrm{Var}(y)}.$$

So there's only one additional quantity required to calculate T_j beyond elements needed for S_j. Similarly,

$$\mathbb{E}\{\widehat{\mathbb{E}^2\{y \mid x_{-j}\}}\} = \frac{1}{N-1}\mathbb{E}\{y \mid M'\}^{\top}\mathbb{E}\{y \mid M'_j\}$$

$$\approx \frac{1}{N-1}\mathbb{E}\{y \mid M\}^{\top}\mathbb{E}\{y \mid M_j\},$$

with the latter approximation saving us the effort of predicting at locations in M' by re-using those on hand at M.

Consider total sensitivity indices on the Friedman data.

```
T <- EE2mj <- rep(NA, 6)
for(j in 1:6) {
  Mj <- M
  Mj[,j] <- Mprime[,j]
  pMj <- predGPsep(gpi, Mj, lite=TRUE, nonug=TRUE)
  EE2mj[j] <- (t(pM$mean) %*% pMj$mean)/(N - 1)
  T[j] <- 1 - (EE2mj[j] - Ey^2)/Vary
}
deleteGPsep(gpi)
```

As shown below, input four again has the highest index, with inputs one and two close behind.

```
T
```

```
## [1] 0.271472 0.247760 0.077858 0.366271 0.080405 0.001955
```

Although the information seems redundant, their difference can be used to order potential for interaction among pairs of input variables ...

```
I <- T-S
I[I < 0] <- 0
I
```

```
## [1] 0.04991 0.04998 0.00000 0.00000 0.00000 0.00000
```

... which we know is correct from the definition of the `fried` function. Notice that any negative differences are thresholded above. This is to compensate for sampling variability in the MC approximation. Theoretically, $T \geq S$ when inputs are independent under U. (All bets are off for dependent inputs.) MC error, leading to negative I, can be exacerbated by one of the big drawbacks of Saltelli/Sobol analysis: first-order and total indices may fail to sum to the total variance. Although both measure a proportion, they don't partition variability. To address this issue, Owen (2014) proposed an alternative sensitivity measure called a Shapely effect, motivated by Shapley values from game theory. Shapley effects always partition variance when inputs are independent under U. Unfortunately, estimating Shapley effects can be cumbersome computationally. Song et al. (2016) suggest one possible approach, as well as provide an excellent survey of the modern landscape of sensitivity estimation when there are many inputs.

So that everything is in one place, Algorithm 8.4 codifies the sequence of steps required to calculate first-order and total sensitivity indices. The set of input locations which must be evaluated under the surrogate for each calculation of indices S_j and T_j

---

Algorithm 8.4 Sobol First-Order and Total Sensitivity Indices

---

Assume m-dimensional inputs x.

Require uncertainty distribution U which may readily be sampled, say with LHS via margins $u_1, \ldots, u_m$; a desired sample size N controlling MC accuracy; surrogate or predictive equations $p(y \mid x)$, say from a GP, generating $\mu(x)$.

Then

1. Draw two $N \times m$ LHSs from U, and denote these as

$$M = \begin{pmatrix} s_{1_1} & s_{1_2} & \cdots & s_{1_m} \\ s_{2_1} & s_{2_2} & \cdots & s_{2_m} \\ \vdots & \vdots & \ddots & \vdots \\ s_{N_1} & s_{N_2} & \cdots & s_{N_m} \end{pmatrix} \quad \text{and} \quad M' = \begin{pmatrix} s'_{1_1} & s'_{1_2} & \cdots & s'_{1_m} \\ s'_{2_1} & s'_{2_2} & \cdots & s'_{2_m} \\ \vdots & \vdots & \ddots & \vdots \\ s'_{N_1} & s'_{N_2} & \cdots & s'_{N_m} \end{pmatrix}.$$

2. Evaluate surrogate (GP) predictive equations at M, saving mean N-vector $\hat{y}$ with i^{th} component $\mu(s_i)$, and approximate

$$\widehat{\mathbb{E}\{y\}} = \frac{1}{N} \sum_{i=1}^{N} \mu(s_i) \quad \text{and} \quad \widehat{\text{Var}(y)} = \frac{\hat{y}^\top \hat{y}}{N} - \widehat{\mathbb{E}^2\{y\}}.$$

For each coordinate x_j of x, i.e., for $j = 1, \ldots, m$, do ...
3. Create column-swapped matrices M'_j and M_j from M and M' as follows

$$M'_j = \begin{pmatrix} s'_{1_1} & \cdots & s_{1_j} & \cdots & s'_{1_m} \\ s'_{2_1} & \cdots & s_{2_j} & \cdots & s'_{2_m} \\ \vdots & \vdots & \vdots & \ddots & \vdots \\ s'_{N_1} & \cdots & s_{N_j} & \cdots & s'_{N_m} \end{pmatrix} \quad \text{and} \quad M_j = \begin{pmatrix} s_{1_1} & \cdots & s'_{1_j} & \cdots & s_{1_m} \\ s_{2_1} & \cdots & s'_{2_j} & \cdots & s_{2_m} \\ \vdots & \vdots & \vdots & \ddots & \vdots \\ s_{N_1} & \cdots & s'_{N_j} & \cdots & s_{N_m} \end{pmatrix}.$$

4. Evaluate surrogate (GP) predictive equations at M'_j and M_j saving mean N-vectors $\hat{y}'_j$ and $\hat{y}_j$ analogous to Step 2, above, and approximate

$$\widehat{\mathbb{E}\{\mathbb{E}^2\{y \mid x_j\}\}} = \frac{\hat{y}^\top \hat{y}'_j}{N-1} \quad \text{and} \quad \widehat{\mathbb{E}\{\mathbb{E}^2\{y \mid x_{-j}\}\}} = \frac{\hat{y}^\top \hat{y}_j}{N-1}.$$

5. Finally, build first-order and total indices by combining quantities calculated above as follows:

$$S_j = \frac{\widehat{\mathbb{E}\{\mathbb{E}^2\{y \mid x_j\}\}} - \widehat{\mathbb{E}^2\{y\}}}{\widehat{\text{Var}(y)}} \quad \text{and} \quad T_j = 1 - \frac{\widehat{\mathbb{E}\{\mathbb{E}^2\{y \mid x_{-j}\}\}} - \widehat{\mathbb{E}^2\{y\}}}{\widehat{\text{Var}(y)}}.$$

End For

Return m-vectors S and T.

---

is $\{M, M', M'_1, M_1, \ldots, M'_m, M_m\}$, which is $N(2m + 2)$ in total. Saltelli's original version involves $N(m + 2)$, almost half as many, but I find the variation in Algorithm 8.4 to be easier to explain and implement. Saltelli also recommends using an alternative estimate $\widehat{\mathbb{E}\{y\}} = \frac{1}{N-1}\mathbb{E}\{y \mid M\}^\top \mathbb{E}\{y \mid M'\}$ in calculating first-order indices, S_j, as this brings the index closer to zero for non-influential variables. In my experience, that trick can induce bias in estimates and suffer from numerical instability. Instead I prefer the simpler, default setup presented above.

LHSs M and M' can be re-used to estimate main effects, at little extra computational cost, as a byproduct of calculations required for Sobol indices. A post-processing one-dimensional nonparametric regression through the scatterplot of $[s_{1_j}, \ldots, s_{N_j}, s'_{1_j}, \ldots, s'_{N_j}]$ vs. $[\mathbb{E}\{y \mid M\}, \mathbb{E}\{y \mid M'\}]$ for each of $j = 1, \ldots, m$ input variables may be snapped to a grid for plotting purposes. I've found this approach to be quite robust to choices of smoother since $2N$ typically represents a very large sample in 1d. For example, lowess works well in R. This technique is utilized by the tgp package (Gramacy and Taddy, 2016) for fully Bayesian sensitivity and main effects, discussed in more detail momentarily.

As a final note before showcasing that library implementation, the numerical integration scheme(s) outlined above extend(s) nicely to other Sobol indices (e.g., second-order, etc.) for particular combinations of inputs. For details and further discussion, the curious reader is invited to explore some of the references offered at the start of this section (§8.2), particularly the edited volume by Saltelli et al. (2000). Such extensions are less common in the surrogate modeling literature.

8.2.4   Bayesian sensitivity

Now those were just point estimates – by which I mean everything we did above on main effects and first-order/total sensitivity – derived from MLE GP fits. We visualized quantiles of main effects, but that offered a higher resolution view of the concept of main effect rather than a quantification of their uncertainty. Like any quantity estimated from data, sensitivity indices have a sampling distribution. In turn, their distribution could help determine, say, which indices are indeed substantially bigger than others or bigger than some baseline like zero.

Considering the MC nature of calculations, closed form derivation of the sampling distribution of sensitivity indices is a nonstarter. One option is the bootstrap. I don't know if this has ever been done for main, first-order and total Sobol indices with GP surrogates. The (parametric) bootstrap has been used with GP surrogates and for other applications, such as KOH-style calibration (Gramacy et al., 2015). Yet use of the bootstrap in this context, where the prevailing view is Bayesian (§5.3.2), feels like a mismatch of technologies at best, and incoherent at worst.

Sampling from the posterior distribution of all unknown quantities when surrogate modeling, including hyperparameters, is no different in principle than MCMC for calibration parameters, e.g., Algorithm 8.1. Augmenting that MCMC with an extra layer of MC over U LHSs is pretty straightforward. (Hopefully it's not too confusing that both are notated by u.) Yet implementing such a method is too cumbersome for Rmarkdown presentation here. Fortunately it's implemented in software, in several packages actually.

Here I shall illustrate the implementation in tgp, as described by Gramacy and Taddy (2010), which is based on a flexible family of (treed) GPs. Simpler, traditional GP formulations may leverage a greater degree of analytic tractability (Oakley and O'Hagan, 2004). See GPMSA software for implementation.

```
library(tgp)
```

More detail on `tgp` is provided in §9.2.2. Similar functionality is provided in `dynaTree` (Gramacy et al., 2017) on CRAN; for more details on *dynamic trees* as an alternative to GPs for surrogate modeling see §9.2.3. For use in optimization and sensitivity analysis see Gramacy et al. (2013). In `tgp`, the function `sens` invokes a sensitivity analysis.[16] Providing `model=bgp` indicates a Bayesian GP surrogate. Argument `nn.lhs` is like N. However the extra layer of MC offered by posterior sampling with MCMC, which by default involves $T = 4000$ iterations, allows smaller N like `nn.lhs=1000` to be used compared to our earlier, pointwise, analysis. The effective number of LHS draws for U derives from their product, which in this case is 4 million.

```
sf <- tgp::sens(data[,1:6], data$Y, nn.lhs=1000, model=bgp, verb=0)
```

The `tgp` package provides a `plot` method for `"tgp"`-class objects, which has an optional argument `layout="sens"` furnishing visuals for a suite of main effects, first-order and total sensitivities. These are shown in Figure 8.14.

```
plot(sf, layout="sens", legendloc="topleft")
```

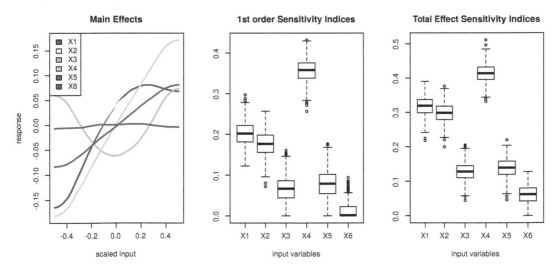

FIGURE 8.14: Summary of Bayesian sensitivity on the Friedman data.

As you can see, boxplots for first-order and total indices offer a window into posterior uncertainty in these calculations. The boxplot corresponding to first-order indices for x_6 indicates not much more than 50% probability of being positive, providing strong indication that this input is not affecting the response. In fact, a common tactic is to deliberately

---

[16]The `tgp` package defines `sens` as an ordinary function, but `dynaTree` treats it as an identically named S3 method. To avoid confusion and runtime errors, explicit `tgp::sens` calls are used for code chunks in this chapter. In your own R session the `tgp::` prefix is only required if `dynaTree` is also loaded.

insert a dummy, useless input like x_6 to get a baseline S and T distribution through which to gauge better what might be an important effect. By default, main effects are visualized without error-bars in order to reduce clutter. An optional argument `maineff` allows the user to specify which inputs to view main effects on, and when that argument is provided 90% error-bars are added to the resulting plot(s). See Figure 8.15.

```
plot(sf, layout="sens", maineff=t(1:5))
```

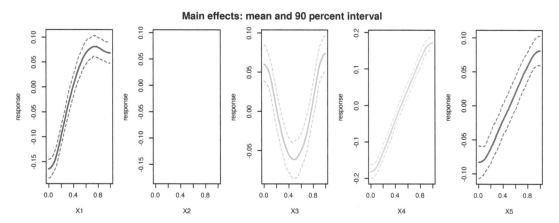

FIGURE 8.15: Bayesian main effects with error-bars on the Friedman data.

Superficially, the view here is quite similar to our mean-and-quantile point estimate-based main effects from Figure 8.13. This time, however, dashed errorbars capture full posterior uncertainty in $Y(x)$ as each coordinate is varied while the others are integrated out.

When searching for interactions, a posterior probability can be calculated as follows.

```
I <- sf$sens$T - sf$sens$S
I[I < 0] <- 0
colMeans(I)
```

```
## [1] 0.11730 0.12255 0.06226 0.05820 0.06160 0.04901
```

Observe that x_1 and x_2 have the highest posterior probability of being involved in an interaction. The others are reasonably high too. One disadvantage to `tgp`'s implementation here is that its sampling scheme doesn't support a `nonug` option. In other words, indices are calculated over $Y(x)$ rather than $\mu(x)$, conditional on training data. Consequently, resulting posterior summaries of sensitivity appear noisier, reflecting greater uncertainty than typically presented. A more favorable assessment would be that, as a result, they offer a conservative view by averaging over a more complete assessment of posterior variability; hence a large probability of (potential) interaction for all variables.

As another illustration, consider the `airquality` data in the base distribution of R. These data contain daily readings of mean ozone in parts per billion (`Ozone`), solar radiation (`Solar.R`), wind speed (`Wind`), and maximum temperature (`Temp`), for New York City between May 1 and September 30, 1973. The `tgp` package supports specification of Beta-distributed marginals for use in sensitivity analysis. Admittedly, this is somewhat restrictive in the landscape of statistical distributions. However, most studies focus on limited ranges for inputs, coded to the unit cube, and uncertainty distributions of interest tend to be unimodal.

Beta marginals offer a relatively flexible, and straightforwardly specified, parametric family meeting that description.

Suppose we take each margin to be scaled beta with shape=2 and mode equal to the average setting for that input. Specifying a beta distribution in this way is thought to be somewhat more intuitive than the typical a and b arguments.

```
X <- airquality[, 2:4]
Z <- airquality$Ozone
rect <- t(apply(X, 2, range, na.rm=TRUE))
mode <- apply(X, 2, mean, na.rm=TRUE)
shape <- rep(2, 3)
```

This dataset has missing values. These are automatically discarded by tgp, however a warning is printed to the screen. To keep our presentation tidy, the code chunk below suppresses those warnings for the sensitivity analysis.

```
s.air <- suppressWarnings(tgp::sens(X=X, Z=Z, nn.lhs=300, rect=rect,
    shape=shape, mode=mode, model=bgp, verb=0))
```

Figure 8.16 shows the default sensitivity layout.

```
plot(s.air, layout="sens")
```

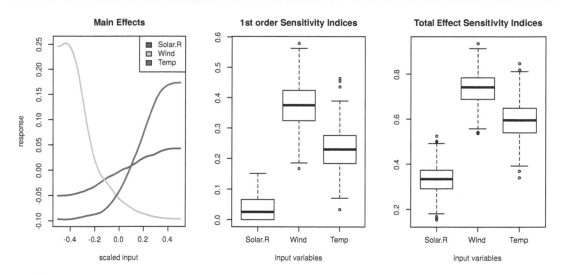

FIGURE 8.16: Sensitivity summary on the air quality data.

Main effects show nonlinear marginal relationships for all three predictors. Solar radiation has the smallest effect on the response, which is echoed in all three measurements. Wind has the largest. Differences in first and total indices indicate modest pairwise interactions among all three variables.

```
I <- s.air$sens$T - s.air$sens$S
I[I < 0] <- 0
colMeans(I)
```

```
## [1] 0.2999 0.3648 0.3640
```

More options for sensitivity analysis in `tgp` are detailed by Gramacy et al. (2013), with supplements in the package documentation.

Hopefully the illustration herein offers a glimpse into what's possible by pairing surrogate modeling with a generous helping of Monte Carlo integration, supporting a curiosity to tinker so long as one has the patience to let it all run. Sensitivity analysis and calibration are numerical procedures which attempt to salvage, or squeeze out, a degree of interpretability from an otherwise opaque nonparametric predictor. Often the great flexibility offered by surrogates, like those based on GPs, thwarts concrete statements derived from inference through optimization and sampling. Critics rightly point to identifiability concerns, for example. Yet much insight can be gleaned from these "beasts" with a few simple tools. That is, until the data get big ...

8.3 Homework exercises

A sample of exercises on calibration and sensitivity analysis follows. Most of these focus on calibration, as there's rather more to explore there – fewer implementations left to libraries.

#1: Calibration with free computer model simulation

Revisit the ball dropping calibration example from §8.1.2 Figure 8.1, with field data in ball.csv[17]. Assume that the computer model `timedrop` is free to evaluate for any height and gravity parameter(s). Specifically, there's no need to fit a surrogate, and thus the setup follows Higdon's special case of the KOH framework (8.1)–(8.2).

a. Develop a calibration apparatus for this situation. In your description, be clear about what quantities are being modeled and how they're being estimated. As in the chapter, provide two versions: (i) one where discrepancy between the computer model and field data is estimated; and (ii) where discrepancy is assumed to be zero.
b. Implement both versions (i–ii) and carry out an analysis that reports on the estimated calibration parameter, $\hat{g}$, in natural units.
c. Provide predictions for time(s) as a function of height for both versions (i–ii) which fully propagates all uncertainties, excepting ones due to hyperparameters (e.g., lengthscales and nuggets to GPs).
d. Finally, re-implement the cross validation (CV) exercise and compare the two methods based on their proper scores. Report the jackknife sampling distribution for $\hat{g}$ in both cases and comment.

---

[17]http://bobby.gramacy.com/surrogates/ball.csv

#2: Calibration and sensitivity in 2d

Consider the following mathematical description of a computer model via inputs $x \in [0,1]^2$ and calibration parameter $u \in [0,1]^2$:

$$y^M(x,u) = \left(1 - e^{-\frac{1}{2x_2}}\right) \frac{1000u_1x_1^3 + 1900x_1^2 + 2092x_1 + 60}{100u_2x_1^3 + 500x_1^2 + 4x_1 + 20}.$$

Now suppose that field data are generated as

$$Y^F(x) = y^M(x,u^\star) + b(x) + \varepsilon, \quad \text{where } b(x) = \frac{10x_1^2 + 4x_2^2}{50x_1x_2 + 10} \quad \text{and } \varepsilon \overset{\text{iid}}{\sim} \mathcal{N}(0, 0.25^2),$$

using $u^\star = (0.2, 0.1)$. See §9.3.6 for implementation in R.

a. Generate an LHS of size 500 in four-dimensional (x,u)-space and evaluate the computer model at those locations. Based on those runs, provide visualizations of main effects, first-order, and total sensitivity indices for each of the four variables.

b. Create a field data design X_{n_F} under a 2d LHS of size 50 and two replicates at each location so that $n_F = 100$. Then obtain $\hat{u}$ estimates under both the bias corrected and nobias modularized KOH calibration apparatus. Use a Beta$(2,2)$ prior independently on the margins of u and GP predictors throughout. How do your $\hat{u}$ estimates compare to the true $u^\star$ value?

c. Generate a testing set of $y^R(x)$-values (i.e., without noise), under 2d LHS of size $N_R = 1000$. Which of your two calibrated predictors, bias corrected and nobias, yield better scores on the testing set?

d. Suppose you were to use the true $u^\star$ value in place of $\hat{u}$, in both bias corrected and nobias alternatives above. How do the resulting "oracle-calibrated" scores compare to the ones you calculated above. Are you at all surprised by the results?

#3: Bayesian version of #2

Revisit exercise #2b with a more fully Bayesian treatment (§8.1.5), ideally using the same data. Compare the posterior distribution for u with the point estimate $\hat{u}$ you obtained in #2. As above, use a Beta$(2,2)$ prior independently on the margins of u.

#4: KOH (Bayesian) prediction

In the KOH framework, ...

a. ... derive the posterior predictive distribution $Y^F(\mathcal{X}) \mid [Y_{n_M}, Y_{n_F}], u$ conditional on all other hyperparameters defining the GPs for surrogate and bias. Use MVN conditional identities similar to those involved in prediction with ordinary GPs (5.2).

b. Implement that predictor in the context of the (four-dimensional (x,u)) exercises #2c and #3 above, simultaneously averaging over all $u^{(t)}$ sampled from the posterior. Or, if you haven't worked on those exercises, similarly extend the Bayesian analysis of the ball drop example (§8.1.5). Report predictive uncertainty on the testing set (#2c, or the grid from the chapter) and compare these to Figure 8.6 obtained from the modular optimization framework. Your comparison should be both qualitative, and quantitative (i.e., compared to the truth).

#5: Sensitivity in high dimension

Oakley and O'Hagan (2004) consider a fifteen dimensional problem whose specification[18] includes five variables having a substantial effect on the response, five with a smaller effect, and five with almost no contribution. Fit a separable GP to responses obtained on a size-250 maximin LHS (§4.3) in fifteen dimensions. Note that your estimated lengthscales should be quite long (if you allow them to be). Use predictive equations from that fit to visualize main effects and calculate first-order sensitivity indices S_j, for $j = 1, \ldots, 15$. (No T_j are required since Oakley & O'Hagan didn't present those.) You may need to increase the MC size N due to the much higher dimensional input space. See if you can partition inputs into three "effect classes" of five based on main effects and first-order indices. Compare your results to Table 1 from Oakley and O'Hagan (2004). *Note that they used $U_i \equiv \mathcal{N}(0,1)$.* Contrast with the fully Bayesian alternative provided by tgp. *Be careful about how you "map" a Gaussian uncertainty distribution to tgp's Beta(s); longer MCMC than the* sens *default may be needed due to the higher dimensional space. You may fix a small nugget by providing* nug.p=0 *and* gd=c(0.0001, 0.1) *to* sens.

[18]https://www.sfu.ca/~ssurjano/oakoh04.html

9

GP Fidelity and Scale

Gaussian processes are fantastic, but they're not without drawbacks. Computational complexity is one. Flops in $\mathcal{O}(n^3)$ for matrix decompositions, furnishing determinants and inverses, is severe. Many practitioners point out that storage, which is in $\mathcal{O}(n^2)$, is the real bottleneck, at least on modern laptops and workstations. Even if you're fine with waiting hours for MVN density calculation for a single likelihood evaluation, chances are you wouldn't have enough high-speed memory (RAM) to store the $n \times n$ matrix Σ_n, let alone its inverse too and any auxiliary space required. There's some truth to this, but usually I find time to be the limiting factor. MLE calculations can demand hundreds of decompositions. Big memory supercomputing nodes are a thing, with orders of magnitude more RAM than conventional workstations. Big time nodes are not. Except when executions can be massively parallelized, supercomputers aren't much faster than high-end laptops.

Although GP inference and prediction (5.2)–(5.3) more prominently feature inverses Σ_n^{-1}, it's actually determinants $|\Sigma_n|$ that impose the real bottleneck. Both are involved in MVN density/likelihood evaluations. Conditional on hyperparameters, only Σ_n^{-1} is required. A clever implementation could `solve` the requisite system of equations for prediction, e.g., $\Sigma_n^{-1}Y_n$ for the mean and $\Sigma_n^{-1}\Sigma(X_n, x)$ for the variance, without explicitly calculating an inverse. Parallelization over elements of $x \in \mathcal{X}$, extending to $\Sigma_n^{-1}\Sigma(X_n, \mathcal{X})$, is rather trivial. All that sounds sensible and actionable, but actually it's little more than trivia. Knowing good hyperparameters (without access to the likelihood) is "rare" to say the least. Let's presume likelihood-based inference is essential and not pursue that line of thinking further. Full matrix decomposition is required.

Most folks would agree that Cholesky decomposition[1], leveraging symmetry in distance-based Σ_n, offers the best path forward.[2] The Cholesky can furnish both inverse and determinant in $\mathcal{O}(n^3)$ time, and in some cases even faster depending on the libraries used. Divide-and-conquer parallelization also helps on multi-core architectures. I strongly recommend Intel MKL[3] which is the workhorse behind Microsoft R Open[4] on Linux and Windows, and the Accelerate Framework[5] on OSX. Both provide conventional BLAS[6]/LAPACK[7] interfaces with modern implementations and customizations under the hood. An example of the potential with GP regression is provided in Appendix A. As explained therein, OpenBLAS[8] is not recommended in this context because of thread safety issues which become relevant in nested, and further parallelized application (e.g., §9.3).

Ok, that's enough preamble on cumbersome calculations. Computational complexity is one barrier; flexibility is another. Stationarity of covariance is a nice simplifying assumption, but

---

[1] https://en.wikipedia.org/wiki/Cholesky_decomposition
[2] LU decomposition is also popular. See, e.g., Ambikasaran et al. (2015).
[3] https://software.intel.com/en-us/articles/using-intel-mkl-with-r
[4] https://mran.microsoft.com/open
[5] https://developer.apple.com/documentation/accelerate
[6] https://en.wikipedia.org/wiki/Basic_Linear_Algebra_Subprograms
[7] https://en.wikipedia.org/wiki/LAPACK
[8] https://www.openblas.net/

it's not appropriate for all data generating mechanisms. There are many ways data can be nonstationary. The LGBB rocket booster (see §2.1), whose simulations are deterministic, exhibits mean nonstationarity along with rather abrupt regime changes from steep to flat dynamics. When data/simulations are noisy, variance nonstationarity or *heteroskedasticity* may be a concern: noise levels which are input-dependent. Changes in either may be smooth or abrupt. GPs don't cope well in such settings, at least not in their standard form. Enhancing fidelity in GP modeling, to address issues like mean and variance nonstationarity, is easy in theory: just choose a nonstationary kernel. In practice it's harder than that. What nonstationary kernel? What happens when regimes change and dynamics aren't smooth?

Often the two issues, speed and flexibility, are coupled together. Coherent high-fidelity GP modeling schemes have been proposed, but in that literature there's a tendency to exacerbate computational bottlenecks. Coupling processes together, for example to warp a nonstationary surface into one wherein simpler stationary dynamics reign (e.g., Schmidt and O'Hagan, 2003; Sampson and Guttorp, 1992) adds layers of additional computational complexity and/or requires MCMC. Consequently, such methods have only been applied on small data by modern standards. Yet observing complicated dynamics demands rich and numerous examples, putting data collection at odds with modeling goals. In a computer surrogate modeling context, where design and modeling go hand in hand, this state of affairs is particularly limiting. Surrogates must be flexible enough to drive sequential design acquisitions towards efficient learning, but computationally thrifty enough to solve underlying decision problems in time far less than it would take to run the actual simulations – a hallmark of a useful surrogate (§1.2.2).

This chapter focuses on GP methods which address those two issues, computational thrift and modeling fidelity, simultaneously. GPs can only be brought to bear on modern big data problems in statistics and machine learning (ML) by somehow skirting full dense matrix decomposition. There are lots of creative ways of doing that, some explicitly creating sparse matrices, some implicitly. Approximation is a given. A not unbiased selection of examples and references is listed below.

- Pseudo-inputs (Snelson and Ghahramani, 2006) or the predictive process (PP; Banerjee et al., 2008); both examples of methods based on *inducing points*
- Iterating over batches (Haaland and Qian, 2011) and sequential updating (Gramacy and Polson, 2011)
- Fixed rank kriging (Cressie and Johannesson, 2008)
- Compactly supported covariances and fast sparse linear algebra (Kaufman et al., 2011; Sang and Huang, 2012)
- Partition models (Gramacy and Lee, 2008a; Kim et al., 2005)
- Composite likelihood (Eidsvik et al., 2014)
- Local neighborhoods (Emery, 2009; Gramacy and Apley, 2015)

The literature on nonstationary modeling is more niche, although growing. Only a couple of the ideas listed above offer promise in the face of both computational and modeling challenges. This will dramatically narrow the scope of our presentation, as will the accessibility of public implementation in software.

There are a few underlying themes present in each of the approaches above: inducing points, sparse matrices, partitioning, and approximation. In fact all four can be seen as mechanisms for inducing sparsity in covariance. But they differ in how they leverage that sparsity to speed up calculations, and in how they offer scope for enhanced fidelity. It's worth noting that you can't just truncate as a means of inducing sparsity. Rounding small entries Σ_n^{ij} to zero will almost certainly destroy positive definiteness.

We begin in this chapter by illustrating a distance-based kernel which guarantees both sparsity and positive definiteness, however as a device that technique underwhelms. Sparse kernels compromise on long-range structure without gaining enhanced local modeling fidelity. Calculations speed up, but accuracy goes down despite valiant efforts to patch up long-range effects.

We then turn instead to implicit sparsity via partitioning and divide-and-conquer. These approaches, separately leveraging two key ideas from computer science – tree data structures on the one hand and transductive learning on the other – offer more control on speed versus accuracy fronts, which we shall see are not always at odds. The downside however is potential lack of smoothness and continuity. Although GPs are famous for their gracefully flowing surfaces and sausage-shaped error-bars, there are many good reasons to eschew that aesthetic when data get large and when mean and variance dynamics may change abruptly.

Finally, Chapter 10 focuses explicitly on variance nonstationarity, or input dependent noise, and low-signal scenarios. Stochastic simulations represent a rapidly growing sub-discipline of computer experiments. In that context, and when response surfaces are essential, tightly coupled active learning and GP modeling strategies (not unlike the warping ideas dismissed above) are quite effective thanks to a simple linear algebra trick, and generous application of replication as a tried and true design strategy for separating signal from noise.

9.1 Compactly supported kernels

A kernel $k_{r_{\max}}(r)$ is said to have *compact support* if $k_{r_{\max}}(r) = 0$ when $r > r_{\max}$. Recall from §5.3.3 that $r = |x - x'|$ for a stationary covariance. We may still proceed component-wise with $r_j = |x_j - x'_j|$ and $r_{j,\max}$ for a separable compactly supported kernel, augment with scales for amplitude adjustments, nuggets for noisy data and embellish with smoothness parameters (Matèrn), etc. Rate of decay of correlation can be managed by lengthscale hyperparameters, a topic we shall return to shortly.

A compactly supported kernel (CSK) introduces zeros into the covariance matrix, so sparse matrix methods may be deployed to aid in computations, both in terms of economizing on storage and more efficient decomposition for inverses and determinants. Recall from §5.3.3 that a product of two kernels is a kernel, so a good way to build a bespoke CSK with certain properties is to take a kernel with those properties and multiply it by a CSK – an example of covariance *tapering* (Furrer et al., 2006).

Two families of CSKs, Bohman and truncated power, offer decent approximations to the power exponential family (§5.3.3), of which the Gaussian (power $\alpha = 2$) is a special case. These kernels are zero for $r > r_{\max}$, and for $r \leq r_{\max}$:

$$k_{r_{\max}}^{\text{B}}(r) = \left(1 - \frac{r}{r_{\max}}\right)\cos\left(\frac{\pi r}{r_{\max}}\right) + \frac{1}{\pi}\sin\left(\frac{\pi r}{r_{\max}}\right)$$

$$k_{r_{\max}}^{\text{tp}}(r; \alpha, \nu) = [1 - (r/r_{\max})^{\alpha}]^{\nu}, \quad \text{where } 0 < \alpha < 2 \text{ and } \nu \geq \nu_m(\alpha).$$

The function $\nu_m(\alpha)$ in the definition of the truncated power kernel represents a restriction necessary to ensure a valid correlation in m dimensions, with $\lim_{\alpha \to 2} \nu_m(\alpha) = \infty$. Although it's difficult to calculate $\nu_m(\alpha)$ directly, there are known upper bounds for a variety of α-values between 1.5 and 1.955, e.g., $v_1(3/2) \leq 2$ and $v_1(5/3) \leq 3$. Chapter 9 of Wendland

(2004) provides several other common CSKs. The presentation here closely follows Kaufman et al. (2011), concentrating on the simpler Bohman family as a representative case.

Bohman CSKs yield a mean-square differentiable process, whereas the truncated power family does not (unless $\alpha = 2$). Notice that $r_{\max}$ plays a dual role, controlling both lengthscale and degree of sparsity. Augmenting with explicit lengthscales, i.e., $r_\theta = r/\sqrt{\theta}$, enhances flexibility but at the expense of identifiability and computational concerns. Since CSKs are chosen over other kernels with computational thrift in mind, fine-tuning lengthscales often takes a back seat.

9.1.1 Working with CSKs

Let's implement the Bohman CSK and kick the tires.

```
kB <- function(r, rmax)
 {
  rnorm <- r/rmax
  k <- (1 - rnorm)*cos(pi*rnorm) + sin(pi*rnorm)/pi
  k <- k*(r < rmax)
 }
```

To have some distances to work with, the code below calculates a rather large 2000×2000 distance matrix based on a dense grid in $[0, 10]$.

```
library(plgp)
X <- matrix(seq(0, 10, length=2000), ncol=1)
D <- distance(X)
```

We can then feed these distances into $k^B_{r_{\max}}(\cdot)$ and check for sparsity under several choices of $r_{\max}$. Careful: kB is defined for ordinary, rather than squared, pairwise distances. For comparison, an ordinary/dense Gaussian covariance is calculated and saved as K.

```
eps <- sqrt(.Machine$double.eps)  ## numerical stability
K <- exp(-D) + diag(eps, nrow(D))
K2 <- kB(sqrt(D), 2)
K1 <- kB(sqrt(D), 1)
K025 <- kB(sqrt(D), 0.25)
c(mean(K > 0), mean(K2 > 0), mean(K1 > 0), mean(K025 > 0))
```

```
## [1] 1.00000 0.35960 0.18955 0.04889
```

Indeed, as $r_{\max}$ is decreased, the proportion of nonzero entries decreases. Observe that Bohman-based correlation matrices do not require jitter along the diagonal. Like Matèrn, Bohman CSKs provide well-conditioned correlation matrices.

Investigating the extent to which those levels of sparsity translate into computational savings requires investing in a sparse matrix library e.g., spam (Furrer, 2018) or Matrix (Bates and Maechler, 2019). Below I choose Matrix as it's built-in to base R, however CSK-GP fitting software illustrated later uses spam.

```
library(Matrix)
c(system.time(chol(K))[3],
   system.time(chol(Matrix(K2, sparse=TRUE)))[3],
   system.time(chol(Matrix(K1, sparse=TRUE)))[3],
   system.time(chol(Matrix(K025, sparse=TRUE)))[3])
```

```
## elapsed elapsed elapsed elapsed
##    1.002    0.267    0.115    0.061
```

As you can see, small r_{max} holds the potential for more than an order of magnitude speedup. Further improvements may be possible if the matrix can be built natively in sparse representation. So where is the catch? Such (speed) gains must come at a cost (to modeling and inference). We want to encourage sparsity because that means speed, but getting enough sparsity requires lots of zeros, and that means sacrificing long range spatial correlation. If local modeling is sufficient, then why bother with a global model? In §9.3 we'll do just that: eschew global modeling all together. For now, let's explore the cost–benefit trade-off with CSK and potential for mitigating compromises on predictive quality and uncertainty quantification (UQ) potential.

Consider a simple 1d random process, observed on a grid.

```
x <- c(1, 2, 4, 5, 6, 8, 9, 10)/11
n <- length(x)
D <- distance(as.matrix(x))
K <- exp(-5*sqrt(D)^1.5) + diag(eps, n)
library(mvtnorm)
y <- t(rmvnorm(1, sigma=K))
```

Here are predictions gathered on a dense testing grid in the input space from the "ideal fit" to that data, using an ordinary GP conditioned on known hyperparameterization. (It's been a while – way back in Chapter 5 – since we entertained such calculations by hand.)

```
xx <- seq(0, 1, length=100)
DX <- distance(as.matrix(x), as.matrix(xx))
KX <- exp(-5*sqrt(DX)^1.5)
Ki <- solve(K)
m <- t(KX) %*% Ki %*% y
Sigma <- diag(1+eps, ncol(KX)) - t(KX) %*% Ki %*% KX
q1 <- qnorm(0.05, m, sqrt(diag(Sigma)))
q2 <- qnorm(0.95, m, sqrt(diag(Sigma)))
```

Before offering a visual, consider the analog of those calculations with a Bohman CSK using $r_{max} = 0.1$.

```
K01 <- kB(sqrt(D), 0.1)
KX01 <- kB(sqrt(DX), 0.1)
Ki01 <- solve(K01)
m01 <- t(KX01) %*% Ki01 %*% y
tau2 <- drop(t(y) %*% Ki01 %*% y)/n
```

```
Sigma01 <- tau2*(1 - t(KX01) %*% Ki01 %*% KX01)
q101 <- qnorm(0.05, m01, sqrt(diag(Sigma01)))
q201 <- qnorm(0.95, m01, sqrt(diag(Sigma01)))
```

Figure 9.1 shows the randomly generated training data and our two GP surrogate fits. "Full", non-CSK, predictive summaries are shown in black, with a solid line for means and dashed lines for 90% quantiles. Red lines are used for the CSK analog.

```
plot(x, y, xlim=c(0, 1.3), ylim=range(q101, q201))
lines(xx, m)
lines(xx, q1, lty=2)
lines(xx, q2, lty=2)
lines(xx, m01, col=2)
lines(xx, q101, col=2, lty=2)
lines(xx, q201, col=2, lty=2)
legend("topright", c("full", "CSK"), lty=1, col=1:2, bty="n")
```

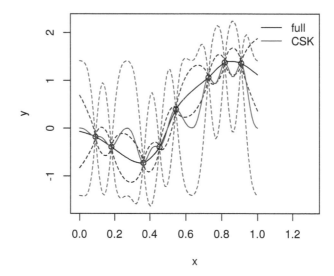

FIGURE 9.1: Predictions under CSK compared to the ideal full GP.

Relative to the ideal baseline, Bohman CSK predictions are too wiggly in the mean and too wide in terms of uncertainty. Predictive means are off because aggregation transpires in a narrower window. Variance is larger simply because sparse K leads to larger K^{-1}. Evidently, inducing sparsity can have deleterious effects on GP prediction equations. What can be done?

9.1.2 Sharing load between mean and variance

A key idea in Kaufman et al. (2011) is to "mop up" long range non-linearity with rich mean structure, leaving a residual that may be modeled by shorter-range (sparse) correlations, via CSKs. In the GP surrogate modeling landscape, where mean-zero processes are the default, this represents somewhat of a paradigm shift. In ML and geostatistics, non-zero mean modeling is more common however simple linear means are the norm. Shifting more of the burden of modeling from covariance to mean is clever because it taps an underutilized resource. There are, of course, many ways that idea could be operationalized. Recall, e.g., that there are covariance kernels that can implement linear means (§5.3.3). Kaufman's partition of focus, shifting between mean and variance, is therefore somewhat arbitrary. Yet human capacity to intuit statistical modeling dynamics generally favors locations over scales.

One alternative that's similar to Kaufman's idea in spirit, but different in form, is *full-scale approximation* (FSA; Sang and Huang, 2012). FSA partitions effort between two spatial processes: a predictive process (PP; Banerjee et al., 2008; Finley et al., 2009) for long range dependence, and a CSK for short-range spatial correlations. To establish a conceptual link between Kaufman and FSA, imagine the PP mapping to a nonparametric (nonlinear) mean structure. That analogy is imperfect because PPs are covariance-centric. Inference remains tractable because PPs emit a reduced rank approximation whose covariance structure can be decomposed quickly through the Sherman–Morrison–Woodbury identity[9], circumventing $\mathcal{O}(n^3)$ matrix manipulation.

A downside to FSA is that implementing PP requires inferring a potentially large set of reference knots whose desired number, and subsequent inference, may become unwieldy as input dimension gets large. Many of the geo-spatial applications targeted by FSA/PP are two-dimensional, where they have enjoyed considerable success in large-n settings. While there are no fundamental or theoretical barriers to applying FSA more widely, e.g., to address larger input spaces more common in computer experiments and ML, I'm unaware of any successful ports to surrogate modeling. Pseudo inputs (Snelson and Ghahramani, 2006), an ML take on PPs, have been applied more widely, but to my knowledge they have not been combined with CSK or an FSA-style analysis.

Kaufman's rich-mean/CSK hybrid more squarely targets modestly higher-dimensional computer surrogate modeling contexts. For that reason, the narrative here focuses on that approach. You may recall from a homework problem in §5.5 that augmenting GPs with unknown linear mean structure can be accommodated analytically: closed-form expressions concentrate out – i.e., marginalize in the Bayesian setting, or replace with MLE for profile likelihoods – unknown regression coefficients, even ones derived from rather large nonlinear bases. The result is an inferential procedure demanding time in $\mathcal{O}(n(n_{\text{sparse}}^2) + p^3)$ where n_{sparse} is the average number of non-zero entries in a row of a CSK matrix, and p is the size of the basis encoding the mean.

There are many reasonable families of bases to choose from. Kaufman et al. found that Legendre polynomials[10] work well. The presentation below reverse engineers some of the computational details from their setup, which is packaged together in an R library called `SparseEm` (for sparse emulation) that can be downloaded from Cari Kaufman's web page[11].

```
library(SparseEm)
```

Unfortunately, Cari's version fails to install in more modern R environments. I've provided a slightly modified version[12] which is more up-to-date in terms of package structure, and should install for most Rs. Code below creates a degree four Legendre polynomial basis over training x values used by the simple 1d example in Figure 9.1.

---

[9] https://en.wikipedia.org/wiki/Sherman-Morrison_formula
[10] https://en.wikipedia.org/wiki/Legendre_polynomials
[11] https://www.stat.berkeley.edu/~cgk/rcode/index.html
[12] http://bobby.gramacy.com/surrogates/SparseEm_0.2-2.tar.gz

```
leg01 <- legFun(0, 1)
degree <- 4
X <- leg01(x, terms=polySet(1, degree, 2, degree))
colnames(X) <- paste0("l", 0:(ncol(X) - 1))
X <- data.frame(X)
X
```

```
##    l0     l1      l2      l3      l4
## 1  1 -1.4171  1.1273 -0.3757 -0.5243
## 2  1 -1.1022  0.2402  0.8210 -1.2784
## 3  1 -0.4724 -0.8686  0.9482  0.3608
## 4  1 -0.1575 -1.0903  0.3558  1.0329
## 5  1  0.1575 -1.0903 -0.3558  1.0329
## 6  1  0.7873 -0.4250 -1.1827 -0.6391
## 7  1  1.1022  0.2402 -0.8210 -1.2784
## 8  1  1.4171  1.1273  0.3757 -0.5243
```

First, let's see how well this basis works on its own, i.e., in a linear regression with iid noise structure (no GP). Notice that `leg01` generates its own intercept column.

```
lfit <- lm(y ~ . -1, data=X)
```

This `leg01` basis must also be evaluated on the `xx` testing grid before it can be fed into `predict.lm`.

```
XX <- leg01(xx, terms=polySet(1, degree, 2, degree))
colnames(XX) <- paste0("l", 0:(ncol(X) - 1))
p <- predict(lfit, newdata=data.frame(XX), interval="prediction", level=0.9)
```

Figure 9.2 augments Figure 9.1. Perhaps with all those lines the plot is a bit busy. Yet it's plain to see that the new `leg01` fit, using `lm` on a Legendre basis, offers a better fit than CSK, but perhaps not as good as the ideal full GP fit.

```
plot(x, y, xlim=c(0, 1.35), ylim=range(q101, q201, p[,2], p[,3]))
lines(xx, m)
lines(xx, q1, lty=2)
lines(xx, q2, lty=2)
lines(xx, m01, col=2)
lines(xx, q101, col=2, lty=2)
lines(xx, q201, col=2, lty=2)
lines(xx, p[,1], col=3)
lines(xx, p[,2], col=3, lty=2)
lines(xx, p[,3], col=3, lty=2)
legend("topright", c("full", "CSK", "leg01"), lty=1, col=1:3, bty="n")
```

The linear/Legendre predictive surface is over-smooth, and consequently its error-bars are everywhere too large compared to ideal. Independent error modeling is a mismatch to our data generating mechanism, being used here to exemplify inherently deterministic computer model simulation. CSK's error-bars can be even wider, but that surface still interpolates due

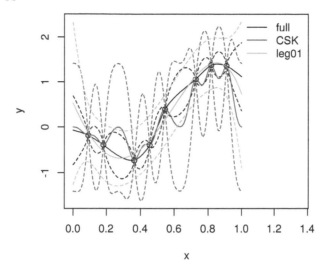

FIGURE 9.2: Legendre-basis linear prediction versus CSK and the ideal GP from Figure 9.1.

to its nugget-free inverse distance-based covariance structure. It may be worth repeating this example in your own R session to observe a diversity of behaviors across data realizations and subsequent fits. You might also try higher **degree** Legendre bases (5 or 6, say).

Separately, neither Legendre basis nor CSK are on par with the ideal full GP, but how about together? As a quick illustration of potential, with a more coherent and Bayesian hybrid on the horizon, consider applying CSK on residuals from Legendre basis-derived fitted values.

```
m2 <- t(KX01) %*% Ki01 %*% lfit$resid
tau22 <- drop(t(lfit$resid) %*% Ki01 %*% lfit$resid)/n
Sigma2 <- tau22*(1 - t(KX01) %*% Ki01 %*% KX01)
```

Now consider predictive summaries formed by combining Legendre basis means with CSK covariances. Uncertainties are not properly managed in this hybrid, but my aim is simply to illustrate potential.

```
m2 <- p[,1] + m2
q12 <- qnorm(0.05, m2, sqrt(diag(Sigma2)))
q22 <- qnorm(0.95, m2, sqrt(diag(Sigma2)))
```

Figure 9.3 shows the resulting predictive surface. In place of green Legendre lines and red CSK lines, blue hybrid lines show how Legendre–CSK predictions compare to the ideal GP fit from earlier.

```
plot(x, y, xlim=c(0, 1.35), ylim=range(q1, q2, q12, q22))
lines(xx,m)
lines(xx, q1, lty=2)
lines(xx, q2, lty=2)
lines(xx, m2, col=4)
```

```
lines(xx, q12, col=4,lty=2)
lines(xx, q22, col=4,lty=2)
legend("topright", c("full", "hybrid"), lty=1, col=c(1,4), bty="n")
```

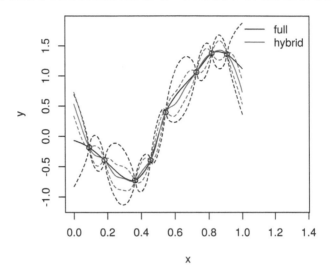

FIGURE 9.3: Comparing hybrid Legendre basis/CSK covariance to the ideal GP.

Although the two surfaces are not identical, they're indeed very similar. Combining a Legendre basis-expanded linear mean with CSK structure on residuals does an excellent job of mimicking an ideal GP fit, at least on this very simple example. Observe that the hybrid predictive surface interpolates. It bears repeating that this illustration doesn't provide full UQ. In particular, we used only the mean of the Legendre basis fit p[,1], ignoring uncertainty stored in p[,2:3]. So our fitted error-bars should actually be even wider. At the same time, we cheated by using true values of the correlation function in our ideal GP fit as plug-ins. So our benchmark is similarly too good to be true. Finally, we didn't estimate the CSK range parameter r_{max}; the only hyperparameter fit from data was scale $\hat{\tau}^2$ ...

```
c(ideal=tau2, CSKresid=tau22)
```

```
##     ideal CSKresid
## 0.73559  0.01439
```

... which is (reasonably) lower for the residual process, compared to the original.

9.1.3 Practical Bayesian inference and UQ

Kaufman et al. argue that the simplest, coherent way to put this hybrid together, and fully account for all relevant uncertainties in prediction while retaining a handle on trade-offs between computational complexity and accuracy (though CSK sparsity), is with Bayesian hierarchical modeling. They describe a prior linking together separable $r_{max,j}$ hyperparameters, for each input dimension j. That prior encourages coordinates to trade off against one another – competing in a manner not unlike in L1 penalization for linear regression using lasso[13] – to produce a covariance matrix with a desired degree of sparsity. That is, some directions yield zero correlations faster than others as a function of coordinate-wise distance. Specifically, Kaufman et al. recommend

$$r_{\max} \text{ uniform in } R_C = \left\{ r_{\max} \in \mathbb{R}^d : r_{\max,j} \geq 0, \sum_{j=1}^{d} r_{\max,j} \leq C \right\}. \tag{9.1}$$

Said another way, this penalty allows some $r_{\max,j}$ to be large to reflect a high degree of correlation in particular input directions, shifting the burden of sparsity to other coordinates. Parameter C determines the level of sparsity in the resulting MVN correlation matrix.

That prior on $r_{\max}$ is then paired with the usual reference priors for scale τ^2 and regression coefficients β through the Legendre basis. Conditional on $r_{\max}$, calculations similar to those required for an exercise in §5.5, analytically integrating out τ^2 and β, yield closed form expressions for the marginal posterior density. For details, see, e.g., Appendix A of Gramacy (2005). Inference for $r_{\max}$ and any other kernel hyperparameters may then be carried out with a conventional mixture of Metropolis and Gibbs sampling steps.

It remains to choose a C yielding enough sparsity for tractable calculation given data sizes present in the problem on hand. Kaufman et al. produced a map, which I shall not duplicate here, relating degree of sparsity to computation time for various data sizes, n. That map helps mitigate search efforts, modulo computing architecture nuances, toward identifying C yielding a specified degree of sparsity. Perhaps more practically, they further provide a numerical procedure which estimates the value of C required, after fixing all other choices for priors and their hyperparameters. I shall illustrate that pre-processing step momentarily.

As one final detail, Kaufman et al. recommend Legendre polynomials up to degree 5 in a "tensor product" form for their motivating cosmology example, including all main effects, and all two-variable interactions in which the sum of the maximum exponent in each interacting variable is constrained to be less than or equal to five. However, in their simpler coded benchmark example[14], which we shall borrow for our illustration below, it would appear they prefer degree 2.

Borehole example

Consider the borehole data[15], which is a classic synthetic computer simulation example (Morris et al., 1993), originally described by Worley (1987). It's a function of eight inputs, modeling water flow through a borehole.

$$y = \frac{2\pi T_u [H_u - H_l]}{\log\left(\frac{r}{r_w}\right) \left[1 + \frac{2LT_u}{\log(r/r_w)r_w^2 K_w} + \frac{T_u}{T_l}\right]}.$$

Input ranges are

$$r_w \in [0.05, 0.15] \quad r \in [100, 5000] \quad T_u \in [63070, 115600]$$
$$T_l \in [63.1, 116] \quad H_u \in [990, 1110] \quad H_l \in [700, 820]$$
$$L \in [1120, 1680] \quad K_w \in [9855, 12045].$$

---

[13]https://en.wikipedia.org/wiki/Lasso_(statistics)
[14]https://www.stat.berkeley.edu/~cgk/rcode/assets/SparseEmExample.R
[15]https://www.sfu.ca/~ssurjano/borehole.html

The function below provides an implementation in coded inputs.

```
borehole <- function(x)
 {
  rw <- x[1]*(0.15 - 0.05) + 0.05
  r <-  x[2]*(50000 - 100) + 100
  Tu <- x[3]*(115600 - 63070) + 63070
  Hu <- x[4]*(1110 - 990) + 990
  Tl <- x[5]*(116 - 63.1) + 63.1
  Hl <- x[6]*(820 - 700) + 700
  L <-  x[7]*(1680 - 1120) + 1120
  Kw <- x[8]*(12045 - 9855) + 9855
  m1 <- 2*pi*Tu*(Hu - Hl)
  m2 <- log(r/rw)
  m3 <- 1 + 2*L*Tu/(m2*rw^2*Kw) + Tu/Tl
  return(m1/m2/m3)
 }
```

Consider the following Latin hypercube sample (LHS; §4.1) training and testing partition *a la* Algorithm 4.1.

```
n <- 4000
nn <- 500
m <- 8
library(lhs)
x <- randomLHS(n + nn, m)
y <- apply(x, 1, borehole)
X <- x[1:n,]
Y <- y[1:n]
XX <- x[-(1:n),]
YY <- y[-(1:n)]
```

Observe that the problem here is bigger than any we've entertained so far in this text. However it's worth noting that $n = 4000$ is not too big for conventional GPs. Following from Kaufman's example, we shall provide a full GP below which (in being fully Bayesian and leveraging a Legendre basis-expanded mean) offers a commensurate look for the purpose of benchmarking computational demands. Yet Appendix A illustrates a thriftier GP implementation leveraging the MKL library which can handle more than $n = 10000$ training data points on this very same borehole problem. But that compares apples with oranges. Therefore we shall press on with the example, whose primary role – by the time the chapter is finished – will anyway be to serve as a straw man against more recent advances in the realm of local–global GP approximation.

The first step is to find the value of C that provides a desired level of sparsity. I chose 99% sparse through an argument `den` specifying density as the opposite of sparsity.

```
C <- find.tau(den=1 - 0.99, dim=ncol(x))*ncol(X)
C
```

```
## [1] 2.708
```

Next, R code below sets up a degree-two Legendre basis with all two-variable interactions, and then collects two thousand samples from the posterior. Compute time is saved for later comparison. Warnings that occasionally come from spam, when challenges arise in solving sparse linear systems, are suppressed in order to keep the document clean.

```
D <- I <- 2
B <- 2000
tic <- proc.time()[3]
suppressWarnings({
  samps99 <- mcmc.sparse(Y, X, mc=C, degree=D, maxint=I,
    B=B, verbose=FALSE)
})
time99 <- as.numeric(proc.time()[3] - tic)
```

Output samps99 is a B × d matrix storing $r_{\max,j}$ samples from the posterior distribution for each input coordinate, $j = 1, \ldots, d$. Trace plots for these samples are shown in Figure 9.4. The left panel presents each $r_{\max,j}$ marginally; on the right is their aggregate as in the definition of R_C (9.1). Observe how that aggregate bounces up against the estimated C-value of 2.71. Even with Legendre basis mopping up a degree of global nonlinearity, the posterior distribution over $r_{\max,j}$ wants to be as dense as possible in order to capture spatial correlations at larger distances. A low C-value is forcing coordinates to trade off against one another in order to induce the desired degree of sparsity.

```
par(mfrow=c(1,2))
matplot(samps99, type="l", xlab="iter")
plot(rowSums(samps99), type="l", xlab="iter", ylab="Rc")
```

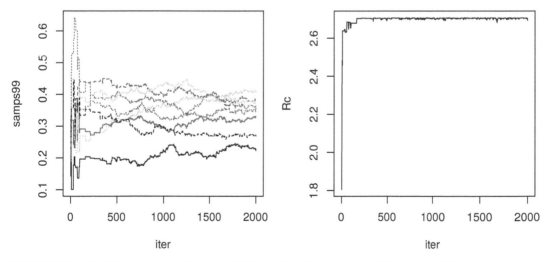

FIGURE 9.4: Trace plots of $r_{\max,j}$ (left) and their aggregate $\sum_j r_{\max,j}$ (right).

Both trace plots indicate convergence of the Markov chain after five hundred or so iterations followed by adequate – certainly not excellent – mixing. Marginal effective sample size calculations indicate that very few "equivalently independent" samples have been obtained from the posterior, however this could be improved with a longer chain.

```
library(coda)
burnin <- 500
apply(samps99[-(1:burnin),], 2, effectiveSize)
```

```
## [1]  1.925 1.932 2.013 2.762 2.142 4.139 3.686 1.632
```

Considering that it has already taken 41 minutes to gather these two thousand samples, it may not be worth spending more time to get a bigger collection for this simple example. (We'll need even more time to convert those samples into predictions.)

```
time99/60
```

```
## [1] 40.72
```

Instead, perhaps let me encourage the curious reader to explore longer chains offline. Pushing on, the next step is to convert those hyperparameters into posterior predictive samples on a testing set. Below, discard the first five hundred iterations as burn-in, then save subsamples of predictive evaluations from every tenth iteration thereafter.

```
index <- seq(burnin+1, B, by=10)
tic <- proc.time()[3]
suppressWarnings({
  p99 <- pred.sparse(samps99[index,], X, Y, XX, degree=D,
    maxint=I, verbose=FALSE)
})
time99 <- as.numeric(time99 + proc.time()[3] - tic)
time99/60
```

```
## [1] 49.09
```

The extra work required to make this conversion depends upon the density of the predictive grid, **XX**. In this particular case it doesn't add substantially to the total compute time, which now totals 49 minutes. The keen reader will notice that we didn't factor in time to calculate C with `find.tau` above. Relative to other calculations, this represents a rather small, fixed-cost pre-processing step.

Before assessing the quality of these predictions, consider a couple alternatives to compare to. Kaufman et al. provide a non-sparse version, that otherwise works identically, primarily for timing and accuracy comparisons. That way we'll be comparing apples to apples, at least in terms of inferential apparatus, when it comes to computation time. Working with dense covariance matrices in this context is really slow. Therefore, the code below collects an order of magnitude fewer MCMC samples from the posterior. (Sampling and predictive stages are combined.) Again, I encourage the curious reader to gather more for a fairer comparison.

```
tic <- proc.time()[3]
suppressWarnings({
  samps0 <- mcmc.nonsparse(Y, X, B=B/3, verbose=FALSE)
})
index <- seq(burnin/3 + 1, D/3, by=10)
suppressWarnings({
```

```
   p0 <- pred.nonsparse(samps0[index,], X, Y, XX, 2, verbose=FALSE)
})
time0 <- as.numeric(proc.time()[3] - tic)
```

As illustrated in Figure 9.5, 667 samples are not nearly sufficient to be confident about convergence. Admittedly, I'm not sure why the mixing here seems so much worse than in the CSK analog. It may have been that Kaufman didn't put as much effort into fine-tuning this straw man relative to their showcase methodology.

```
matplot(samps0, type="l", xlab="iter")
```

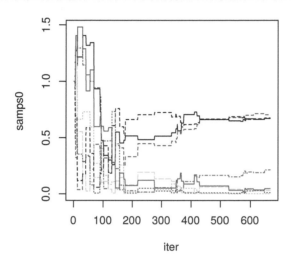

FIGURE 9.5: Trace plot of lengthscales under the full (non CSK) GP posterior; compare to the left panel of Figure 9.4.

Considering the amount of time it took to get as many samples as we did, we'll have to be content with extrapolating a bit to make a proper timing comparison.

```
time0/60
```

```
## [1] 51.12
```

Gathering the same number of samples as in the 99% sparse CSK case would've required more than 2 hours of total compute time. For a comparison on accuracy grounds, consider pointwise proper scores via Eq. (27) from Gneiting and Raftery (2007). Also see §8.1.4. Higher scores are better.

```
scorep <- function(YY, mu, s2) { mean(-(mu - YY)^2/s2 - log(s2)) }
scores <- c(sparse99=scorep(YY, p99$mean, p99$var),
  dense=scorep(YY, p0$mean, p0$var))
scores
```

```
## sparse99     dense
##   -1.918    -6.966
```

CSK is both faster and more accurate on this example. It's reasonable to speculate that the dense/ordinary GP results could be improved with better MCMC. Perhaps MCMC proposals, tuned to work well for short $r_{\max,j}$ on residuals obtained from a rich Legendre basis, work rather less well for ordinary lengthscales θ_j applied directly on the main process. In §9.3.4 we'll show how a full GP under MLE hyperparameters can be quite accurate on these data.

To add one final comparator into the mix, let's see how much faster and how much less accurate a 99.9% sparse version is. Essentially cutting and pasting from above ...

```
C <- find.tau(den=1 - 0.999, dim=ncol(x))*ncol(x)
tic <- proc.time()[3]
suppressWarnings({
  samps999 <- mcmc.sparse(Y, X, mc=C, degree=D, maxint=I,
    B=B, verbose=FALSE)
})
index <- seq(burnin+1, B, by=10)
suppressWarnings({
  p999 <- pred.sparse(samps999[index,], X, Y, XX, degree=D,
    maxint=I, verbose=FALSE)
})
time999 <- as.numeric(proc.time()[3] - tic)
```

In terms of computing time, the 99.9% sparse version is almost an order of magnitude faster.

```
times <- c(sparse99=time99, dense=time0, sparse999=time999)
times
```

```
##  sparse99     dense sparse999
##    2945.4    3066.9     351.7
```

In terms of accuracy, it's just a little bit worse than the 99% analog and much better than the slow and poorly mixing full GP.

```
scores <- c(scores, sparse999=scorep(YY, p999$mean, p999$var))
scores
```

```
##  sparse99     dense sparse999
##    -1.918    -6.966    -2.105
```

To summarize this segment on CSKs, consider the following notes. Sparse covariance matrices decompose faster compared to their dense analogs, but the gap in execution time is only impressive when matrices are very sparse. In that context, intervention is essential to mop up long-range structure left unattended by all those zeros. A solution entails hybridization between processes targeting long- and short-distance correlation. Kaufman et al. utilize a rich mean structure; Sang and Huang's FSA stays covariance-centric. Either way, both agree that Bayesian posterior sampling is essential to average over competing explanations. We have seen that the MCMC required can be cumbersome: long chains "eat up" computational savings offered by sparsity. Nevertheless, both camps offer dramatic success stories. For example, Kaufman fit a surrogate to more than twenty thousand runs of a photometric redshift simulation – a cosmology example – in four input dimensions, and predict with full

UQ at more then eighty thousand sites. Results are highly accurate, and computation time is reasonable.

One big downside to these ideas, at least in the context of our motivation for this chapter, is that neither approach addresses nonstationarity head on. Computational demands are eased, somewhat, but modeling fidelity has not been substantially increased. I qualify with "substantially" because the introduction of a basis-expanded mean does hold the potential to enhance even though it was conceived to compensate. Polynomial bases, when dramatically expanded both by location and degree, do offer a degree of nonstationary flexibility. Smoothing splines[16] are a perfect example; also see the splines supplement[17]. But such techniques break down computationally when input dimension is greater than two. Exponentially many more knots are required as input dimension grows. A more deliberate and nonparametric approach to obtaining local variation in a global landscape could represent an attractive alternative.

9.2 Partition models and regression trees

Another way to induce sparsity in the covariance structure is to *partition* the input space into independent regions, and fit separate surrogates therein. The resulting covariance matrix is block-diagonal after row–column reordering. In fact, you might say it's implicitly block-diagonal because it'd be foolish to actually build such a matrix. In fact, even "thinking" about the covariance structure on a global scale, after partitioning into multiple local models, can be a hindrance to efficient inference and effective implementation.

The trouble is, it's hard to know just how to split things up. Divide-and-conquer is almost always an effective strategy computationally. But dividing haphazardly can make conquering hard. When statistical modeling, it's often sensible to let the data decide. Once we've figured that out – i.e., how to let data say how it "wants" to be partitioned for independent modeling – many inferential and computational details naturally suggest themselves.

One happy consequence of partitioning, especially when splitting is spatial in nature, is a cheap nonstationary modeling mechanism. Independent latent processes and hyperparameterizations, thinking particularly about fitting GP surrogates to each partition element, kills two birds with one stone: 1) disparate spatial dynamics across the input space; 2) smaller matrices to decompose for faster local inference and prediction. The downside is that all bets for continuity are off. That "bug" could be a "feature", e.g., if the data generating mechanism is inherently discontinuous, which is not as uncommon as you might think. But more often a scheme for smoothing, or averaging over all (likely) partitions is desired.

Easy to say, hard to do. I know of only two successful attempts involving GPs on partition elements: 1) with Voronoi tessellations (Kim et al., 2005); 2) with trees (Gramacy and Lee, 2008a). In both cases, those references point to the original attempts. Other teams of authors have subsequently refined and extended these ideas, but the underlying themes remain the same. Software is a whole different ballgame; I know of only one package for R.

Tessellations are easy to characterize mathematically, but a nightmare computationally. Trees are easy mathematically too, and much friendlier in implementation. Although no walk

---

[16]https://en.wikipedia.org/wiki/Smoothing_spline
[17]http://bobby.gramacy.com/surrogates/splines.html

in the park, tree data structures are well developed from a software engineering perspective. Translating tree data structures from their home world of computer science textbooks over to statistical modeling is relatively straightforward, but as always the devil is in the details.

Figure 9.6 shows a partition tree $\mathcal{T}$ in two views. The left-hand drawing offers a graph view, illustrating a tree with two *internal*, or splitting nodes, and three *leaf* or terminal nodes without splits. Internal nodes are endowed with splitting criteria, which in this case refers to conditional splits in a two-dimensional x-space. Internal nodes have two *children*, called *siblings*. All nodes have a *parent* except the *root*, paradoxically situated at the top of the tree.

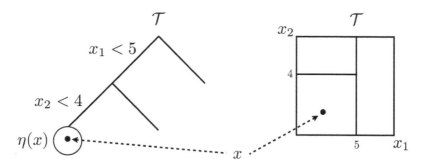

FIGURE 9.6: Tree graph (left) and partition of a 2d input space (right). Borrowed from Chipman et al. (2013) with many similar variations elsewhere; used with permission from Wiley.

The right-hand drawing illustrates the recursive nature of those splits geographically in the input space, creating an axis-aligned partition. A generic 2d input coordinate x would land in one of the three leaf nodes, depending on the setting of its two coordinates x_1 and x_2. Leaf node $\eta(x)$ resides in the lower-left partition of the input space.

For statistical modeling, the idea is that recursive, axis-aligned, splits represent a simple yet effective way to divvy up the input space $\mathcal{X}$ into independent predictive models for responses y. Predictions $\hat{y}(x)$ are dictated by tree structure, $\mathcal{X}$ and *leaf model*: historically, a simple prediction rule tailored to the subset of data residing at each terminal node. My plan is to showcase GPs at the leaves, but let's take a step back first.

9.2.1 Divide-and-conquer regression

Use of trees in regression dates back to *AID (automatic interaction detection)* by Morgan and Sonquist (1963). *Classification and regression trees* (CART; Breiman et al., 1984), a suite of methods obtaining fitted partition trees, popularized the idea. The selling point was that trees facilitate parsimonious divide-and-conquer, leading to flexible yet interpretable modeling.

Fitting partition structure (depth, splits, etc.) isn't easy, however. You need a leaf model/prediction rule, goodness-of-fit criteria, and a search algorithm. And there are lots of very good ways to make choices in that arena. In case it isn't yet obvious, I prefer likelihood whenever possible. Although other approaches are perhaps more common in the trees literature, with as many/possibly more contributions from computer science as statistics, likelihoods rule the roost as a default in modern statistics.

Given a particular tree, $\mathcal{T}$, the (marginal) likelihood factorizes into a product form.

$$p(y^n \mid \mathcal{T}, x^n) \equiv p(y_1, \ldots, y_n \mid \mathcal{T}, x_1, \ldots, x_n) = \prod_{\eta \in \mathcal{L}_{\mathcal{T}}} p(y^{\eta} \mid x^{\eta})$$

Above, shorthand $x^n = x_1, \ldots, x_n$ and $y^n = y_1, \ldots, y_n$ represents the full dataset of n pairs $(x, y)^n = \{(x_i, y_i) : i = 1, \ldots, n\}$. Analogously, superscript η represents those elements of the data which fall into leaf node η. That is, $(x, y)^{\eta} = \{(x_i, y_i) : x_i \in \eta, i = 1, \ldots, n\}$. The product arises from independent modeling, and by indexing over all leaf nodes in $\mathcal{T}$, $\eta \in \mathcal{L}_{\mathcal{T}}$, we cover all indices $i \in \{1, \ldots, n\}$. All that remains in order to complete the specification is to choose a model for $p(y^{\eta} \mid x^{\eta})$ to apply at leaf nodes η.

Usually such models are specified parametrically, via θ_{η}, but calculations are simplified substantially if those parameters can be integrated out. Hence the parenthetical "marginal" above. The simplest leaf model for regression is the constant model with unknown mean and variance $\theta_{\eta} = (\mu_{\eta}, \sigma_{\eta}^2)$:

$$p(y^{\eta} \mid \mu_{\eta}, \sigma_{\eta}^2, x^{\eta}) \propto \sigma_{\eta}^{-|\eta|} \exp\left\{-\frac{1}{2\sigma_{\eta}^2} \sum_{y \in \eta} (y - \mu_{\eta})^2\right\}$$

$$\text{so that} \quad p(y^{\eta} \mid x^{\eta}) = \frac{1}{(2\pi)^{\frac{|\eta|-1}{2}}} \frac{1}{\sqrt{|\eta|}} \left(\frac{s_{\eta}^2}{2}\right)^{-\frac{|\eta|-1}{2}} \Gamma\left(\frac{|\eta|-1}{2}\right)$$

upon taking reference prior $p(\mu_{\eta}, \sigma_{\eta}^2) \propto \sigma_{\eta}^{-2}$. Above, $|\eta|$ is a count of the number of data points in leaf node η, and $s_{\eta}^2 \equiv \hat{\sigma}_{\eta}^2$ is the typical residual sum of squares from $\bar{y}_{\eta} \equiv \hat{\mu}_{\eta}$. Concentrated analogs, i.e., not committing to a Bayesian approach, are similar. However the Bayesian view is natural from the perspective of coherent regularization.

Clearly some kind of penalty on complexity is needed for inference, otherwise marginal likelihood is maximized when there's one leaf for each observation. The original CART family of methods relied on minimum leaf-size and other heuristics, paired with a cross validation (CV) pruning stage commencing after greedily growing a deep tree. The fully Bayesian approach, which is more recent, has a more natural feel to it although at the expense of greater computation through MCMC. A silver lining, however, is that the Monte Carlo can smooth over hard breaks and lend a degree of continuity to an inherently "jumpy" predictive surface.

Completing the Bayesian specification requires a prior over trees, $p(\mathcal{T})$. There were two papers, published at almost the same time, proposing a so-called *Bayesian CART* model, or what is known as the *Bayesian treed constant model* in the regression (as opposed to classification) context. Denison et al. (1998) were looking for a light touch, and put a Poisson on the number of leaves, but otherwise specified a uniform prior over other aspects such as tree depth. Chipman et al. (1998) called for a more intricate class of priors which allowed heavier regularization to be placed on tree depth. Time says they won the argument, although the reasons for that are complicated. Almost everyone has since adopted the so-called *CGM prior*, although that's not evidence of much except popularity. (VHS beat out BetaMax in the videotape format war[18], but not because the former is better.) If the last twenty years have taught us nothing, we've at least learned that a hearty dose of regularization – even when not strictly essential – is often a good default.

CGM's prior is based on the following tree growing stochastic process. A tree $\mathcal{T}$ may *grow* from one of its leaf nodes η, which might be the root, with a probability that depends

---

[18]https://en.wikipedia.org/wiki/Videotape_format_war

on the depth D_η of that node in the tree. The family of probabilities preferred by CGM, dictating terms for when such a leaf node might split into two new children, is provided by the expression below.

$$p_{\text{split}}(\eta, \mathcal{T}) = \alpha(1 + D_\eta)^{-\beta}$$

One can use this probability to simulate the tree growing process, generating recursively starting from a null, single leaf/root tree, and stopping when all leaves refuse to draw a split. CGM studied this distribution under various choices of hyperparameters $0 < \alpha < 1$ and $\beta \geq 0$ in order to provide insight into characteristics of trees which are typical under this prior.

Of course, the primary aim here isn't to generate trees *a priori*, but to learn trees via partitions of the data and the subsequent patchwork of regressions they imply. For that, a density on $\mathcal{T}$ is required. It's simple to show that this prior process induces a prior density for tree $\mathcal{T}$ through the probability that internal nodes $\mathcal{I}_\mathcal{T}$ split and leaves $\mathcal{L}_\mathcal{T}$ do not:

$$p(\mathcal{T}) \propto \prod_{\eta \in \mathcal{I}_\mathcal{T}} p_{\text{split}}(\eta, \mathcal{T}) \prod_{\eta \in \mathcal{L}_\mathcal{T}} [1 - p_{\text{split}}(\eta, \mathcal{T})].$$

As in the *DMS prior* (Denison et al., 1998), CGM retains uniformity on everything else: splitting location/dimension, number of leaf node observations. Note that a minimum number of observations must be enforced in order to ensure proper posteriors at the leaves. That is, under the reference prior $p(\mu_\eta, \sigma_\eta^2) \propto \sigma_\eta^{-2}$, we must have at least $|\eta| \geq 2$ observations in each leaf node $\eta \in \mathcal{L}_\mathcal{T}$.

Inference

Posterior inference proceeds by MCMC. Note that there are no parameters except tree $\mathcal{T}$ when leaf-node θ_η are integrated out. Here is how a single iteration of MCMC would go. Randomly choose one of a limited number of stochastic tree modification operations (grow, prune, change, swap, rotate; more below), and conditional on that choice, randomly select a node $\eta \in \mathcal{T}$ on which that proposed modification would apply. Those two choices comprise proposal $q(\mathcal{T}, \mathcal{T}')$ for generating a new tree $\mathcal{T}'$ from $\mathcal{T}$, taking a step along a random walk in tree space. Accept the move with Metropolis–Hastings (MH) probability:

$$\frac{p(\mathcal{T}' \mid y^n, x^n)}{p(\mathcal{T} \mid y^n, x^n)} \times \frac{q(\mathcal{T}', \mathcal{T})}{q(\mathcal{T}, \mathcal{T}')} = \frac{p(y^n \mid \mathcal{T}', x^n)}{p(y^n \mid \mathcal{T}, x^n)} \times \frac{p(\mathcal{T}')}{p(\mathcal{T})} \times \frac{q(\mathcal{T}', \mathcal{T})}{q(\mathcal{T}, \mathcal{T}')}.$$

There's substantial scope for computational savings here with local moves $q(\mathcal{T}, \mathcal{T}')$ in tree space, since many terms in the big product over $\eta \in \mathcal{L}_\mathcal{T}$ in the denominator marginal likelihood, and over $\eta' \in \mathcal{L}_{\mathcal{T}'}$ in the numerator one, would cancel for unaltered leaves in $\mathcal{T} \to \mathcal{T}'$.

What do tree proposals q look like? Well, they can be whatever you like so long as they're reversible[19], which is required by the ergodic theorem for MCMC convergence, providing samples from the target distribution $p(\mathcal{T} \mid y^n, x^n)$. That basically means proposals must be matched with an opposite, undo proposal. Figure 9.7 provides an example of the four most popular tree moves, converting tree $\mathcal{T}$ from Figure 9.6 to $\mathcal{T}'$ shown in the same two views.

---

[19]https://en.wikipedia.org/wiki/Markov_chain#Reversible_Markov_chain

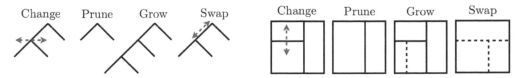

FIGURE 9.7: Random-walk proposals in tree space graphically (left four) and as partitions (right four). Borrowed from Chipman et al. (2013) with many similar variations elsewhere; used with permission from Wiley.

Observe that these moves have reversibility built in. *Grow* and *prune* are the undo of one another, and *change* and *swap* are the undo of themselves. Grow and prune are extremely local moves, acting only on leaf nodes and parents thereof, respectively. Swap and change are slightly more global as they may be performed on any internal node, or adjacent pair of nodes, respectively. Consequently, they may shuffle the contents of all their descendant leaves. Such "high up" proposals can have low MH acceptance rates because they tend to create $\mathcal{T}'$ far from $\mathcal{T}$.

Several new moves have been introduced to help. Gramacy and Lee (2008a) provide *rotate* which, like swap, acts on pairs of nodes which might reside anywhere in the tree. However, no matter how high up a rotation is, leaf nodes always remain unchanged. Thus acceptance is determined only by the prior. The idea comes from tree re-balancing in the computer science literature, for example as applied for red–black trees[20]. Wu et al. (2007) provide a *radical restructure* move targeting similar features, but with greater ambition. Pratola (2016) offers an alternative *rotate* targeting local moves which traverse disparate regions of partition space along contours of high, rather than identical likelihood.

To illustrate inference under the conventional move set, consider the motorcycle accident data in the `MASS` library for R. These data are derived from simulation of the acceleration of the helmet of a motorcycle rider before and after an impact. The pre- and post-whiplash effect, which we shall visualize momentarily, is extremely hard to model, whether using parametric (linear) models, or with GPs.

```
library(MASS)
library(tgp)
```

We shall utilize a *Bayesian CART* (BCART) implementation from the `tgp` package (Gramacy and Taddy, 2016) on CRAN. One quirk of `tgp` is that you must provide a predictive grid, **XX** below, at the time of fitting. Trees are complicated data structures, which makes saving samples from lots of MCMC iterations cumbersome. It's far easier to save predictions derived from those trees, obtained by dropping elements of $x \in \mathcal{X} \equiv$ **XX** down to leaf(s). In fact, it's sufficient to save the average means and quantiles accumulated over MCMC iterations. For each $x \in \mathcal{X}$, one can separately aggregate $\hat{\mu}_{\eta(x)}$ and quantiles $\hat{\mu}_{\eta(x)} + 1.96\hat{\sigma}_{\eta(x)}$ for all $\eta(x) \in \mathcal{L}_{\mathcal{T}}$, for every tree $\mathcal{T}$ visited by the Markov chain, normalizing at the end. That quick description is close to what `tgp` does by default.

```
XX <- seq(0, max(mcycle[,1]), length=1000)
out.bcart <- bcart(X=mcycle[,1], Z=mcycle[,2], XX=XX, R=100, verb=0)
```

---

[20]https://en.wikipedia.org/wiki/Red-black_tree

Another peculiarity here is the argument R=100, which calls for one hundred restarts of the MCMC. CGM observed that mixing in tree space can be poor, resulting in chains becoming stuck in local posterior maxima. Multiple restarts can help alleviate this. Whereas posterior mean predictive surfaces require accumulating predictive draws over MCMC iterations, the most probable predictions can be extracted after the fact. Code below utilizes predict.tgp to extract the predictive surface from the most probable tree, i.e., the maximum *a posteriori* (MAP) tree $\hat{\mathcal{T}}$.

```
outp.bcart <- predict(out.bcart, XX=XX)
```

It's worth reiterating that this way of working is different from the typical fit-then-predict scheme in R. The main prediction vehicle in tgp is driven by providing XX to bcart and similar methods. Still both surfaces, posterior mean and MAP, offer instructive visualizations. To that end, the R code below establishes a macro that I shall reuse, in several variations, to visualize predictive output from fitted tree models.

```
plot.moto <- function(out, outp)
  {
  plot(outp$XX[,1], outp$ZZ.km, ylab="accel", xlab="time",
    ylim=c(-150, 80), lty=2, col=1, type="l")
  points(mcycle)
  lines(outp$XX[,1], outp$ZZ.km + 1.96*sqrt(outp$ZZ.ks2), col=2, lty=2)
  lines(outp$XX[,1], outp$ZZ.km - 1.96*sqrt(outp$ZZ.ks2), col=2, lty=2)
  lines(out$XX[,1], out$ZZ.mean, col=1, lwd=2)
  lines(out$XX[,1], out$ZZ.q1, col=2, lwd=2)
  lines(out$XX[,1], out$ZZ.q2, col=2, lwd=2)
  }
```

Observe in the macro that solid bold (lwd=2) lines are used to indicate posterior mean predictive; thinner dashed lines (lty=2) indicate the MAP. On both, black (col=1) shows the center (mean of means or MAP mean) and red (col=2) shows 95% quantiles. Figure 9.8 uses this macro for the first time in the context of our Bayesian CART fit, also for the first time showing the training data.

```
plot.moto(out.bcart, outp.bcart)
```

What can be seen in these surfaces? Organic nonstationarity and heteroskedasticity, that's what. The rate of change of outputs is changing as a function of inputs, and so is the noise level. Training data exhibit these features, and predictive surfaces are coping well, albeit not gracefully. Variances, exhibited by quantiles, may be too high (wide) at the end. The whiplash effect in the middle of the data appears overly dampened by forecasts on the testing grid.

The MAP surface (dashed lines) in the figure exemplifies an "old CART way" of regression. Hard breaks abound, being both unsightly and a poor surrogate for what are likely smooth physical dynamics. Posterior mean predictive summaries (solid lines) are somewhat more smooth. Averaging over the posterior for $\mathcal{T}$ with MCMC smooths over abrupt transitions that come in disparate form with each individual sample from the chain. Yet the surface, even after aggregation, is still blocky: like a meandering staircase with rounded edges. A

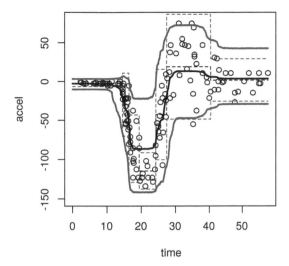

FIGURE 9.8: Bayesian treed constant model fit to the motorcycle accident data in terms of means and 95% quantiles. Posterior means are indicated by solid lines; MAP dashed.

longer MCMC chain, or more restarts, could smooth things out more, but with diminishing return.

Other leaf models

One of the cool things about this setup is that any data type/leaf model may be used without extra computational effort if $p(y^\eta \mid x^\eta)$ is analytic; that is, as long as we can evaluate the marginal likelihood, integrating out parameters θ_η in closed form. Fully conjugate, scale-invariant, default (non-informative) priors on θ_η make this possible for a wide class of models for response y, even conditional on x. A so-called *Bayesian treed linear model* (BTLM; Chipman et al., 2002) uses

$$p(y^\eta \mid \beta_\eta, \sigma_\eta^2, x^\eta) \propto \sigma_\eta^{-|\eta|} \exp\{(y^\eta - X^\eta \beta_\eta)^2/2\sigma_\eta^2\} \quad \text{and} \quad p(\beta_\eta, \sigma_\eta^2) \propto \sigma_\eta^{-2}.$$

In that case we have

$$p(y^\eta \mid x^\eta) = \frac{1}{(2\pi)^{\frac{|\eta|-d-1}{2}}} \left(\frac{|\mathcal{G}_\eta^{-1}|}{|\eta|}\right)^{\frac{1}{2}} \left(\frac{s_\eta^2 - \mathcal{R}_\eta}{2}\right)^{-\frac{|\eta|-m-1}{2}} \Gamma\left(\frac{|\eta|-d-1}{2}\right),$$

where $\mathcal{G}_\eta = \bar{X}_\eta^\top \bar{X}_\eta$, $\mathcal{R}_\eta = \hat{\beta}_\eta^\top \mathcal{G}_\eta \hat{\beta}_\eta$ and intercept-adjusted $(m+1)$-column $\bar{X}_\eta$ is a centered X_η.

Without getting too bogged down in details, how about a showcase of BTLM in action through it's `tgp` implementation? Again, we must specify **XX** during the fitting stage for full posterior averaging in prediction.

```
out.btlm <- btlm(X=mcycle[,1], Z=mcycle[,2], XX=XX, R=100, verb=0)
outp.btlm <- predict(out.btlm, XX=XX)
```

As before, a MAP predictor may be extracted after the fact. Figure 9.9 reuses the plotting macro in order to view the result, and qualitatively compare to the earlier BCART fit.

```
plot.moto(out.btlm, outp.btlm)
```

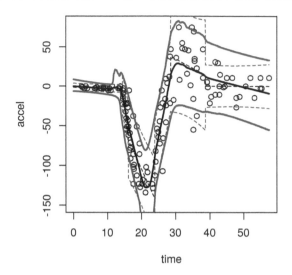

FIGURE 9.9: Bayesian treed linear model fit to the motorcycle accident data; compare to Figure 9.8.

The MAP surface (dashed) indicates fewer partitions compared to the BCART analog, but the full posterior average (solid) implies greater diversity in trees and linear-leaves over MCMC iterations. In particular, there's substantial posterior uncertainty in timing of the impact, when acceleration transitions from level at zero into whiplash around `time=14`. For the right third of inputs there's disagreement about both slope and noise level. Both BTLM surfaces, MAP and posterior mean, dampen the whiplash to a lesser extent compared to BCART. Which surface is better likely depends upon intended use.

If responses y are categorical, then a multinomial leaf model and Dirichlet prior pair leads to an analytic marginal likelihood (Chipman et al., 1998). Other members of the exponential family proceed similarly: Poisson, exponential, negative binomial... Yet to my knowledge none of these choices – besides multinomial – have actually been implemented in software as leaf models in a Bayesian setting.

Technically, any leaf model can be deployed by extending the MCMC to integrate over leaf parameters θ_η too; in other words, replace analytic integration to calculate marginal likelihoods, in closed form, with a numerical alternative. Since the dimension of the parameter space is changing when trees grow or prune, reversible jump MCMC (Richardson and Green, 1997) is required. Beyond that technical detail, a more practical issue is that deep trees/many leaves can result in a prohibitively large parameter space. An important exception is GPs. GPs offer a parsimonious take on nonlinear nonparametric regression, mopping up much of the variability left to the tree with simpler leaf models. GP leaves encourage shallow trees with fewer leaf nodes. At the same time, treed partitioning enables (axis aligned) regime changes in mean stationarity and skedasticity.

Before getting into further detail, let's look at a stationary GP fit to the motorcycle data. The `tgp` package provides a Bayesian GP fitting method that works similarly to `bcart` and `btlm`. Since GP MCMC mixes well, fewer restarts need be entertained. (Even then default of `R=1` works well.)

```
out.bgp <- bgp(X=mcycle[,1], Z=mcycle[,2], XX=XX, R=10, verb=0)
outp.bgp <- predict(out.bgp, XX=XX)
```

Although the code above executes an order of magnitude fewer MCMC iterations, runtimes (not quoted here) are much slower for BGP due to the requisite matrix decompositions. Figure 9.10, again using our macro, shows predictive surfaces which result.

```
plot.moto(out.bgp, outp.bgp)
```

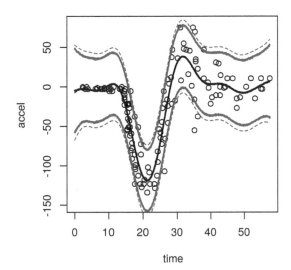

FIGURE 9.10: Bayesian GP fit to the motorcycle data; compare to Figures 9.8–9.9.

Both are nice and smooth, but lead to a less than ideal fit especially as regards variance. In fact you might say that GP and tree-based predictors are complementary. Where one is good the other is bad. Can they work together in harmony?

9.2.2 Treed Gaussian process

Bayesian *treed Gaussian process (TGP)* models (Gramacy and Lee, 2008a) can offer the best of both worlds, marrying the smooth global perspective of an infinite basis expansion, via GPs, with the thrifty local adaptivity of trees. Their divide-and-conquer nature means faster computation from smaller matrix decompositions, and nonstationary and heteroskedasticity effects as conditionally independent leaves allow for disparate spatial dependencies. Perversely, the two go hand in hand. The more the training data exhibit nonstationary/heteroskedastic features, the more treed partitioning and the faster it goes!

There are too many modeling and implementation details to introduce here. References shall be provided – in addition to the original methodology paper cited above – in due course. For now the goal is to illustrate potential and then move on to more ambitious enterprises with TGP. The program is the same as above, using `tgp` from CRAN, but with `btgp` instead.

Bayesian posterior sampling is extended to cover GP hyperparameters (lengthscales and nuggets) at the leaves.

```
out.btgp <- btgp(X=mcycle[,1], Z=mcycle[,2], XX=XX, R=30,
  bprior="b0", verb=0)
```

Previous calls to `tgp`'s suite of `b*` functions specified `verb=0` to suppress MCMC progress output printed to the screen by default. The call above is no exception. That output was suppressed because it was either excessive (`bcart` and `btlm`) or boring (`bgp`). Situated in-between on the modeling landscape, `btgp` progress statements are rather more informative, and less excessive, providing information about accepted tree moves and giving an online indication of trade-offs navigated between smooth and abrupt dynamics. I recommend trying `verb=1`.

Argument `bprior="b0"`, above, is optional. By default, `tgp` fits a linear mean GP at the leaves, unless `meanfn="constant"` is given. Specifying `bprior="b0"` creates a hierarchical prior linking β_η and σ_η^2, for all $\eta \in \mathcal{L}_{\mathcal{T}}$, together. That makes sense for the motorcycle data because it starts and ends flat. Under the default setting of `bprior="bflat"`, β_η and σ_η^2 parameters of the linear mean are unrestricted. Results are not much different in that case.

As before, the MAP predictor may be extracted for comparison.

```
outp.btgp <- predict(out.btgp, XX=XX)
```

Figure 9.10, generated with our macro, provides a summary of both mean and MAP predictive surfaces.

```
plot.moto(out.btgp, outp.btgp)
```

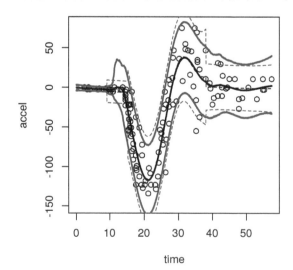

FIGURE 9.11: Bayesian treed GP fit to the motorcycle data; compare to Figures 9.8–9.10.

It's hard to imagine a better compromise. Both surfaces offer excellent fits on their own, but the posterior mean clearly enjoys greater smoothness, which is warranted by the physics under study. Both surfaces, but particularly the posterior mean, reflect uncertainty in the location of the transition between zero-acceleration and whiplash dynamics for time $\in (10, 15)$. The posterior over trees supports many transition points in that region more-or-less equally.

Often having a GP at all leaves is overkill, and this is the case with the motorcycle accident data. The response is flat for the first third of inputs, and potentially flat in the last third too. Sometimes spatial correlation is only expressed in some input coordinates; linear may be sufficient in others. Gramacy and Lee (2008b) explain how a *limiting linear model (LLM)* can allow the data to determine the flexibility of the leaf model, offering a more parsimonious fit and speed enhancements when training data determine that a linear model is sufficient to explain local dynamics.

For now, consider how LLMs work in the simple 1d case offered by mcycle.

```
out.btgpllm <- btgpllm(X=mcycle[,1], Z=mcycle[,2], XX=XX, R=30,
  bprior="b0", verb=0)
outp.btgpllm <- predict(out.btgpllm, XX=XX)
```

Figure 9.12, showcasing btgpllm, offers a subtle contrast to the btgp fit shown in Figure 9.11.

```
plot.moto(out.btgpllm, outp.btgpllm)
```

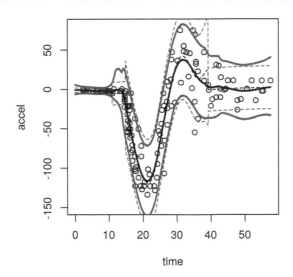

FIGURE 9.12: Bayesian treed GP with jumps to the limiting linear model (LLM) on the motorcycle data; compare to Figure 9.11.

Observe how the latter third of inputs enjoys a slightly tighter predictive interval in this setting, borrowing strength from the obviously linear (actually completely flat) fit to the first third of inputs. Transition uncertainty from zero-to-whiplash is also somewhat diminished.

For a two-dimensional example, revisit the exponential data first introduced in §5.1.2. In fact, that data was created to showcase subtle nonstationarity with TGP. A data-generating shorthand is included in the tgp package.

```
exp2d.data <- exp2d.rand(n1=30, n2=70)
X <- exp2d.data$X
Z <- exp2d.data$Z
XX <- exp2d.data$XX
```

The `exp2d.rand` function works with a grid in the input space and allows users to specify how many training data points should come from the interesting, lower-left quadrant of the input space versus the other three flat quadrants. The call targets slightly higher sampling in the interesting region, taking remaining grid elements as testing locations. Consider an ordinary (Bayesian) GP fit to these data as a warm up. To illustrate some of the alternatives offered by `tgp`'s GP capability, the call below asks for isotropic Gaussian correlation with `corr="exp"`.

```
out.bgp <- bgp(X=X, Z=Z, XX=XX, corr="exp", verb=0)
```

The `tgp` package provides a somewhat elaborate suite of `plot` methods defined for `"tgp"`-class objects. Figure 9.13 utilizes a paired `image` layout for mean and variance (actually 90% quantile gap) surfaces.

```
plot(out.bgp, pc="c")
```

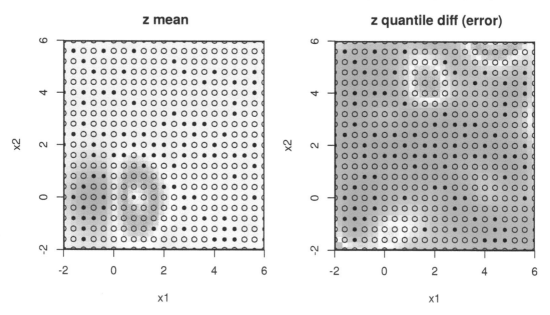

FIGURE 9.13: Bayesian GP fit to the 2d exponential data (§5.1.2) via mean (left) and uncertainty (right; difference between 95% and 5% quantiles).

Occasionally the predictive mean surface (left panel) is exceptionally poor, depending on the random design and response generated by `exp2d.data`. The predictive variance (right) almost always disappoints. That's because the GP is stationary which implies, among other things, uniform uncertainty in distance. Consequently, the uncertainty surface is unable to reveal what is intuitively obvious from the pictures: that the interesting quadrant is harder to predict than the other flat ones. The uncertainty surface is sausage-shaped: higher where training data is scarce. To learn otherwise requires building in a degree of nonstationary flexibility, which is what the tree in TGP facilitates. Consider the analogous btgp fit, with modest restarting to avoid the Markov chain becoming stuck in local posterior modes. With GPs at the leaves, rather than constant or linear models, trees are less deep so tree movement is more fluid.

```
out.btgp <- btgp(X=X, Z=Z, XX=XX, corr="exp", R=10, verb=0)
```

Analogous plots of posterior predictive mean and uncertainty reveal a partition structure that quarantines the interesting region away from the rest, and learns that uncertainty is indeed higher in the lower-left quadrant in spite of denser sampling there. Dashed lines in Figure 9.14 correspond to the MAP treed partition found during posterior sampling.

```
plot(out.btgp, pc="c")
```

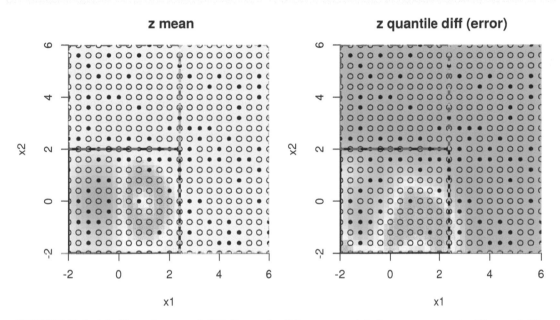

FIGURE 9.14: Bayesian treed GP fit to the 2d exponential data; compare to Figure 9.13.

Recursive axis-aligned partitioning is both a blessing and curse in this example. Notice that one of the flat quadrants is needlessly partitioned away from the other two. But the view in the figure only depicts one, highly probable tree. Posterior sampling averages over many other trees. In fact, a single accepted swap move would result in the diametrically opposed quadrant being isolated instead. This averaging over disparate, highly probable partitions explains why variance is about the same in these two regions. Unfortunately, there's no support for viewing all of these trees at once, which would anyways be a mess.

Increased uncertainty for the lower-left quadrant in the right panel of Figure 9.14 is primarily due to the shorter lengthscale and higher nugget estimated for data in that region, as supported by many of the trees sampled from the posterior, particularly the MAP. The diagram in Figure 9.15 provides another visual of the MAP tree, relaying a count of the number of observations in each leaf and estimated marginal variance therein.

```
tgp.trees(out.btgp, heights="map")
```

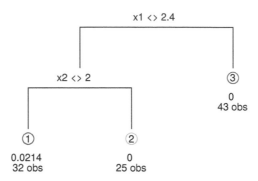

FIGURE 9.15: MAP tree under Bayesian treed GP. Leaf information includes an estimate of local scale (equivalent to $\hat{\tau}^2$) and number of observations.

Again, the tree indicates that only the lower-left quadrant has substantial uncertainty. Having a regional notion of model inadequacy is essential to sequential design efforts which utilize variance-based acquisition. Examples are ALM/C, IMSPE, etc., from Chapter 6. All those heuristics are ultimately space-filling unless the model accommodates nonstationary flexibility. A homework exercise in §9.4 targets exploration of these ideas on the motivating NASA rocket booster data (§2.1). In an earlier §6.4 exercise we manually partitioned these data to affect sequential design decisions and direct acquisition towards more challenging-to-model regimes. TGP can take the human out of that loop, automating iteration between flexible learning and adaptive design.

Revisiting LGBB (rocket booster) data

TGP was invented for the rocket booster data. NASA scientists knew they needed to partition modeling, and accompanying design, to separate subsonic and supersonic speeds. They had an idea about how the partition might go, but thought it might be better if the data helped out. Below we shall explore that potential with data collected from a carefully implemented, sequentially designed, computer experiment conducted on NASA's Columbia supercomputer[21].

```
lgbb.as <- read.table("lgbb/lgbb_as.txt", header=TRUE)
lgbb.rest <- read.table("lgbb/lgbb_as_rest.txt", header=TRUE)
```

Those files contain training input–output pairs obtained with ALC-based sequential design (§6.2.2) selected from a dense candidate grid (Gramacy and Lee, 2009). Un-selected elements from that grid form a testing set on which predictions are desired. Here our illustration centers on depicting the final predictive surface, culminating after a sequential design effort on the lift output, one of six responses. The curious reader may wish to repeat this analysis and subsequent visuals with one of the other five output columns. Code below sets up the data we shall use for training and testing.

---

[21]https://en.wikipedia.org/wiki/Columbia_(supercomputer)

```
X <- lgbb.as[,2:4]
Y <- lgbb.as$lift
XX <- lgbb.rest[2:4]
c(X=nrow(X), XX=nrow(XX))
```

```
##     X    XX
##   780 37128
```

The training set is modestly sized and the testing set is big. Fitting isn't speedy, so we won't do any restarts (using default R=1). Even better results can be obtained with larger R and with more MCMC iterations, which is controlled with the BTE argument ("B"urn-in, "T"otal and thinning level to save "E"very sample). Defaults used here target fast execution, not necessarily ideal inferential or predictive performance. CRAN requires all coded examples in documentation files finish in five seconds. This larger example has no hope of achieving that speed, however results with the defaults are acceptable as we shall see.

```
t1 <- system.time(fit <- btgpllm(X=X, Z=Y, XX=XX, bprior="b0", verb=0))[3]
t1/60
```

```
## elapsed
##   59.37
```

A fitting time of 59 minutes is quite a wait, but not outrageous. Compared to CSK timings from earlier, these btgpllm calculations are slower even though the training data entertained here is almost an order of magnitude smaller. The reason is that our "effective" covariance matrices from treed partitioning aren't nearly as sparse. We entertained CSKs at 99% and 99.9% sparsity but our btgpllm MAP tree yields effective sparsity closer to 30%, as we illustrate below. Also, keep in mind that tgp bundles fitting and prediction, and our predictive set is huge. CSK examples entertained just five-hundred predictive locations.

Figure 9.16 provides a 2d slice of the posterior predictive surface where the third input, side-slip angle (beta), is fixed to zero. The plot.tgp method provides several hooks that assist in 2d visualization of higher dimensional fitted surfaces through slices and projections. More details can be found in package documentation.

```
plot(fit, slice=list(x=3, z=0), gridlen=c(100, 100),
  layout="surf", span=0.01)
```

Observe that the predictive mean surface is able to capture the ridge nearby low speeds (mach) and for high angles of attack (alpha), yet at the same time furnish a more slowly varying surface at higher speeds. That would not be possible under a stationary GP: one of the two regimes (or both) must compromise, and the result would be an inferior fit.

The MAP tree, visualized in Figure 9.17, indicates a two-element partition.

```
tgp.trees(fit, heights="map")
```

The number of data points in each leaf implies an effective (global) covariance matrix that's about 30% sparse, with the precise number depending on the random seed used to generate this Rmarkdown build. Code for a more precise calculation – for a more interesting case with

z mean, with (beta) fixed to (0)

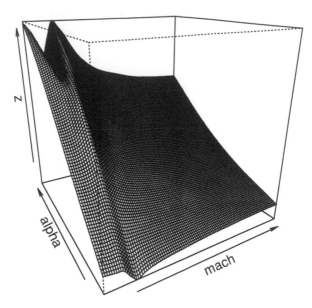

FIGURE 9.16: Slice of the Bayesian TGP with LLM predictive surface for the LGBB lift response.

height=2, log(p)=3309.44

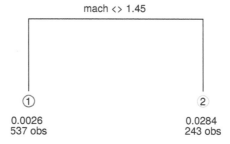

FIGURE 9.17: MAP tree for the lift response.

more leaves – is provided momentarily. Note this applies only for the MAP tree; the other thousands of trees visited by the Markov chain would likely be similar but seldom identical.

By default, bt* fitting functions begin with a null tree/single leaf containing all of the data, implying a 100% dense 780 × 780 covariance matrix. So the first several hundred iterations of MCMC, before the first grow move is accepted, may be particularly slow. Even after successfully accepting a grow, subsequent prunes entertain a full 780 × 780, covariance matrix when evaluating MH acceptance ratios. Thus the method is still in $\mathcal{O}(N^3)$, pointing to very little improvement on computation, at least in terms of computational order. A more favorable assessment would be that we get enhanced fidelity at no extra cost compared to a (dense covariance) stationary GP. More aggressive use of the tree is required when speed is a priority. But let's finish this example first.

A slightly tweaked plot.tgp call can provide the predictive variance surface. See Figure 9.18. As in our visuals for the 2d exponential data (e.g., Figure 9.14), training inputs (dots) and testing locations (open circles) are shown automatically. A dense testing grid causes lots of open circles to be drawn, which unfortunately darkens the predictive surface. Adding `pXX=FALSE` makes for a prettier picture, but leaves the testing grid to the imagination. (Here, that grid is pretty easy to imagine in the negative space.) The main title says "z ALM stats", which should be interpreted as "variance of the response(s)". Recall that ALM sequential design from §6.2.1 involves a maximizing variance heuristic.

```
plot(fit, slice=list(x=3,z=0), gridlen=c(100,100), layout="as", as="alm",
  span=0.01, pXX=FALSE)
```

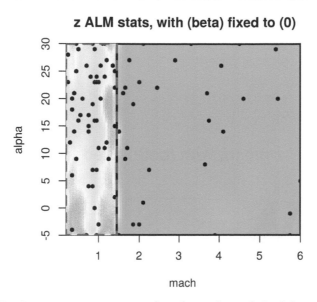

FIGURE 9.18: Predictive uncertainty surface for a slice of the lift response with MAP partition and training data (in slice) overlayed.

Several notable observations can be drawn from this surface. First, the partition doesn't split the input space equally in a geographic sense. However it does partition the 780 inputs somewhat more equally. This is because the design is non-uniform, emphasizing low-speed inputs. Sequential design was based on ALC, not ALM, but since both focus on variance they would recommend similar acquisitions. Observe that predictive uncertainty is much higher in the low-speed regime, so future acquisitions would likely demand even heavier sampling in that region. In the next iteration of an ALC/M scheme, one might select a new run from the lower-left (low `mach`, low `alpha`) region to add into the training data. Finally, notice how high uncertainty bleeds across the MAP partition boundary – a relic of uncertainty in the posterior for $\mathcal{T}$.

The `tgp` package provides a number of "knobs" to help speed things up at the expense of faithful modeling. One way is through the prior. For example, p_{split} arguments α (bigger) and β (smaller) can encourage deeper trees and consequently smaller matrices and faster execution.

Another way is through MCMC initialization. Providing `linburn=TRUE` will burn-in treed GP MCMC with a treed linear model, and then switch-on GPs at the leaves before collecting samples. That facilitates two economies. For starters it shortcuts an expensive full GP burn-in while waiting for accepted grow moves to organically partition up the input space into smaller GPs. More importantly, it causes the tree to over-grow, ensuring smaller partitions, initializing the chain in a local mode of tree space that's hard to escape out of even after GPs are turned on. Once that happens some pruning is typical, but almost never entirely back to where the chain should be under the target distribution. You might say that `linburn=TRUE` takes advantage of poor tree mixing, originally observed by CGM, to favor speed.

Consider a `btgpllm` call identical to the one we did above, except with `linburn=TRUE`.

```
t2 <- system.time(
  fit2 <- btgpllm(X=X, Z=Y, XX=XX, bprior="b0", linburn=TRUE, verb=0))[3]
```

As you can see in Figure 9.19, our new visual of the `beta=0` slice is not much different than Figure 9.16's ideal fit. Much of the space is plausibly piecewise linear anyway, but it helps to smooth out rough edges with the GP.

```
plot(fit2, slice=list(x=3, z=0), gridlen=c(100, 100),
  layout="surf", span=0.01)
```

z mean, with (beta) fixed to (0)

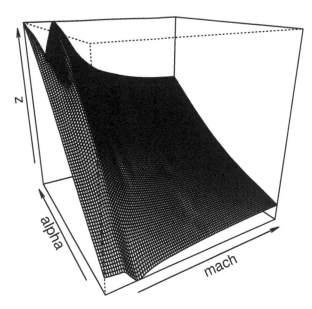

FIGURE 9.19: Approximate LGBB lift posterior mean slice under linear model burn-in; compare to Figure 9.16.

It's difficult to speculate on the exact nature of the aesthetic difference between this new `linburn=TRUE` surface and the earlier ideal one. Sometimes the thriftier surface over-smooths. Sometimes it under-smooths revealing kinks or wrinkles. Sometimes it looks pretty much the same. What is consistent, however, is that nothing looks entirely out of place while execution time is about 22 times faster than otherwise, taking about 3 minutes in this build.

```
c(full=t1, linburn=t2)
```

```
##     full.elapsed linburn.elapsed
##           3562.4           162.6
```

To explain the speedup, code below calculates the effective degree sparsity realized by the MAP tree.

```
map.height <-fit2$post$height[which.max(fit2$posts$lpost)]
leafs.n <- fit2$trees[[map.height]]$n
1 - sum(leafs.n^2)/(sum(leafs.n)^2)
```

```
## [1] 0.9139
```

Approximately 91% sparse – much closer to the 99% of CSK. Maybe its predictive surface isn't as smooth as it could be, or as would be ideal, say for a final visualization in a published article. For most other purposes, however, `linburn` fits possess all of the requisite ingredients.

One such purpose is sequential design, where turning around acquisitions quickly can be essential to applicability. We already talked about ALM: an estimate of predictive variance at **XX** locations comes for free. Usually, although not in all Rmarkdown builds, the ALM/variance surface plotted in Figure 9.20 demands future acquisitions in the interesting part of the space, for low speeds and high angles of attack. Even when that's not the case, the maximizing location is usually sensible. An important thing to keep in mind about sequential design, especially when individual decisions seem erratic, is that the long run of many acquisitions is what really counts.

```
plot(fit2, slice=list(x=3,z=0), gridlen=c(100,100), layout="as", as="alm",
  span=0.01, pXX=FALSE)
```

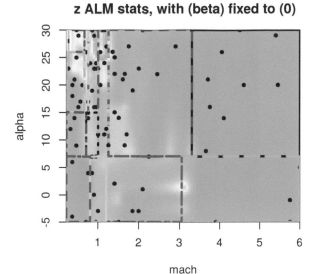

FIGURE 9.20: Approximate LGBB lift posterior mean slice under linear model burn-in; compare to Figure 9.16.

Having to specify XX locations in advance rules out continuous optimization of acquisition criteria, say with optim. Gramacy and Lee (2009) describe a sequential *treed maximum entropy (ME)* candidate-based scheme for coping with this drawback. The essence is to seek space-filling candidates separately in each region of the MAP partition. ALC (§6.2.2) is available via Ds2x=TRUE, signaling calculations of integrated change in variance $\Delta\sigma^2(x)$ for all $x \in$ XX. Testing XX locations are also used as reference sites in a sum approximating the ALC integral (6.7). This makes the computational expense quadratic in nrow(XX), which can be a tall order for testing sets sized in the tens of thousands ($780^3 \ll 37128^2$). Treed ME XX thus prove valuable as reference locations as well. Finally, additional computational savings may be realized by undoing some of tgp's defaults, such as automatic predictive sampling at training X locations. Those aren't really necessary for acquisition and can be skipped by providing pred.n=FALSE.

Often simple leaf models, e.g., constant or linear, lead to great sequential designs. Variance estimates can be quite good, on relative terms, even if predictive means are unfaithful to dynamics exhibited by the response surface. Consequently ALM/C heuristics work quite well. You don't need smooth prediction to find out where model uncertainty (predictive variance) is high. Consider the treed linear model ...

```
t3 <- system.time(
  fit3 <- btlm(X=X, Z=Y, XX=XX, BTE=c(2000, 7000, 10), R=10, verb=0))[3]
```

... and resulting ALM surface provided in Figure 9.21. Treed partitioning can be quite heavy, so option pparts=FALSE suppresses those colorful rectangles to provide a clearer view of spatial variance.

```
plot(fit3, slice=list(x=3,z=0), gridlen=c(100,100), layout="as", as="alm",
  span=0.01, pparts=FALSE, pXX=FALSE)
```

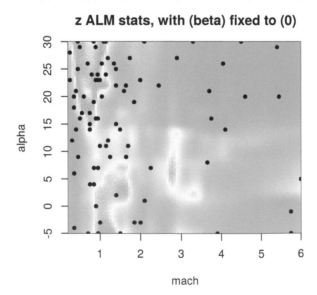

FIGURE 9.21: Even more approximate predictive uncertainty surface under a Bayesian treed LM for a slice of the lift response; compare to Figures 9.18 and 9.20.

That surface indicates a sufficiently localized sense of uncertainty despite having, what would most certainly be, a highly discontinuous (and thus potentially quite inaccurate) predictive mean analog. Despite a factor of R=10 more MCMC sampling, that fit takes a fraction of the time: 1 minute(s), which is 52 times faster than a full TGP.

```
c(full=t1, linburn=t2, btlm=t3)
```

```
##    full.elapsed linburn.elapsed    btlm.elapsed
##        3562.44          162.57           68.15
```

Sometimes treed LMs provide even better acquisitions than treed GPs do. In my experience, this only happens in the latter stages of sequential design. Early on, GP smoothness – at least in part – is key to sensible acquisition. Linear models only have high variance at the edges, i.e., at partition boundaries, which can cause self-reinforcing acquisitions and trigger a vicious cycle. Boundary-targeting sequential design puts a heavy burden on tree mixing in the MCMC, which CGM remind is problematic for linear and constant leaves. If speed is a concern, my preference is for a treed GP with `linburn=TRUE` shortcuts over wholesale treed linear or constant models. Treed LM burn-in offers a nice compromise: fast predictions and design estimates; smoothed out rough edges, reducing spurious variances due to "over-quilted" input spaces arising from deep trees compensating for crude linear fits at the leaves.

9.2.3   Regression tree extensions, off-shoots and fix-ups

The `tgp` package also supports expected improvement (EI) for Bayesian optimization (BO; §7.2). Providing `improv=TRUE` to any `b*` function causes samples to be gathered from the posterior mean of improvements, converted from raw improvement $I(x)$ using $Y(x)$ sampled from the posterior predictive distribution. So you get a fully Bayesian EI, averaged over tree and leaf model uncertainty, for a truly Bayesian BO. (Before the "B" in BO meant marginalizing over a GP latent field; see §5.3.2. Here it means averaging over all uncertainties through posterior integration over the full set of unknown parameters, including trees.) A downside is that candidates, as with other Bayesian TGP sampling of active learning heuristics (ALC/M), must be specified in advance of sampling, thwarting acquisition by local optimization. The upside is that such sampling naturally extends to powered up improvements, encouraging exploration, and ranking by improvement for batch sequential optimization (Taddy et al., 2009).

For more details on `tgp`, tutorials and examples, see

- Gramacy (2007): a beginner's primer, including instruction on custom compilation for fast linear algebra and threaded prediction for large XX, e.g., for LGBB; also see `vignette("tgp")` in the package;
- Gramacy and Taddy (2010): advanced topics like EI, categorical inputs (Broderick and Gramacy, 2011), sensitivity analysis, and importance tempering (Gramacy et al., 2010) to improve MCMC mixing; also see `vignette("tgp2")` in the package.

Several authors have further extended TGP capability. Konomi et al. (2017) demonstrate a nonstationary calibration framework (§8.1) based on TGP. MATLAB® code is provided as supplementary material supporting their paper. Classification with TGP, utilizing a logit linked multinomial response model paired with latent (treed) GP random field, has been explored (Broderick and Gramacy, 2011). However synergy between tree and GP is weaker here than in the regression context. Notions of smoothness are artificial when classification labels are involved. A GP latent field can mimic tree-like partitioning features without the

help of the tree. Yet the authors were able to engineer some examples where their hybrid
was successful.

Another popular approach to tree regression is to combine many simple trees additively. This
idea was first introduced in ML as boosting decision stumps[22]. Perhaps the most popular
such implementation is gradient tree boosting[23]. See the gbm package (Greenwell et al.,
2019), originally from Ridgeway (2007), and xgboost (Chen et al., 2019) on CRAN. More
recently, Bayesian additive regression trees (BART) from CGM (Chipman et al., 2010) has
gained traction as a sampling-based alternative offering better UQ which can be key in
surrogate modeling applications such as BO. Although BART has a built-in additive error
model, tweaks can be applied in order to mimic sausage-shaped error-bars when modeling
deterministic computer simulations (Chipman et al., 2012). Several R packages support
BART, including BayesTree (Chipman and McCulloch, 2016) and BART (McCulloch et al.,
2019), with the latter accommodating a multitude of response types. The space of BART
research is quite active and I'd expect many new developments in future. Perhaps the only
downside to BART is that predictive surfaces are pathologically non-smooth. Its additive
structure does however lend great flexibility and reactiveness to the mean surface which is a
substantial asset.

Dynamic trees (DTs; Taddy et al., 2011) were developed specifically to target sequential
applications, such as arise in computer simulation, active learning, and BO, with additional
support for input importance and sensitivity analysis (§8.2) applications (Gramacy et al.,
2013). An implementation is provided by dynaTree (Gramacy et al., 2017) on CRAN. DT
development revisits the tree prior as a process evolving sequentially in time, and extends
that prior to posterior updates as new data arrive. A new data point (x_{t+1}, y_{t+1}) may
support tree growth or pruning, or neither in favor of the status quo. Inference for DT
processes is facilitated by the sequential Monte Carlo[24] method of *particle learning* (Carvalho
et al., 2010). Dynamic trees have been applied on a wide variety of computer surrogate
modeling tasks such as computer code semantic translation and autotuning (Balaprakash
et al., 2013b,a), stochastic control for epidemic management, financial options pricing, and
autonomous vehicle tracking (Gramacy and Ludkovski, 2015). Streaming applications, which
deploy data point *retirement* in order to work in fixed memory and *forgetting factors* in the
face of *concept drift* (a target response surface that's evolving in time), are described by
Anagnostopoulos and Gramacy (2013).

```
library(dynaTree)
```

For applications which are not inherently sequential in nature, DTs can be applied to random
data orderings. Randomization over the "arrival-time" of data has a simultaneous bootstrap[25]
and likelihood annealing[26] effect. The results are, at least in some cases, astounding. Consider
the following multiple DTs (dynaTrees) fit to mcycle data (§9.2.1), separately under constant
and linear leaves.

```
XX <- seq(0,max(mcycle[,1]), length=1000)
out.dtc <- dynaTrees(X=mcycle[,1], y=mcycle[,2], XX=XX, verb=0, pverb=0)
```

[22]https://en.wikipedia.org/wiki/Boosting_(machine_learning)
[23]https://en.wikipedia.org/wiki/Gradient_boosting#Gradient_tree_boosting
[24]https://en.wikipedia.org/wiki/Particle_filter
[25]https://en.wikipedia.org/wiki/Bootstrapping_(statistics)
[26]https://en.wikipedia.org/wiki/Simulated_annealing

```
out.dtl <- dynaTrees(X=mcycle[,1], y=mcycle[,2], XX=XX, model="linear",
  verb=0, pverb=0)
```

Unfortunately the structure of the output objects doesn't align with `tgp` so we can't use the plotting macro from earlier. Similar code, evaluated below, aggregates means and quantiles extracted from 1000 random re-passes through the data.

```
plot(out.dtc$XX[,1], rowMeans(out.dtc$mean), type="l", ylim=c(-160, 110),
  ylab="accel", xlab="time")
points(mcycle)
lines(out.dtc$XX[,1], rowMeans(out.dtc$q1), lty=2)
lines(out.dtc$XX[,1], rowMeans(out.dtc$q2), col=1, lty=2)
lines(out.dtl$XX[,1], rowMeans(out.dtl$mean), col=2)
lines(out.dtl$XX[,1], rowMeans(out.dtl$q1), col=2, lty=2)
lines(out.dtl$XX[,1], rowMeans(out.dtl$q2), col=2, lty=2)
legend("topleft", legend=c("DT constant", "DT linear"), bty="n",
  lty=1, col=1:2)
```

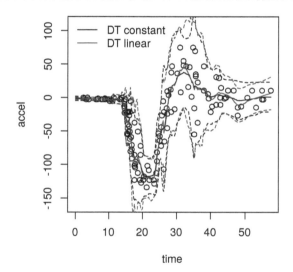

FIGURE 9.22: Dynamic trees posterior predictive surfaces on the motorcycle accident data.

The first time I saw these plots I was blown away. Aesthetically, constant leaves seem to offer better compromise between dynamic reactivity and smoothing. But how can piecewise constant models look so smooth, on average, but still mimic abrupt features in the data? Bootstrap aggregation, or bagging[27], manifest as random data orderings in DT, is a powerful tool. Bagging is the workhorse behind random forest (RF)[28] models, supported by several packages including **randomForest** (Breiman et al., 2018) on CRAN. Indeed, RFs make similarly strong surrogates, however extracting full predictive uncertainty can be challenging. A review paper by Chipman et al. (2013) provides more in-depth qualitative and quantitative comparison between Bayesian (and classical) tree-based regression methods. In fairness, not

---

[27]https://en.wikipedia.org/wiki/Bootstrap_aggregating
[28]https://en.wikipedia.org/wiki/Random_forest

all examples (whether via `dynaTree`, `bcart`, `btlm` or `BayesTree`) look as good as the `mcycle` ones presented here. GPs are hard to beat in bakeoffs spanning diverse data-generating mechanisms, input dimensions and sizes (so long as the calculations remain computationally tractable).

Finally, tessellation-based partition modeling has been revisited from several angles in recent literature. Rushdi et al. (2017) provide new data structures and new computational approaches to shortcut expensive search subroutines required for prediction under Voronoi piecewise surrogates. Park and Apley (2018) consider patching together piecewise fits to furnish a degree of smoothness across partition boundaries. Rullière et al. (2018) propose a nested approach to divide-and-conquer modeling. Those are just a few, non-representative examples; I doubt we've heard the last word on partition-based regression and surrogate modeling. Divide-and-conquer remains an attractive device for marrying computational thrift with modeling fidelity.

9.3   Local approximate GPs

A *local approximate Gaussian process* (LAGP), which I've been so excited to tell you about for hundreds of pages, has aspects in common with partition based schemes, in the sense that it creates sparsity in the covariance structure in a geographically local way. In fact, LAGP is a partitioning scheme in a limiting sense, although delving too deeply into that connection is counterproductive because the approach is quite different from partitioning in spirit. The core LAGP innovation is reminiscent of what Cressie (1992, pp. 131–134) called "ad hoc local kriging neighborhoods". Perhaps in 1992 the basic idea, which at face value isn't mind blowing but might have gotten less credit than it deserves, was simply a little ahead of its time. I think that the geostatistical community of that era may not have anticipated the scale of modern data, a ubiquity of applications to computer simulation and ML (with inputs other than longitude and latitude), and the architecture of contemporary supercomputers. Multi-core/cluster parallelization begs for divide-and-conquer.

All in all, LAGP's building blocks and their synthesis are more modern, both technologically and culturally, than could have been anticipated thirty-odd years ago. Technology wise, it draws on recent findings for approximate likelihoods in spatial data (e.g., Stein et al., 2004), and active learning techniques for sequential design (e.g., Cohn, 1994). But the big divergence, particularly from geostatistics, is cultural. Interest in LAGP lies squarely in prediction, which is the primary goal in computer experiments and ML applications.

Although designed for large-scale statistical surrogate modeling, the LAGP mindset is distinctly ML. The methodology is an example of *transductive learning* (Vapnik, 2013), with training tailored to predictive goals. This is as opposed to the more familiar inductive sort, where model fitting and prediction transpire in serial, usually in two distinct stages. A transductive learner utilizes training data differently depending on where prediction is required and to what end predictions might be used. Consequently the enterprise is more about reaction, decision and adaptation than it is about inference.

9.3.1   Local subdesign

For the next little bit, focus on prediction at a single testing location x. Coordinates encoded by x are arbitrary; it's only important that it be a single location in the input space $\mathcal{X}$. Let's think about the properties of a GP surrogate at x. Training data far from x have vanishingly small influence on GP predictions, especially when correlation is measured as an inverse of exponentiated Euclidean distances. This is what motivates a CSK approach to inducing sparsity (§9.1), but the difference here is that we're thinking about a particular x, not the entire spatial field.

The crux of LAGP is a search for the most useful training data points – a *subdesign* relative to x – for predicting at x, without considering/handling large matrices. One option is a *nearest neighbor (NN)* subset. Specifically, fill $X_n(x) \subset X_N$ with local-$n \ll$ full-N closest locations to x. Notice that I've tweaked notation a bit to have big N represent the size of a potentially enormous training set, unwieldy for conventional GPs, and now little n denotes a much smaller, more manageable size. Derive GP predictive equations under $Y(x) \mid D_n(x)$ where $D_n(x) = (X_n, Y_n)$, pretending that no other data exist. The best reference for this idea is Emery (2009). This prediction rule is as simple to implement as it is to describe, and it's very fast on relative terms when $n \ll N$. Costs are in $\mathcal{O}(n^3)$ and $\mathcal{O}(n^2 + N)$ for decomposition(s) and storage, respectively; and NNs can be found in $\mathcal{O}(n \log N)$ time with k-d trees[29] after an up-front $\mathcal{O}(N \log N)$ build cost. In practice, one can choose local-n as large as computational constraints allow, although there may be reasons to prefer smaller n on reactivity grounds. Predictors may potentially be more accurate at x if they're not burdened by information from training data far from x.

This is different, and much simpler than, what other authors have recently dubbed *nearest neighbor GP* regression (NNGP; Datta et al., 2016), which is a potential source of confusion. NNGP is clever, but targets global rather than local inference and prediction. The way in which neighbors are used is not akin to canonical NN[30], i.e., nonparametric regression and classification where each testing prediction conditions only on a small set of very closest training points. Neighborhood sets in NNGP, rather, anchor an approximate Cholesky decomposition leading to a joint distribution similar to what could be obtained at greater computational expense under a full conditioning set. This so-called Vecchia approximation (Vecchia, 1988; Stroud et al., 2017) induces sparsity in the inverse covariance structure. After this fashion, NNGP might be more aptly named "Bayesian Vecchia". Also see Katzfuss and Guinness (2018) for a more general treatment of conditioning sets toward that end. Both groups of authors provide implementations on CRAN; see spBayes (Finley and Banerjee, 2019) and GpGp (Guinness and Katzfuss, 2019), respectively. Empirical performance with these packages, tackling large geospatial data and furnishing accurate predictions and estimates of uncertainty, is impressive. As far as I know, they're untested in (higher dimensional) computer surrogate modeling and ML contexts.

Ok, apologies for the short digression. The essence of NN-based local GP approximation, using as training data that $D_n(x)$ which is closest to predictive location x, is embodied by the cartoon in Figure 9.23. Gridded black dots represent a massive training design X_N. There are mere thousands of such dots in the grid, but imagine hundreds of thousands or millions. Five solid, colored dots in the figure represent potential x sites. Open circles of the same color indicate NN subdesigns $X_n(x) \subset X_N$ corresponding to those predictive locations.

Notice how topology of the global design X_N impacts the shape of local designs $X_n(x)$.

---

[29]https://en.wikipedia.org/wiki/K-d_tree
[30]https://en.wikipedia.org/wiki/Nearest_neighbor_search

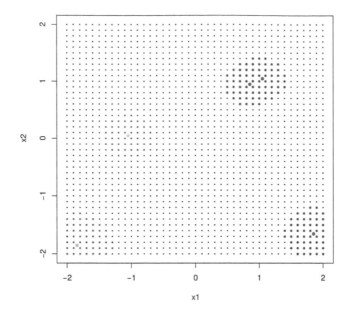

FIGURE 9.23: Local neighborhoods (colored open circles) based on NN subdesign for predictive locations (colored filled dots) as selected from a large design (small black dots).

When two predictive locations are nearby, as illustrated in pink and red, training data sites may be shared by subdesigns. There are no hard boundaries whereby adjacent, arbitrarily close predictive locations might be trained on totally disjoint data subsets. It's even possible to have two very close predictive locations $x \neq x'$ with the same subdesign $X_n(x) = X_n(x')$ when they share the same NN sets.

What can be said about this NN-based GP approximation? Is it sensible? Under fixed hyperparameterization and some regularity conditions that would be a distraction to review, one can show that as $n \to N$ predictions $Y(x) \mid D_n \to Y(x) \mid D_N$. In other words, larger local design size means better approximation. Being endowed with the label "approximate" requires some notion of accuracy relative to an exact alternative. It can also be shown, again under some regularity conditions, that $V(x) \mid D_n \gg V(x) \mid D_N$, reflecting uncertainties inflated by the smaller design, where $\sigma^2(x) = \hat{\tau}^2 V(x)$.

Is it good? Empirically, yes, but it's not optimal given computational limits, n (Vecchia, 1988; Stein et al., 2004). Of all size-n subsets of N training data sites, those residing closest to x in terms of Euclidean distance are not optimal for predicting at x. Clearly it's good to have some, perhaps many, nearby training sites. However, some farther out sites – not included because they're not part of the NN set – may be useful as anchors, providing long-range spatial dependence information. That information is potentially of greater value because it's less correlated/more independent than that which is provided by closer-in points. (If we already have a bunch of nearby points, the marginal value of another one has diminishing returns. It could be better to have independent information farther out.) That being said, finding the optimal n of N, of which there are $\binom{N}{n}$ alternatives, could be a combinatorially huge undertaking.

So that begs the question: can we do better than NN (in terms of prediction accuracy) without much extra effort (in terms of computational cost)? More precisely, n-NN GP prediction requires computation in $\mathcal{O}(n^3)$. So can we find a dataset $D_n(x)$, using time in

$\mathcal{O}(n^3)$ combining search, hyperparameter inference, and prediction, where accuracy at x based on that set is no worse than under $D_n^{(NN)}(x)$, the NN special case? Of course by "no worse" I really mean "hopefully much better", but choose to manage expectations by phrasing things conservatively.

The answer to that question is a qualified "Yes!", with a greedy[31]/forward stepwise[32] scheme. For a particular predictive location x, solve a sequence of easy decision problems.

For $j = n_0, \dots, n$:

1. given $D_j(x)$, choose x_{j+1} according to some criterion;
2. augment the design $D_{j+1}(x) = D_j(x) \cup (x_{j+1}, y(x_{j+1}))$ and update the GP approximation.

Optimizing the criterion (1), and updating the GP (2), must not exceed $\mathcal{O}(j^2)$ so the total scheme remains in $\mathcal{O}(n^3)$. Initialize with a small $D_{n_0}(x)$ comprised of NNs.

Gramacy and Apley (2015), G&A below, proposed the following criterion for sequential subdesign. Given $D_j(x)$ for particular x, search for $x_{j+1} \in X_N \setminus X_n(x)$ considering its impact on predictive variance $V_j(x) \equiv V(x) \mid D_j(x)$, while taking into account uncertainty in hyperparameters θ, by minimizing empirical Bayes *mean-squared prediction error*:

$$
\begin{aligned}
J(x_{j+1}, x) &= \mathbb{E}\{[Y(x) - \mu_{j+1}(x; \hat{\theta}_{j+1})]^2 \mid D_j(x)\} \\
&\approx V_j(x \mid x_{j+1}; \hat{\theta}_j) + \left(\frac{\partial \mu_j(x; \theta)}{\partial \theta} \Big|_{\theta = \hat{\theta}_j} \right)^2 \Big/ \mathcal{G}_{j+1}(\hat{\theta}_j).
\end{aligned} \tag{9.2}
$$

The approximation stems from Gaussian instead of Student-t predictive equations, and plugging in estimated kernel hyperparameters $\hat{\theta}_j$ instead of $\hat{\theta}_{j+1}$. Student-t equations arise upon estimating, or integrating out, the covariance scale τ^2. This detail is glossed over in our Chapter 5 introduction; since predictive equations presented therein were Gaussian, these were technically an approximation as well. As N, or in this context n, gets larger ($\gg 30$, say), the approximation error is small. Plugging in $\hat{\theta}_j$ for $\hat{\theta}_{j+1}$ avoids entertaining how estimated lengthscales might change as new y_{j+1} are incorporated, depending on the selected x_{j+1} location. In practice θ is fixed throughout sequential design iterations; more implementation details will be covered later.

Let's break down elements of the MSPE criterion $J(x_{j+1}, x)$. Apparently it combines variance and rate of change of the mean at x. G&A's presentation, and indeed the original laGP package implementation (Gramacy and Sun, 2018), emphasized isotropic lengthscale parameters θ. Our summary here follows that simplified setup. For extensions to vectorized θ for separable, coordinate-wise, lengthscales see the appendix to the original paper. A subsequently updated version of laGP supports separable lengthscales, as detailed by our empirical work below.

The first part of J, namely $V_j(x \mid x_{j+1}; \theta)$, is our old friend: an estimate of the new variance that will result after adding x_{j+1} into D_j, treating θ as known.

$$
V_j(x \mid x_{j+1}; \theta) = \frac{\psi_j}{j-2} v_{j+1}(x; \theta),
$$

where $\quad v_{j+1}(x; \theta) = [K_{j+1}(x, x) - k_{j+1}^\top(x) K_{j+1}^{-1} k_{j+1}(x)] \quad$ and $\quad \psi_j = j\hat{\tau}_j^2$.

Integrating $V_j(x|x_{j+1}; \theta)$ over x yields the ALC acquisition criterion for approximate global

---

A-optimal design (Seo et al., 2000; Cohn, 1994). §6.3 showed how v_{j+1} can be updated from v_j, and other quantities available at iteration j, in $\mathcal{O}(j^2)$ time.

Next, $\frac{\partial \mu_j(x;\theta)}{\partial \theta}$ is the partial derivative of the predictive mean at x, given D_j, with respect to lengthscale:

$$\frac{\partial \mu_j(x;\theta)}{\partial \theta} = K_j^{-1}[\dot{k}_j(x) - \dot{K}_j K_j^{-1} k_j(x)]^\top Y_j,$$

where $\dot{k}_j(x)$ is a length-j column vector of derivatives of kernel correlations $K(x, x_k)$, for $k = 1, \ldots, j$, taken with respect to θ. So it's the rate of change of predictive mean with respect to changes in lengthscale(s). These can be updated in $\mathcal{O}(j^2)$ too; see G&A for more details.

Finally, $\mathcal{G}_{j+1}(\theta)$ is also an old friend (§6.2.3): the Fisher information (FI) from D_j, including an expected component from future Y_{j+1} at x_{j+1}:

$$\mathcal{G}_{j+1}(\theta) = F_j(\theta) + \mathbb{E}\left\{-\frac{\partial^2 \ell_j(y_{j+1};\theta)}{\partial \theta^2} \,\Big|\, Y_j;\theta\right\}$$

$$\approx F_j(\theta) + \frac{1}{2V_j(x_{j+1};\theta)^2} \times \left(\frac{\partial V_j(x_{j+1};\theta)}{\partial \theta}\right)^2 + \frac{1}{V_j(x_{j+1};\theta)}\left(\frac{\partial \mu_j(x_{j+1};\theta)}{\partial \theta}\right)^2,$$

where $F_j(\theta) = -\ell''(Y_j;\theta)$, and with $\dot{\mathcal{K}}_j = \dot{K}_j K_j^{-1}$ and $\tilde{k}_j(x) = K_j^{-1} k_j(x)$,

$$\frac{\partial V_j(x;\theta)}{\partial \theta} = \frac{Y_j^\top K_j^{-1} \dot{\mathcal{K}}_j Y_j}{j-2} \left(K(x,x) - k_j^\top(x)\tilde{k}_j(x)\right)$$

$$- \psi_j \left[\dot{k}_j(x)\tilde{k}_j(x) + \tilde{k}_j(x)^\top(\dot{k}_j(x) - \dot{\mathcal{K}}_j k_j(x))\right].$$

G&A similarly detail how the derivative of V_j may be updated in $\mathcal{O}(j^2)$ time which is nearly identical to our development for v_j provided in §6.3.

Observe how inverse FI, which for vectorized θ would be a matrix inverse and applied in full quadratic form as $(\partial \mu_j)^\top \mathcal{G}_{j+1}^{-1}(\partial \mu_j)$, serves as a weight in a trade-off between variance reduction and sensitivity of predictive mean to hyperparameters θ. Referring back to our global FI-based sequential design illustration from §6.2.3, FI generally increases as more samples are added. Therefore the weight applied to the second term in the MSPE criterion J in Eq. (9.2) diminishes as j increases, effectively up-weighting reduction in variance.

Although MSPE nests an ALC-like criteria (§6.2.2), importantly we don't need to integrate (or in practice sum) over reference locations. The single testing location x is our (only) reference location. Reducing future variance is a sensible criterion in its own right, and considering that the FI-based weight acts most strongly for low j, one may wonder whether the extra complication of calculating first and second derivatives is "worth it" for full MSPE? Perhaps ALC on its own – basing sequential subdesign decisions on $V_j(x \mid x_{j+1}; \theta)$ only – is sufficient to beat the simple NN set. Let's see how these two alternatives, referred to as MSPE and ALC below, compare qualitatively and empirically in a simple example.

9.3.2 Illustrating LAGP: ALC v. MSPE

Consider a design of size $N \approx 40,000$ on a 2d grid in $[-2, 2]^2$.

```
xg <- seq(-2, 2, by=0.02)
X <- as.matrix(expand.grid(xg, xg))
nrow(X)
```

```
## [1] 40401
```

Technically, greedy subdesign search (using ALC or MSPE) doesn't require a response. Conditional on hyperparameters θ, calculations involve X_N and x only. Still, being engineered to furnish predictions, laGP wants responses Y_N too, i.e., the full D_N. We'll be looking at those predictions shortly anyways, so this is as good a time as any to introduce a challenging test problem.

The function below was chosen by G&A because it's sufficiently complicated to warrant a large training set despite being in relatively small input dimension. Some have taken to calling this "Herbie's tooth"; it was dreamed up as a challenging Bayesian optimization test problem by Herbie Lee (Lee et al., 2011), my PhD advisor, and looks like a molar when plotted in 2d. We used it in §7.3.7 to illustrate an augmented Lagrangian BO method. Although most uses, including the original, are in 2d, the function is sometimes mapped to higher input dimension. Let

$$g(z) = \exp\left(-(z-1)2\right) + \exp\left(-0.8(z+1)2\right) - 0.05\sin\left(8(z+0.1)\right), \qquad (9.3)$$

be defined for scalar inputs z. Then, for inputs x with m coordinates $x_1, \ldots, x_m$, the response is $f(x) = -\prod_{j=1}^{m} g(x_j)$. Some variations in the literature use $-f(x)$ instead, depending on whether the application (e.g., optimization) emphasizes minima or maxima. The un-negated version is easier to visualize with perspective plots, as we do shortly.

An implementation is provided below, modified from the Chapter 7 version to handle generic input dimension.

```
herbtooth <- function(X)
  {
  g <- function(z)
    return(exp(-(z - 1)^2) + exp(-0.8*(z + 1)^2) - 0.05*sin(8*(z + 0.1)))
  return(-apply(apply(X, 2, g), 1, prod))
  }
```

In spite of that upgrade in generality, we shall apply Herbie's tooth here, over our large gridded $X \equiv X_N$, in the easy-to-visualize 2d case. For a four-dimensional application, see Section 4.3 of Sun et al. (2019a).

```
Y <- herbtooth(X)
```

Now consider predictive location x, denoted by Xref in the code below, via local designs constructed greedily based on MSPE and ALC.

```
library(laGP)
Xref <- matrix(c(-1.725, 1.725), nrow=1)
p.mspe <- laGP(Xref, 6, 50, X, Y, d=0.1, method="mspe")
p.alc <- laGP(Xref, 6, 50, X, Y, d=0.1, method="alc")
```

As specified by the calls above, both design searches initialize with $n_0 = 6$ NNs, make greedy selections until $n = 50$ locations are chosen, and use an isotropic d $\equiv \theta = 0.1$ hyperparameterized Gaussian kernel. The goal is to interpolate these deterministic evaluations. Although one can specify a zero nugget, a more conservative default setting of g=1e-4 is used to ensure numeric positive-definiteness. When numerics are more favorable, smaller g often leads to better results, as we shall illustrate later in a more ambitious setting. Figure 9.24 shows the resulting selections.

```
plot(X[p.mspe$Xi,], xlab="x1", ylab="x2", type="n",
  xlim=range(X[p.mspe$Xi,1]), ylim=range(X[p.mspe$Xi,2]))
text(X[p.mspe$Xi,], labels=1:length(p.mspe$Xi), cex=0.7)
text(X[p.alc$Xi,], labels=1:length(p.alc$Xi), cex=0.7, col=2)
points(Xref[1], Xref[2],pch =19, col=3)
legend("right", c("mspe", "alc"), text.col=c(1,2), bty="n")
```

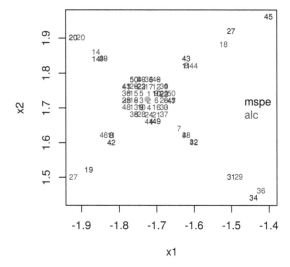

FIGURE 9.24: Comparing LAGP subdesigns under MSPE (9.2) and the ALC special case. The predictive location x is a filled-green dot. Numbers indicate the order in which each training-data location was greedily selected for the local subdesign.

The green dot is $x \equiv$ Xref. Each number plotted, whether in red for ALC or black for MSPE, indicates the order in which greedy selections were made. Although the grid of possible candidates $X_N \setminus X_{n_0}(x)$ is not shown, it's easy to mentally visualize them filling in the negative space. Notice that there are early selections which are not NNs, and there are late selections which are. The order in which an NN or non-NN is chosen is hard to predict. Non-NN subdesign selections are sometimes called *satellite points*. Although these satellites are not NNs, they aren't that far from x in the entire space, which is $[-2,2]^2$. (Observe that the plotting domain is smaller than the full span of the design X_N.) Finally, any differences between ALC and MSPE are slight at best, at least aesthetically speaking.

Is the result in the figure surprising? Why do the criteria not prefer only the closest possible points, i.e., NNs? An exponentially decaying correlation should substantially devalue locations far from x. Gramacy and Haaland (2016) explain that the form of the correlation has very little to do with it. Consider (scale free) reduction in variance, an expression we've seen before (6.9):

$$v_j(x; \theta) - v_{j+1}(x; \theta) = k_j^\top(x)G_j(x_{j+1})v_j(x_{j+1})k_j(x) + \cdots + K(x_{j+1}, x)^2/v_j(x_{j+1}).$$

Although quadratic in inverse distance $K(x_{j+1}, x)^2$, terms are also quadratic in *inverse* inverse distance, e.g.,

$$G_j(x') \equiv g_j(x')g_j^\top(x') \quad \text{where} \quad g_j(x') = K_j^{-1}k_j(x')/v_j(x').$$

So the criteria make a trade-off: minimize scaled distance to x while maximizing distance (or minimizing inverse distance) to the existing design $X_j(x)$. Or in other words, the potential value of new design element (x_{j+1}, y_{j+1}) depends not just on its proximity to x, but also on how potentially different that information is to where we already have (lots of) it, at $X_j(x)$.

Both MSPE and ALC provide a mixture of NNs and satellite points. What about the rays, emanating from x, that satellite points seem to arrange themselves around? Those are due to the radial structure of our isotropic kernel. Satellite points would like to be even more radial except that a discrete, gridded X_N is thwarting that outcome. Gramacy and Haaland show some of the interesting patterns, manifest as ribbons and rings, that materialize in local designs depending on kernel and hyperparameterization.

How about prediction? The two local subdesigns are qualitatively similar. Predictions are, empirically speaking, nearly identical:

```
p <- rbind(c(p.mspe$mean, p.mspe$s2, p.mspe$df),
  c(p.alc$mean, p.alc$s2, p.alc$df))
colnames(p) <- c("mean", "s2", "df")
rownames(p) <- c("mspe", "alc")
p
```

```
##           mean        s2 df
## mspe -0.3725 2.519e-06 50
## alc  -0.3725 2.445e-06 50
```

Despite being built under greedy criteria for fixed lengthscale $\theta = 0.1$, the predictive equations output by laGP utilize local MLEs based on data subset $D_n(x)$. That is, after n selections, $\hat{\theta}_n(x) \mid D_n(X)$ is calculated before applying the usual GP predictive equations (5.2).

```
mle <- rbind(p.mspe$mle, p.alc$mle)
rownames(mle) <- c("mspe", "alc")
mle
```

```
##             d dits
## mspe 0.3589    7
## alc  0.3378    7
```

These are also nearly identical for the two sequential subdesign criteria. Very few Newton iterations (`dits`) are required for convergence to $\hat{\theta}_n(x)$. Finally, both local subdesign searches are fast.

```
ts <- c(mspe=p.mspe$time, alc=p.alc$time)
ts
```

TABLE 9.1: Summarizing calculation times in local lengthscale estimation.

	mean	s2	df	d	dits	ts
mspe	-0.3725	0	50	0.3589	7	0.177
alc	-0.3725	0	50	0.3378	7	0.052
nn	-0.3726	0	50	0.2096	6	0.005

```
## mspe.elapsed  alc.elapsed
##         0.177        0.052
```

ALC is about 3× faster because it bypasses derivative calculations, motivating that choice as the default in the package. For a point of reference, inverting a 4000×4000 matrix takes about five seconds on the same machine, but this can be improved upon with customized linear algebra (Appendix A); an improvement which would be enjoyed by laGP as well. Never mind a $40,000 \times 40,000$ decomposition – impossible on ordinary workstations.

Calculating an NN subdesign is faster even though the computational order is the same. Constant and lower order (quadratic and linear terms) add substantially to the work required to make greedy ALC selections. Yet calculating MLE $\hat{\theta}_n(x)$ at the end is what dominates. Table 9.1 summarizes predictive quantities and calculations involved in local lengthscale estimation.

```
p.nn <- laGP(Xref, 6, 50, X, Y, d=0.1, method="nn")
p <- rbind(p, nn=c(p.nn$mean, p.nn$s2, p.nn$df))
mle <- rbind(mle, nn=p.nn$mle)
ts <- c(ts, nn=p.nn$time)
kable(cbind(p, mle, ts),
   caption="Summarizing calculation times in local lengthscale estimation.")
```

NN is the exception in lengthscale hyperparameter $d \equiv \hat{\theta}_n(x)$ and speed. NN's shorter lengthscale can be explained by its more compact subdesign: longer lengthscales require support from (absent) longer pairwise distances. Considering speed differences, it's worth asking if the extra effort of ALC or MSPE local subdesign is worth it compared to NN. Until we're able to collect predictions at a suite of testing locations, for many $x \in \mathcal{X}$, it's not worth commenting in finer detail on relative accuracies. If you want to take my word for it: for fixed subdesign size n, ALC and MSPE outperform NN. But a more natural comparator is NN based on a bigger $n' > n$, one requiring a commensurate amount of computational time to compute. That's a difficult comparison to make, or at least to make fair. See, for example, Section 3.1 of Gramacy (2016) which illustrates that, in at least one view on the borehole data, $n = 50$ ALC-based subdesigns yield predictions that are more accurate than ones based on much larger $n' = 200$ NN local subsets in about the same amount of time.

Complicating things further, there are several mechanisms built into laGP targeting search speedups at the expense of (possibly) less faithful greedy selection. A parameter close adjusts the scope of candidates entertained for local acquisition. Rather than search over all $X_N \setminus X_j(x)$, one can safely entertain only close $\equiv N' \ll N$ NNs, say close=1000 (the default), without any effect on the chosen subdesign compared to one obtained under an exhaustive search. Sung et al. (2018) show how an even smaller, but more dynamically determined, set can be entertained by ruling out large swaths of $X_N \setminus X_j(x)$ as "noncompetitive for selection". Gramacy et al. (2014) demonstrate how greedy selection among tens of thousands of candidates can be off-loaded to a GPU for essentially instantaneous execution; Gramacy and Haaland (2016) illustrate how exhaustive discrete searches over $X_N \setminus X_j(x)$ can be replaced by simple 1d line searches, along rays emanating from x, whose solutions may be snapped back to the original (un-selected) design locations; finally Sun et al. (2019b) provide a similar – perhaps more exact but also more involved – approach using gradients.

In a nutshell, one can dramatically speed up greedy ALC and MSPE searches with little impact on accuracy, furnishing predictions more accurate than NN in about the same amount of time. All of the methods listed above are built into laGP for ALC (and some also for MSPE). An exercise in §9.4 entertains some of these alternatives alongside defaults. Perhaps the best reason to prefer ALC and MSPE over NN is that G&A show empirically that those methods lead to better numerical conditioning of the resulting $n \times n$ local covariance matrices. When training data are too close together in the design space, the covariance structure becomes difficult to decompose. Local GP approximation via NN can exacerbate this numerical challenge; satellite points from ALC and MSPE offer some relief. Numerically stable local prediction is a crucial engineering detail to consider in repeated application, say over a dense testing set – which is our next subject.

9.3.3 Global LAGP surrogate

How can this local strategy be extended to predict on a big testing/predictive set $\mathcal{X}$? One simple option is to serialize: tack on a `for` loop over each $x \in \mathcal{X}$. But why serialize when you can parallelize? Each $D_n(x)$ is obtained independently of other x's, so they may be constructed simultaneously. In laGP's C implementation, that's as simple as a `parallel for` OpenMP[33] pragma.

```
#ifdef _OPENMP
  #pragma omp parallel for private(i)
#endif
  for(i = 0; i < npred; i++) { ...
```

It really is that simple. The original implementation in laGP used exactly that pragma. Subsequent versions have upgraded to more elaborate families of pragmas in order to trim overheads inherent in `parallel for`.

With modern laptops having two hyperthreaded cores (meaning they can run four threads in parallel), and many desktops having eight (16 threads) or more, parallelization in implementation is essential to taking full advantage of contemporary architectures. OpenMP is the simplest way to accomplish shared memory parallelization (SMP)[34] in C and Fortran. Statistically speaking, leveraging SMP means exploiting or introducing statistical independence. Local approximation with LAGP imposes exactly this kind of statistical, and thus algorithmic independence structure already. No further approximation is required in order to predict with laGP in parallel for many x.

To illustrate, consider the following $\approx$ 10K-element predictive grid in $[-2, 2]^2$, spaced to avoid the original $N \approx 40$K design.

```
xx <- seq(-1.97, 1.95, by=0.04)
```

---

[33]https://www.openmp.org/
[34]https://en.wikipedia.org/wiki/Parallel_programming_model#Shared_memory

```
XX <- as.matrix(expand.grid(xx, xx))
YY <- herbtooth(XX)
```

The aGP function automates "iteration" over elements of $\mathcal{X} = $ XX, and its omp.threads argument controls the number of OpenMP threads. Here we'll use 8 threads, even though my desktop is hyperthreaded (can do 16).

```
nth <- 8
P.alc <- aGP(X, Y, XX, omp.threads=nth, verb=0)
P.alc$time
```

```
## elapsed
##    70.71
```

Observe that the compute time reflects almost linear scaling by comparison to the time extrapolated for our earlier singleton ALC run.

```
p.alc$time*nrow(XX)/8
```

```
## elapsed
##    63.71
```

Figure 9.25 offers a view of the predictive mean surface thus obtained. That surface is negated to ease visibility in this perspective.

```
persp(xx, xx, -matrix(P.alc$mean, ncol=length(xx)), phi=45, theta=45,
    xlab="x1", ylab="x2", zlab="yhat(x)")
```

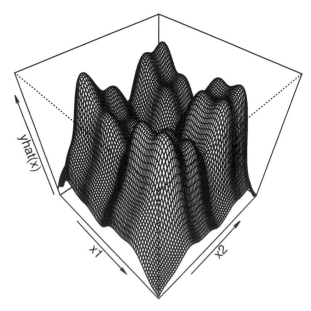

FIGURE 9.25: Global approximation by independent and parallel LAGP predictions on Herbie's tooth.

Perhaps you can see why this surface is called "Herbie's tooth". Although input dimension is low, input–output relationships are nuanced and merit a dense design to fully map. For a closer look, consider a slice through that predictive surface at $x_2 = 0.51$. R code below sets up that slice ...

```
med <- 0.51
zs <- XX[,2] == med
sv <- sqrt(P.alc$var[zs])
r <- range(c(-P.alc$mean[zs] + 2*sv, -P.alc$mean[zs] - 2*sv))
```

... followed by its visualization in Figure 9.26. A 1d view allows us to inspect error-bars and compare to the truth all in the same plotting window.

```
plot(XX[zs,1], -P.alc$mean[zs], type="l", lwd=2, ylim=r,
  xlab="x1", ylab="y")
lines(XX[zs,1], -P.alc$mean[zs] + 2*sv, col=2, lty=2, lwd=2)
lines(XX[zs,1], -P.alc$mean[zs] - 2*sv, col=2, lty=2, lwd=2)
lines(XX[zs,1], -YY[zs], col=3, lwd=2, lty=3)
legend("bottom", c("mean", "95% interval", "truth"), lwd=2, lty=1:3,
  col=1:3, bty="n")
```

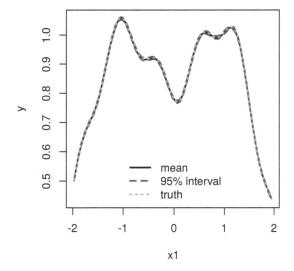

FIGURE 9.26: Slice $x_2 = 0.51$ through Figure 9.25, augmented with predictive interval and true response.

What do we see? Prediction is accurate; had the green-dotted truth line been plotted first, with black mean coming after, the black would've completely covered the green, rendering it invisible. Error bars are very tight on the scale of the response; again, had black come after red we'd barely see red peeking out around black. Despite no continuity being enforced – calculations at nearby locations are independent and potentially occur in parallel – the resulting surface looks smooth to the eye.

What don't we see? Accuracy, despite generally being high, is not uniform. Consider discrepancy with the truth, measured out-of-sample.

```
diff <- P.alc$mean - YY
```

Figure 9.27 uses the same slice as above. We're over-predicting. Systematic pattern and persistent positive bias is evident, although extremely small, judging by the scale of the y-axis.

```
plot(XX[zs,1], diff[zs], type="l", lwd=2, xlab="x1", ylab="yhat-y")
```

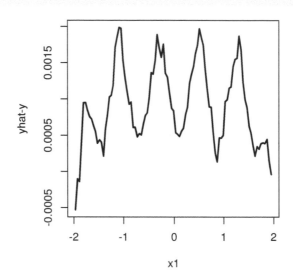

FIGURE 9.27: Illustrating bias through differences between predictions and truth in the slice from Figure 9.26.

Remember that we're using a zero-mean GP. Evidently in this slice, mean reversion is pulling us up a little too far, away from the training data and consequently away from the true response at these testing locations. Having an explanation helps, but it's still unsatisfying to be leaving systematic predictive potential on the table. Also notice the lack of smoothness in the discrepancy surface. Training and testing responses are both smooth (and observed without noise), so it must be that the predictor – the LAGP surrogate – is not smooth. In fact it's discontinuous, but on a small scale and not everywhere. Nearby predictive locations eventually share the same subdesign (as $x' \to x$) and thus inherit GP smoothness properties on a fine local scale.

Focusing on patterns evident in the bias plotted in Figure 9.27, what limits LAGP's dynamic ability? Why does it leave such strong signal in the residuals? Considering the density of the input design in 2d, perhaps the fit is not flexible enough to characterize fast-moving changes in input–output relationships. Although an approximation, the local nature of modeling means that, from a global perspective, the predictor is more flexible than a full-N stationary GP predictor. So we can rest assured that an ordinary full GP wouldn't fare any better in this regard, assuming the calculations were computationally tractable. Statistically independent fits in space, across elements of a vast predictive grid, lends aGP a degree of nonstationarity. By default, the laGP calls inside aGP go beyond that by learning separate $\hat{\theta}_n(x)$ local to each $x \in \mathcal{X}$ by maximizing local likelihoods.

In fact local lengthscales vary spatially, and relatively smoothly. Figure 9.28 plots $\hat{\theta}_n(x)$ as x varies along our 1d slice. A loess smoothed alternative is overlayed in dashed-red.

```
plot(XX[zs,1], P.alc$mle$d[zs], type="l", lwd=2, xlab="x1",
  ylab="thetahat(x)")
df <- data.frame(y=log(P.alc$mle$d), XX)
lo <- loess(y ~ ., data=df, span=0.01)
lines(XX[zs,1], exp(lo$fitted)[zs], col=2, lty=2, lwd=2)
legend("topleft", "loess smoothed", col=2, lty=2, lwd=2, bty="n")
```

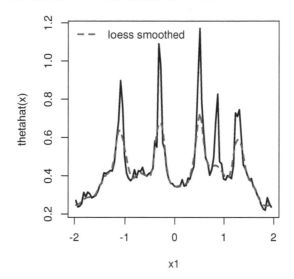

FIGURE 9.28: Locally estimated and smoothed lengthscales along the slice from Figures 9.26–9.27.

Spatial signal in rate of decay of inverse exponentiated squared Euclidean distance-based correlation is distinct in Figure 9.28, but the pattern doesn't perfectly align with spatial bias along the slice shown in Figure 9.26. Smoothing $\hat{\theta}_n(x)$ may help downplay numerical instabilities, but can in general represent a problem just as computationally difficult as the original one: an implicitly large-N spatial regression. Exceptions arise when working with grids, which is not uncommon if predicting for the purpose of visualization. (In other words, the tail doesn't necessarily wag the dog when smoothing over $\hat{\theta}_n(x)$'s spatially in x.)

Although the spatial field is locally isotropic, tacitly assuming stationarity to a certain extent, globally characteristics are less constrained. Nevertheless, the extra degree of flexibility afforded by spatially varying $\hat{\theta}_n(x)$ is not enough to entirely mitigate the small amount of bias we saw above. Several enhancements offer potential for improved performance. Perhaps the first, and most obvious, entails deploying an anisotropic/separable correlation structure. That's a mere implementation detail, so hold that thought for a moment.

Another option is to put those $\hat{\theta}_n(x)$ to better use. You see, the local subdesign process must start somewhere – condition on "known" quantities – and laGP utilizes a fixed $\theta_0 \equiv$ d to that end. By default, aGP invokes a built-in function called darg to choose a d to use identically for all $x \in \mathcal{X} \equiv$ XX. Alternatively, a two-stage scheme, re-designing $X_n(x) \mid \hat{\theta}_n(x)$, could help soften the influence of that ultimately arbitrary initialization. Below, subdesign search is based on the smoothed lengthscales from the first stage.

```
P.alc2 <- aGP(X, Y, XX, d=exp(lo$fitted), omp.threads=nth, verb=0)
```

Basically, we're redoing the whole local analysis, globally over all $x \in \mathcal{X}$, but with a lengthscale unique to x and thus a subdesign tailored to dynamics nearby x. Previously only the topology of the global design X_N mattered nearby x to determine $X_n(x)$. Now conditioning on $\hat{\theta}_n(x)$ from an earlier run gets responses Y_n involved too. Comparing predictions from the first iteration to those from the second in terms of RMSE ...

```
rmse <- data.frame(alc=sqrt(mean((P.alc$mean - YY)^2)),
  alc2=sqrt(mean((P.alc2$mean - YY)^2)))
rmse
```

```
##            alc        alc2
## 1 0.0006453 0.0003161
```

... reveals a degree of improvement, albeit perhaps not by an impressive amount (a factor of around 2 in this case). Such a two-stage analysis offers consistent, if slight improvement across a large swath of examples. Yet that thinking represents the tip of an iceberg: there are lots of ways to prime the pump, as it were. Before we get to those, let's transition to a more ambitious example. By going back to the borehole (§9.1.3) we'll be able to draw comparisons to a wider range of alternatives, coming full circle to CSK. (Objects x and y are unchanged from our earlier **borehole** examples, but X, Y and XX must be rebuilt from those.)

```
X <- x[1:n,]
Y <- y[1:n]
XX <- x[-(1:n),]
YY <- y[-(1:n)]
```

The borehole example benefits from long lengthscales, so aGP calls below adjust the default maximum lengthscale upwards to twenty. As with the default nugget g=1e-4, darg's are conservative, but hold that thought for a moment. Long lengthscales are likely to yield poorly conditioned covariance matrices. Below an ordinary aGP run is invoked first, followed by a run primed with lengthscales $\hat{\theta}_n \equiv$ out1mled.

```
out1 <- aGP(X, Y, XX, d=list(max=20), omp.threads=nth, verb=0)
out2 <- aGP(X, Y, XX, d=list(start=out1$mle$d, max=20),
  omp.threads=nth, verb=0)
```

Recall that we used pointwise proper scores to compare CSK methods. The table below collects new timing and score results, combining with those obtained earlier for CSKs.

```
times <- c(times, aGP=as.numeric(out1$time), aGP2=as.numeric(out2$time))
scores <- c(scores, aGP=scorep(YY, out1$mean, out1$var),
  aGP2=scorep(YY, out2$mean, out2$var))
rbind(times, scores)
```

```
##         sparse99     dense sparse999      aGP     aGP2
## times 2945.376 3066.946   351.746   4.9170   4.7390
## scores   -1.918    -6.966    -2.105  -0.6545  -0.6149
```

That `aGP` is faster than our earlier CSK calculations is an understatement; the factor of improvement on time is a whopping 599. Local surrogates are also quite a bit more accurate. Perhaps this is surprising because `aGP`, being based on many independent `laGP` fits, utilizes far less data to make predictions. That can translate into over-estimates of variance which is penalized by score. Since training and testing data are noise free, one might instead compare using Mahalanobis distance, as explained in §5.2.1 nearby Eqs. (5.6)–(5.7). However, `aGP` only provides point-wise variance, necessitating a Mahalanobis approximation like Nash–Sutcliffe efficiency[35] or out-of-sample R^2. The curious reader may wish to update our comparison with those values, or consult the original G&A paper.

Now back to that separable lengthscale "implementation detail". We're able to beat CSK on two fronts, in time and score, with an isotropic correlation function. But CSK had the benefit of learning decay of correlation (and sparsity) in a coordinate-wise fashion. A function `aGPsep`, which is the separable analog of `aGP`, estimates separate lengthscales $\theta_k(x)$ for each coordinate $k = 1, \ldots, m$.

```
outs <- aGPsep(X, Y, XX, d=list(max=20), omp.threads=nth, verb=0)
```

Using separable lengthscales represents a bigger implementation feat than at first it may seem, requiring an additional four thousand lines of C and R code in `laGP`. Working with a higher dimensional hyperparameter space, gradients, etc., demands extra scaffolding. One particular challenge centers around R's built-in BFGS solver which, due to an antique implementation utilizing static variables[36], is not thread safe[37]. Parallelization with OpenMP requires an intricate blocking mechanism to ensure that threaded BFGS calculations for $\hat{\theta}_n(x)$ and $\hat{\theta}_n(x')$ don't interfere with one another. In the simpler scalar θ setting `laGP` is able to avoid BFGS via R's `optimize` (rather than `optim`) back-end.

Augmenting our table, notice below that utilizing separable lengthscales results in a similarly speedy execution, but yields substantially more accurate prediction.

```
times <- c(times, aGPs=as.numeric(outs$time))
scores <- c(scores, aGPs=scorep(YY, outs$mean, outs$var))
rbind(times, scores)
```

```
##         sparse99     dense sparse999      aGP     aGP2      aGPs
## times 2945.376 3066.946   351.746   4.9170   4.7390  4.92000
## scores   -1.918    -6.966    -2.105  -0.6545  -0.6149  0.03639
```

A second pass, priming a new `aGPsep` run with $\hat{\theta}_n$'s, offers slight improvements. Even better results may be obtained by taking a step back for more global perspective.

---

[35]https://en.wikipedia.org/wiki/Nash-Sutcliffe_model_efficiency_coefficient

[36]https://stackoverflow.com/questions/572547/what-does-static-mean-in-c

[37]https://en.wikipedia.org/wiki/Thread_safety

9.3.4   Global/local multi-resolution effect

Surely something is lost on this local approach to GP approximation. Kaufman et al. astutely observed that, especially when inducing sparsity in the covariance structure, it can be important to "put something global back in". Recall that they partition modeling between trend (global/nonstationary) and residual (local/stationary), with the former being basis-expanded linear and the latter being spatial. That's not easily mapped to the LAGP setup, which is the other way around: the local part is where the nonstationary effect comes in.

Towards an appropriate analog in the LAGP world, consider not a partition between trend and residual, but rather between lengthscales: global and local. Liu (2014) showed that a "consistent" estimator of global (separable) lengthscale can be estimated through more manageably sized random data subsets if those subsets are generated with a block-bootstrap Latin hypercube sampling (BLHS) scheme. Also see Zhao et al. (2018). Rather than dig much under the surface of what Liu meant by "consistent", instead let's take it as "similar to what you'd get by maximizing the full data log likelihood", i.e., the MLE. Basically, you don't need to directly manipulate training data from a big design to estimate the lengthscale you would get from the big design. Now lengthscale hyperparameter estimates aren't particularly useful on their own. A missing ingredient in that work is obtaining predictions, given those lengthscales, comparable to ones that would've been obtained from the big design. Those would still require big-matrix decomposition.

The idea is to convert global lengthscales into local subdesigns, subsequent local refinement of lengthscale, and ultimately tractable and accurate prediction. Here we take Liu's BLHS as inspiration and use a simpler random subset analog. A random subset could not guarantee a similar distribution of pairwise distances compared to the original. From that perspective, BLHS accomplishes a feat akin to betadist designs introduced in §6.2.3, but for subdesign. See Sun et al. (2019a) for more detail on BLHS, an implementation, and an empirical illustration of why it's better than simple random subsampling, which nevertheless works very well as a quick-and-dirty alternative. A homework exercise in §9.4 invites readers to entertain a `blhs` (from `laGP`) variation of the following.

Consider a random 1000-sized subset in our running borehole example.

```
tic <- proc.time()[3]
nsub <- 1000
d2 <- darg(list(mle=TRUE, max=100), X)
subs <- sample(1:nrow(X), nsub, replace=FALSE)
gpsi <- newGPsep(X[subs,], Y[subs], rep(d2$start, m), g=1/1000, dK=TRUE)
that <- mleGPsep(gpsi, tmin=d2$min, tmax=d2$max, ab=d2$ab, maxit=200)
psub <- predGPsep(gpsi, XX, lite=TRUE)
deleteGPsep(gpsi)
toc <- as.numeric(proc.time()[3] - tic)
that$d
```

```
## [1]  0.4489 34.5091 36.3133  5.9696 31.7979  5.7055  2.3457 11.5340
```

Values $\hat{\theta}^{(g)} \equiv$ `that` quoted above don't matter in and of themselves, but there they are. We can repeat that a bunch of times, in a bootstrap-like fashion, but that's overkill. As a sanity check, observe that global random subset GP prediction is pretty good on its own, because the response is super smooth and pretty stationary.

```
times <- c(times, sub=toc)
scores <- c(scores, sub=scorep(YY, psub$mean, psub$s2))
rbind(times, scores)
```

```
##          sparse99    dense sparse999     aGP    aGP2    aGPs      sub
## times   2945.376 3066.946   351.746  4.9170  4.7390 4.92000 110.1060
## scores    -1.918   -6.966    -2.105 -0.6545 -0.6149 0.03639   0.6384
```

In fact, that result almost makes you wonder what was going on with CSK. All that sparsity-inducing structure and it'd have been sufficient (even better) to take a random subset and run with that, in a fraction of the time. Subset-based GP surrogates even beat laGP, but at the expense of greater computation. Keep in mind this is a small example, and in particular that $n_{sub} = 1000$ is a sizable portion of the total, $N = 4000$. It's easy to engineer much larger-N examples where laGP is the only viable option in terms of accuracy per unit flop. The real power comes from combining global and local estimates.

A trick from Sun et al. (2019a) is to use estimated global lengthscales $\hat{\theta}^{(g)}$ to pre-scale inputs X_N, and any testing locations $\mathcal{X}$, so that afterwards the effective global lengthscales are 1. Careful, θ-values modulate squared distance, so a square root must be taken before applying these back on the original scale of X_N.

```
scale <- sqrt(that$d)
Xs <- X
XXs <- XX
for(j in 1:ncol(Xs)) {
   Xs[,j] <- Xs[,j]/scale[j]
   XXs[,j] <- XXs[,j]/scale[j]
}
```

Next fit LAGPs on these scaled inputs, stretching and compressing the input space, achieving a "multi-resolution effect". With scaled inputs, an initial setting of d=1 makes sense. Otherwise the call below is the same as our aGP above. Notice that Y-values are the same as before. We're not modeling a residual obtained from the "sub"-predictor, although we could. Local structure is isotropic (via aGP), but we could similarly do separable (via aGPsep). Both have been done before – even together – with success. The LAGP comparator included in a competition summarized by Heaton et al. (2018) was set up in that way, using aGPsep on residuals and pre-scaled inputs obtained from a random subset-trained GP predictor.

```
out3  <- aGP(Xs, Y, XXs, d=list(start=1, max=20), omp.threads=nth, verb=0)
```

Before making yet another table with one new column, how about we do one more thing first (for two more columns instead)? The default nugget value in laGP/aGP is too large for most deterministic computer experiment applications. It was chosen conservatively so new users don't get frustrated by inscrutable error messages. But we can safely dial it way down for this borehole example.

```
out4  <- aGP(Xs, Y, XXs, d=list(start=1, max=20), g=1/10000000,
   omp.threads=nth, verb=0)
```

TABLE 9.2: Comparing CSK and LAGP.

	times	scores
sparse99	2945.376	-1.918
dense	3066.946	-6.966
sparse999	351.746	-2.105
aGP	4.917	-0.654
aGP2	4.739	-0.615
aGPs	4.920	0.036
sub	110.106	0.638
aGPsm	4.746	1.008
aGPsmg	4.718	5.288

Table 9.2 finishes off the comparison with a prettier presentation, including the two new columns.

```
times <- c(times, aGPsm=as.numeric(out3$time),
  aGPsmg=as.numeric(out4$time))
scores <- c(scores, aGPsm=scorep(YY, out3$mean, out3$var),
  aGPsmg=scorep(YY, out4$mean, out4$var))
kable(t(rbind(times, scores)), digits=3,
  caption="Comparing CSK and LAGP.")
```

Wow, that's lots better without lots more time! Technically, some of the times in the table above should incorporate time accrued by the earlier calculations they condition on. For example, a full accounting of compute times for aGPsm and aGPsmg would be 5 seconds longer. A caveat is that global random subset-based GP training of $\hat{\theta}^{(g)}$ represents a fixed startup cost no matter how large the predictive set $|\mathcal{X}|$ is.

9.3.5 Details and variations

Before turning to more realistic examples, including our motivating satellite drag application from §2.3 and CRASH calibration from §2.2, let's codify LAGP algorithmically. Then I shall introduce some of the rather newer variations developed in order to support those challenging applications.

Algorithm

Pseudocode in Algorithm 9.1 details aGPsep, although its application with a singleton $\mathcal{X} = \{x\}$, skipping Steps 1 and 3 (focusing only on 2), provides the essence of laGPsep. Local approximate GP regression at x is deployed as a subroutine applied identically over elements of a vast predictive set $\mathcal{X}$. Isotropic aGP/laGP arise as trivial simplifications. The pseudocode is agnostic to a choice of active learning heuristic $J(\cdot, \cdot)$ and covariance kernel $K(\cdot, \cdot)$ except that the latter be distance-based as NNs are used in several places, e.g., as a means of priming and of short-cutting unnecessarily huge exhaustive searches through pre selection of nearby candidates.

Algorithm 9.1 (Local) Approximate GP Regression

Assume criterion $J(x_{j+1}, x; \theta)$, e.g., MSPE (9.2) or ALC, on distance-based covariance through hyperparameters θ which are vectorized below; any priors/restrictions on θ may be encoded through a log posterior/penalty addendum to GP log likelihood $\ell(\theta; D)$.

Require large-N training data $D_N = (X_N, Y_N)$ and predictive/testing set $\mathcal{X}$ which might be singleton $\mathcal{X} = \{x\}$; local design size $n \ll N$ with NN init size $n_0 < n$ and NN search window size $n \le N' \ll N$.

Then

1. Optional multi-resolution modeling: **additionally require** subset size $n < n_{\text{sub}} \ll N$ and solve for global MLE hyperparameters; **otherwise require** initial θ_0 and set $\theta \leftarrow \theta_0$.

   a. Draw n indices $I_n \subset \{1, \ldots, N\}$ without replacement and form global data subset $D_n = (X_n, Y_n)$ where $X_n = X_N[I_n, :]$ and $Y_n = Y_N[I_n]$

   b. Calculate $\hat{\theta}^{(g)} = \text{argmin}_\theta - \ell(\theta)$.

   c. Scale training and testing inputs. For each of $k = 1, \ldots, m$:

   $$X_N[:, k] \leftarrow X_N[:, k] \Big/ \sqrt{\hat{\theta}_k^{(g)}} \quad \text{and} \quad \mathcal{X}[:, k] \leftarrow \mathcal{X}[:, k] \Big/ \sqrt{\hat{\theta}_k^{(g)}}.$$

   d. Set $\theta \leftarrow 1$.

2. **(Parallel) For** each predictive location $x \in \mathcal{X}$ calculate GP surrogate with data subset $D_n(x)$.

   a. Build candidate set $X_{N'}^{\text{cand}}(x)$ of N' NNs in X_N to x.

   b. Initialize $X_{n_0}(x)$ with n_0 nearest $X_{N'}^{\text{cand}}(x)$ to x and remove these from the candidate set, yielding $X_{N'-n_0}^{\text{cand}}(x)$.

   c. **For** $j = n_0, \ldots, n - 1$, acquire the next local design element.

      i. Optimize criterion J to select

      $$x_{j+1} = \text{argmin}_{x' \in X_{N'-j}^{\text{cand}}} J(x', x; \theta) \quad (\text{exhaust., approx., or parallel}).$$

      ii. Update $X_{j+1}(x) \leftarrow X_n \cup \{x_{j+1}\}$ and $X_{N'-j-1}^{\text{cand}}(x) \leftarrow X_{N'-j}^{\text{cand}}(x) \setminus x_j$.

   End For

   d. Pair $X_n(x)$ with Y_n-values to form local data $D_n(x)$.

   e. Optionally update hyperparameters $\theta \leftarrow \hat{\theta}_n(x)$ where

   $$\hat{\theta}_n(x) = \text{argmin}_\theta - \ell(\theta; D_n(x)).$$

   f. Record predictions, e.g., pointwise means and variances (5.2) given θ and $D_n(x)$.

   End For

3. Optionally repeat Step 2 with $\theta \leftarrow \hat{\theta}_n(x)$-values, separately for each x, potentially after smoothing.

Return predictions on $\mathcal{X}$, e.g., as $|\mathcal{X}| \times 2$ matrix of means and variances.

Several details, caveats and variations are worth explaining. Implementation in the `laGP` package calculates Student-t predictive equations under a reference prior for local scale parameter τ^2. Consequently, an additional degrees-of-freedom parameter is returned. Optionally, when `Xi.ret=TRUE` an $n \times |\mathcal{X}|$ matrix of indices into X_N is returned to enable rebuilding of local designs and GP surrogates. Local MLE calculations (Step 2e) can be turned off with modifications to the `d` argument; incorporation of nuggets is facilitated by modifications to the default `g` argument. Only ALC is fully implemented by all variations: separable, isotropic, and otherwise. MSPE and Fisher information (FI) are provided as options for isotropic implementations only. Fully NN local approximation, which is accommodated by specifying $n_0 = n$, could skip many of the lettered sub-steps of Step 2 in the algorithm. These are superfluous as NN calculation is not influenced by hyperparameters θ.

Step 1, creating a multi-resolution effect through pre-estimation of global lengthscale, assumes random subsampling; a BLHS alternative is similar, but would require an additional `for` loop. See `blhs` in the package documentation for more detail on this variation. Working with residuals from a global subset predictor (Heaton et al., 2018) is a potential variation that may be worth exploring. Step 2 may be SMP parallelized (on a single multi-core compute node) through the `omp.threads` argument to `aGP/aGPsep`. For distributed parallelization over the nodes of a cluster, see `aGP.parallel`. At this time there's no `aGPsep.parallel` analog although a bespoke translation would be rather straightforward.

Step 2, implementing `laGP/laGPsep` local subdesign search as a subroutine, is initialized with `start` $\equiv n_0 = 6$ NNs by default. Smaller `start` is not recommended on numerical stability grounds. Remaining local design selections, up to `end` $\equiv n = 50$, follow the greedy search criterion optimized in Step 2c. Rather than search over all $X_N \setminus X_j(x)$, which might be an enormous set, Algorithm 9.1 utilizes local candidates $X_{N'}^{\text{cand}}(x)$ of N' NNs to x. The number `close` $\equiv N'$ of such candidates offers a compromise between full enumeration and an more limited approximate search. Package default `close=1000` yields identical results compared to more exhaustive alternatives on a wealth of benchmark problems. Choosing $N' = n$ would facilitate a clumsy reduction to fully NN local approximation.

Several alternative Step 2c implementations offer potential for further speedups from shortcuts under ALC. Providing `method="alcray"` replaces iteration over candidates with a 1d line search over rays emanating from x (Gramacy and Haaland, 2016), with solutions snapped back to candidates. Derivatives offer another way to replace discrete with continuous search. Sun et al. (2019a) provide details, with implementation as `method="alcopt"`. In both cases the number of candidates N', to which continuous solutions are snapped, can safely be increased without taking a substantial computational hit. The `laGP` package automatically uses `close=10000` for such cases. Gramacy et al. (2014) describe an exhaustive GPU-based search implemented in CUDA[38]. Arguments including `num.gpus`, `parallel="gpu"`, and `alc.gpu` support this interface, however special compilation of the original sources is required to enable these features. Speedups up to a factor of seventy have been observed in benchmarking exercises.

Finally, Step 3 provides the option of multiple passes through local design with refinements of hyperparameterization $\hat{\theta}_n(x)$ learned in earlier stages. One re-pass can yield minor improvements in terms of predictive accuracy; more re-passes are not usually beneficial. When stretching and compressing inputs with globally estimated lengthscales $\hat{\theta}^{(g)}$ from Step 1, re-passes offer limited additional benefit. (That is, typically one would not engage both of Steps 1 and 3 simultaneously.)

---

[38]https://www.geforce.com/hardware/technology/cuda

Joint path sampling

A downside of Algorithm 9.1 – equations furnished by `aGP`/`aGPsep`, or cluster analog `aGP.parallel` – is that predictive summaries are point-wise. A consequence of statistical independence, from which parallelization advantages stem, is a lack of predictive covariance across $\mathcal{X} \equiv \mathsf{XX}$. Sun et al. (2019a) describe how to extend LAGP to sets of points, which they call *joint path sampling*, motivated by a desire for joint predictive equations along a trajectory of orbits in their/our motivating satellite drag application (§2.3).

Consider a fixed, discrete and finite set of input locations $W \subset \mathcal{X}$. A natural extension of the ALC criterion, the essence of which is reduction in variance $v_j(x) - v_{j+1}(x)$, is

$$v_j(W) - v_{j+1}(W) = \frac{1}{|W|} \sum_{w \in W} \{v_j(w) - v_{j+1}(w)\},$$

$$= \frac{1}{|W|} \sum_{w \in W} \left\{ k_j^\top(w) G_j(x_{j+1}) v_j(x_{j+1}) k_j(w) + 2 k_j^\top(w) g_j(x_{j+1}) K(x_{j+1}, w) + \frac{K(x_{j+1}, w)^2}{v_j(x_{j+1})} \right\}.$$

Notation here, including g_j and G_j, etc., is borrowed from partitioned inverse-based updating equations (6.8) from §6.3. Observe that $v_j(W) - v_{j+1}(W)$ is a scalar, measuring average reduction in predictive variance over W. We wish to maximize over this criterion to greedily determine new x_{j+1} and augment local design $X_n(W)$. Otherwise the setup is the same as $X_n(x)$ with ALC, which arises as a special case under the degenerate path $W = \{x\}$.

To illustrate, consider again Herbie's tooth in 2d. Whenever you provide a set of reference locations to `laGP`/`laGPsep`, via `Xref` comprised of multiple rows, a joint path ALC criterion is automatically engaged. Joint path sampling is not implemented in the package for other criteria, J, like MSPE. Code below recreates training data from earlier and designs a path W in 2d.

```
x <- seq(-2, 2, by=0.02)
X <- as.matrix(expand.grid(x, x))
Y <- herbtooth(X)
wx <- seq(-0.85, 0.45, length=100)
W <- cbind(wx - 0.75, wx^3 + 0.51)
```

Next, three variations on joint path sampling are entertained. The first is based on ALC. When using candidates $X_{N'}^{\text{cand}}(W)$ to shortcut local search it makes sense to allow N' to grow with the number of points (or length) of the path, $|W|$. The second is NN: ignoring J and gathering samples based on nearest neighbors calculated in terms of aggregated Euclidean distance for all elements of W. No candidates are required for this comparator. The third one approximates the first using derivative-based/continuous search (Sun et al., 2019a), with snaps back to $X_{N'}^{\text{cand}}(W)$.

```
p.alc <- laGPsep(W, 6, 100, X, Y, close=10000, lite=FALSE)
p.nn <- laGPsep(W, 6, 100, X, Y, method="nn", close=10000, lite=FALSE)
p.alcopt <- laGPsep(W, 6, 100, X, Y, method="alcopt", lite=FALSE)
```

Providing `lite=FALSE` causes full predictive covariance matrices (5.3) to be returned. Visuals for these local joint path designs follow in Figure 9.29. Points selected by ALC are indicated as red open circles and denoted as "ALC-ex". Suffix "-ex" reminds us that they're based on exhaustive, discrete search. Analogous ones from NN (green diamonds) and derivative-based optimization (blue diamonds/"ALC-opt") have been shifted down-right and up-left, respectively, to ease visualization. Path W, in all three cases (shifted as appropriate), is shown as a solid black line.

```
plot(W, type="l", xlab="x1", ylab="x2", xlim=c(-2.25,0),
  ylim=c(-0.75,1.25), lwd=2)
points(X[p.alc$Xi,], col=2, cex=0.6)
lines(W[,1] + 0.25, W[,2] - 0.25, lwd=2)
points(X[p.nn$Xi,1] + 0.25, X[p.nn$Xi,2] - 0.25, pch=22, col=3, cex=0.6)
lines(W[,1] - 0.25, W[,2] + 0.25, lwd=2)
points(X[p.alcopt$Xi,1] - 0.25, X[p.alcopt$Xi,2] + 0.25, pch=23,
  col=4, cex=0.6)
legend("bottomright", c("ALC-opt", "ALC-ex", "NN"),
  pch=c(22, 21, 23), col=c(4, 2, 3))
```

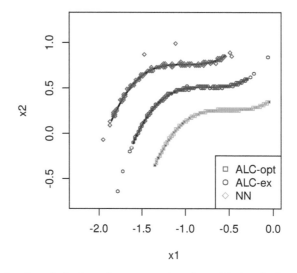

FIGURE 9.29: Joint local design along a path of predictive/reference locations under three criteria. Two paths are shifted from the original (middle) to ease visualization.

Observe how "ALC-ex" sacrifices some NNs for satellite points, primarily selected along rays emanating from the ends of the line W. Some are quite far out. Derivative-based search, exemplified by "ALC-opt", exhibits similar behavior but to a lesser extent. Instead it prefers satellite points off the middle of W, which are likely inferior. Value in "ALC-opt" is primarily computational.

```
c(alc=p.alc$time, alcopt=p.alcopt$time, nn=p.nn$time)
```

```
##     alc.elapsed alcopt.elapsed    nn.elapsed
##          10.836          0.221         0.044
```

Assessing these comparators out-of-sample would only be fair upon averaging over representative, but random, such W. See `randLine` in the package. Sun et al. provide several such comparisons, including pointwise benchmarks. To summarize the outcome of those experiments: ALC-ex is the most accurate but also the slowest; NN is the fastest but least accurate. ALC-opt offers a nice compromise. As a testament to the value of joint path sampling, Figure 9.30 shows samples from the joint predictive distribution (5.3) provided by "ALC-ex".

```
YY <- rmvnorm(100, p.alc$mean, p.alc$Sigma)
matplot(wx, t(YY), col="gray", type="l", lty=1,
  xlab="Indices wx of W", ylab="Y(W)")
```

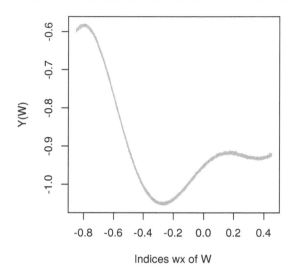

FIGURE 9.30: Full-covariance-based draws from the posterior predictive distribution along the local path from Figure 9.29.

Such samples would not be possible with an ordinary, pointwise (aGP/aGPsep) alternative. Note that although this example featured a connected W, sampled discretely on a grid, the joint path sampling methodology, and its implementation in laGP/laGPsep, is applicable to any Xref $\equiv W$ characterized as a discrete set of points and stored in a matrix. Recall that we looked at a similar setting in the context of reference-based IMSPE design in §6.1.2.

That concludes our methodological description and demonstration of LAGP surrogate modeling. For further examples, consult package documentation or see Gramacy (2016), which is also available as vignette("laGP") with the package. We turn now to two motivating applications where LAGP methods excel: large-scale satellite drag emulation (§2.3) and Kennedy and O'Hagan (KOH) style modularized calibration (§8.1.2) of radiative shock hydrodynamics computer simulations and field measurements (§2.2).

9.3.6 Two applications

Satellite drag

Let's revisit the Hubble Space Telescope (HST) atmospheric drag simulations introduced in §2.3. Recall that the goal was to be able to predict drag coefficients (our response), globally in low Earth orbit (LEO), to within 1% root mean-squared percentage error (RMSPE). In Chapter 2 we restricted our analysis to a small training dataset with limited ranges of yaw and pitch angles. Here data from a larger simulation campaign is entertained over the full range of those angles. There are eight inputs, including HST's panel angle, and supplementary material linked from the book web page[39] contains files with 2 million runs obtained on LHS designs, separately for each chemical species (O, O_2, N, N_2, He, H). Our focus here will be on helium (He) runs, one of the harder species, whose design is built as two separate 1 million-sized LHSs.

Consider the first 1-million run LHS only.

```
hstHe <- read.table("tpm-git/data/HST/hstHe.dat", header=TRUE)
nrow(hstHe)
```

```
## [1] 1000000
```

Inputs are in natural units, so begin by coding these to the unit cube.

```
m <- ncol(hstHe) - 1
X <- hstHe[,1:m]
Y <- hstHe[,m+1]
maxX <- apply(X, 2, max)
minX <- apply(X, 2, min)
for(j in 1:ncol(X)) {
  X[,j] <- X[,j] - minX[j]
  X[,j] <- X[,j]/(maxX[j] - minX[j])
}
range(Y)
```

```
## [1] 0.3961 9.1920
```

The range of output Y-values is relatively tame, so no need to re-scale these. It can help to center them, so that the mean-zero assumption in our (local approximate) GPs is not too egregiously violated, but that's not required to get good results in this exercise. Sun et al. (2019a) considered ten-fold CV on these data, and similarly for the other five species.

```
cv.folds <- function (n, folds=10)
  split(sample(1:n), rep(1:folds, length=n))
f <- cv.folds(nrow(X))
```

Our illustration here repeats one such random fold. Completing with the other nine folds is easily automated with a `for(i in 1:length(f))` loop wrapped around the chunks of code entertained below.

```
i <- 1 ## potentially replace with for(i in 1:length(f))) { ...
o <- f[[i]]
Xtest <- X[o,]
Xtrain <- X[-o,]
Ytest <- Y[o]
Ytrain <- Y[-o]
c(test=length(Ytest), train=length(Ytrain))
```

[39]http://bobby.gramacy.com/surrogates

```
##    test   train
## 100000 900000
```

Using these data, the first inferential step is to fit a random subset GP. That simple surrogate serves as a benchmark on its own, and as the basis of a multi-resolution global/local lengthscale approach (§9.3.4). All together, the plan is to entertain three comparators: global subset GP, local approximate GP, and multi-resolution local/global approximate GP. Recall that drag simulations are not deterministic. They're the outcome of a Monte Carlo (MC) solver called tpm. A large MC size was used to generate these data, so noise on outputs is quite low. To acknowledge that I've hard coded a small, but nontrivial nugget $g = 10^{-3}$ below. The curious reader may wish to try a (e.g., $10\times$) smaller nugget here. It doesn't help, but it doesn't hurt either. Out-of-sample mean accuracy results in this example are not much effected by the nugget as long as g is chosen to be reasonably small on the scale of range(Y).

```
da.orig <- darg(list(mle=TRUE), Xtrain, samp.size=10000)
sub <- sample(1:nrow(Xtrain), 1000, replace=FALSE)
gpsi <- newGPsep(Xtrain[sub,], Ytrain[sub], d=0.1, g=1/1000, dK=TRUE)
mle <- mleGPsep(gpsi, tmin=da.orig$min, tmax=10*da.orig$max, ab=da.orig$ab)
psub <- predGPsep(gpsi, Xtest, lite=TRUE)
deleteGPsep(gpsi)
rmspe <- c(sub=sqrt(mean(((100*(psub$mean - Ytest)/Ytest)^2)))
rmspe
```

```
##   sub
## 11.4
```

These results are similar to ones we obtained with a pilot exercise in §2.3: not even close to the 1% target. How about a separable local GP?

```
alcsep <- aGPsep(Xtrain, Ytrain, Xtest, d=da.orig, omp.threads=nth, verb=0)
rmspe <- c(rmspe, alc=sqrt(mean(((100*(alcsep$mean - Ytest)/Ytest)^2)))
rmspe
```

```
##    sub    alc
## 11.398  5.829
```

Much better, but not quite to 1%. Our final comparator is the global/local multi-resolution approximation based on local GPs applied to inputs stretched and compressed by the mle calculated on the subset above.

```
for(j in 1:ncol(Xtrain)) {
  Xtrain[,j] <- Xtrain[,j]/sqrt(mle$d[j])
  Xtest[,j] <- Xtest[,j]/sqrt(mle$d[j])
}
```

Before setting things running it helps to re-construct a default prior appropriate for scaled inputs.

```
da.s <- darg(list(mle=TRUE), Xtrain, samp.size=10000)
da.s$start <- 1
```

Finally, fit locally on globally scaled inputs.

```
alcsep.s <- aGPsep(Xtrain, Ytrain, Xtest, d=da.s, omp.threads=nth, verb=0)
rmspe <- c(rmspe, alcs=sqrt(mean((100*(alcsep.s$mean - Ytest)/Ytest)^2)))
rmspe
```

```
##    sub    alc   alcs
## 11.398  5.829  0.779
```

Although results here are somewhat sensitive to the random CV fold, more often than not this local–global surrogate beats 1%. In addition to being more accurate, the multi-resolution fit is actually faster than its pure-local alternative.

```
round(c(alc=alcsep$time/60, alcs=alcsep.s$time/60))
```

```
##  alc.elapsed alcs.elapsed
##          51           40
```

This speedup, of about 22%, stems from two factors. One is that local MLE calculations for $\hat{\theta}(x)$ require fewer iterations after scaling inputs and initializing search from d=1.

```
itrat <- alcsep$mle$dits/alcsep.s$mle$dits
c(better=mean(itrat > 1), x2=mean(itrat > 2))
```

```
## better      x2
## 0.7947  0.2486
```

As you can see above, 79% of testing locations $x \in \mathcal{X} \equiv$ Xtest required fewer BFGS iterations for the latter comparator, corresponding to the multi-resolution case, than for the former. About 25% of the time the latter was two times better. Fewer iterations means faster execution. It also means less OpenMP blocking to circumvent static variable shortcomings in R's internal BFGS implementation, and thus higher engagement of running threads on idle cores. The first, alcsep run had 400% engagement (of a total 800%) on my 8-core machine; alcsep.s had about 500% thanks to lower latency in OpenMP blocking from faster lbfgsb calculations. Finally, a similar aGPsep run from an laGP package linked to Intel MKL R (and executed off-line of this Rmarkdown build) gets consistently above 700% because that latency is further reduced by fast matrix decompositions in local MLE calculations. Total runtime for alcsep.s is reduced from 40 down to about 21 minutes with MKL. That's pretty amazing for 100 thousand predictions on 900 thousand training points.

How may even better results be obtained? Several ways: use more data (augmenting with the other 1 million LHS); use BLHS rather than random subsampling for estimating $\hat{\theta}^{(g)}$; perform a second stage of local analysis, conditioned upon $\hat{\theta}_n(x)$. Keen readers are encouraged to pursue these alternatives in a homework exercise combining all six pure-species predictors (i.e., H, N, N_2, O, and O_2 in addition to He, above) in order to obtain accurate ensemble predictors (§2.3.2) out-of-sample.

Large-scale calibration

Modularized KOH calibration from §8.1.2 relied on computer surrogate model predictions at a small number n_F of field data sites X_{n_F} paired with "promising" values of the calibration parameter u. In other words, surrogate GP prediction is only required at a relatively small set of locations, determined on-line as optimization over u proceeds in search of $\hat{u}$, regardless of the (potentially massive) size of the computer experiment $N_M \gg n_F$. Local GPs couldn't be more ideal for this situation, furnishing quick approximations to $y^M(x, \cdot)$ nearby testing sites, e.g., for $x \in X_{n_F}$ paired with u-values along a trajectory toward $\hat{u}$. Algorithm 9.2 makes that explicit by re-defining Steps 1–2 of our earlier (full GP/agnostic surrogate) version in Algorithm 8.2 from §8.1.3.

Algorithm 9.2 LAGP-based Modifications to Modularized KOH Calibration via Optimization (Algorithm 8.2)

Assume and **Require** are unchanged except that `laGP`-based methods are used for surrogate modeling $y^M(\cdot, \cdot)$.

Then

1. Let $[X_{n_F}, u^\top]$ denote calibration-parameter augmented field data locations obtained by concatenating $u^\top$ identically to each of n_F rows of field data inputs X_{n_F}.
2. Form the objective `obj(u)` as follows:
   a. Obtain surrogate predictive mean values $\hat{Y}_{n_F}^{M|u} = \mu(X_{n_F}, u)$ from local surrogate $\hat{y}^M(\cdot, u)$, e.g., via `aGP/aGPsep` evaluations with `XX` $\equiv [X_{n_F}, u^\top]$ and training data `X` $\equiv (X_{N_M}, U_{N_M})$ and `Y` $\equiv Y_{N_M}$.
   b. – d. unchanged
3. Solve $\hat{u} = \mathrm{argmin}_u$ `obj`(u) with a derivative-free solver,
   • e.g., `snomadr` in the `crs` package for R.
4. unchanged

Return unchanged.

Several variations may be worth entertaining. LAGP surrogate fits, via `aGP/aGPsep`, could be enhanced with multi-resolution global/local or multi-pass local refinement of lengthscale. Steps 2b–d, which detail fitting discrepancy between computer model predictions and field data responses can utilize GP regression with a nugget, as favored in §8.1.3. Alternatively, something entirely different could be entertained, including a "nobias" alternative (§8.1.4). Local GPs only substitute for computer model surrogate. While they could equally be deployed for discrepancy, which might be helpful if n_F is also large, I'm not aware of any successful such applications in the literature.

It's easy to overlook Step 3 in Algorithm 9.2 as an insignificant modification. It calls for derivative-free optimization (Larson et al., 2019) over u, whereas in Algorithm 8.2 `optim` was recommended. The trouble here is the discrete nature of independent local design searches for $\hat{y}^M(x_j^F, u)$, for each index $j = 1, \dots, n_F$ into X_{n_F}. These ensure an objective (8.4), quoted here as

$$p(u) \left[\max_{\theta_b} p_b(\theta_b \mid D_{n_F}^b(u)) \right],$$

that's not continuous in u. Discontinuous objectives thwart derivative (BFGS) or continuity-based (Nelder-Mead) solvers, although some are more robust to such nuances than others. Gramacy et al. (2015) suggest *mesh adaptive direct search* (MADS; Audet and Dennis Jr, 2006) as implemented in NOMAD[40]. An interface for R is furnished by subroutines in crs (Racine and Nie, 2018) on CRAN.

As MADS is a local solver, NOMAD requires initialization. Gramacy et al. suggest choosing starting $u^{(0)}$-values from the best value(s) of the objective found on a small space-filling design. The laGP package contains several functions that automate the objective explained in Algorithm 9.2: fcalib is like calib from Chapter 8, returning $-\text{obj}(u)$ evaluations (in log space) for minimization; discrep.est is like bhat.fit; special cases for nobias calibration are also implemented. For more details see Section 4.2 of Gramacy (2016), or as vignette("laGP").

To illustrate, revisit the setup from exercise #2 in §8.3 with synthetic computer model

$$y^M(x, u) = \left(1 - e^{-\frac{1}{2x_2}}\right) \frac{1000u_1 x_1^3 + 1900x_1^2 + 2092x_1 + 60}{100u_2 x_1^3 + 500x_1^2 + 4x_1 + 20},$$

borrowed from Goh et al. (2013), but originally due to Bastos and O'Hagan (2009). Design inputs x and calibration parameters u are both unit two-dimensional, which eases visualization. An implementation in R is provided below.

```
M <- function(x, u)
 {
  x <- as.matrix(x)
  u <- as.matrix(u)
  out <- (1 - exp(-1/(2*x[,2])))
  out <- out*(1000*u[,1]*x[,1]^3 + 1900*x[,1]^2 + 2092*x[,1] + 60)
  out <- out/(100*u[,2]*x[,1]^3 + 500*x[,1]^2 + 4*x[,1] + 20)
  return(out)
 }
```

Again slightly simplifying from Goh et al. (2013), the field data are generated as

$$y^F(x) = y^M(x, u^\star) + b(x) + \varepsilon, \quad \text{where} \quad b(x) = \frac{10x_1^2 + 4x_2^2}{50x_1 x_2 + 10}$$

$$\text{and} \quad \varepsilon \stackrel{\text{iid}}{\sim} \mathcal{N}(0, 0.5^2),$$

using $u^\star = (0.2, 0.1)$. In R $b(\cdot)$ may be implemented as follows.

```
bias <- function(x)
 {
  x <- as.matrix(x)
  out <- (10*x[,1]^2 + 4*x[,2]^2)/(50*x[,1]*x[,2] + 10)
  return(out)
 }
```

Compared to the homework, our example here will follow Gramacy et al. (2015) for a slightly bigger field experiment with $n_F = 100$ runs formed of two replicates of 50-sized 2d LHS of

---

[40]https://www.gerad.ca/nomad/Project/Home.html

x-values; and a much bigger simulation experiment with $N_M = 10500$ runs comprised of a 4d LHS of size 10000 of (x, u)-values, augmented with runs at the 50 unique field data sites x randomly paired with a 500-sized 2d LHS of u-values. Code immediately below generates the field experiment component of these data, where Zu holds intermediate computer model evaluations at $u^\star$.

```
ny <- 50
X <- randomLHS(ny, 2)
u <- c(0.2, 0.1)
Zu <- M(X, matrix(u, nrow=1))
sd <- 0.5
reps <- 2
Y <- rep(Zu, reps) + rep(bias(X), reps) + rnorm(reps*length(Zu), sd=sd)
length(Y)
```

```
## [1] 100
```

Next, the computer experiment component is designed and simulations are run with a Z object storing output Y_{N_M}-values.

```
nz <- 10000
XU <- randomLHS(nz, 4)
XU2 <- matrix(NA, nrow=10*ny, ncol=4)
for(i in 1:10) {
  I <- ((i-1)*ny + 1):(ny*i)
  XU2[I, 1:2] <- X
}
XU2[,3:4] <- randomLHS(10*ny, 2)
XU <- rbind(XU, XU2)
Z <- M(XU[,1:2], XU[,3:4])
length(Z)
```

```
## [1] 10500
```

The following code chunk sets default priors and specifies details of the laGP-based surrogate and (full) GP bias.

```
bias.est <- TRUE
methods <- rep("alc", 2)
da <- d <- darg(NULL, XU)
g <- garg(list(mle=TRUE), Y)
```

Changing bias.est <- FALSE causes estimation of $\hat{b}(\cdot)$ to be skipped. Instead only the level of noise between computer model and field data is estimated, setting up a nobias calibrator. The methods vector specifies the nature of search and number of passes through simulation data for local design and inference. We'll be doing two passes of isotropic ALC local design search where the second pass is primed with $\hat{\theta}_n(x)$'s from the first one. The model is completed with a (log) prior density on calibration parameter u, chosen to be independent Beta with mode in the middle of the space. This choice is made primarily to avoid boundary-focused $\hat{u}$ as discussed in §8.1.5 and revisited in homework exercises.

```
lprior <- function(u, shape1=2, shape2=2)
  sum(dbeta(u, shape1, shape2, log=TRUE))
```

Now evaluate the objective on a space-filling grid to search for good starting values for subsequent NOMAD optimization. Our "grid" is actually a maximin LHS (§4.3) focused away from the edges of the input space.

```
initsize <- 10*ncol(X)
uinit <- maximinLHS(initsize, 2)
uinit <- 0.9*uinit + 0.05
```

The for loop below collects objective evaluations on that "grid" through fcalib calls.

```
llinit <- rep(NA, nrow(uinit))
for(i in 1:nrow(uinit)) {
  llinit[i] <- fcalib(uinit[i,], XU, Z, X, Y, da, d, g, lprior,
    methods, NULL, bias.est, nth, verb=0)
}
```

Finally we're ready to optimize. The snomadr interface allows a number of options to be passed to NOMAD. Those provided below have been found to work well in a number of laGP-based calibration examples. The NOMAD user guide[41] can be consulted for more detail and information on further options.

```
library(crs)
imesh <- 0.1
opts <- list("MAX_BB_EVAL"=1000, "INITIAL_MESH_SIZE"=imesh,
  "MIN_POLL_SIZE"="r0.001", "DISPLAY_DEGREE"=0)
```

Unfortunately snomadr doesn't provide any mechanism for saving progress information, which will be handy later for visualization. So fcalib has an optional save.global argument for specifying in which R environment (e.g., .GlobalEnv) to save such relevant information. R code below invokes snomadr on the best input(s) found on the starting grid, looping over them until a minimum number of NOMAD iterations has been reached. Usually just one pass through this outer loop is sufficient, i.e., using only the very best starting location. In situations where NOMAD may prematurely converge, having a backup starting location (or two) can be handy.

```
its <- 0
i <- 1
out <- NULL
o <- order(llinit)
while(its < 10) {
  outi <- snomadr(fcalib, 2, c(0,0), 0, x0=uinit[o[i],], lb=c(0,0),
    ub=c(1,1), opts=opts, XU=XU, Z=Z, X=X, Y=Y, da=da, d=d, g=g,
    methods=methods, M=NULL, verb=0, bias=bias.est, omp.threads=nth,
```

---

[41]https://www.gerad.ca/nomad/Downloads/user_guide.pdf

```
      uprior=lprior, save.global=.GlobalEnv)
    its <- its + outi$iterations
    if(is.null(out) || outi$objective < out$objective) out <- outi
    i <- i + 1
}
```

```
##
## iterations: 16
## time:        59
```

What came out? Here's some code that reads the `fcalib.save` object stored in `.GlobalEnv` to prepare a plot for the forthcoming figure. Objective (`fcalib`) evaluations from both grid and `NOMAD` optimization are combined and used as the basis of a reconstruction of a 2d surface in u-space.

```
Xp <- rbind(uinit, as.matrix(fcalib.save[,1:2]))
Zp <- c(-llinit, fcalib.save[,3])
wi <- which(!is.finite(Zp))
if(length(wi) > 0) {
  Xp <- Xp[-wi, ]
  Zp <- Zp[-wi]
}
library(akima)
surf <- interp(Xp[,1], Xp[,2], Zp, duplicate="mean")
u.hat <- out$solution
```

The `interp` command above, from `akima` (Akima et al., 2016) on CRAN, projects these "z" coordinates onto an "x–y" mesh used by `image` in Figure 9.31. Lighter/whiter colors correspond to higher (log) objective evaluations. The grid of initial runs is indicated by open circles; `NOMAD` evaluations are shown as green diamonds, and the best of those ($\hat{u}$) as a solitary blue diamond. The true data-generating value of the calibration parameter $u^\star$ is indicated by an intersecting pair of horizontal and vertical dashed lines.

```
image(surf, xlab="u1", ylab="u2", col=heat.colors(128),
  xlim=c(0,1), ylim=c(0,1))
points(uinit)
points(fcalib.save[,1:2], col=3, pch=18)
points(u.hat[1], u.hat[2], col=4, pch=18)
abline(v=u[1], lty = 2)
abline(h=u[2], lty = 2)
```

So true $u^\star$ is far from the $\hat{u}$ that we found. Indeed, the surface is fairly peaked around $\hat{u}$, giving very little support to regions nearby the true value. Admittedly, very few u-evaluations were tried nearby $u^\star$. It may be worth checking if our scheme missed an area of high likelihood.

```
obju <- fcalib(u, XU, Z, X, Y, da, d, g, lprior, methods, NULL, bias.est,
  nth, verb=0)
c(u.hat=out$objective, u.star=obju)
```

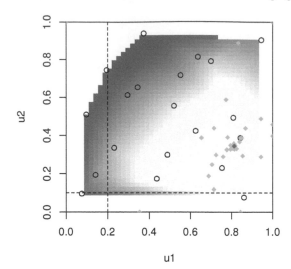

FIGURE 9.31: Log posterior surface for u built by linearly interpolating evaluations under an initial maximin LHS (open circles) and `NOMAD` evaluations (green diamonds). MAP estimate $\hat{u}$ is indicated by a blue diamond. Cross-hairs show true $u^\star$.

```
##   u.hat u.star
##   127.6   130.7
```

Nope: our $\hat{u}$ is better than the true $u^\star$. (Recall that `fcalib` is designed for minimization, so smaller is better.) Perhaps a better question is: which ($\hat{u}$ or $u^\star$) leads to better prediction out-of-sample? Obtaining a predictor that can be used at new testing locations, the following code goes step-by-step through calls automated by `fcalib`.

```
Xu <- cbind(X, matrix(rep(u, ny), ncol=2, byrow=TRUE))
Mhat.u <- aGP.seq(XU, Z, Xu, da, methods, ncalib=2,
  omp.threads=nth, verb=0)
cmle.u <- discrep.est(X, Y, Mhat.u$mean, d, g, bias.est, FALSE)
cmle.u$ll <- cmle.u$ll + lprior(u)
-cmle.u$ll
```

```
## [1] 130.7
```

The final line above serves as a sanity check that indeed that code duplicates the process behind `fcalib`: same as `obju` above. Entry `cmle.u$gp` holds the bias-correcting GP reference, which we shall use momentarily to make predictions. First, build up a testing set with computer model evaluations on 1000 new space-filling sites x, all paired with true data-generating value $u^\star$, followed by true discrepancy. No noise is added to these out-of-sample validation responses.

```
nny <- 1000
XX <- randomLHS(nny, 2)
ZZu <- M(XX, matrix(u, nrow=1))
YYtrue <- ZZu + bias(XX)
```

Now consider prediction and subsequent RMSE calculation with true $u^\star$.

```
XXu <- cbind(XX, matrix(rep(u, nny), ncol=2, byrow=TRUE))
Mhat.oos.u <- aGP.seq(XU, Z, XXu, da, methods, ncalib=2,
  omp.threads=nth, verb=0)
YYm.pred.u <- predGP(cmle.u$gp, XX)
YY.pred.u <- YYm.pred.u$mean + Mhat.oos.u$mean
rmse.u <- sqrt(mean((YY.pred.u - YYtrue)^2))
deleteGP(cmle.u$gp)
```

For estimated $\hat{u}$ we must first backtrack through what `fcalib` automated earlier and save surrogate predictions and estimated bias-corrections.

```
Xu <- cbind(X, matrix(rep(u.hat, ny), ncol=2, byrow=TRUE))
Mhat <- aGP.seq(XU, Z, Xu, da, methods, ncalib=2, omp.threads=nth, verb=0)
cmle <- discrep.est(X, Y, Mhat$mean, d, g, bias.est, FALSE)
cmle$ll <- cmle$ll + lprior(u.hat)
```

Here's a sanity check that this gives the same objective evaluation as what came out of snomadr.

```
c(-cmle$ll, out$objective)
```

```
## [1] 127.6 127.6
```

Finally, repeat what we did above with the true $u^\star$ value, but with estimated $\hat{u}$ instead.

```
XXu <- cbind(XX, matrix(rep(u.hat, nny), ncol=2, byrow=TRUE))
Mhat.oos <- aGP.seq(XU, Z, XXu, da, methods, ncalib=2,
  omp.threads=nth, verb=0)
YYm.pred <- predGP(cmle$gp, XX)
YY.pred <- YYm.pred$mean + Mhat.oos$mean
rmse <- sqrt(mean((YY.pred - YYtrue)^2))
deleteGP(cmle$gp)
```

How do these RMSEs compare?

```
c(u.hat=rmse, u=rmse.u)
```

```
##   u.hat      u
## 0.1126 0.1189
```

Indeed, our estimated $\hat{u}$ version leads to better predictions too. Clearly there's an identifiability issue – something other than the true data-generating parameter works better – in this supremely flexible calibration apparatus; but it does a good job of predicting. Other merits include that the framework is computationally thrifty. Although no timing results are quoted by the code above, this entire segment on large-scale `laGP`-based calibration takes but a few minutes to run. Our illustrative example was synthetic, but served to highlight many of the pertinent features of the methodology and its implementation. Porting that setup to the motivating CRASH data from §2.2 is rather straightforward. A homework exercise in §9.4 takes the reader through the details.

The core idea of tailoring GP prediction to the application goal, in a transductive fashion, is catching on as a means of gaining computational tractability and modeling fidelity. For example, Gul (2016) developed a so-called *in situ emulator* where the surrogate is tailored to a UQ task focused around a nominal input setting. Huang et al. (2020) develop a similar framework of *on-site surrogates (OSSs)* for a large-run (big N), large input dimension (m), computer model calibration problem. This is similar in spirit to laGP-based KOH calibration except that design of new computer model runs is incorporated directly into surrogate construction in lieu of sub-selecting points from an existing design.

To close out this chapter, it's worth acknowledging that the methods described herein are at best representative and at worst barely scratch the surface of the state-of-the-art of thrifty surrogate modeling. Pace of development of new methodology for large-scale GP approximation is feverish. My goal was to provide some depth, but at the expense of clear bias toward methodology I know well. Nevertheless we covered all of the important pillars: sparsity, divide-and-conquer, parallel computing and hybridization (e.g., global and local). A downside to all of these approaches, but especially those leveraging divide-and-conquer, is that they sacrifice smoothness and force compromise to be struck between local/dynamic behavior and global/long range features. For situations where, for example, mean and variance evolve smoothly throughout the input space – a feature common to many modern stochastic computer model simulations – a different perspective is warranted.

9.4   Homework exercises

Exercises focusing primarily on `tgp` and `laGP` follow, in several cases circling back to motivating examples from Chapter 2 with a proper treatment.

#1: Sequential design with `tgp`

Revisit exercise #5 from §6.4, on sequential design for a nonstationary process, but this time using treed Gaussian processes. In particular, consider #5b with ALM and ALC heuristics under `btgp`, using any approximations or options you deem necessary for the method to be fast enough on your machine in order to complete the full exercise. Repeat multiple times (say thirty), report average RMSE-based progress and the distribution of final RMSEs measured on the full dataset. As in the original description of the problem, restrict yourself to the `side` output, and two-dimensional input spaces in the `beta=2` slice. Make an explicit comparison between your new TGP-based results and those you previously obtained for the stationary GP (1) and manual 2-partition GP (2) in your solution to #5 from §6.4. For brownie points, extend the comparison to the full 3d input space.

#2: SARCOS robotics data

The SARCOS data[42] features as a prime example in Rasmussen and Williams (2006). These data come pre-partitioned as a set of about 44 thousand training runs, and 4.4 thousand testing runs, both having twenty-one inputs and seven outputs. Consider here just the first

---

[42]http://www.gaussianprocess.org/gpml/data/

of the seven outputs. For more details, see the link(s) above. MATLAB files storing these data have been converted to plain text and are linked here: train[43] and test[44]. Entertain the following comparators in terms of compute time (combining learning and prediction) and out-of-sample RMSE.

i. NN-based local approximate GP (LAGP) with isotropic and separable alternatives (aGP/aGPsep with `method=nn"`), local design size $n \equiv$ `end=50` (default) and $n \equiv$ `end=200`;
ii. ALC-based LAGP, both isotropic and separable (aGP/aGPsep with default `method="alc"`);
iii. ALC-based LAGP with approximate ray search, both isotropic and separable (aGP/aGPsep with `method="alcray"`);
iv. Separable GP trained on a random subsample of 1000 input–output pairs;
v. Revisit #i–iii with inputs stretched and compressed by the MLE lengthscales from #iv for a multi-resolution effect (§9.3.4).

You should have a total of seventeen comparators. Code inputs for #i–iv into $[0, 1]^{21}$, fix the nugget at $g = 1/1000000$ and use priors on the lengthscale built with `darg` using `samp.size=10000`. Base predictors for #v variations on scaled versions of coded inputs and appropriately modified priors. Provide visuals capturing time-versus-accuracy trade-offs, possibly on log scales.

#3: Ensemble satellite drag benchmarking

Revisit exercise #3 from §2.6 combining surrogates for pure-species simulations under a realistic mixture of chemical species in LEO, but this time use the big simulation sets over the entire range of pitch and yaw angles. Specifically, choose one of the following:

- HST via 2-million run (8d) LHS pure-species training sets, which may be found in `tpm-git/data/HST/hst[H,He,He2,N,N2,O,O2].dat`, and ensemble testing set residing in `tpm-git/data/HST/hstEns.dat`. Note that you'll need to combine He and He2 to obtain the full 2-million training runs for pure helium;
- or GRACE via 1-million run (7d) LHS pure-species training sets residing in `tpm-git/data/GRACE/grace[H,He,He2,N,N2,O,O2].dat` and ensemble testing set which can be found in `tpm-git/data/GRACE/graceEns.dat`.

Work with inputs coded to $[0, 1]^m$. Subsequently stretch and compress those coded inputs using MLE lengthscales obtained from a surrogate fit to a 1000-sized random subset to achieve a multi-resolution effect (§9.3.4). *Hint: When fitting a global subset GP to GRACE simulations, you might find it helpful to initialize MLE calculation for lengthscales with* `c(rep(1,5), 0.01, 0.1)`. Obtain predictions at coded and scaled ensemble inputs under the six multi-resolution global–local (§9.3.4) surrogates, and combine them by the appropriate weighted average (§2.3.2). Confirm that your out-of-sample RMSPE is below LANL's 1% target. You may wish to compare your results to BLHS-based pre-scaling instead (§9.3.4), using the `blhs` function in `laGP`.

#4: CRASH calibration

Consider the radiative shock hydrodynamics (CRASH) example from §2.2, which is described as a calibration problem. There are nine design variables (x), two calibration parameters

---

[43]http://bobby.gramacy.com/surrogates/sarcos_inv_train.csv
[44]http://bobby.gramacy.com/surrogates/sarcos_inv_test.csv

(u), two computer experiments with a combined five thousand or so runs, and a field data experiment with twenty observations. Combining the two computer experiments necessitates an expansion of the dataset vis-à-vis one of the calibration parameters, so that there are more like twenty thousand runs. That data setup is given in the code below.

```
## read in computer model runs
rs12 <- read.csv("crash/RS12_SLwithUnnormalizedInputs.csv")
rs13 <- read.csv("crash/RS13Minor_SLwithUnnormalizedInputs.csv")
## a correction
rs13$ElectronFluxLimiter <- 0.06

## read in the physical data
crash <- read.csv("crash/CRASHExpt_clean.csv")
## a correction
crash$BeThickness <- 21

## create expanded computer model data
sfmin <- rs13$EffectiveLaserEnergy/5000
sflen <- 10
rs13.sf <- matrix(NA, nrow=sflen*nrow(rs13), ncol=ncol(rs13)+2)
for(i in 1:sflen) {
  sfi <- sfmin + (1 - sfmin)*(i/sflen)
  rs13.sf[(i-1)*nrow(rs13) + (1:nrow(rs13)),] <-
    cbind(as.matrix(rs13), sfi, rs13$EffectiveLaserEnergy/sfi)
}
rs13.sf <- as.data.frame(rs13.sf)
names(rs13.sf) <- c(names(rs13), "EnergyScaleFactor", "LaserEnergy")

## merge the data.frames
rsboth <- rbind(rs12, rs13.sf[,names(rs12)])

## extract out Xs and Ys
XU.orig <- rsboth[,-which(names(rsboth) %in%
  c("FileNumber", "ShockLocation"))]
Z.orig <- rsboth$ShockLocation
X.orig <- crash[,names(XU.orig)[1:(ncol(XU.orig) - 2)]]
Y.orig <- crash$ShockLocation

## scale to coded outputs
minZ <- min(Z.orig)
maxZ <- max(Z.orig)
Z <- Z.orig - minZ
Z <- Z/(maxZ - minZ)
Y <- Y.orig - minZ
Y <- Y/(maxZ - minZ)

## scale to coded inputs
maxX <- apply(XU.orig, 2, max)
minX <- apply(XU.orig, 2, min)
XU < XU.orig
```

```
X <- X.orig
for(j in 1:ncol(XU)) {
  XU[,j] <- XU[,j] - minX[j]
  XU[,j] <- XU[,j]/(maxX[j] - minX[j])
  if(j <= ncol(X)) {
    X[,j] <- X[,j] - minX[j]
    X[,j] <- X[,j]/(maxX[j] - minX[j])
  }
}
```

Your task is to use `laGP` to perform a computer model calibration on these data. Report on the estimate of $\hat{u}$ that you obtained from a modularized KOH setup (estimate a bias correction; no need to entertain a nobias alternative). Finally, gather predictions at the following x-values: Be thick 21, laser energy 3800, Xe fill 1.15, tube diam 1150, taper length 500, nozzle length 500, aspect ratio 2, time 26. Note that this example is entertained in Gramacy et al. (2015), which you may use as guidance/inspiration. Although inputs and outputs are coded, report predictions, etc., back on the natural scale.

10

Heteroskedasticity

Historically, design and analysis of computer experiments focused on deterministic solvers from the physical sciences via Gaussian process (GP) interpolation (Sacks et al., 1989). But nowadays computer modeling is common in the social (Cioffi-Revilla, 2014, Chapter 8), management (Law, 2015) and biological (Johnson, 2008) sciences, where stochastic simulations abound. Queueing systems[1] and agent-based models[2] replace finite elements and simple Monte Carlo (MC) with geometric convergence (Picheny and Ginsbourger, 2013). Data in geostatistics (Banerjee et al., 2004) and machine learning (ML; Rasmussen and Williams, 2006) aren't only noisy but frequently involve signal-to-noise ratios that are low or changing over the input space. Noisier simulations/observations demand bigger experiments/studies to isolate signal from noise, and more sophisticated GP models – not just adding nuggets to smooth over noise, but variance processes to track changes in noise throughout the input space in the face of that *heteroskedasticity*.

Modeling methodology for large simulation efforts with intrinsic stochastic dynamics has lagged until recently. Partitioning is one option (§9.2.2), but leaves something to be desired when underlying processes are smooth and signal-to-noise ratios are low. A theme in this chapter is that *replication*, a tried and true design strategy for separating signal from noise, can play a crucial role in computationally efficient heteroskedastic modeling, especially with GPs. To be concrete, replication here means repeated y-observations at the same input x setting. In operations research[3], stochastic kriging (SK; Ankenman et al., 2010) offers approximate methods that exploit large degrees of replication. Its independent method of moments[4]-based inferential framework can yield an efficient heteroskedastic modeling capability. However the setup exploited by SK is highly specialized; moment-based estimation can strain coherency in a likelihood dominated surrogate landscape. Here focus is on a heteroskedastic GP technology (Binois et al., 2018) that offers a modern blend of SK and ideas from ML (largely predating SK: Goldberg et al., 1998; Quadrianto et al., 2009; Boukouvalas and Cornford, 2009; Lazaro-Gredilla and Titsias, 2011).

Besides leveraging replication in design, the idea is: what's good for the mean is good for the variance. If it's sensible to model the latent (mean) field with GPs (§5.3.2), why not extend that to variance too? Unfortunately, latent variance fields aren't as easy to integrate out. (Variances must also be positive, whereas means are less constrained.) Numerical schemes are required for variance latent learning, and there are myriad strategies – hence the plethora of ML citations above. Binois et al. (2018)'s scheme remains within our familiar class of library-based (e.g., BFGS) likelihood-optimizing methodology, and is paired with tractable strategies for sequential design (Binois et al., 2019) and Bayesian optimization (BO; Chapter 8). Other approaches achieving a degree of input-dependent variance include quantile kriging (QK; Plumlee and Tuo, 2014); use of pseudoinputs (Snelson and Ghahramani, 2006) or

---

[1]https://en.wikipedia.org/wiki/Queueing_theory
[2]https://en.wikipedia.org/wiki/Agent-based_model
[3]https://en.wikipedia.org/wiki/Operations_research
[4]https://en.wikipedia.org/wiki/Method_of_moments_(statistics)

predictive processes (Banerjee et al., 2008); and (non-GP-based) tree methods (Pratola et al., 2017a). Unfortunately, none of those important methodological contributions, to my knowledge, pair with design, BO or open source software.

```
library(hetGP)
```

Illustrations here leverage `hetGP` (Binois and Gramacy, 2019) on CRAN, supporting a wide class of alternatives in the space of coupled-GP heteroskedastic (and ordinary homoskedastic) modeling, design and optimization.

10.1 Replication and stochastic kriging

Replication can be a powerful device for separating signal from noise, offering a pure look at noise not obfuscated by signal. When modeling with GPs, replication in design can yield substantial computational savings as well. Continuing with N-notation from Chapter 9, consider training data pairs $(x_1, y_1), \ldots, (x_N, y_N)$. These make up the "full-N" dataset (X_N, Y_N). Now suppose that the number n of unique x_i-values in X_N is far smaller than N, i.e., $n \ll N$. Let $\bar{Y}_n = (\bar{y}_1, \ldots, \bar{y}_n)^\top$ collect averages of a_i replicates at unique locations $\bar{x}_i$, and similarly let $\hat{\sigma}_i^2$ collect normalized residual sums of squares for those replicate measurements:

$$\bar{y}_i = \frac{1}{a_i} \sum_{j=1}^{a_i} y_i^{(j)} \quad \text{and} \quad \hat{\sigma}_i^2 = \frac{1}{a_i - 1} \sum_{j=1}^{a_i} (y_i^{(j)} - \bar{y}_i)^2.$$

Unfortunately, $\bar{Y}_n = (\bar{y}_1, \ldots, \bar{y}_n)^\top$ and $\hat{\sigma}_{1:n}^2$ don't comprise of a set of sufficient statistics for GP prediction under the full-N training data. But these are nearly sufficient: only one quantity is missing and will be provided momentarily in §10.1.1. Nevertheless "unique-n" predictive equations based on $(\bar{X}_n, \bar{Y}_n)$ are a best linear unbiased predictor (BLUP):

$$\mu_n^{\mathrm{SK}}(x) = \nu k_n^\top(x)(\nu C_n + S_n)^{-1} \bar{Y}_n$$
$$\sigma_n^{\mathrm{SK}}(x)^2 = \nu K_\theta(x, x) - \nu^2 k_n^\top(x)(\nu C_n + S_n)^{-1} k_n(x), \qquad (10.1)$$

where $k_n(x) = (K_\theta(x, \bar{x}_1), \ldots, K_\theta(x, \bar{x}_n))^\top$ as in Chapter 5,

$$S_n = [\hat{\sigma}_{1:n}^2] A_n^{-1} = \mathrm{Diag}(\hat{\sigma}_1^2/a_1, \ldots, \hat{\sigma}_n^2/a_n),$$

$C_n = \{K_\theta(\bar{x}_i, \bar{x}_j)\}_{1 \le i,j \le n}$, and $a_i \gg 1$. This is the basis of *stochastic kriging* (SK; Ankenman et al., 2010), implemented as an option in `DiceKriging` (Roustant et al., 2018) and `mlegp` (Dancik, 2018) packages on CRAN. SK's simplicity is a virtue. Notice that when $n = N$, all $a_i = 1$ and $\hat{\sigma}_i^2 = \nu g$, a standard GP predictor (§5) is recovered when mapping $\nu = \tau^2$.[5] Eq. (10.1) has intuitive appeal as an application of ordinary kriging equations (5.2) on (almost) sufficient statistics. An independent moments-based estimate of variance is used in lieu of the more traditional, likelihood-based (hyperparametric) alternative. This could

---

[5]Letter ν is preferred here to the familiar τ^2 from Chapter 5 to avoid double squaring: $\nu^2 \equiv (\tau^2)^2$.

be advantageous if variance is changing in the input space, as discussed further in §10.2. Computational expedience is readily apparent even in the usual homoskedastic setting: $S_n = \text{Diag}(\frac{1}{n}\sum \hat{\sigma}_i^2)$. Only $\mathcal{O}(n^3)$ matrix decompositions are required, which could represent a huge savings compared with $\mathcal{O}(N^3)$ if the degree of replication is high.

Yet independent calculations have their drawbacks. Thinking heteroskedastically, this setup lacks a mechanism for specifying a belief that variance evolves smoothly over the input space. Lack of smoothness thwarts a pooling of variance that's essential for predicting uncertainties out-of-sample at novel $x' \in \mathcal{X}$. Basically we don't know $K_\theta(x, x)$ in Eq. (10.1), whereas homoskedastic settings use $1 + g$ for all x. Ankenman et al. (2010) suggest smoothing $\hat{\sigma}_i^2$-values with a second GP, but that two-stage approach will feel ad hoc compared to what's presented momentarily in §10.2. More fundamentally, numbers of replicates a_i must be relatively large in order for the $\hat{\sigma}_i^2$-values to be reliable. Ankenman et al. recommend $a_i \geq 10$ for all i, which can be prohibitive. Thinking homoskedastically, the problem with this setup is that it doesn't emit a likelihood for inference for other hyperparameters, such as lengthscale(s) θ and scale ν. Again, this is because $(\bar{Y}_n, S_n)$ don't quite constitute a set of sufficient statistics for hyperparameter inference.

10.1.1 Woodbury trick

The fix, ultimately revealing the full set of sufficient statistics for prediction and likelihood-based inference, involves Woodbury linear algebra identities[6]. These are not unfamiliar to the spatial modeling community (e.g., Opsomer et al., 1999; Banerjee et al., 2008; Ng and Yin, 2012). However, their application toward efficient GP inference under replicated design is relatively recent (Binois et al., 2018). Let D and B be invertible matrices of size $N \times N$ and $n \times n$, respectively, and let U and $V^\top$ be matrices of size $N \times n$. The Woodbury identities are

$$(D + UBV)^{-1} = D^{-1} - D^{-1}U(B^{-1} + VD^{-1}U)^{-1}VD^{-1} \qquad (10.2)$$

$$|D + UBV| = |B^{-1} + VD^{-1}U| \times |B| \times |D|. \qquad (10.3)$$

Eq. (10.3) is also known as the matrix determinant lemma[7], although it's clearly based on the same underlying principles as its inversion cousin (10.2). I refer to both as "Woodbury identities", in part because GP likelihood and prediction application requires both inverse and determinant calculations. To build $K_N = UC_nV^\top + D$ for GPs under replication, take $U = V^\top = \text{Diag}(1_{a_1}, \dots, 1_{a_n})$, where 1_k is a k-vector of ones, $D = g\mathbb{I}_N$ with nugget g, and $B = C_n$. Later in §10.2, we'll take $D = \Lambda_n$ where Λ_n is a diagonal matrix of latent variances.

Before detailing how the "Woodbury trick" maps to prediction (kriging) and likelihood identities for $\mathcal{O}(n^3)$ rather than $\mathcal{O}(N^3)$ calculations in §10.1.2, consider how it helps operate on a full covariance structure K_N through its unique counterpart $K_n = C_n + gA_n^{-1}$. The example below creates a matrix X_n with $n = 50$ unique rows and then builds X_N identical to X_n except that four of its rows have been replicated a number of times, leading to a much bigger $N = 150$-sized matrix.

```
n <- 50
Xn <- matrix(runif(2*n), ncol=2)
```

---

[6]https://en.wikipedia.org/wiki/Woodbury_matrix_identity
[7]https://en.wikipedia.org/wiki/Matrix_determinant_lemma

```
Xn <- Xn[order(Xn[,1]),]
ai <- c(8, 27, 38, 45)
mult <- rep(1, n)
mult[ai] <- c(25, 30, 9, 40)
XN <- Xn[rep(1:n, times=mult),]
N <- sum(mult)
```

Code below calculates covariance matrices K_n and K_N corresponding to X_n and X_N, respectively. Rather than use our usual **distance** or **covar.sep** from **plgp** (Gramacy, 2014), covariance matrix generation subroutines from **hetGP** are introduced here as an alternative. An arbitrary lengthscale of $\theta = 0.2$ and (default) nugget of $g = \epsilon$ is used with an isotropic Gaussian kernel. Separable Gaussian, and isotropic and separable Matèrn $\nu \in \{3/2, 5/2\}$ are also supported.

```
KN <- cov_gen(XN, theta=0.2)
Kn <- cov_gen(Xn, theta=0.2)
```

Next, build U for use in the Woodbury identity (10.2)–(10.3).

```
U <- c(1, rep(0, n - 1))
for(i in 2:n){
  tmp <- rep(0, n)
  tmp[i] <- 1
  U <- rbind(U, matrix(rep(tmp, mult[i]), nrow=mult[i], byrow=TRUE))
}
U <- U[,n:1]
```

Figure 10.1 shows these three matrices side-by-side, illustrating the mapping $K_n \to U \to K_N$ with choices of B and $U = V^\top$ described after Eq. (10.3), above. Note that in this case $C_n = K_n$ since nugget g is essentially zero.

```
cols <- heat.colors(128)
layout(matrix(c(1, 2, 3), 1, 3, byrow=TRUE), widths=c(2, 1, 2))
image(Kn, x=1:n, y=1:n, main="uniq-n: Kn", xlab="1:n", ylab="1:n", col=cols)
image(t(U), x=1:n, y=1:N, asp=1, main="U", xlab="1:n", ylab="1:N", col=cols)
image(KN, x=1:N, y=1:N, main="full-N: KN", xlab="1:N", ylab="1:N", col=cols)
```

Storage of K_n and U, which may be represented sparsely, or even implicitly, is not only more compact than K_N, but the Woodbury formulas show how to calculate requisite inverse and determinants by acting on $\mathcal{O}(n^2)$ quantities rather than $\mathcal{O}(N^2)$ ones.

10.1.2 Efficient inference and prediction under replication

Here my aim is to make SK simultaneously more general, exact, and prescriptive, facilitating full likelihood-based inference and conditionally (on hyperparameters) exact predictive equations with the Woodbury trick. Pushing the matrix inverse identity (10.2) through to

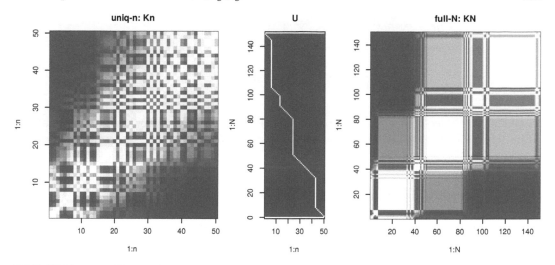

FIGURE 10.1: Example mapping $K_n \to U \to K_N$ through Woodbury identities (10.2)–(10.3).

predictive equations, mapping $\nu C_N + S_N = \nu(C_N + \Lambda_N) \equiv \nu(C_N + g\mathbb{I}_N)$ between SK and more conventional Chapter 5 notation, yields the following predictive identities:

$$\nu k_N^\top(x)(\nu C_N + S_N)^{-1}Y_N = k_n^\top(x)(C_n + \Lambda_n A_n^{-1})\bar{Y}_n \qquad (10.4)$$
$$vk - \nu^2 k_N^\top(x)(\nu C_N + S_N)^{-1}k_N(x) = vk - \nu k_n^\top(x)(C_n + \Lambda_n A_n^{-1})^{-1}k_n(x),$$

where $vk \equiv \nu K_\theta(x,x)$ is used as a shorthand to save horizontal space.

In words, typical full-N predictive quantities may be calculated identically through unique-n counterparts, potentially yielding dramatic savings in computational time and space. The unique-n predictor (10.4), implicitly defining $\mu_n(x)$ and $\sigma_n^2(x)$ on the right-hand side(s) and updating SK (10.1), is unbiased and minimizes mean-squared prediction error (MSPE) by virtue of the fact that those properties hold for the full-N predictor on the left-hand side(s). No asymptotic or frequentist arguments are required. Crucially, no minimum data or numbers of replicates (e.g., $a_i \geq 10$ for SK) are required to push asymptotic arguments through, although replication can still be helpful from a statistical efficiency perspective. See §10.3.1.

The same trick can be played with the concentrated log likelihood (5.8). Recall that $U_n = C_n + A_n^{-1}\Lambda_n$, where for now $\Lambda_n = g\mathbb{I}_n$ encoding homoskedasticity. Later I shall generalize Λ_n for the heteroskedastic setting. Using these quantities and Eqs. (10.2)–(10.3) simultaneously,

$$\ell = c + \frac{N}{2}\log\hat{\nu}_N - \frac{1}{2}\sum_{i=1}^{n}[(a_i - 1)\log\lambda_i + \log a_i] - \frac{1}{2}\log|K_n|, \qquad (10.5)$$

where $\hat{\nu}_N = \frac{1}{N}(Y_N^\top \Lambda_N^{-1} Y_N - \bar{Y}_n^\top A_n \Lambda_n^{-1} \bar{Y}_n + \bar{Y}_n^\top K_n^{-1} \bar{Y}_n).$

Notice additional terms in $\hat{\nu}_N$ compared with $\hat{\nu}_n = n^{-1}\bar{Y}_n^\top K_n^{-1} Y_n$. Thus $N^{-1}Y_N^\top \Lambda_N^{-1} Y_N$ is our missing statistic for sufficiency. Since Λ_N is diagonal, evaluation of ℓ requires just $\mathcal{O}(n^3)$ operations beyond the $\mathcal{O}(N)$ required for $\bar{Y}_n$. Closed form derivative evaluations are available in $\mathcal{O}(n^3)$ time too, facilitating library-based optimization, e.g., with BFGS.

$$\frac{\partial \ell}{\partial \cdot} = \frac{N}{2} \frac{\partial (Y_N^\top \Lambda_N^{-1} Y_N - \bar{Y}_n A_n \Lambda_n^{-1} \bar{Y}_n + n \hat{\nu}_n)}{\partial \cdot} \times (N \hat{\nu}_N)^{-1}$$

$$- \frac{1}{2} \sum_{i=1}^{n} \left[(a_i - 1) \frac{\partial \log \lambda_i}{\partial \cdot} \right] - \frac{1}{2} \mathrm{tr} \left(K_n^{-1} \frac{\partial K_n}{\partial \cdot} \right) \qquad (10.6)$$

Above, "$\cdot$" is a place-holder for hyperparameter(s) of interest. Fast likelihood and derivative evaluation when $N \gg n$ can be computationally much more efficient, compared to schemes introduced in Chapter 5, yet still reside under an otherwise identical numerical umbrella.

Example: computational advantage of replication

To demonstrate the potential benefit of this Woodbury mapping, consider the following modestly-sized example based on our favorite 2d test function (§5.1.2). You may recall we leveraged replication in an extension of this example (§5.2.4) in order to control signal-to-noise behavior in repeated Rmarkdown builds. Here the response is observed at $n = 100$ unique input locations, each having a random number of replicates $a_i \sim \mathrm{Unif}\{1, 2, \ldots, 50\}$. Otherwise the setup is identical to previous (noisy) uses of this data-generating mechanism.

```
library(lhs)
Xbar <- randomLHS(100, 2)
Xbar[,1] <- (Xbar[,1] - 0.5)*6 + 1
Xbar[,2] <- (Xbar[,2] - 0.5)*6 + 1
ytrue <- Xbar[,1]*exp(-Xbar[,1]^2 - Xbar[,2]^2)
a <- sample(1:50, 100, replace=TRUE)
N <- sum(a)
X <- matrix(NA, ncol=2, nrow=N)
y <- rep(NA, N)
nf <- 0
for(i in 1:100) {
  X[(nf+1):(nf+a[i]),] <- matrix(rep(Xbar[i,], a[i]), ncol=2, byrow=TRUE)
  y[(nf+1):(nf+a[i])] <- ytrue[i] + rnorm(a[i], sd=0.01)
  nf <- nf + a[i]
}
```

The code below invokes `mleHomGP` from the `hetGP` package in two ways. One cripples automatic pre-processing subroutines that would otherwise calculate sufficient statistics for unique-n modeling, forcing a more cumbersome full-N calculation. Another does things the default, more thrifty unique-n way. In both cases, `mleHomGP` serves as a wrapper automating calls to `optim` with `method="L-BFGS-B"`, furnishing implementations of log concentrated likelihood (10.5) and derivative (10.6).

```
Lwr <- rep(sqrt(.Machine$double.eps), 2)
Upr <- rep(10, 2)
fN <- mleHomGP(list(X0=X, Z0=y, mult=rep(1, N)), y, Lwr, Upr)
un <- mleHomGP(X, y, Lwr, Upr)
```

Repeating search bounds Lwr and Upr to match input dimension invokes a separable kernel, which is Gaussian by default. Execution times saved by the `mleHomGP` calls above show a dramatic difference.

```
c(fN=fN$time, un=un$time)
```

```
## fN.elapsed un.elapsed
##     168.90       0.03
```

Calculations on unique-n quantities using Woodbury identities results in execution that's 5630 times faster than an otherwise equivalent full-N analog. The outcome of those calculations, exemplified below through a reporting of estimated lengthscales, is nearly identical in both versions.

```
rbind(fN=fN$theta, un=un$theta)
```

```
##      [,1]  [,2]
## fN 1.092 1.941
## un 1.092 1.942
```

Excepting a user-triggered SK feature offered by `mlegp` and `DiceKriging` packages on CRAN, I'm not aware of any other software package, for R or otherwise, that automatically pre-process data into a format leveraging Woodbury identities to speed up GP calculations under heavy replication. Repeated sampling is a common design tactic in the face of noisy processes, and the example above demonstrates potential for substantial computational benefit when modeling with GPs. Replication is essential when signal-to-noise features may exhibit heterogeneity across the input space. Whenever two things are changing simultaneously, a means of pinning down one – in this case variance, if only locally – can be a game-changer.

10.2 Coupled mean and variance GPs

GP kernels based on Euclidean distance (e.g., Chapter 5 or above) emit stationary processes where input–output behavior exhibits highly regular dynamics throughout the input space. Yet we saw in §9.2.2–9.3 that many data-generating mechanisms are at odds with stationarity. A process can diverge from stationary in various ways, but few are well accommodated by computationally viable modeling methodology – exceptions in Chapter 9 notwithstanding. Input-dependent variance, or heteroskedasticity, is a particular form of nonstationarity that's increasingly encountered in stochastic simulation. A fine example is the motorcycle accident data introduced in §9.2.1. Our TGP fit to these data, shown in Figure 9.11, nicely captures noise regime changes from low, to high, to medium as inputs track from left to right. Accommodating those shifts required partitioning/hard breaks in the predictive surface which (we hoped) could be smoothed over by Markov chain Monte Carlo (MCMC) sampling from the Bayesian posterior on trees. Entertaining a smoother alternative might be worthwhile.

To lay the groundwork for a heteroskedastic GP-based alternative, and remind the reader of the setting and challenges involved, consider the following ordinary homoskedastic GP fit to these data, followed by prediction on a testing grid.

```
library(MASS)
hom <- mleHomGP(mcycle$times, mcycle$accel)
Xgrid <- matrix(seq(0, 60, length=301), ncol=1)
p <- predict(x=Xgrid, object=hom)
df <- data.frame(table(hom$mult))
colnames(df) <- c("reps", "howmany")
rownames(df) <- NULL
df
```

```
##   reps howmany
## 1    1      66
## 2    2      22
## 3    3       3
## 4    4       2
## 5    6       1
```

Considering the importance of separating signal from noise in this simulation experiment, perhaps it's not surprising that the design includes moderate replication. Figure 10.2 shows the resulting predictive surface overlaid on a scatterplot of the data. The solid-black line is the predictive mean, and dashed-red lines trace out a 90% predictive interval. Output from predict.homGP separates variance in terms of epistemic/mean (p$sd2) and residual (p$nugs) estimates, which are combined in the figure to show full predictive uncertainty.

```
plot(mcycle)
lines(Xgrid, p$mean)
lines(Xgrid, qnorm(0.05, p$mean, sqrt(p$sd2 + p$nugs)), col=2, lty=2)
lines(Xgrid, qnorm(0.95, p$mean, sqrt(p$sd2 + p$nugs)), col=2, lty=2)
```

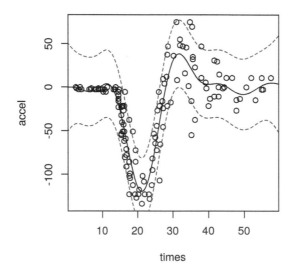

FIGURE 10.2: Homoskedastic GP fit to the motorcycle data via mean (solid-black) and 90% error-bars (dashed-red).

As in Figure 9.10, showing a (Bayesian) stationary GP, this fit is undesirable on several fronts. Not only is the variance off, but the mean is off too: it's way too wiggly in the left-hand, pre-impact regime. Perhaps this is due to noise in the right-hand regime being incorrectly interpreted as signal, which then bleeds into the rest of the fit because the covariance structure is stationary. Learning how variance is changing is key to learning about mean dynamics. Replication is key to learning about variance.

10.2.1 Latent variance process

Heteroskedastic GP modeling targets learning the diagonal matrix Λ_N, or its unique-n counterpart Λ_n; see Eq. (10.4). Allowing λ_i-values to exhibit heterogeneity, i.e., not all $\lambda_i = g/\nu$ as in the homoskedastic case, is easier said than done. Care is required when performing inference for such a high-dimensional parameter. SK (10.1) suggests taking $\Lambda_n A_n^{-1} = S_n = \text{Diag}(\hat{\sigma}_1^2, \ldots, \hat{\sigma}_n^2)$, but that requires large numbers of replicates a_i and is anyways useless out of sample. By fitting each $\hat{\sigma}_i^2$ separately there's no pooling effect for interpolation/smoothing. Ankenman et al. (2010) suggest the quick fix of fitting a second GP to the variance observations with "data":

$$(\bar{x}_1, \hat{\sigma}_1^2), (\bar{x}_2, \hat{\sigma}_2^2), \ldots, (\bar{x}_n, \hat{\sigma}_n^2),$$

to obtain a smoothed variance for use out of sample.

A more satisfying approach that's similar in spirit, coming from ML (Goldberg et al., 1998) and actually predating SK by more than a decade, applies regularization that encourages a smooth evolution of variance. Goldberg et al. introduce latent (log) variance variables under a GP prior and develop an MCMC scheme performing joint inference for all unknowns, including hyperparameters for both mean and noise GPs. The overall method, which is effectively on the order of $\mathcal{O}(TN^4)$ for T MCMC samples, is totally impractical but works well on small problems. Several authors have economized on aspects of this framework (Kersting et al., 2007; Lazaro-Gredilla and Titsias, 2011) with approximations and simplifications of various sorts, but none to my knowledge have resulted in public R software.[8] The key ingredient in these works, of latent variance quantities smoothed by a GP, has merits and can be effective when handled gingerly. Binois et al. (2018) introduced the methodology described below.

Let $\delta_1, \delta_2, \ldots, \delta_n$ denote latent variance variables (or equivalently latent nuggets), each corresponding to one of the n unique design sites $\bar{x}_i$ under study. It's important to introduce latents only for the unique-n locations. A similar approach in the full-N setting, i.e., without exploiting Woodbury identities, is fraught with identifiability and numerical stability challenges. Store these latents diagonally in a matrix Δ_n and place them under a GP prior:

$$\Delta_n \sim \mathcal{N}_n(0, \nu_{(\delta)}(C_{(\delta)} + g_{(\delta)}A_n^{-1})).$$

Inverse exponentiated squared Euclidean distance-based correlations in $n \times n$ matrix $C_{(\delta)}$ are hyperparameterized by novel lengthscales $\theta_{(\delta)}$ and nugget $g_{(\delta)}$. Smoothed λ_i-values can be calculated by plugging Δ_n into GP mean predictive equations:

$$\Lambda_n = C_{(\delta)}K_{(\delta)}^{-1}\Delta_n, \quad \text{where} \quad K_{(\delta)} = (C_{(\delta)} + g_{(\delta)}A_n^{-1}). \tag{10.7}$$

Smoothly varying Λ_n generalize both $\Lambda_n = g\mathbb{I}_n$ from a homoskedastic setup, and moment-estimated S_n from SK. Hyperparameter $g_{(\delta)}$ is a nugget of nuggets, controlling the smoothness

---

[8]A partial implementation is available for Python in GPy[9].

of λ_i's relative to δ_i's; when $g_{(\delta)} = 0$ the λ_i's interpolate δ_i's. Although the formulation above uses a zero-mean GP, a positive nonzero mean $\mu_{(\delta)}$ may be preferable for variances. The hetGP package automates a classical plugin estimator $\hat{\mu}_{(\delta)} = \Delta_n^\top K_{(\delta)}^{-1} \Delta_n (1_n^\top K_{(\delta)}^{-1} 1_n)^{-1}$. See exercise #2 from §5.5 and take intercept-only β. This seemingly simple twist unnecessarily complicates the remainder of the exposition, so I shall continue to assume zero mean throughout.

Variances must be positive, and the equations above give nonzero probability to negative δ_i and λ_i-values. One solution is to threshold latent variances at zero. Another is to model $\log \Delta_n$ in this way instead. The latter option, guaranteeing positive variance after exponentiating, is the default in hetGP but both variations are implemented. Differences in empirical performance and mathematical development are slight. Modeling $\log \Delta_n$ makes derivative expressions a little more complicated after applying the chain rule, so our exposition here shall continue with the simpler un-logged variation. Logged analogs are left to an exercise in §10.4.

So far the configuration isn't much different than what Goldberg et al. (1998) described more than twenty years ago, except for an emphasis here on unique-n latents. Rather than cumbersome MCMC, Binois et al. describe how to stay within a (Woodbury) MLE framework, by defining a joint log likelihood over both mean and variance GPs:

$$\tilde{\ell} = c - \frac{N}{2} \log \hat{\nu}_N^2 - \frac{1}{2} \sum_{i=1}^{n} [(a_i - 1) \log \lambda_i + \log a_i] - \frac{1}{2} \log |K_n| \tag{10.8}$$

$$- \frac{n}{2} \log \hat{\nu}_{(\delta)} - \frac{1}{2} \log |K_{(\delta)}|.$$

Maximization is assisted by closed-form derivatives which may be evaluated with respect to all unknown quantities in $\mathcal{O}(n^3)$ time. For example, the derivative with respect to latent Δ_n may be derived as follows.

$$\frac{\partial \tilde{\ell}}{\partial \Delta_n} = \frac{\partial \Lambda_n}{\partial \Delta_n} \frac{\partial \tilde{\ell}}{\partial \Lambda_n} = C_{(\delta)} K_{(\delta)}^{-1} \frac{\partial \tilde{\ell}}{\partial \Lambda_n} - \frac{K_{(\delta)}^{-1} \Delta_n}{\hat{\nu}_{(\delta)}}$$

$$\text{where} \quad \frac{\partial \tilde{\ell}}{\partial \lambda_i} = \frac{N}{2} \times \frac{\frac{a_i \hat{\sigma}_i^2}{\lambda_i^2} + \frac{(K_n^{-1} \bar{Y}_n)_i^2}{a_i}}{\hat{\nu}_N} - \frac{a_i - 1}{2\lambda_i} - \frac{1}{2a_i} (K_n)_{i,i}^{-1}$$

Recall that $\hat{\sigma}_i^2 = \frac{1}{a_i} \sum_{j=1}^{a_i} (y_i^{(j)} - \bar{y}_j)^2$. So an interpretation of Eq. (10.8) is as an extension of SK estimates $\hat{\sigma}_i^2$ at $\bar{x}_i$. In contrast with SK, GP smoothing provides regularization needed in order to accommodate small numbers of replicates, even $a_i = 1$ in spite of $\hat{\sigma}_i^2 = 0$. In this single replicate case, y_i still contributes to local variance estimates through the rest of Eq. (10.8). Note that Eq. (10.8) is not constant in δ_i; in fact it depends (10.7) on all of Δ_n. Smoothing may be entertained on other quantities, e.g., $\Lambda_n \hat{\nu}_N = C_{(\delta)} K_{(\delta)}^{-1} S_n^2$ presuming $a_i > 1$, resulting in smoothed moment-based variance estimates in the style of SK (Kamiński, 2015; Wang and Chen, 2016). There may similarly be scope for bypassing a latent GP noise process with a *SiNK predictor* (Lee and Owen, 2015) by taking

$$\Lambda_n = \rho(\bar{X}_n)^{-1} C_{(\delta)} K_{(\delta)}^{-1} \Delta_n \quad \text{with} \quad \rho(x) = \sqrt{\hat{\nu}_{(\delta)} c_{(\delta)}(x)^\top K_{(\delta)}^{-1} c_{(\delta)}(x)},$$

or alternatively through $\log \Lambda_n$.

Binois et al. (2018) show that $\tilde{\ell}$ in Eq. (10.8) is maximized when $\Delta_n = \Lambda_n$ and $g_{(\delta)} = 0$.

In other words, smoothing latent nuggets (10.7) is unnecessary at the MLE. However, intermediate smoothing is useful as a device in three ways: 1) connecting SK to Goldberg's latent representation; 2) annealing[10] to avoid local minima; and 3) yielding a smooth solution when the numerical optimizer is stopped prematurely, which may be essential in big data (large unique-n) contexts.

It's amazing that a simple `optim` call can be brought to bear in such a high dimensional setting. Thoughtful initialization helps. Residuals from an initial homoskedastic fit can prime Δ_n-values. For more implementation details and options, see Section 3.3 of Binois and Gramacy (2018). $\mathcal{O}(n)$ latent variables can get big, especially as higher dimensional examples demand training datasets with more unique inputs. This is where replication really shines, inducing a likelihood that locks in high-replicate latents early on in the optimization. The Woodbury trick (§10.1.1) is essential here. Beyond simply being inefficient, having multiple latent $\delta_i^{(j)}$ at unique $\bar{x}_i$ corrupts optimization by inserting numerical weaknesses into the log likelihood.

Goldberg et al. (1998)'s MCMC is more forgiving in this regard. MC is a robust, if (very) slowly converging, numerical procedure. Besides that nuance favoring a Woodbury unique-n formulation, the likelihood surface is exceedingly well behaved. Critics have suggested expectation maximization (EM)[11] to integrate out latent variances. But this simply doesn't work except on small problems; EM represents nearly as much work as MCMC. Kersting et al. (2007) and Quadrianto et al. (2009) keyed into this fact a decade ago, but without the connection to SK, replication, and the Woodbury trick.

10.2.2 Illustrations with `hetGP`

Passages below take the reader through a cascade of examples illustrating `hetGP`: starting simply by returning to the motorcycle accident data, and progressing up though real-simulation examples from epidemiology to Bayesian model selection and inventory management.

Motorcycle accident data

The code below shows how to fit a heteroskedastic, coupled mean and variance GP, with smoothed initialization and Matèrn covariance structure. This is but one of several potential ways to obtain a good fit to these data using the methods provided by `hetGP`.

```
het <- mleHetGP(mcycle$times, mcycle$accel, covtype="Matern5_2",
  settings=list(initStrategy="smoothed"))
het$time
```

```
## elapsed
##   0.354
```

Time required to perform the requisite calculations is trivial, although admittedly this isn't a big problem. Built-in `predict.hetGP` works the same as its `.homGP` cousin, illustrated earlier.

---

[10]https://en.wikipedia.org/wiki/Simulated_annealing
[11]https://en.wikipedia.org/wiki/Expectation-maximization_algorithm

```
p2 <- predict(het, Xgrid)
ql <- qnorm(0.05, p2$mean, sqrt(p2$sd2 + p2$nugs))
qu <- qnorm(0.95, p2$mean, sqrt(p2$sd2 + p2$nugs))
```

Figure 10.3, shows the resulting predictive surface in two views. The first view, in the left panel, complements Figure 10.2. Observe how the surface is heteroskedastic, learning the low-variance region in the first third of inputs and higher variance for the latter two-thirds. As a consequence of being able to better track signal-to-noise over the input space, extraction of signal (particularly for the first third of **times**) is better than in the homoskedastic case.

```
par(mfrow=c(1,2))
plot(mcycle, ylim=c(-160, 90), ylab="acc", xlab="time",
  main="predictive surface")
lines(Xgrid, p2$mean)
lines(Xgrid, ql, col=2, lty=2)
lines(Xgrid, qu, col=2, lty=2)
plot(Xgrid, p2$nugs, type="l", ylab="s2", xlab="times",
  main="variance surface", ylim=c(0, 2e3))
points(het$X0, sapply(find_reps(mcycle[,1], mcycle[,2])$Zlist, var),
  col=3, pch=20)
```

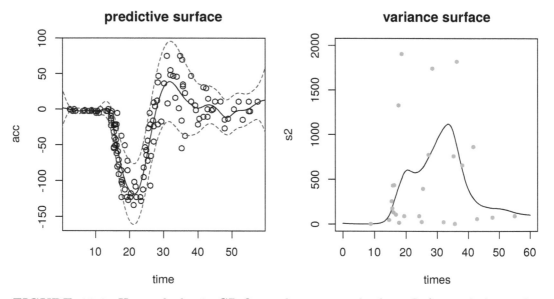

FIGURE 10.3: Heteroskedastic GP fit to the motorcycle data. Left panel shows the predictive distribution via mean (solid-black) and 90% error-bars (dashed-red); compare to Figure 10.2. Right panel shows the estimated variance surface and moment-based estimates of variance (green dots).

The second view, in the right panel of the figure, provides more detail on latent variances. Predictive uncertainty is highest for the middle third of **times**, which makes sense because this is where the whiplash effect is most prominent. Green dots in that panel indicate moment-based estimates of variance obtained from a limited number of replicates $a_i > 1$ available for some inputs. (There are nowhere near enough for an SK-like approach.) Observe how the black curve, smoothing latent variance values Δ_n, extracts the essence of the pattern of those values. This variance fit uses the full dataset, leveraging smoothness of the GP prior to incorporate responses at inputs with only one replicate, which in this case represents most of the data.

Susceptible, infected, recovered

Hu and Ludkovski (2017) describe a 2d simulation arising from the spread of an epidemic in a susceptible, infected, recovered (SIR) setting, a common stochastic compartmental model. Inputs are (S_0, I_0), the numbers of initial susceptible and infected individuals. Output is the total number of newly infected individuals which may be calculated by approximating

$$f(x) := \mathbb{E}\left\{ S_0 - \lim_{T \to \infty} S_T \mid (S_0, I_0, R_0) = x \right\} = \gamma \mathbb{E}\left\{ \int_0^\infty I_t \, dt \mid x \right\}$$

under continuous time Markov dynamics with transitions $S + I \to 2I$ and $I \to R$. The resulting surface is heteroskedastic and has some high noise regions. Parts of the input space represent volatile regimes wherein random chance intimately affects whether or not epidemics spread quickly or die out.

A function generating the data for standardized inputs in the unit square (corresponding to coded S_0 and I_0) is provided by `sirEval` in the `hetGP` package. Output is also standardized. Consider a space-filling design of $n = 200$ unique runs under random replication $a_i \sim$ Unif$\{1, \ldots, 100\}$. Coordinate x_1 represents the initial number of infecteds I_0, and x_2 the initial number of susceptibles S_0.

```
Xbar <- randomLHS(200, 2)
a <- sample(1:100, nrow(Xbar), replace=TRUE)
X <- matrix(NA, ncol=2, nrow=sum(a))
nf <- 0
for(i in 1:nrow(Xbar)) {
  X[(nf+1):(nf+a[i]),] <- matrix(rep(Xbar[i,], a[i]), ncol=2, byrow=TRUE)
  nf <- nf + a[i]
}
nf
```

```
## [1] 10407
```

The result is a full dataset with about ten thousand runs. Code below gathers responses – expected total number of infecteds at the end of the simulation – at each input location in the design, including replicates.

```
Y <- apply(X, 1, sirEval)
```

Code below fits our `hetGP` model. By default, lengthscales for the variance GP ($\theta_{(\delta)}$) are linked to those from the mean GP (θ), requiring that the former be a scalar multiple $k > 1$ of the latter. That feature can be switched off, however, as illustrated below. Specifying scalar `lower` and `upper` signals an isotropic kernel.

```
fit <- mleHetGP(X, Y, settings=list(linkThetas="none"),
  covtype="Matern5_2", maxit=1e4)
fit$time
```

```
## elapsed
##   3.274
```

Around 3.3 seconds are needed to train the model. To visualize the resulting predictive surface, code below creates a dense grid in 2d and calls `predict.hetGP` on the `fit` object.

```
xx <- seq(0, 1, length=100)
XX <- as.matrix(expand.grid(xx, xx))
psir <- predict(fit, XX)
vsir <- psir$sd2 + psir$nugs
```

The left panel of Figure 10.4 captures predictive means; the right panel shows predictive standard deviation. Text overlaid on the panels indicates the location of the training data inputs and the number of replicates observed thereupon.

```
par(mfrow=c(1,2))
image(xx, xx, matrix(psir$mean, ncol=100), xlab="S0", ylab="I0",
  col=cols, main="mean infected")
text(Xbar, labels=a, cex=0.5)
image(xx, xx, matrix(sqrt(vsir), ncol=100), xlab="S0", ylab="I0",
  col=cols, main="sd infected")
text(Xbar, labels=a, cex=0.5)
```

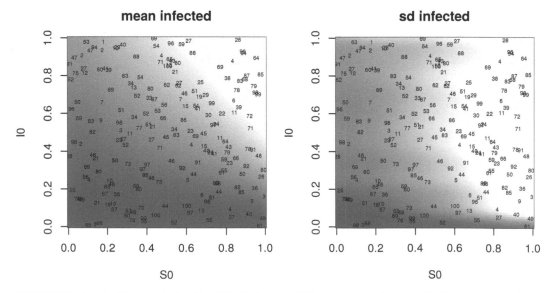

FIGURE 10.4: Heteroskedastic GP fit to the SIR data showing predictive mean surface (left) and estimated standard deviation (right). Text in both panels shows numbers of replicates in training.

Notice in the figure how means are changing more slowly than variances. Un-linking their lengthscales was probably the right move. Interpreting the mean surface, the number of infecteds is maximized when initial S_0 and/or I_0 is high. That makes sense. Predictive uncertainty in total number of infecteds is greatest when there's a large number of initial susceptible individuals, but a moderate number of initial infecteds. That too makes sense. A moderate injection of disease into a large, vulnerable population could swell to spell disaster.

Bayesian model selection

Model selection by Bayes factor (BF) is known to be sensitive to hyperparameter choice in hierarchical models, which is further complicated and obscured by MC evaluation injecting a substantial source of noise. To study BF surfaces in such settings, Franck and Gramacy (2020) propose treating expensive BF evaluations, with MCMC say, as a (stochastic) computer simulation experiment. BF calculations at a space-filling design in the input/hyperparameter space can be used to map and thus better understand sensitivity of model selection to those settings.

As a simple warm-up example, consider an experiment described in Sections 3.3–3.4 of Gramacy and Pantaleo (2010) where BF calculations determine whether data is leptokurtic (Student-t errors) or not (simply Gaussian). Here we study BF approximation as a function of prior parameterization on the Student-t degrees of freedom parameter ν, which Gramacy and Pantaleo took as $\nu \sim \text{Exp}(\theta = 0.1)$. Their intention was to be diffuse, but ultimately they lacked an appropriate framework for studying sensitivity to this choice. Franck and Gramacy (2020) designed a grid of θ-values, evenly spaced in $\log_{10}$ space from 10^{-3} to 10^6 spanning "solidly Student-t" (even Cauchy) to "essentially Gaussian" prior specifications for $\mathbb{E}\{\nu\}$. Each θ_i gets its own reversible jump MCMC simulation (Richardson and Green, 1997), as automated by `blasso` in the `monomvn` package (Gramacy, 2018a) on CRAN, taking about 36 minutes on a 3.20GHz Intel Core i7 processor. Converting those posterior draws into $\text{BF}^{(i)}_{\text{St}\mathcal{N}}$ evaluations follows a post-processing scheme described by Jacquier et al. (2004). To better understand MC variability in those calculations, ten replicates of BFs under each hyperparameter setting θ_i were simulated. Collecting 200 BFs in this way takes about 120 hours. These data are saved in the `hetGP` package for later analysis.

```
data(bfs)
thetas <- matrix(bfs.exp$theta, ncol=1)
bfs <- as.matrix(t(bfs.exp[,-1]))
```

BF evaluation by MCMC is notoriously unstable. We shall see that even in log-log space, the $\text{BF}_{\text{St}\mathcal{N}}$ process is not only heteroskedastic but also has heavy tails. Consequently Franck and Gramacy (2020) fit a *heteroskedastic Student-t process*. Details are provided in Section 3.2 of Binois and Gramacy (2018) and Chung et al. (2019) who describe several other extensions including an additional layer of latent variables to handle missing data, and a scheme to enforce a monotonicity property. Both are motivated by a challenging class of inverse problems involved in characterizing spread of influenza.

```
bfs1 <- mleHetTP(X=list(X0=log10(thetas), Z0=colMeans(log(bfs)),
  mult=rep(nrow(bfs), ncol(bfs))), Z=log(as.numeric(bfs)), lower=1e-4,
  upper=5, covtype="Matern5_2")
```

Predictive evaluations may be extracted on a grid in the input space …

```
dx <- seq(0, 1, length=100)
dx <- 10^(dx*4 - 3)
p <- predict(bfs1, matrix(log10(dx), ncol=1))
```

... and visualized in Figure 10.5. Each open circle is a $\mathrm{BF}_{\mathrm{St}\mathcal{N}}$ evaluation, plotted in $\log_{10}$–$\log_e$ space.

```
matplot(log10(thetas), t(log(bfs)), col=1, pch=21, ylab="log(bf)")
lines(log10(dx), p$mean)
lines(log10(dx), p$mean + 2*sqrt(p$sd2 + p$nugs), col=2, lty=2)
lines(log10(dx), p$mean - 2*sqrt(p$sd2 + p$nugs), col=2, lty=2)
legend("topleft", c("hetTP mean", "hetTP interval"), col=1:2, lty=1:2)
```

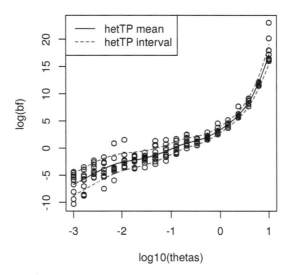

FIGURE 10.5: Heteroskedastic TP fit to Bayes factor data under exponential hyperprior.

It bears repeating that the $\mathrm{BF}_{\mathrm{St}\mathcal{N}}$ surface is heteroskedastic, even after log transform, and has heavy tails. A take-home message from these plots is that BF surfaces can be extremely sensitive to hierarchical modeling hyperparameterization. When θ is small, the Student-t ($\mathrm{BF}_{\mathrm{St}\mathcal{N}} < 1$) is essentially a foregone conclusion, whereas if θ is large the Gaussian ($\mathrm{BF}_{\mathrm{St}\mathcal{N}} > 1$) is. This is a discouraging result for model selection with BFs in this setting: a seemingly innocuous hyperparameter is essentially determining the outcome of selection.

Although the computational burden involved in this experiment – 120 hours – is tolerable, extending the idea to higher dimensions is problematic. Suppose one wished to entertain $\nu \sim \mathrm{Gamma}(\alpha, \beta)$, where the $\alpha = 1$ case reduces to $\nu \sim \mathrm{Exp}(\beta \equiv \theta)$ above. Over a similarly dense hyperparameter grid, runtime would balloon to more than one hundred days which is clearly unreasonable. Instead it makes sense to build a surrogate model from a more limited space-filling design and use the resulting posterior predictive surface to understand variability in BFs in the hyperparameter space. Five replicate responses on a size $n = 80$ LHS in $\alpha \times \beta$-space were obtained for a total of $N = 400$ MCMC runs. These data are provided as `bfs.gamma` in the `bfs` data object in `hetGP`.

```
D <- as.matrix(bfs.gamma[,1:2])
bfs <- as.matrix(t(bfs.gamma[,-(1:2)]))
```

A similar **hetTP** fit may be obtained with ...

```
bfs2 <- mleHetTP(X=list(X0=log10(D), Z0=colMeans(log(bfs)),
  mult=rep(nrow(bfs), ncol(bfs))), Z=log(as.numeric(bfs)),
  lower=rep(1e-4, 2), upper=rep(5, 2), covtype="Matern5_2")
```

... followed by predictions on a dense grid in the 2d input space.

```
dx <- seq(0, 1, length=100)
dx <- 10^(dx*4 - 3)
DD <- as.matrix(expand.grid(dx, dx))
p <- predict(bfs2, log10(DD))
```

Figure 10.6 shows the outcome of that experiment: mean surface in the left panel and standard deviation in the right. Numbers overlaid indicate average $BF_{St\mathcal{N}}$ obtained for the five replicates at each input location. Contours in the left panel correspond to levels in the so-called Jeffrey's scale[12].

```
par(mfrow=c(1,2))
mbfs <- colMeans(bfs)
image(log10(dx), log10(dx), t(matrix(p$mean, ncol=length(dx))), col=cols,
  xlab="log10 alpha", ylab="log10 beta",  main="mean log BF")
text(log10(D[,2]), log10(D[,1]), signif(log(mbfs), 2), cex=0.5)
contour(log10(dx), log10(dx), t(matrix(p$mean, ncol=length(dx))),
  levels=c(-5,-3,-1,0,1,3,5), add=TRUE, col=4)
image(log10(dx), log10(dx), t(matrix(sqrt(p$sd2 + p$nugs),
  ncol=length(dx))), co =cols, xlab="log10 alpha",
  ylab="log10 beta", main="sd log BF")
text(log10(D[,2]), log10(D[,1]), signif(apply(log(bfs),2,sd), 2), cex=0.5)
```

The story here is much the same as before in terms of β, which maps to θ in the earlier experiment, especially near $\alpha = 1$, or $\log_{10}\alpha = 0$ where equivalence is exact. The left panel shows that along that slice one can get just about any model selection conclusion one wants. Smaller α-values tell a somewhat more nuanced story, however. A rather large range of smaller α-values leads to somewhat less sensitivity in the outcome, except when β is quite large. Apparently, having a small α setting is essential if data are going to have any influence on model selection with BFs. The right panel shows that variances are indeed changing over the input space, justifying the heteroskedastic surrogate.

Inventory management

The assemble-to-order (ATO) problem (Hong and Nelson, 2006) involves a queuing simulation targeting inventory management scenarios. At its heart it's an optimization or reinforcement

---

[12]https://en.wikipedia.org/wiki/Bayes_factor#Interpretation

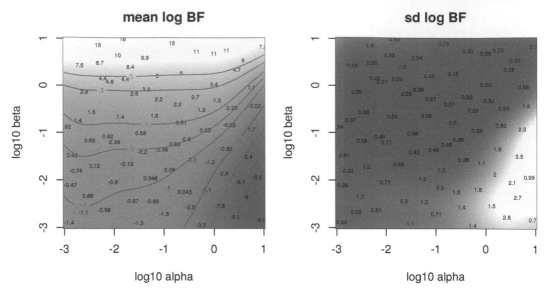

FIGURE 10.6: Heteroskedastic TP fit to the Bayes factor data under Gamma hyperprior.

learning problem, however here I simply treat it as a response surface to be learned. Although the signal-to-noise ratio is relatively high, ATO simulations are known to be heteroskedastic, e.g., as illustrated by documentation for the MATLAB® library utilized for simulations (Xie et al., 2012).

The setting is as follows. A company manufactures m products. Products are built from base parts called items, some of which are "key" in that the product can't be built without them. If a random request comes in for a product that's missing a key item, a replenishment order is executed and filled after random delay. Holding items in inventory is expensive, so balance must be struck between inventory costs and revenue. Binois et al. (2018) describe an experiment under target stock vector inputs $b \in \{0, 1, \dots, 20\}^8$ for eight items.

Code below replicates results from that experiment, which entail a uniform design of size $n_{\text{tot}} = 2000$ in 8d with ten replicates for a total design size of $N_{\text{tot}} = 20000$. A training–testing partition was constructed by randomly selecting $n = 1000$ unique locations and replicates $a_i \sim \text{Unif}\{1, \dots, 10\}$ thereupon. The `ato` data object in `hetGP` contains one such random partition, which is subsequently coded into the unit cube $[0, 1]^8$. Further detail is provided in package documentation for the `ato` data object.

Actually that object provides two testing sets. One is a genuine out-of-sample set, where testing sites are exclusive of training locations. The other is replicate based, involving those $10 - a_i$ runs not selected for training. The training set is large, making MLE calculations slow, so the `ato` object additionally provides a fitted model for comparison. Examples in the `ato` documentation file provide code which may be used to reproduce that fit or to create a new one based on novel training–testing partitions.

```
data(ato)
c(n=nrow(Xtrain), N=length(unlist(Ztrain)), time=out$time)
```

```
##              n            N time.elapsed
##           1000         5594         8584
```

Storing these objects facilitates fast illustration of prediction and out-of-sample comparison. It also provides a benchmark against which a more thoughtful sequential design scheme, introduced momentarily in §10.3.1, can be judged. Code below performs predictions at the held-out testing sites and then calculates a pointwise proper score (Eq. (27) from Gneiting and Raftery, 2007) with ten replicates observed at each of those locations. Higher scores are better.

```
phet <- predict(out, Xtest)
phets2 <- phet$sd2 + phet$nugs
mhet <- as.numeric(t(matrix(rep(phet$mean, 10), ncol=10)))
s2het <- as.numeric(t(matrix(rep(phets2, 10), ncol=10)))
sehet <- (unlist(t(Ztest)) - mhet)^2
sc <- - sehet/s2het - log(s2het)
mean(sc)
```

```
## [1] 3.396
```

A similar calculation may be made for held-out training replicates, shown below. These are technically out-of-sample, but accuracy is higher since training data were provided at these sites.

```
phet.out <- predict(out, Xtrain.out)
phets2.out <- phet.out$sd2 + phet.out$nugs
s2het.out <- mhet.out <- Ztrain.out
for(i in 1:length(mhet.out)) {
  mhet.out[[i]] <- rep(phet.out$mean[i], length(mhet.out[[i]]))
  s2het.out[[i]] <- rep(phets2.out[i], length(s2het.out[[i]]))
}
mhet.out <- unlist(t(mhet.out))
s2het.out <- unlist(t(s2het.out))
sehet.out <- (unlist(t(Ztrain.out)) - mhet.out)^2
sc.out <- - sehet.out/s2het.out - log(s2het.out)
mean(sc.out)
```

```
## [1] 5.085
```

Those two testing sets may be combined to provide a single score calculated on the entire corpus of held-out data. The result is a compromise between the two score statistics calculated earlier.

```
mean(c(sc, sc.out))
```

```
## [1] 3.926
```

Binois et al. (2018) repeated that training–testing partition one hundred times to provide score boxplots which may be compared to simpler (e.g., homoskedastic) GP-based approaches. I shall refer the curious reader to Figure 2 in that paper for more details. To make a long story short, fits accommodating heteroskedasticity in the proper way – via coupled GPs and fully likelihood-based inference – are superior to all other (computationally tractable) ways entertained.

10.3   Sequential design

A theme from Chapter 6 is that design for GPs has a chicken-or-egg problem. Model-based designs are hyperparameter dependent, and data are required to estimate parameters. Space-filling designs may be sensible but are often sub-optimal. It makes sense to proceed sequentially (§6.2). That was for the ordinary, homoskedastic/stationary GPs of Chapter 5. Now noise is high and possibly varying regionally. Latent variables must be optimized. Replication helps, but how much and where?

The following characterizes the state of affairs as regards replication in experiment design for GP surrogates. In low/no-noise settings and under the assumption of stationarity (i.e., constant stochasticity), replication is of little value. Yet no technical result precludes replication except in deterministic settings. Conditions under which replicating is an optimal design choice are, until recently, unknown. When replicating, as motivated perhaps by an abundance of caution, one spot is sufficient because under stationarity the variance is the same everywhere.

Even less is known about heteroskedastic settings. Perhaps more replicates may be needed in high-variance regions? Ankenman et al. (2010) show that once you have a design, under a degree of replication "large enough" to trust its moment-based estimates of spatial variance, new replicates can be allocated optimally. But what use are they if you already have "enough", and what about the quagmire when you learn that you need fewer than you already have? (More on that in §10.3.2.) On the other hand, if means are changing quickly – but otherwise exhibit stationary dynamics – might it help to concentrate design acquisitions where noise is lower so that signal-learning is relatively cheap in replication terms? One thing's for sure: we'd better proceed sequentially. Before data are collected, regions of high or low variance are unknown.

Details driving such trade-offs, requisite calculations and what can be learned from them, depend upon design criteria. Choosing an appropriate criterion depends upon the goal of modeling and/or prediction. §10.3.1–10.3.3 concentrate on predictive accuracy under heteroskedastic models, extending §6.2.2. §10.3.4 concludes with pointers to recent work in Bayesian optimization (extending §7.2), level set finding, and inverse problems/calibration (§8.1).

10.3.1   Integrated mean-squared prediction error

Integrated mean-squared prediction error (IMSPE) is a common model-based design criterion, whether for batch design (§6.1.2) or in sequential application as embodied by ALC (§6.2.2). Here our focus is sequential, however many calculations, and their implementation in `hetGP`, naturally extend to the batch setting as already illustrated in §6.1.2. IMSPE is predictive variance averaged over the input space, which must be minimized.

$$I_{n+1}(x_{n+1}) \equiv \text{IMSPE}(\bar{x}_1, \ldots, \bar{x}_n, x_{n+1}) = \int_{x \in \mathcal{X}} \breve{\sigma}_{n+1}^2(x) \, dx \qquad (10.9)$$

Recall that $\breve{\sigma}^2(x)$ is nugget-free predictive variance (7.5), capturing only epistemic uncertainty about the latent random field (§5.3.2). IMSPE above is expressed in terms of unique n

inputs for reasons that shall be clarified shortly. Following the right-hand side of Eq. (10.4), we have

$$\breve{\sigma}_{n+1}^2(x) = \nu(1 - k_{n+1}^\top(x)(C_{n+1} + \Lambda_{n+1}A_{n+1}^{-1})^{-1}k_{n+1}(x)). \tag{10.10}$$

Re-expression for full-N is straightforward. The next design location, x_{N+1}, may be a new unique location ($\bar{x}_{n+1}$) or a repeat of an existing one (i.e., one of $\bar{x}_1, \ldots, \bar{x}_n$).

Generally speaking, integral (10.9) requires numerical evaluation (Seo et al., 2000; Gramacy and Lee, 2009; Gauthier and Pronzato, 2014; Gorodetsky and Marzouk, 2016; Pratola et al., 2017b). §6.2.2 offers several examples; or Ds2x=TRUE in the tgp package (Gramacy and Taddy, 2016) – see §9.2.2. Both offer approximations based on sums over a reference set, with the latter providing additional averaging over MCMC samples from the posterior of hyperparameters and trees. Conditional on GP hyperparameters, and when the study region $\mathcal{X}$ is an easily integrable domain such as a hyperrectangle, requisite integration may be calculated in closed form. Although examples exist elsewhere in the literature (e.g., Ankenman et al., 2010; Anagnostopoulos and Gramacy, 2013; Burnaev and Panov, 2015; Leatherman et al., 2017) for particular cases, hetGP provides the only implementation on CRAN that I'm aware of. Binois et al. (2019) extend many of those calculations to the Woodbury trick, showing for the first time how choosing a replicate can be optimal in active learning/sequential design.

The closed form solution leverages that IMSPE is similar to an expectation over covariance functions.

$$I_{n+1}(x_{n+1}) = \mathbb{E}\{\breve{\sigma}_{n+1}^2(X)\} = \mathbb{E}\{K_\theta(X, X) - k_{n+1}^\top(X)K_{n+1}^{-1}k_{n+1}(X)\} \tag{10.11}$$
$$= \mathbb{E}\{K_\theta(X, X)\} - \text{tr}(K_{n+1}^{-1}W_{n+1}),$$

where $W_{ij} = \int_{x \in \mathcal{X}} k(x_i, x)k(x_j, x)\,dx$. Closed forms for W_{ij} exist with $\mathcal{X}$ being a hyperrectangle. Notice that K_{n+1} depends on the number of replicates per unique design element, so this representation includes a tacit dependence on noise level and replication counts $a_1, \ldots, a_n$. Binois et al. (2019) provide forms for several popular covariance functions, including the Matèrn. In the case of the separable Gaussian:

$$W_{ij} = \prod_{k=1}^m \frac{\sqrt{2\pi\theta_k}}{4} \exp\left\{-\frac{(x_{ik} - x_{jk})^2}{2\theta_k}\right\} \left[\text{erf}\left\{\frac{2 - (x_{ik} + x_{jk})}{\sqrt{2\theta_k}}\right\} + \text{erf}\left\{\frac{x_{ik} + x_{jk}}{\sqrt{2\theta_k}}\right\}\right].$$

Gradients are also available in closed form, which is convenient for fast library-based optimization as a means of solving for acquisitions. Partitioned inverse equations (6.8) make evaluation of $I_{n+1}(x_{n+1})$ and its derivative quadratic in n. See Binois et al. for details.

To investigate how replication can be favored by IMSPE, consider the following scenario(s). Let $r(x)$ denote a belief about extrinsic variance. Choice of $r(x)$ is arbitrary. In practice we shall use estimated $r(x) = \breve{\sigma}_n(x)$, as above. However this illustration contrives two simpler $r(x)$, primarily for pedagogical purposes, based on splines that agree at five knots. Knot locations could represent design sites $\bar{x}_i$ where potentially many replicate responses have been observed.

```
rn <- c(6, 4, 5, 6.5, 5)
X0 <- matrix(seq(0.2, 0.8, length.out=length(rn)))
```

```
X1 <- matrix(c(X0, 0.3, 0.4, 0.9, 1))
Y1 <- c(rn, 4.7, 4.6, 6.3, 4.5)
r1 <- splinefun(x=X1, y=Y1, method="natural")
X2 <- matrix(c(X0, 0.0, 0.3))
Y2 <- c(rn, 7, 2)
r2 <- splinefun(x=X2, y=Y2, method="natural")
```

Figure 10.7 provides a visual of these two hypotheses evaluated on a testing grid. Below I shall refer to these surfaces as "green" and "blue", respectively, referencing the colors from the figure. Knots are shown as red open circles.

```
xx <- matrix(seq(0, 1, by=0.005))
plot(X0, rn, xlab="x", ylab="r(x)", xlim=c(0,1), ylim=c(2,8), col=2)
lines(xx, r1(xx), col=3)
lines(xx, r2(xx), col=4)
```

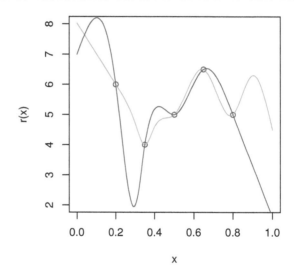

FIGURE 10.7: Two example $r(x)$ variance surfaces based on splines with knots as open red circles.

The code below implements Eq. (10.11) for generic variance function **r**, like one of our splines from above. It uses internal **hetGP** functions such as **Wij** and **cov_gen**. (We shall illustrate the intended hooks momentarily; these low-level functions are of value here in this toy example.)

```
IMSPE.r <- function(x, X0, theta, r) {
  x <- matrix(x, nrow = 1)
  Wijs <- Wij(mu1=rbind(X0, x), theta=theta, type="Gaussian")
  K <- cov_gen(X1=rbind(X0, x), theta=theta)
  K <- K + diag(apply(rbind(X0, x), 1, r))
  return(1 - sum(solve(K)*Wijs))
}
```

Next, apply this function on a grid for each $r(x)$, green and blue ...

```
imspe1 <- apply(xx, 1, IMSPE.r, X0=X0, theta=0.25, r=r1)
imspe2 <- apply(xx, 1, IMSPE.r, X0=X0, theta=0.25, r=r2)
xstar1 <- which.min(imspe1)
xstar2 <- which.min(imspe2)
```

... and visualize in Figure 10.8. The x-locations of the knots – our hypothetical unique design sites $\bar{x}_1, \ldots, \bar{x}_n$ – are indicated as dashed red vertical bars.

```
plot(xx, imspe1, type="l", col=3, ylab="IMSPE", xlab="x", ylim=c(0.6, 0.7))
lines(xx, imspe2, col=4)
abline(v=X0, lty=3, col='red')
points(xx[xstar1], imspe1[xstar1], pch=23, bg=3)
points(xx[xstar2], imspe2[xstar2], pch=23, bg=4)
```

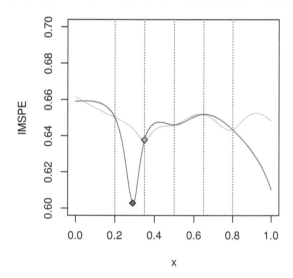

FIGURE 10.8: IMSPE surfaces from the two $r(x)$ in Figure 10.7; knots indicated as red vertical dashed lines.

In the figure, the blue variance hypothesis is minimized at a novel $\bar{x}_{n+1}$ location, not coinciding with any of the previous design sites. But the green hypothesis is minimized at $\bar{x}_2$, counting vertical red-dashed lines from left to right. IMSPE calculated on the green $r(x)$ prefers replication. That's not a coincidence or fabrication. Binois et al. (2019) showed that the next point x_{N+1} will be a replicate, i.e., one of the existing unique locations $\bar{x}_1, \ldots, \bar{x}_n$ rather than a new $\bar{x}_{n+1}$, when

$$r(x_{N+1}) \geq \frac{k_n^\top(x_{N+1})K_n^{-1}W_nK_n^{-1}k_n(x_{N+1}) - 2w_{n+1}^\top K_n^{-1}k_n(x_{N+1}) + w_{n+1,n+1}}{\text{tr}(B_{k^*}W_n)}$$
$$- \sigma_n^2(x_{N+1}), \tag{10.12}$$

where $k^* = \text{argmin}_{k \in \{1,\ldots,n\}} \text{IMSPE}(\bar{x}_k)$ and

$$B_k = \frac{(K_n^{-1})_{.,k}(K_n^{-1})_{k,.}}{\frac{\nu\lambda_k}{a_k(a_k+1)} - (K_n)_{k,k}^{-1}}.$$

To describe that result from thirty thousand feet: IMSPE will prefer replication when predictive variance is "everywhere large enough." To illustrate, code below utilizes hetGP internals to enable evaluation of the right-hand side of inequality (10.12) above.

```
rx <- function(x, X0, rn, theta, Ki, kstar, Wijs) {
  x <- matrix(x, nrow=1)
  kn1 <- cov_gen(x, X0, theta=theta)
  wn <- Wij(mu1=x, mu2=X0, theta=theta, type="Gaussian")
  a <- kn1 %*% Ki %*% Wijs %*% Ki %*% t(kn1) - 2*wn %*% Ki %*% t(kn1)
  a <- a + Wij(mu1=x, theta=theta, type="Gaussian")
  Bk <- tcrossprod(Ki[,kstar], Ki[kstar,])/(2/rn[kstar] - Ki[kstar, kstar])
  b <- sum(Bk*Wijs)
  sn <- 1 - kn1 %*% Ki %*% t(kn1)
  return(a/b - sn)
}
```

Evaluating on **XX** commences as follows ...

```
bestk <- which.min(apply(X0, 1, IMSPE.r, X0=X0, theta=0.25, r=r1))
Wijs <- Wij(X0, theta=0.25, type="Gaussian")
Ki <- solve(cov_gen(X0, theta=0.25, type="Gaussian") + diag(rn))
rx.thresh <- apply(xx, 1, rx, X0=X0, rn=rn, theta=0.25, Ki=Ki,
  kstar=bestk, Wijs=Wijs)
```

... which may be used to augment Figure 10.7 with a gray-dashed line in Figure 10.9. Since the green hypothesis is everywhere above that threshold in this instance, replication is recommended by the criterion. Observe that the point of equality coincides with the selection minimizing IMSPE.

```
plot(X0, rn, xlab="x", ylab="r(x)", xlim=c(0,1), ylim=c(2,8), lty=2, col=2)
lines(xx, r1(xx), col=3)
lines(xx, r2(xx), col=4)
lines(xx, rx.thresh, lty=2, col="darkgrey")
```

That green hypothesis is, of course, just one instance of a variance function above the replication threshold. Although those hypotheses were not derived from GP predictive equations, the example illustrates potential. So here's what we know: replication is 1) good for GP calculations ($n^3 \ll N^3$); 2) preferred by IMSPE under certain (changing) variance regimes; and 3) intuitively helps separate signal from noise. But how often is IMSPE going to ask for replications in practice? The short answer: not often enough.

One challenge is numerical precision in optimization when mixing discrete and continuous search. Given a continuum of potential new locations, up to floating-point precision, a particular setting corresponding to a finite number of replicate sites is not likely to be preferred over all other settings, such as ones infinitesimally nearby. Another issue, which is more fundamental, is that IMSPE is myopic. Value realized by the current selection, be it

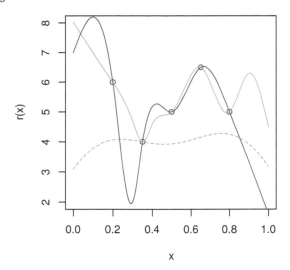

FIGURE 10.9: Variance hypotheses from Figure 10.7 with replicating threshold added in gray.

a replicate or new unique location, is not assessed through its impact on a future decision landscape.

10.3.2   Lookahead over replication

Entertaining acquisition decision spaces that are forward looking is hard. Several examples may be found in the literature, primarily emphasizing BO applications (Chapter 7). (See, e.g., Ginsbourger and Le Riche, 2010; Gonzalez et al., 2016; Lam et al., 2016; Huan and Marzouk, 2016.) To my knowledge, closed form solutions only exist for the one-step-ahead case. Many approaches, like IECI from §7.2.3, leverage numerics to a degree. More on optimization is provided, in brief, in §10.3.4. In the context of IMSPE, Binois et al. (2019) found that a replication-biased lookahead, correcting myopia in the context of separating signal from noise, is manageable.

Consider a horizon h into the future, over which the goal is to plan for up to one new, unique site $\bar{x}_{n+1}$, and h or $h+1$ replicates, each at n or $n+1$ existing unique locations. The first can be a new, unique site, and the remaining h replicates. Or replicates can come first, with h ways to assign a new unique site later. Figure 10.10 provides a flow diagram corresponding to horizon $h = 3$ case. Each node in the graph represents a lookahead state. Red arrows denote potential timings for a new unique site graphically, as a transition between states. Replicates are indicated by black arrows.

Finding each new unique location involves a continuous, potentially derivative-based IMSPE search. Entertaining replicates is discrete. In total, looking ahead over horizon h requires $h+1$ continuous searches, and $(h+1)(h+2)/2 - 1$ discrete ones. Searching over discrete alternatives, involving replicates at one of the existing $\bar{x}_k$ locations, $k = 1, \ldots, n, n+1$, may utilize a simplified IMSPE (10.9) calculation:

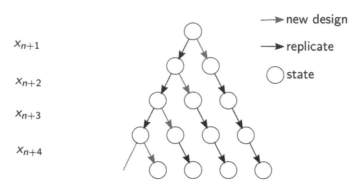

x_{n+1}

x_{n+2}

x_{n+3}

x_{n+4}

FIGURE 10.10: Flow chart of lookahead over replication. A similar chart may be found in Binois et al. (2019).

$$I_{n+1}(\bar{x}_k) = \nu(1 - \text{tr}(B'_k W_n)), \quad \text{with} \tag{10.13}$$

$$B'_k = \frac{\left(\left(C_n + A_n^{-1}\Lambda_n\right)^{-1}\right)_{.,k} \left(\left(C_n + A_n^{-1}\Lambda_n\right)^{-1}\right)_{k,.}}{a_k(a_k + 1)/\lambda_k - \left(C_n + A_n^{-1}\Lambda_n\right)^{-1}_{k,k}}.$$

Implementation in `hetGP` considers $h \in \{0, 1, 2, \ldots\}$ with $h = 0$ representing ordinary myopic IMSPE search. Although larger values of h entertain future sequential design decisions, the goal (for any h) is to determine what to do *now* as $N \to N + 1$. Toward that end, $h + 1$ decision paths are entertained spanning alternatives between exploring sooner and replicating later, or vice versa. During each iteration along a given path through Figure 10.10, either Eq. (10.9) or (10.13), but not simultaneously, is taken up as the hypothetical action. In the first iteration, if a new $\bar{x}_{n+1}$ is chosen by optimizing Eq. (10.9), that location is combined with existing $(\bar{x}_1, \ldots, \bar{x}_n)$ and considered as candidates for future replication over the remaining h lookahead iterations (when $h \geq 1$). If instead a replicate is chosen in the first iteration, the lookahead scheme recursively searches over choices of which of the remaining h iterations will pick a new $\bar{x}_{n+1}$, with others optimizing over replicates. Recursion is resolved by moving to the second iteration and again splitting the decision path into a choice between replicate-explore-replicate-... and replicate-replicate-..., etc. After fully unrolling the recursion in this way, optimizing up to horizon h along $h + 1$ paths, the ultimate IMSPE for the respective hypothetical design with size $N + 1 + h$ is computed, and the decision path yielding the smallest IMSPE is stored. Selection x_{N+1} is a new location if the explore-first path was optimal, and is a replicate otherwise.

Horizon is a tuning parameter. It determines the extent to which replicates are entertained in lookahead, and therefore larger h somewhat inflates chances for replication. As h grows there are more decision paths that delay exploration to a later iteration; if any of them yield smaller IMSPE than the explore-first path, the immediate action is to replicate. Although larger h allows more replication before committing to a new, unique $\bar{x}_{n+1}$, it also magnifies the value of an $\bar{x}_{n+1}$ chosen in the first iteration, as that location could potentially accrue its own replicates in subsequent lookahead or real design iterations. Therefore, although in practice larger h leads to more replication in the final design, this association is weak.

Before considering criteria for how to set horizon, consider the following illustration with fixed $h = 5$. Take $f(x) = (6x - 2)^2 \sin(12x - 4)$ from Forrester et al. (2008), implemented as `f1d` in `hetGP`, and observe $Y(x) \sim \mathcal{N}(f(x), r(x))$, where the noise variance function is $r(x) = (1.1 + \sin(2\pi x))^2$.

```
fr <- function(x) { (1.1 + sin(2*pi*x)) }
fY <- function(x) { f1d(x) + rnorm(length(x), sd=fr(x)) }
```

This example was engineered to be similar to the motorcycle accident data (§10.2), but allow for bespoke evaluation. Begin with an initial uniform design of size $N = n = 10$, i.e., without replicates, and associated `hetGP` fit based on a Gaussian kernel.

```
X <- seq(0, 1, length=10)
Y <- fY(X)
mod <- mleHetGP(X=X, Z=Y, lower=0.0001, upper=10)
```

Using that fit, calculate IMSPE with horizon $h = 5$ lookahead over replication. The `IMSPE_optim` call below leverages closed form IMSPE (10.11) and derivatives using an `optim` call with `method="L-BFGS-B"`, mixed with discrete evaluations (10.13).

```
opt <- IMSPE_optim(mod, h=5)
X <- c(X, opt$par)
X
```

```
##  [1] 0.0000 0.1111 0.2222 0.3333 0.4444 0.5556 0.6667 0.7778 0.8889
## [10] 1.0000 0.8889
```

Whether or not the chosen location, in position eleven above (0.889), is a replicate depends on the random Rmarkdown build, challenging precise narrative here. The `hetGP` package provides an efficient partitioned inverse-based updating method (6.8), utilizing $\mathcal{O}(n^2)$ or $\mathcal{O}(n)$ updating calculations for new data points depending on whether that point is unique or a replicate, respectively. Details are provided by Binois et al. (2019), extending those from §6.3 to the heteroskedastic case.

```
ynew <- fY(opt$par)
Y <- c(Y, ynew)
mod <- update(mod, Xnew=opt$par, Znew=ynew, ginit=mod$g*1.01)
```

That's the basic idea. Let's continue and gather a total of 500 samples in this way, in order to explore the aggregate nature of sequential design. Periodically, every 25 iterations in the code below, it can help to restart MLE calculations to "unstick" any solutions found in local modes of the likelihood surface. Gathering 500 points is overkill for this simple 1d problem, but it helps create a nice visualization.

```
for(i in 1:489) {

  ## find the next point and update
  opt <- IMSPE_optim(mod, h=5)
  X <- c(X, opt$par)
  ynew <- fY(opt$par)
```

```
Y <- c(Y, ynew)
mod <- update(mod, Xnew=opt$par, Znew=ynew, ginit=mod$g*1.01)

## periodically attempt a restart to try to escape local modes
if(i %% 25 == 0){
  mod2 <- mleHetGP(X=list(X0=mod$X0, Z0=mod$Z0, mult=mod$mult), Z=mod$Z,
    lower=0.0001, upper=1)
  if(mod2$ll > mod$ll) mod <- mod2
}
}
nrow(mod$X0)
```

```
## [1] 67
```

Of the $N = 500$ total acquisitions, the final design contains $n = 67$ unique locations. To help visualize and assess the quality of the final surface with that design, the code below gathers predictive quantities on a dense grid in the input space.

```
xgrid <- seq(0, 1, length=1000)
p <- predict(mod, matrix(xgrid, ncol=1))
pvar <- p$sd2 + p$nugs
```

Figure 10.11 shows the resulting predictive surface in red, with true analog in black. Gray vertical bars help visualize relative degrees of replication at each unique input location.

```
plot(xgrid, f1d(xgrid), type="l", xlab="x", ylab="y", ylim=c(-8,18))
lines(xgrid, qnorm(0.05, f1d(xgrid), fr(xgrid)), col=1, lty=2)
lines(xgrid, qnorm(0.95, f1d(xgrid), fr(xgrid)), col=1, lty=2)
points(X, Y)
segments(mod$X0, rep(0, nrow(mod$X0)) - 8, mod$X0,
  (mod$mult - 8)*0.5, col="gray")
lines(xgrid, p$mean, col=2)
lines(xgrid, qnorm(0.05, p$mean, sqrt(pvar)), col=2, lty=2)
lines(xgrid, qnorm(0.95, p$mean, sqrt(pvar)), col=2, lty=2)
legend("top", c("truth", "estimate"), col=1:2, lty=1)
```

Both degree of replication and density of unique elements is higher in the high-noise region than where noise is lower. In a batch design setting and in the unique situation where relative noise levels are somehow known, a rule of thumb of more samples or replicates in higher noise regimes is sensible. Such knowledge is unrealistic, and even so the optimal density differentials and degrees of replication aren't immediate. Proceeding sequentially allows learning to adapt to design and vice versa.

A one-dimensional input case is highly specialized. In higher dimension, where volumes are harder to fill, a more delicate balance may need to be struck between the value of information in high- versus low-noise regions. Given the choice between high- and low-noise locations, which are otherwise equivalent, a low-noise acquisition is clearly preferred. Determining an appropriate lookahead horizon can be crucial to making such trade-offs.

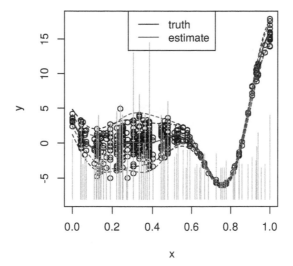

FIGURE 10.11: Sequential design with horizon $h = 5$: truth in black; predictive in red; vertical gray line segments indicate relative degrees of replication.

Tuning the horizon

Horizon $h = 5$ is rather arbitrary. Although it's difficult to speculate on details regarding the quality of the surface presented in Figure 10.11, due to the random nature of the Rmarkdown build, scope for improvement may be apparent. Chances are that uncertainty is overestimated in some regions and underestimated in others. In the version I'm looking at, the right-hand/low-noise region is over-sampled. A solution lies in tuning lookahead horizon h online. Binois et al. (2019) proposed two simple schemes. The first adjusts h in order to *target* a desired ratio $\rho = n/N$ and thus manage surrogate modeling computational cost through the Woodbury trick (§10.1.1):

$$
\text{Target:} \quad h_{N+1} \leftarrow \begin{cases} h_N + 1 & \text{if } n/N > \rho \text{ and a new } \bar{x}_{n+1} \text{ is chosen} \\ \max\{h_N - 1, -1\} & \text{if } n/N < \rho \text{ and a replicate is chosen} \\ h_N & \text{otherwise.} \end{cases}
$$

Horizon h is nudged downward when the empirical degree of replication is lower than the desired ratio; or vice versa.[13] The second scheme attempts to *adapt* and minimize IMSPE regardless of computational cost:

$$
\text{Adapt:} \quad h_{N+1} \sim \text{Unif}\{a_1', \dots, a_n'\} \quad \text{with} \quad a_i' := \max(0, a_i^* - a_i).
$$

Ideal replicate values

$$
a_i^* \propto \sqrt{r(\bar{x}_i)(K_n^{-1} W_n K_n^{-1})_{i,i}} \tag{10.14}
$$

come from a criterion in the SK literature (Ankenman et al., 2010). See `allocate_mult` in

---

[13]Horizon $h = -1$ corresponds a software implementation detail eliminating (setting to zero) thresholds applied to identify as replicates the results of continuous derivative-based searches residing nearby existing $\bar{x}_k$. Setting these to zero reduces selection of replicates, but doesn't entirely preclude them since Eq. (10.13) outcomes are still entertained.

hetGP and homework exercises in §10.4 for more details. Here the idea is to stochastically nudge empirical replication toward an optimal setting derived by fixing unique-n locations $\bar{x}_1, \ldots, \bar{x}_n$, and allocating any remaining design budget entirely to replication. An alternative, deterministic analog, could be $h_{N+1} = \max_i a_i'$.

Code below duplicates the example above with the *adapt* heuristic. Alternatively, `horizon` calls can be provided `target` and `previous_ratio` arguments in order to implement the *target* heuristic instead. Begin by reinitializing the design.

```
X <- X[1:10]
Y <- Y[1:10]
mod.a <- mleHetGP(X=X, Z=Y, lower=0.0001, upper=10)
h <- rep(NA, 500)
```

Next, loop to obtain $N = 500$ observations under an adaptive horizon scheme.

```
for(i in 1:490) {

  ## adaptively adjust the lookahead horizon
  h[i] <- horizon(mod.a)

  ## find the next point and update
  opt <- IMSPE_optim(mod.a, h=h[i])
  X <- c(X, opt$par)
  ynew <- fY(opt$par)
  Y <- c(Y, ynew)
  mod.a <- update(mod.a, Xnew=opt$par, Znew=ynew, ginit=mod.a$g*1.01)

  ## periodically attempt a restart to try to escape local modes
  if(i %% 25 == 0){
    mod2 <- mleHetGP(X=list(X0=mod.a$X0, Z0=mod.a$Z0, mult=mod.a$mult),
      Z=mod.a$Z, lower=0.0001, upper=1)
    if(mod2$ll > mod.a$ll) mod.a <- mod2
  }
}
```

Once that's done, predict on the grid.

```
p.a <- predict(mod.a, matrix(xgrid, ncol=1))
pvar.a <- p.a$sd2 + p.a$nugs
```

The left panel of Figure 10.12 shows adaptively selected horizon over iterations of sequential design. Observe that a horizon of $h = 5$ is not totally uncommon, but is also higher than typically preferred by the adaptive scheme. In total, $n = 104$ unique sites were chosen – more than in the fixed $h = 5$ case, but in the same ballpark compared to the full size of $N = 500$ acquisitions. The right panel of the figure shows the final design and predictions versus the truth, matching closely that of Figure 10.11 corresponding to the fixed horizon ($h = 5$) case.

```
par(mfrow = c(1,2))
plot(h, main="horizon", xlab="iteration")
plot(xgrid, f1d(xgrid), type="l", xlab="x", ylab="y",
  main="adaptive horizon design", ylim=c(-8,18))
lines(xgrid, qnorm(0.05, f1d(xgrid), fr(xgrid)), col=1, lty=2)
lines(xgrid, qnorm(0.95, f1d(xgrid), fr(xgrid)), col=1, lty=2)
points(X, Y)
segments(mod$X0, rep(0,nrow(mod$X0)) - 8, mod$X0,
  (mod$mult - 8)*0.5,  col="gray")
lines(xgrid, p$mean, col=2)
lines(xgrid, qnorm(0.05, p$mean, sqrt(pvar.a)), col=2, lty=2)
lines(xgrid, qnorm(0.95, p$mean, sqrt(pvar.a)), col=2, lty=2)
```

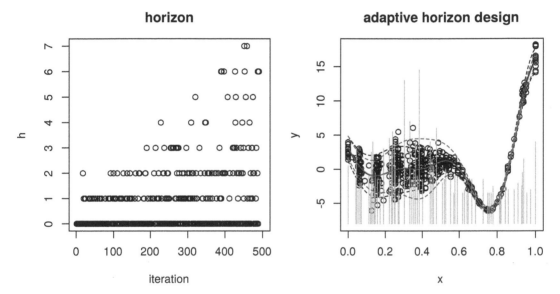

FIGURE 10.12: Horizons chosen per iteration (left); final design and predictions versus the truth (right) as in Figure 10.11.

The code below offers an out-of-sample comparison with RMSE (lower is better).

```
ytrue <- f1d(xgrid)
rmse <- c(h5=mean((ytrue - p$mean)^2), ha=mean((ytrue - p.a$mean)^2))
rmse
```

```
##      h5      ha
## 0.01781 0.01351
```

Although it varies somewhat from one Rmarkdown build to the next, a typical outcome is that the adaptive scheme yields slightly lower RMSE. Being able to adjust lookahead horizon enables the adaptive scheme to better balance exploration versus replication in order to efficiently learn a smoothly changing signal-to-noise ratio exhibited by the data-generating mechanism.

10.3.3   Examples

Here we revisit SIR and ATO examples from §10.2.2 from a sequential design perspective.

Susceptible, infected, recovered

Begin with a limited seed design of just twenty unique locations n and five replicates upon each for $N = 100$ total runs.

```
X <- randomLHS(20, 2)
X <- rbind(X, matrix(rep(t(X), 4), ncol=2, byrow=TRUE))
Y <- apply(X, 1, sirEval)
```

A fit to that initial data uses options similar to the batch fit from §10.2.2, augmenting with a fixed mean and limits on $\theta_{(\delta)}$. Neither is essential in this illustration, but help to limit sensitivity to a small seed design in this random Rmarkdown build.

```
fit <- mleHetGP(X, Y, covtype="Matern5_2", settings=list(linkThetas="none"),
   known=list(beta0=0), noiseControl=list(upperTheta_g=c(1,1)))
```

To visualize predictive surfaces and progress over acquisitions out-of-sample, R code below completes the 2d testing input grid XX from the batch example in §10.2.2 with output evaluations for benchmarking via RMSE and score.

```
YY <- apply(XX, 1, sirEval)
```

Code below then allocates storage for RMSEs, scores and horizons, for a total of 900 acquisitions.

```
N <- 1000
score <- rmse <- h <- rep(NA, N)
```

The acquisition sequence here, under an adaptive horizon scheme, looks very similar to our previous examples, modulo a few subtle changes. When reinitializing MLE calculations, for example, notice that `fit$used_args` is used below to maintain consistency in search parameters and modeling choices.

```
for(i in 101:N) {

  ## find the next point and update
  h[i] <- horizon(fit)
  opt <- IMSPE_optim(fit, h=h[i])
  ynew <- sirEval(opt$par)
  fit <- update(fit, Xnew=opt$par, Znew=ynew, ginit=fit$g*1.01)

  ## periodically attempt a restart to try to escape local modes
  if(i %% 25 == 0){
```

```
   fit2 <- mleHetGP(X=list(X0=fit$X0, Z0=fit$Z0, mult=fit$mult), Z=fit$Z,
     maxit=1e4, upper=fit$used_args$upper, lower=fit$used_args$lower,
     covtype=fit$covtype, settings=fit$used_args$settings,
     noiseControl=fit$used_args$noiseControl)
   if(fit2$ll > fit$ll) fit <- fit2
 }


 ## track progress
 p <- predict(fit, XX)
 var <- p$sd2 + p$nugs
 rmse[i] <- sqrt(mean((YY - p$mean)^2))
 score[i] <- mean(-(YY - p$mean)^2/var - log(var))
}
```

Figure 10.13 shows the resulting predictive surfaces and design selections. Compare to Figure 10.4 for a batch version. Similarity in these surfaces is high, especially for the mean (left panel), despite an order of magnitude smaller sequential design compared to the batch analog, which was based on more than ten thousand runs. Depending on the random Rmarkdown build, the standard deviation (right) may mis-locate the area of high noise in the I_0 direction, over-sampling in the upper-right corner. This is eventually resolved with more active learning.

```
par(mfrow=c(1,2))
xx <- seq(0, 1, length=100)
image(xx, xx, matrix(p$mean, ncol=100), xlab="S0", ylab="I0",
  col=cols, main="mean infected")
text(fit$X0, labels=fit$mult, cex=0.5)
image(xx, xx, matrix(sqrt(var), ncol=100), xlab="S0", ylab="I0",
  col=cols, main="sd infected")
text(fit$X0, labels=fit$mult, cex=0.5)
```

Density of unique locations, indicated by the coordinates of the numbers plotted in both panels, appears uniform to the naked eye. However, degree of replication is greater in the right-half of inputs corresponding to $S_0 > 0.5$ under the adaptive horizon IMSPE scheme. All double-digit multiplicities are in this right-hand half-plane. As a summary of progress over sequential design iterations, Figure 10.14 shows a progression of RMSEs (left), proper scores (middle) and horizon (right). The first two panels have the batch analog overlaid in dashed-red.

```
par(mfrow=c(1,3))
plot(rmse, type="l", xlab="n: sequential iterate", ylim=c(0.06, 0.07))
abline(h=sqrt(mean((YY - psir$mean)^2)), col=2, lty=2)
legend("topright", c("sequential", "batch"), col=1:2, lty=1:2)
plot(score, type="l", xlab="n: sequential iterate", ylim=c(4, 5))
abline(h=mean(-(YY - psir$mean)^2/vsir - log(vsir)), col=2, lty=2)
plot(h, xlab="n: sequential iterate", ylab="lookahead horizon")
```

In spite of the much larger batch training set, our sequential design performs very similarly out-of-sample. Note that both solid and dashed lines depend on random initializations/designs and responses through sirEval. Thus they have a distribution that could be averaged-over

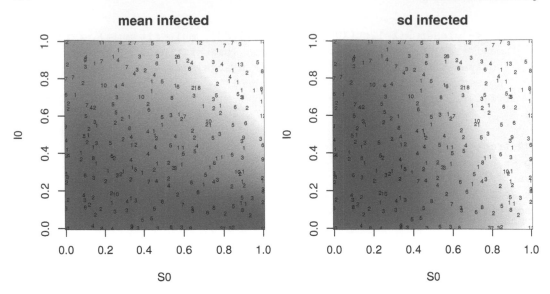

FIGURE 10.13: Heteroskedastic GP fit to sequentially designed SIR data showing predictive mean surface (left) and estimated standard deviation (right). Compare to batch analog in Figure 10.4.

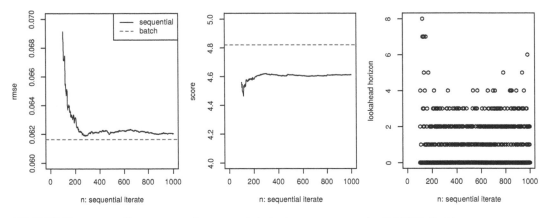

FIGURE 10.14: Summary of sequential design progress via RMSE (left), proper score (middle) and selected lookahead horizon (right) under the adapt scheme.

in a MC fashion. That exercise is left to the curious reader. To wrap up, observe in the right-hand panel that horizon favors low numbers, emphasizing $h = 0$ and $h = 1$, but more than twenty percent of selections are larger than that.

```
mean(h > 1, na.rm=TRUE)
```

```
## [1] 0.2822
```

Limited lookahead horizon is deemed to be beneficial for separating signal from noise in this example. It's worth reiterating that the IMSPE criterion (10.9) and lookahead horizon (10.14) emphasize mean-accuracy not variance accuracy or minimization. Nevertheless they perform well on score which combines measurements of both. This is because they – especially longer horizons – encourage replication which offers a pure look at noise in order to pin down local variance.

Inventory management

ATO is a much larger example. §10.2.2 leveraged a fit stored in the `ato` data object. Fitting live in Rmarkdown would've represented a prohibitive undertaking. That same data object contains an `"hetGP"`-class model `out.a` that was trained under an adaptive horizon IMSPE-based sequential design scheme. The size of that design and the time it took to train are quoted by the output of the code below.

```
c(n=nrow(out.a$X0), N=length(out.a$Z), time=out.a$time)
```

```
##            n          N time.elapsed
##         1194       2000        38738
```

Recall that the earlier experiment used $n = 1000$ unique sites with an average of five replicates upon each, for a total of about $N \approx 5000$ training data points. The actively selected training set here is much smaller, having $N = 2000$ at $n = 1194$ unique locations. So the sequential design has more unique locations but still a nontrivial degree of replication, resulting in many fewer overall runs of the ATO simulator. Utilizing the same out-of-sample testing set from the previous score-based comparison, code below calculates predictions and pointwise scores with this new sequential design.

```
phet.a <- predict(out.a, Xtest)
phets2.a <- phet.a$sd2 + phet.a$nugs
mhet.a <- as.numeric(t(matrix(rep(phet.a$mean, 10), ncol=10)))
s2het.a <- as.numeric(t(matrix(rep(phets2.a, 10), ncol=10)))
sehet.a <- (unlist(t(Ztest)) - mhet.a)^2
sc.a <- - sehet.a/s2het.a - log(s2het.a)
c(batch=mean(sc), adaptive=mean(sc.a))
```

```
##     batch adaptive
##     3.396    3.615
```

So active learning leads to a more accurate predictor than the earlier batch alternative, despite the former having been trained on about 60% fewer runs. Illustrating those acquisitions requires "rebuilding" the `out.a` object. To keep `hetGP` compact for CRAN, $\mathcal{O}(n^2)$ covariance matrices and inverses have been deleted by providing `return.matrices=FALSE` to the `mleHetGP` command used to build `out.a`.

```
out.a <- rebuild(out.a)
```

The calculation sequence for acquisitions involved in this sequential design begins by determining the horizon, and then searching with IMSPE while looking ahead over that horizon. In code, that amounts to the following:

```
Wijs <- Wij(out.a$X0, theta=out.a$theta, type=out.a$covtype)
h <- horizon(out.a, Wijs=Wijs)
```

```
control <- list(tol_dist=1e-4, tol_diff=1e-4, multi.start=30)
opt <- IMSPE_optim(out.a, h, Wijs=Wijs, control=control)
```

Pre-calculating `Wijs` (W) economizes a little on execution time since these are needed by both `horizon` and `IMSPE_optim`. A `control` argument to `IMSPE_optim` allows users to fine-tune fidelity of search over the criterion. Above, the number of restarts and tolerances on distance to existing sites (i.e., replicates) in search is adjusted upwards of their default settings. For restarts, our adjustments to the default are intended to make search more global. Regarding tolerances, higher values make selection of replicates more likely, determining that `optim` outputs within `1e-4` of input x or output $I_{N+1}(x)$ should snap to existing sites.

```
opt$par
```

```
##         [,1]    [,2]    [,3]    [,4]    [,5]    [,6]    [,7]   [,8]
## [1,] 0.202  0.8664  0.8689  0.2666  0.4139  0.4508  0.8075  0.892
```

Acquisition is completed by feeding that 8d location into the ATO simulator. Subsequently, the chosen input–output pair would be used to `update` the model fit. An indication of whether or not the new location is unique (i.e., actually new) or a replicate is provided by the `path` field in `IMSPE_optim` output. The `path` list contains the best sequence of points via elements `par`, `value`, and `new` of the acquisition followed by h others computed as lookahead selections.

```
opt$path[[1]]$new
```

```
## [1] TRUE
```

ATO inputs are actually on a grid, so evaluation would require "snapping" this continuous solution to that grid after undoing any coding of inputs. Further replication could be induced as part of that discretization process.

10.3.4   Optimization, level sets, calibration and more

Throughout this chapter we've focused on reducing mean-squared prediction error, a global criterion. Targeting specific regions of interest, such as global minima (BO; Chapter 7) or level sets is also of interest. Bogunovic et al. (2016) provide a treatment unifying criteria for these settings. Consider BO first, having already treated the noisy case in §7.2.4. Picheny et al. (2012) provide an important set of benchmarks specifically for this setting. Although there are many suitable criteria for BO with GPs, expected improvement (EI) emerges as perhaps the most appealing because it organically balances exploitation and exploration without any tuning parameters, doesn't require numerical approximation, and has a closed-form derivative. In the face of noise, and particularly heteroskedasticity, simplicity and computational tractability are important features when adapting a method originally designed for noiseless/deterministic settings.

As mentioned in the lead-in to §10.3.2, lookahead versions of EI have been studied (see, e.g., Ginsbourger and Le Riche, 2010; Gonzalez et al., 2016; Lam et al., 2016; Huan and Marzouk, 2016), but closed-form expressions exist only for one-step-ahead versions. Unlike IMPSE, future values of the criterion depend on future function evaluations. Inspired by

Lam et al. (2016) and IMSPE-based lookahead presented in §10.3.2, Binois and Gramacy (2018) introduce a replication-biased lookahead for EI which circumvents unknown future function values through simple imputation: $y_{N+1} \leftarrow \mu_N(x_{N+1})$, which is also known as a kriging "believer" approach (Ginsbourger et al., 2010). Examples are left to Section 4.1 in that paper, which is also available as a vignette in the `hetGP` package.

Active learning for the related problem of *contour finding*, or level set estimation, proceeds similarly. The objective is to identify a region of inputs defined by a threshold on outputs

$$\Gamma = \{x \in \mathbb{R}^m : Y(x) > T\},$$

where T can be zero without loss of generality. As with EI, the canonical development is for deterministic settings. In the presence of noise, take $Y(x) \equiv F(x)$, the latent random field (§5.3.2) with nugget-free epistemic uncertainty $\breve{\sigma}(x)$ from Eq. (10.10). Criteria defined by Lyu et al. (2018) are implemented for homoskedastic and heteroskedastic GPs and TPs in `hetGP`. Several other criteria can be found in the literature (Chevalier et al., 2013; Bogunovic et al., 2016; Azzimonti et al., 2016) with a selection of implementations provided by the `KrigInv` package (Chevalier et al., 2018, 2014a,b) on CRAN.

The simplest criterion for active learning in this context is *maximum contour uncertainty (MCU)*, implemented as `crit_MCU` in `hetGP`. MCU is based on local probability of misclassification, namely that the function is incorrectly predicted to be below or above T (Bichon et al., 2008; Ranjan et al., 2008). A second criterion, *contour stepwise uncertainty reduction (cSUR)*, implemented by `crit_cSUR` in `hetGP`, amounts to calculating a one-step-ahead reduction of MCU. A more computationally intensive, but more global alternative entails integrating cSUR over the domain in a manner similar to IMSPE (10.9) for variance reduction or IECI (Gramacy and Lee, 2011) for BO (§7.2.3). In practice, the integral is approximated by a finite sum, which is the approach taken by `crit_ICU` in `hetGP`. Finally, *targeted mean-squared error* (tMSE; Picheny et al., 2010), is provided by `crit_tMSE` and designed to reduce variance close to the target contour. Again, examples are provided in Section 4.2 of Binois and Gramacy (2018).

As a final topic, consider computer model calibration (§8.1) and inverse problems with a noisy, possibly heteroskedastic simulator. I'm not aware of SK or `hetGP`-like methodology yet being deployed in this setting, but Fadikar et al. (2018) consider quantile kriging (QK; Plumlee and Tuo, 2014) to good effect. Pairing a massive agent-based epidemiological simulation campaign leveraging social networks and data on an Ebola outbreak in Liberia, Fadikar et al. built a QK surrogate for functional output (counts of infecteds over time) via principal components (Higdon et al., 2008) in order to back out a distribution on simulator tuning parameters most likely to have generated an observed (real) sequence of infecteds. Entertaining alternatives to QK, like `hetGP`, in Rmarkdown on this specific example are complicated by access to a supercomputer scale simulation capability and MATLAB (rather than R) libraries for working with functional output (Gattiker et al., 2016).

In lieu of that, consider the following setup involving similar features without the complications of functional output, and ready access to simulations. Herbei and Berliner (2014) describe a Feynman-Kac simulator[14], whose source may be downloaded from GitHub[15], that models the concentration of a tracer within a given spatial domain. One application is modeling oxygen concentration in a thin water layer deep in the ocean (McKeague et al., 2005). The simulator is stochastic, approximating the solution of an advection–diffusion

---

[14]https://en.wikipedia.org/wiki/Feynman-Kac_formula
[15]https://github.com/herbei/FK_Simulator

with MC, and is highly heteroskedastic. There are four real-valued inputs: two spatial coordinates (longitude and latitude) and two diffusion coefficients. Interest is in learning, i.e., inverting, the diffusion coefficients. In the notation of Chapter 8, the spatial coordinates are x and diffusion coefficients are calibration parameters u. So the goal is to use the simulator $Y^M(x, u)$ to learn $\hat{u}$ from measurements $Y^F(x)$ of oxygen concentration in the field, e.g., at x's in the Southern Ocean.

In some regions of the input space the simulator response is multimodal, i.e., non-Gaussian, but this can be resolved through minimal (and automatic) replication and averaging. While simulation code is available for C/CUDA, R and MATLAB on the GitHub page, code linked from the book web page provides the essentials; see fksim.R[16] and configuration files fkset.RData[17] read therein, with several enhancements. Coded inputs and sensible defaults for the diffusion coefficients (calibration parameters u) promote plug-and-playability; automatic degree-6 replication multiplicity with averaging eliminates multimodality.

With those codes, consider simulations along a longitudinal slice in the input space ...

```
source("fksim.R")
x2 <- seq(0, 1, length=100)
y <- sapply(x2, function(x2) { fksim(c(0.8, x2)) })
```

... with visual following in Figure 10.15.

```
plot(x2, y, xlab="coded longitude", ylab="oxygen concentration")
```

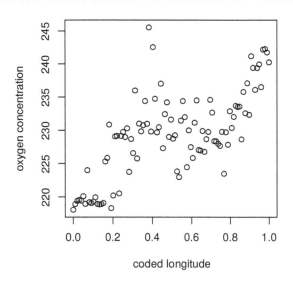

FIGURE 10.15: Simulated oxygen concentration along a slice in longitude, with the other three parameters fixed.

Clear signal is evident in the figure, but also clear input-dependent variance. It seems that a much higher degree of replication would be required to effectively separate signal and noise, at least in some parts of the input space, in addition to more exploration to fill out the

---

[16]http://bobby.gramacy.com/surrogates/fksim.R
[17]http://bobby.gramacy.com/surrogates/fkset.RData

4d study area. Two homework exercises in §10.4 take the curious reader through surrogate modeling, design, and inversion with this simulator. For inversion, calibration methods from §8.1 can be applied. Alternatively the log posterior coded below, provided in order to keep our development here somewhat more self-contained, can be used instead.

```
library(mvtnorm)
lpost.invert <- function(theta, XF, yF, GP)
  {
  ## input processing and checking
  if(length(theta) != ncol(GP$X0) - ncol(XF) + 1)
    stop("length(theta), ncol(XF), ncol(GP$X0) mismatch")
  u <- theta[-length(theta)]
  s2 <- theta[length(theta)]

  ## prior checking
  if(any(u < 0 | u > 1)) return (-Inf)
  if(s2 < 0) return(-Inf)

  ## derive predictive distribution for XF paired with u
  XFU <- cbind(XF, matrix(rep(u, nrow(XF)), ncol=length(u), byrow=TRUE))
  p <- predict(GP, XFU, xprime=XFU)
  C <- s2*diag(nrow(p$cov)) + (p$cov + t(p$cov))/2

  ## gaussian log density evaluation for yF under that predictive
  return(dmvnorm(yF, p$mean, C, log=TRUE) - log(s2))
  }
```

Notice that `lpost.invert`, above, is coded to work with a (possibly homoskedastic) `"hetGP"`-class object `GP`. Parameter `theta` combines diffusion coefficients u, in its first several coordinates, with iid noise variance parameter σ^2 in its last coordinate. Implicit in `lpost.invert` is a uniform prior $u \sim \text{Unif}(0, 1)^{m_u}$ and scale-invariant Jeffrey's prior $\sigma^2 \sim 1/\sigma^2$. Arguments `XF` and `YF` are field data pairs (X_{n_F}, Y_{n_F}), observed under the correct but unknown $u^\star$. That likelihood could be optimized, or used in a Bayesian analysis after placing priors on u. It measures the surrogate predictive probability density of Y_{n_F}-values when paired with X_{n_F} and candidate u-settings. This is more or less equivalent to Kennedy and O'Hagan's likelihood (8.3) in the "nobias" case.

10.4 Homework exercises

Prompts below explore `hetGP` variations, sequential design and replication, and application in a calibration (§8.1) setting.

#1: Latent log variances

In §10.2.1 inferential expressions are developed for Λ_n as a diagonal storing smoothed variances for GP modeling with covariance structure $\nu(C_N + \Lambda_N)$. Unfortunately, such

$\Lambda_n \mid \Delta_n$ given latent Δ_n in Eq. (10.7) can be negative, which is bad for a variance. Describe how the scheme can be adapted to work with log variances instead. *Hint: the answer to this question is relatively short, but getting it right requires a careful inspection of likelihood and derivatives.*

#2: Compensating for mean nonstationarity

§5.3.3 discussed an equivalence between mean and kernel/covariance characterization for GPs. What happens when you fit `hetGP`, designed for variance nonstationarity, to data which is homoskedastic but exhibits mean nonstationarity? Consider two examples.

a. Begin with our favorite 2d data: $x_1 \exp(-x_1^2 - x_2^2) + \varepsilon \sim \mathcal{N}(0, 0.1^2)$, where $x \in [-2, 4]^2$. In §9.2.2 we used trees to partition up the input space and focus on the interesting lower-left quadrant. We also used it above in §10.1.2. Train homoskedastic and heteroskedastic separable Gaussian kernel GPs to outputs collected on a design composed of a uniform 21×21 grid in the input space, with three replicates on each site. Plot the predictive mean and standard deviation surfaces using a 101×101 grid and comment. Compare pointwise proper scores [Gneiting and Raftery (2007), Eq. (27); also see §9.1.3] with testing data simulated on that uniform grid.

b. Now investigate a 1d response with a regime shift: $f(x) = \sin(\pi(10x + 0.5)) + 10(-1)^{\mathbb{I}(x < 0.5)}$, for $x \in [0, 1]$ and observed under standard normal noise. Train homoskedastic and heteroskedastic separable Gaussian kernel GPs to data observed on a uniform length $n = 30$ grid. Visualize the predictive surfaces on a uniform 100-length testing grid and comment. Compare RMSEs and pointwise proper scores using testing evaluations on that grid.

#3: Optimal replication?

Here the goal is to compare SK's ideal replication (10.14) numbers a_i at existing training inputs $\bar{x}_i$, $i = 1, \ldots, n$, with lookahead based IMSPE analogs limited to those same sites.

a. Create a uniform design in $[0, 1]$ of length $n = 21$ and evaluate five replicates of each under `fY` from §10.3.2 and fit a `hetGP` surface using a Matèrn $\nu = 5/2$ kernel. (Otherwise the `fit <- mleHetGP(...)` defaults are fine.) Obtain predictions on a fine grid in the input space and compare your predicted nuggets `predict(...)$nugs` to the truth `fr`.

b. Suppose the plan was to gather a total of $N = 201$ simulations from `fY`. You've gathered $n = 21$ already, so there are $N_{\text{rest}} = 180$ more to go. Use `allocate_mult` with `Ntot = N` to calculate the optimal a_i, for $i = 1, \ldots, n$.

c. Similarly use `opt <- IMSPE_optim(...)` with argument `h =` N_{rest} candidates `Xcand=fit$X0` from existing unique sites. Record all `opt$path[[j]]$par` as selected x locations and extract from those a multiplicity count, i.e., number of replicates, corresponding to each unique setting.

d. Plot your results from #a–c on common axes and comment.

#4: Ocean sequential design

Consider the ocean/oxygen simulator introduced in §10.3.4. Here we shall treat only the spatial inputs x, leaving diffusion coefficients u at their default values.

a. Build a design in 2d comprised of $n = 25$ unique LHS (§4.1) locations with twenty replicates on each. Obtain simulation responses at these $N = 500$ sites. Fit heteroskedastic and homoskedastic GPs to these data and provide a visual comparison between the two on a 100×100 predictive grid, and on a slice of that grid fixing the latitude input near 80% in coded units. Overlay runs from §10.3.4 on that slice and comment. You might find it helpful, but not essential, to use the following settings in your fit(s).

```
lower <- rep(0.01, 2)
upper <- rep(30, 2)
covtype <- "Matern5_2"
noiseControl <- list(g_min=1e-6, g_bounds=c(1e-6, 1),
  lowerDelta=log(1e-6))
settings <- list(linkThetas="none", initStrategy="smoothed",
  return.hom=TRUE)
```

b. Now consider sequential design under the heteroskedastic GP only. Begin with $n_{\text{init}} = 25$ locations, ideally the same ones from #a, but only four replicates on each so that $N_{\text{init}} = 100$. Acquire 400 more simulations under an adaptive lookahead IMSPE scheme. At the end, augment your visuals from #a, including horizon over the iterations, and comment. You might find it helpful, but not essential, to utilize the following in your IMSPE searches.

```
control <- list(tol_dist=1e-4, tol_diff=1e-4, multi.start=30)
```

c. Finally, repeat #a–b at least ten times and save RMSEs and pointwise proper scores calculated on out-of-sample testing data observed on the predictive grid. (No need to repeat the visuals, just the designs and fits.) Provide boxplots of RMSEs and scores and comment.

#5: Ocean calibration

Data in ocean_field.txt[18] contains $n_F = N_F = 150$ observations of oxygen concentration at sites in the Southern Ocean matching baseline diffusion coefficients $u^\star$ encoded in the default **fksim** formals. See §10.3.4 and #4 for details. Separately, build homoskedastic and heteroskedastic GP surrogates from **fksim** simulation campaign output provided by ocean_runs.txt[19] combining a 4d LHS of size 500, over latitude and longitude x-sites and two diffusion coefficients u, with a 2d maximin LHS of u-values paired with X_{n_F} from the field data. Ten replicate runs at each location are gathered for a grand total of $N_M = 6000$ simulations. Use **lpost.invert** from §10.3.4 in order to sample from a Bayesian posterior distribution of u and σ^2, separately for both surrogates. Compare your posterior distribution(s) for u with true $u^\star$-values.

---

[18]http://bobby.gramacy.com/surrogates/ocean_field.txt
[19]http://bobby.gramacy.com/surrogates/ocean_runs.txt

A

Numerical Linear Algebra for Fast GPs

This appendix is in two parts. §A.1 illustrates the value of linking against fast linear algebra libraries, yielding 10× speedups and sometimes better, without any coding changes. §A.2 provides pointers to recent developments in MVN likelihood evaluation and prediction via stochastic approximations to log determinants and solves of linear systems.

A.1 Intel MKL and OSX Accelerate

Throughout this text, Rmarkdown builds leveraged an ordinary R installation – one downloaded from CRAN and run without modification. R ships with a rather vanilla, but highly portable, numerical linear algebra library (i.e., implementation of BLAS[1] and LAPACK[2]).[3] This section is an exception. Here an R linked against Intel's Math Kernel Library (MKL)[4] is used,[5] following instructions here for Linux/Unix[6]. Another good resource tied specifically to .deb based systems such as Ubuntu[7] can be found here[8]. A set of instructions from Berkeley's Statistics Department[9], similar to Intel's for MKL, explains how to link against Apple OSX's Accelerate Framework[10] instead. Microsoft R Open[11] provides binary installs for Linux and Windows linked against MKL, and for OSX linked against the Accelerate framework. These are nice options for out-of-the-box speedups, as will be demonstrated momentarily. One reason MATLAB® often outperforms R on linear algebra-intensive benchmarks is that MATLAB ships with Intel MKL linear algebra. R does not. Microsoft R Open fills that gap, although it's not hard to do-it-yourself.

Illustration on borehole

Recall the borehole function[12] introduced in §9.1.3 and revisited throughout Chapter 9. The borehole is a classic synthetic computer simulation example (Morris et al., 1993). It's

---

[1]https://en.wikipedia.org/wiki/Basic_Linear_Algebra_Subprograms

[2]https://en.wikipedia.org/wiki/LAPACK

[3]Even when compiling R from source, or pulling binaries from standard repositories, R links against a vanilla reference BLAS and LAPCK by default.

[4]https://software.intel.com/en-us/mkl

[5]Actually, results from an MKL linked R were performed off-line and read in after-the-fact because RStudio doesn't make it easy for two separate Rs to build a single document.

[6]https://software.intel.com/en-us/articles/quick-linking-intel-mkl-blas-lapack-to-r

[7]https://ubuntu.com/

[8]https://github.com/eddelbuettel/mkl4deb

[9]http://statistics.berkeley.edu/computing/blas

[10]https://developer.apple.com/documentation/accelerate

[11]https://mran.microsoft.com/open

[12]https://www.sfu.ca/~ssurjano/borehole.html

well-fit by a GP, but obtaining accurate predictions requires a relatively large sampling. First pasting ...

```
borehole <- function(x)
  {
    rw <- x[1]*(0.15 - 0.05) + 0.05
    r <-  x[2]*(50000 - 100) + 100
    Tu <- x[3]*(115600 - 63070) + 63070
    Hu <- x[4]*(1110 - 990) + 990
    Tl <- x[5]*(116 - 63.1) + 63.1
    Hl <- x[6]*(820 - 700) + 700
    L <-  x[7]*(1680 - 1120) + 1120
    Kw <- x[8]*(12045 - 9855) + 9855
    m1 <- 2*pi*Tu*(Hu - Hl)
    m2 <- log(r/rw)
    m3 <- 1 + 2*L*Tu / (m2*rw^2*Kw) + Tu/Tl
    return(m1/m2/m3)
  }
```

... then code below generates random training and testing sets in a manner identical to §9.1.3 with two exceptions. Both sets are much bigger: ten thousand elements each from a combined Latin hypercube sample (LHS, §4.1); and these are observed with noise so that nuggets must be estimated alongside lengthscales.

```
library(lhs)
Npred <- N <- 10000
x <- randomLHS(N + Npred, 8)
y <- apply(x, 1, borehole)
y <- y + rnorm(length(y), sd=1)
X <- x[1:N,]
Y <- y[1:N]
XX <- x[-(1:N),]
YY <- y[-(1:N)]
```

Consider the full, non-approximate GP capability offered in `laGP` (Gramacy and Sun, 2018). Code below generates appropriate priors, insuring stable search for MAP estimates of hyperparameters $\theta_1, \ldots, \theta_8$ and g.

```
library(laGP)
ga <- garg(list(mle=TRUE), Y)
da <- darg(list(mle=TRUE, max=100), X)
```

Such a setup, and code below solving for $(\hat{\theta}, \hat{g})$ through the concentrated log likelihood (5.8) and furnishing predictions (5.2), is identical to previous uses in this text. The `laGP` library is identical too. The only difference is that R's BLAS and LAPACK are linked against Intel MKL, a testament to modularity in this presentation.

```
tic <- proc.time()[3]
```

```
## GP initialization and MAP calculation
gpsepi <- newGPsep(X, Y, da$start, g=ga$start, dK=TRUE)
that <- mleGPsep(gpsepi, param="both", tmin=c(da$min, ga$min),
  tmax=c(da$max, ga$max), ab=c(da$ab, ga$ab), maxit=1000)

## predict out of sample
p <- predGPsep(gpsepi, XX, lite=TRUE)
deleteGPsep(gpsepi)

## timing end
toc <- proc.time()[3]
```

Here are the timing results. For the record, my workstation has an eight-core hyperthreaded Intel i7-6900K CPU running at 3.20GHz with 128GB of RAM. It was purchased in 2016 – a mid-range desktop workstation from that era, optimized to be as quiet as possible, not for speed.

```
toc - tic
```

```
## elapsed
##     667
```

So it takes about 0.2 hours which, while not instantaneous, is still well within the realm of tolerable. Running the same example on a two-core laptop, such as my 2015 hyperthreaded Intel i7-5600U running at 2.60GHz with 8GB of RAM, takes about 25% longer. A problem of this size is all but out of reach without optimized linear algebra. I tried, letting it run overnight in ordinary R on the same workstation, and still it had not yet finished. Very conservatively, Intel MKL gives a 10× speedup on this example.

How does it work?

Intel MKL and OSX Accelerate offer a three-pronged approach to faster basic linear algebra (matrix–vector multiplications, etc.), matrix decompositions (determinants, LU, Cholesky), and solves of linear systems. Prong one leverages highly specialized machine instructions supported by modern processor hardware. Intel MKL has a huge advantage here. Experts designing processor pipelines, instructions, and compilers/options work closely together for the specific purpose of optimizing linear algebra and other benchmarks in order to show-up competitors and market to industrial clients.

Prong two involves tuning myriad alternatives to problem sizes and representative spans of use cases. Over the years, beginning in the 1960s, many numerical linear algebra codes have been developed, extended and refined. Some codes and combinations thereof are better than others depending on how big the problem is, and on other features describing common situations. For example, algorithms requiring time in $\mathcal{O}(n^3)$ are sometimes faster than $\mathcal{O}(n^{5/2})$ methods for small n. Determining which n goes to which algorithm depends upon architecture details and can only be determined by extensive experimentation. This tuning enterprise is itself a computer experiment and Bayesian optimization (BO) of sorts. For more details and an open source scripting of such experimentation and search, see the ATLAS project[13].

Prong three is symmetric multiprocessor (SMP) parallelization, which is intimately linked to the first two prongs. Divide-and-conquer multi-threaded calculation features in many modern linear algebra libraries, and the optimal way to divvy up tasks for parallel evaluation depends upon problem size n, numbers of processor cores, cache size, high-speed memory (RAM) capacity and performance, etc.

Swapping in fast linear algebra benefits more than just GP regression. Speedups can be expected in any linear algebra-intensive implementation of statistical methodology, or otherwise. Not every application will see substantial gains. For example, building the entirety of Chapter 9 takes 57 minutes with MKL and 98 minutes without. That may not seem impressive, but keep in mind that a large portion of the R code in that chapter doesn't involve linear algebra (e.g., plotting), uses non-BLAS/LAPACK solvers (e.g., sparse matrix CSK examples using `spam`), or is already heavily parallelized (`tgp` and `laGP` examples). There are even examples where Intel MKL is slower than R's default BLAS and LAPACK. Those are few and far between in my experience. Typical speedups for linear algebra-intensive tasks are between 2× and 10×, and the bigger the problem the bigger the time savings. I run both an MKL-linked and ordinary R on my machine. Development is more straightforward on that latter, because that gives me a good sense of what most users experience. Big problems get run with MKL.

Most university/lab-level supercomputing services offer an R linked against MKL. Virginia Tech's Advanced Research Computing (ARC) service[14] goes one step further by compiling all of R with Intel's C and Fortran compilers[15], in addition to linking with MKL. Differences compared to ordinary (GCC[16]) compiling are slight, as you can see on that page. The big winner is MKL. It's not hard to get similar capabilities on your own laptop or workstation. VT ARC doesn't offer an R linked against the default BLAS/LAPACK as provided by CRAN.

A caution on OpenBLAS

The Berkeley page[17] explains how to link to OpenBLAS[18] on Linux, and VT ARC also provides an OpenBLAS option. But I strongly caution against OpenBLAS when Intel MKL and OSX Accelerate alternatives are available. OpenBLAS is fast, but has thread safety issues which means it doesn't play well with some of the methods in this text, particularly those from Chapter 9 providing additional, bespoke divide-and-conquer parallelization. The `tgp` package (Gramacy and Taddy, 2016) uses pthreads[19] to parallelize prediction across leaves; see Appendix C.2 of Gramacy (2007). The `laGP` package uses OpenMP to parallelize over elements of the testing set. These are incompatible with OpenBLAS. Related messages on discussion boards can be found by googling "OpenBLAS is not thread safe" and adding "with pthreads" or "with OpenMP".

Thread safety[20] means that two independent calculations can run in parallel without

---

[13]http://math-atlas.sourceforge.net/
[14]https://www.arc.vt.edu/userguide/r/#blas
[15]https://software.intel.com/en-us/compilers
[16]https://gcc.gnu.org/
[17]http://statistics.berkeley.edu/computing/blas
[18]https://www.openblas.net/
[19]https://en.wikipedia.org/wiki/POSIX_Threads
[20]https://en.wikipedia.org/wiki/Thread_safety

interfering with one another. Unfortunately, two GP predictive or MLE subroutines (e.g., via `aGP` in §9.3) could not occur in parallel, without error, when the underlying linear algebra is off-loaded to OpenBLAS. When thread safety cannot be guaranteed, calculations which should transpire independently under the two processor threads carrying out their instructions will instead corrupt one another, without any indication that something is wrong. Results output are garbage.

There are ways to cripple OpenBLAS, making it slower but ensuring thread safety by limiting to a single thread. That defeats the purpose, especially on large problems where fast modern divide-and-conquer methods shine most. Intel MKL and OSX Accelerate are both thread safe and their compatibility spans more than 99% of architectures in modern use. There's almost no reason to entertain OpenBLAS, in my opinion, as long as thread safety remains an issue. That said, I've had little traction getting OpenBLAS removed as a default option at VT ARC. OpenBLAS is common in spite of thread safety issues, so beware.

A.2 Stochastic approximation

Many exciting inroads are being made in the realm of stochastic approximation of the sorts of matrix solves and decompositions required for GP inference and prediction. Specifically, the important calculations are $K_n^{-1} Y_n$ and $\log |K_n|$. Both are prohibitive for larger n. These passages are not intended to review the breadth of related methodology, but instead to showcase one modern approach as a representative. In contrast to Appendix A.1, which emphasizes a modular and generic fast linear algebra capability, pairing existing high-level GP code with fast low-level computations provided by libraries, the methods here are more tailored to GPs. They involve re-implementing calculations core to GP likelihood (and derivative) evaluation, and to predictive equations.

Gardner et al. (2018) proposed combining linear conjugate gradients[21] for matrix solves $(K_n^{-1} Y_n)$ with stochastic Lanczos quadrature (Ubaru et al., 2017) to approximate log determinant evaluations $(\log |K_n|)$. They carefully describe many engineering details that make this work, including preconditioning, pivoted Cholesky decomposition, and considerations specific to GP inference and prediction. The methods are approximate and stochastic, which may seem like downsides, but they make the method highly parallelizable, enabling trivial SMP distribution. Another advantage is that they don't require storage of large K_n, but rather only access kernel evaluations $k(\cdot, \cdot)$. This dramatically reduces communication overheads in offloading data and calculations to customized hardware, such as to graphical processing units (GPUs). Other attempts to leverage GPU linear algebra in GP inference have, by contrast, yielded lukewarm results (Franey et al., 2012).

Gardner et al. describe a Python implementation, GPyTorch[22] based on the PyTorch[23] toolkit for distributed computing. They report tractable inference on large-scale GP regression problems, including up to $n = 10^6$ (Wang et al., 2019). I'm not aware of any R implementations at this time, but would anticipate similar developments coming online soon. Such techniques will be key to applying GP regression at scale, especially when assumptions

---

[21]https://en.wikipedia.org/wiki/Conjugate_gradient_method
[22]https://gpytorch.ai/
[23]https://pytorch.org/

of smoothness and stationarity are key to effective application, i.e., meaning that many of the Chapter 9 divide-and-conquer alternatives are less than ideal.

B

An Experiment Game

In-class games are a fun way to engage students in difficult and seemingly esoteric material (Lee, 2007). Out-of-class games, played over an entire semester say, are less common but can effectively synthesize real-life settings such as those encountered in response surface methodology (RSM; Chapter 3) and Bayesian optimization (BO; Chapter 7). One such game was proposed, and first played over forty years ago (Mead and Freeman, 1973). Today, few are aware of that contribution in spite of a citation featuring prominently in a canonical RSM text (Box and Draper, 2007). Perhaps this is because Mead and Freeman were ahead of their time. Their setup required a computing environment with student access, and so on, decades before ubiquitous desktop and laptop computing. Today with R, Rmarkdown and `shiny` (Chang et al., 2017) web interfaces, barriers have come way down.

B.1 A shiny update to an old game

Mead and Freeman's original game centered around blackbox evaluation of agricultural yield as a function of six nutrient levels, following a form borrowed from Nelder (1966):

```
yield <- function(N, P, K, Na, Ca, Mg)
 {
   l1 <- 0.015 + 0.0005*N + 0.001*P + 1/((N+5)*(P+2)) + 0.001*K + 0.1/(K+2)
   l2 <- 0.001*((2 + K + 0.5*Na)/(Ca+1)) + 0.004*((Ca+1)/(2 + K + 0.5*Na))
   l3 <- 0.02/(Mg+1)
   return(1/(l1 + l2 + l3))
 }
```

Players were asked to supply settings for inputs across five campaigns, simulating crop years. In each campaign, after observing yield under a Gaussian noise regime determined by additive block and plot-within-block effects, players could use data to update fits and revise strategies for future campaigns. The ultimate goal was to maximize yield, primarily with Chapter 3-like tools such as steepest ascent and ridge analysis. In a modern landscape of computer experiments, the Mead and Freeman (1973) game seems somewhat antiquated, harking back to Fisher's 1920s work at Rothamsted Experimental Station[1].

Gramacy (2018b, hereafter "I") described a revised variation motivated by modern technology/application, and a more sophisticated methodological toolkit, such as for BO in Chapter 7. Other enhancements target friendly competition through leaderboards and benchmarks,

---

[1]https://en.wikipedia.org/wiki/Ronald_Fisher#Rothamsted_Experimental_Station,_1919-1933

carrots to encourage regular engagement, and partial solutions to catch-up straggling students. Perhaps the biggest innovation in this reboot is the use of modern web interfaces. Game play is facilitated by an R `shiny` app, shown in Figure B.1, which serves as both a multi-player portal and an interface to the back-end database of player(s) records. Code supporting the app is linked from the book web page.

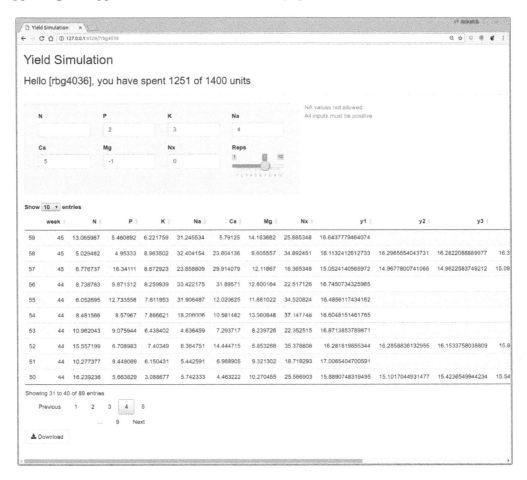

FIGURE B.1: Interactive yield simulation session. Duplicated from Gramacy (2018b).

Once logged in, the player is presented with three blocks of game content. Players are identified by their initials, as in the leaderboard discussed in §B.2, and a secret four-digit pin. The view in the figure is for my personal session; my initials are "rbg" and I chose my office number "403G" as a pin. A greeting block at the top of the page in the figure provides details on my spent and total budget for experimental runs. Details on how budget replenishes weekly, and how a schedule of run costs encourages including replication in the design of runs (§10.3.2), are left to Section 2.2 of Gramacy (2018b). As long as the player has not over-spent their budget, new runs may be performed by entering coordinates and a number of replicates into the second block on the page shown in the figure. Once all entries are valid, a "Run" button appears alongside a warning that there are no do-overs.

Performing a run causes the table in the final block of the page to be updated. All together, the table has 18 columns, recording run week, 7 input coordinates, and up to ten outputs. It's primary purpose is visual confirmation that new runs have been successfully incorporated into the player's database file. It's not intended as the main data-access vehicle. A "Download" button at the bottom saves a text-formatted table to the player's `~/Downloads` directory.

One modern twist in the game's construction encourages players to think about signal-to-noise trade-offs, and nudges them to spend units regularly rather than save them all until the final week of game play. I was worried that students would procrastinate, and wanted to devise a scheme that encouraged rather than mandated engagement. So variance of the additive noise on yield simulations changes weekly, following a smooth process in time. Starting in week w_s, noise in week w follows

$$\sigma^2(w) = 0.1 + 0.05(\cos(2\pi(w - w_s)/10) + 1).$$

Noise peaks in this setup during the first and tenth weeks, although players are warned that variance may be monotonically increasing, substantially devaluing late semester binges. A second modern innovation involves the introduction of a seventh input, `Nx`, that is deliberately unrelated to the response.

Game setup is optimized for play during a fifteen-week semester. I played the game with the class during the Fall semester of 2016. Although the main goal is to optimize yield, a final project assignment prompted students to think about ancillary goals such as main effects (§8.2.2) and sensitivity indices (§8.2.3), and asked them to report on how variance evolves over time (§10.2). Homework exercises encouraged students to try certain specific methods, and were timed with lecture material: beginning with steepest ascent and ridge analysis (Chapter 3) and culminating in BO (Chapter 8). Details and more specific pointers are provided in Gramacy (2018b); all materials are linked from the book web page[2].

B.2 Benchmarking play in real-time

Among those materials is an Rmarkdown script compiling four "leaderboard" style views into student performance over the weeks of game play. Students could visit the leaderboard any time. It was hosted along with the game interface (Figure B.1) on shinyapps.io[3]. I hoped that friendly competition would spur interest. As a peek into one of the four views provided, code below recreates (de-noised) maximum yield progress over thirteen weeks of game play.

Text files storing each player's database of runs may be read in as follows.

```
files <- list.files(path="yield/leaderboard", pattern="txt")
files
```

```
##  [1] "ame4794.txt" "fs0930.txt"  "hm1113.txt"  "ic2997.txt"
##  [5] "jbl1003.txt" "jh0702.txt"  "jtf1020.txt" "mds6266.txt"
##  [9] "rbg4036.txt" "ss0720.txt"  "wt4512.txt"  "ww2222.txt"
```

Names of the files concatenate player initials and pins. Each records a table of inputs and noisy outputs, so these inputs must be run back through `yield` for a de-noised view. De-noising helps identify the true ranking of players, rather than ones corrupted by spurious

[2]http://bobby.gramacy.com/surrogates
[3]https://www.shinyapps.io

noise. The function below vectorizes `yield` in order to streamline that de-noising process. Notice that the seventh, `Nx` input isn't used.

```
yield.fn <- function(X)
  yield(X[,1], X[,2], X[,3], X[,4], X[,5], X[,6])
```

Next, code below loops over each player's database file, building up a `data.frame` of best results by week. The game was run during weeks 38–51 in 2016, and these results were captured during the final, 51st week.

```
wk <- 51
start.wk <- 38
weeks <- (start.wk-1):wk
Ybest <- matrix(-Inf, nrow=length(weeks), ncol=length(files))
for(i in 1:length(files)) {
  data <- read.table(paste0("yield/leaderboard/", files[i]), header=TRUE)
  wk <- data[,1]
  xs <- data[,2:8]
  for(j in 1:length(weeks)) {
    wi <- wk == weeks[j]
    if(j > 1) Ybest[j,i] <- Ybest[j-1,i]
    if(sum(wi) == 0) next
    ybnew <- max(as.matrix(yield.fn(xs[wi,-7])), na.rm=TRUE)
    if(j == 1 || Ybest[j-1,i] < ybnew) Ybest[j,i] <- ybnew
  }
}
```

In order to mask true outputs, lest players learn the actual (non-noisy) value of their best response over the weeks, visuals of de-noised yields were provided on a normalized scale.

```
minY <- min(Ybest)
Ybest <- Ybest/minY - 1
```

Finally, player pin information is scrubbed to leave only initials for presentation in the legend of the leaderboard.

```
initials <- files
for(i in 1:length(initials)) {
  initials[i] <- sub("[0-9]+.txt", "", files[i])
}
```

Figure B.2 shows the result. On the x-axis is the week of game play, and on the y-axis is normalized yield. Each player has a line in the plot.

```
matplot(weeks-start.wk+1, Ybest, type="l", ylab="max yield", xlab="week",
   col=1:length(initials), lty=1:length(initials), lwd=2, xlim=c(0, 19))
legend("bottomright", initials, col=1:length(initials),
   lty=1:length(initials), lwd=2)
```

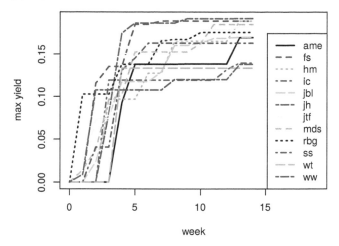

FIGURE B.2: De-noised view into real-time progress on yield optimization captured during the final week of game play.

Observe from the figure that about half of all players' progress is made in the first five weeks, spanning around forty runs. This is a testament to the prowess of simple, classical RSM from Chapter 3. Most subsequent refinement transpired using more modern, Chapter 7 techniques. Students "jh" and "fs" made rapid progress, whereas "hm" ends up at the same place in the end, but with more steady increments. My own progress ("rbg") placed me fifth by this measure. I favored replication over unique runs in hopes of obtaining better main effects, sensitivity indices, and estimates of variance over time.

Three other views are provided by the Rmarkdown file `leader.Rmd` residing in an archive linked from the book web page[4]. One is similar to Figure B.2, presenting de-noised best results over run number instead of by week. Since some students performed many more unique runs than others who favored heavier replication, this view is harder to interpret. Two others present the analog of the first two but without de-noising, and back on the original un-normalized scale.

---

[4]http://bobby.gramacy.com/surrogates

Bibliography

Abrahamsen, P. (1997). A review of Gaussian random fields and correlation functions. Norsk Regnesentral/Norwegian Computing Center Oslo, https://www.nr.no/directdownload/917_Rapport.pdf.

Adler, R. (2010). *The Geometry of Random Fields*. SIAM.

Akima, H., Gebhardt, A., Petzoldt, T., and Maechler, M. (2016). `akima`: *Interpolation of Irregularly and Regularly Spaced Data*. R package version 0.6-2.

Allaire, J., Xie, Y., McPherson, J., Luraschi, J., Ushey, K., Atkins, A., Wickham, H., Cheng, J., and Chang, W. (2018). `rmarkdown`: *Dynamic Documents for R*. R package version 1.10.

Ambikasaran, S., Foreman-Mackey, D., Greengard, L., Hogg, D., and O'Neil, M. (2015). Fast direct methods for Gaussian processes. *IEEE Transactions on Pattern Analysis and Machine Intelligence*, 38(2):252–265.

Anagnostopoulos, C. and Gramacy, R. (2013). Information-theoretic data discarding for dynamic trees on data streams. *Entropy*, 15(12):5510–5535.

Andrianakis, I. and Challenor, P. (2012). The effect of the nugget on Gaussian process emulators of computer models. *Computational Statistics & Data Analysis*, 56(12):4215–4228.

Ankenman, B., Nelson, B., and Staum, J. (2010). Stochastic kriging for simulation meta-modeling. *Operations Research*, 58(2):371–382.

Audet, C. and Dennis Jr, J. (2006). Mesh adaptive direct search algorithms for constrained optimization. *SIAM Journal on Optimization*, 17(1):188–217.

Azzimonti, D., Ginsbourger, D., Chevalier, C., Bect, J., and Richet, Y. (2016). Adaptive design of experiments for conservative estimation of excursion sets. *Preprint on arXiv:1611.07256*.

Ba, S. (2015). `SLHD`: *Maximin-Distance (Sliced) Latin Hypercube Designs*. R package version 2.1-1.

Ba, S. and Joseph, V. (2012). Composite Gaussian process models for emulating expensive functions. *The Annals of Applied Statistics*, 6(4):1838–1860.

Ba, S. and Joseph, V. (2018). `MaxPro`: *Maximum Projection Designs*. R package version 4.1-2.

Ba, S., Myers, W., and Brenneman, W. (2015). Optimal sliced latin hypercube designs. *Technometrics*, 57(4):479–487.

Balaprakash, P., Gramacy, R., and Wild, S. (2013a). Active-learning-based surrogate models for empirical performance tuning. In *2013 IEEE International Conference on Cluster Computing (CLUSTER)*, pages 1–8. IEEE.

Balaprakash, P., Rupp, K., Mametjanov, A., Gramacy, R., Hovland, P., and Wild, S. (2013b). Empirical performance modeling of GPU kernels using active learning. In *ParCo*, pages 646–655. Citeseer.

Banerjee, S., Carlin, B., and Gelfand, A. (2004). *Hierarchical Modeling and Analysis for Spatial Data*. Chapman and Hall/CRC.

Banerjee, S., Gelfand, A., Finley, A., and Sang, H. (2008). Gaussian predictive process models for large spatial data sets. *Journal of the Royal Statistical Society: Series B (Statistical Methodology)*, 70(4):825–848.

Bastos, L. and O'Hagan, A. (2009). Diagnostics for Gaussian process emulators. *Technometrics*, 51(4):425–438.

Bates, D. and Maechler, M. (2019). `Matrix`*: Sparse and Dense Matrix Classes and Methods*. R package version 1.2-17.

Bect, J., Bachoc, F., and Ginsbourger, D. (2016). A supermartingale approach to Gaussian process based sequential design of experiments. *Preprint on arXiv:1608.01118*.

Berger, J., Bernardo, J., and Sun, D. (2009). The formal definition of reference priors. *The Annals of Statistics*, 37(2):905–938.

Berger, J., De Oliveira, V., and Sansó, B. (2001). Objective Bayesian analysis of spatially correlated data. *Journal of the American Statistical Association*, 96(456):1361–1374.

Bertsekas, D. (2014). *Constrained Optimization and Lagrange Multiplier Methods*. Academic Press.

Bichon, B., Eldred, M., Swiler, L., Mahadevan, S., and McFarland, J. (2008). Efficient global reliability analysis for nonlinear implicit performance functions. *AIAA Journal*, 46(10):2459–2468.

Binois, M. and Gramacy, R. (2018). `hetGP`: Heteroskedastic Gaussian process modeling and sequential design in R. *available as a vignette in the R package*.

Binois, M. and Gramacy, R. (2019). `hetGP`*: Heteroskedastic Gaussian process modeling and design under replication*. R package version 1.1.1.

Binois, M., Gramacy, R., and Ludkovski, M. (2018). Practical heteroscedastic Gaussian process modeling for large simulation experiments. *Journal of Computational and Graphical Statistics*, 27(4):808–821.

Binois, M., Huang, J., Gramacy, R., and Ludkovski, M. (2019). Replication or exploration? Sequential design for stochastic simulation experiments. *Technometrics*, 27(4):808–821.

Bisgaard, S. and Ankenman, B. (1996). Standard errors for the eigenvalues in second-order response surface models. *Technometrics*, 38(3):238–246.

Bogunovic, I., Scarlett, J., Krause, A., and Cevher, V. (2016). Truncated variance reduction: a unified approach to Bayesian optimization and level-set estimation. In *Advances in Neural Information Processing Systems*, pages 1507–1515.

Bompard, M., Peter, J., and Desideri, J. (2010). Surrogate models based on function and derivative values for aerodynamic global optimization. In *V European Conference on Computational Fluid Dynamics ECCOMAS CFD 2010*.

Booker, A., Dennis, J., Frank, P., Serafini, D., Torczon, V., and Trosset, M. (1999). A rigorous framework for optimization of expensive functions by surrogates. *Structural Optimization*, 17(1):1–13.

Bornn, L., Shaddick, G., and Zidek, J. (2012). Modeling nonstationary processes through dimension expansion. *Journal of the American Statistical Association*, 107(497):281–289.

Boukouvalas, A. and Cornford, D. (2009). Learning heteroscedastic Gaussian processes for complex datasets. Technical report, Aston University, Neural Computing Research Group.

Box, G. and Draper, N. (1987). *Empirical Model-Building and Response Surfaces*. John Wiley & Sons.

Box, G. and Draper, N. (2007). *Response Surfaces, Mixtures, and Ridge Analyses*, volume 649. John Wiley & Sons.

Breiman, L. (2001). Random forests. *Machine Learning*, 45(1):5–32.

Breiman, L., Cutler, A., Liaw, A., and Wiener, M. (2018). `randomForest`: *Breiman and Cutler's Random Forests for Classification and Regression*. R package version 4.6-14.

Breiman, L., Friedman, J., Olshen, R., and Stone, C. (1984). *Classification and Regression Trees*. Wadsworth Int.

Broderick, T. and Gramacy, R. (2011). Classification and categorical inputs with treed Gaussian process models. *Journal of Classification*, 28(2):244–270.

Brynjarsdóttir, J. and O'Hagan, A. (2014). Learning about physical parameters: The importance of model discrepancy. *Inverse Problems*, 30(11):114007.

Buchanan, M. (2013). *Forecast: What Physics, Meteorology, and the Natural Sciences Can Teach Us about Economics*. Bloomsbury Publishing USA.

Bull, A. (2011). Convergence rates of efficient global optimization algorithms. *Journal of Machine Learning Research*, 12(Oct):2879–2904.

Burnaev, E. and Panov, M. (2015). Adaptive design of experiments based on Gaussian processes. In *Statistical Learning and Data Sciences*, pages 116–125. Springer-Verlag.

Byrd, R., Qiu, P., Nocedal, J., , and Zhu, C. (1995). A limited memory algorithm for bound constrained optimization. *Journal on Scientific Computing*, 16(5):1190–1208.

Carlin, B. and Polson, N. (1991). Inference for nonconjugate Bayesian models using the Gibbs sampler. *Canadian Journal of Statistics*, 19(4):399–405.

Carmassi, M. (2018). `CaliCo`: *Code Calibration in a Bayesian Framework*. R package version 0.1.1.

Carnell, R. (2018). `lhs`: *Latin Hypercube Samples*. R package version 0.16.

Carvalho, C., Johannes, M., Lopes, H., and Polson, N. (2010). Particle learning and smoothing. *Statistical Science*, 25(1):88–106.

Chang, W., Cheng, J., Allaire, J., Xie, Y., and McPherson, J. (2017). shiny: *Web Application Framework for R.* R package version 1.0.5.

Chen, T., He, T., Benesty, M., Khotilovich, V., Tang, Y., Cho, H., Chen, K., Mitchell, R., Cano, I., Zhou, T., Li, M., Xie, J., Lin, M., Geng, Y., and Li, Y. (2019). xgboost: *Extreme Gradient Boosting.* R package version 0.82.1.

Chevalier, C., Ginsbourger, D., Bect, J., and Molchanov, I. (2013). Estimating and quantifying uncertainties on level sets using the Vorob'ev expectation and deviation with Gaussian process models. In Ucinski, D., Atkinson, A. C., and Patan, M., editors, *mODa 10 - Advances in Model-Oriented Design and Analysis*, Contributions to Statistics, pages 35–43. Springer-Verlag.

Chevalier, C., Ginsbourger, D., and Emery, X. (2014a). Corrected kriging update formulae for batch-sequential data assimilation. In *Mathematics of Planet Earth*, pages 119–122. Springer-Verlag.

Chevalier, C., Picheny, V., and Ginsbourger, D. (2014b). KrigInv: An efficient and user-friendly implementation of batch-sequential inversion strategies based on kriging. *Computational Statistics & Data Analysis*, 71:1021–1034.

Chevalier, C., Picheny, V., Ginsbourger, D., and Azzimonti, D. (2018). KrigInv: *Kriging-Based Inversion for Deterministic and Noisy Computer Experiments.* R package version 1.4.1.

Chipman, H., George, E., Gramacy, R., and McCulloch, R. (2013). Bayesian treed response surface models. *Wiley Interdisciplinary Reviews: Data Mining and Knowledge Discovery*, 3(4):298–305.

Chipman, H., George, E., and McCulloch, R. (2002). Bayesian treed models. *Machine Learning*, 48(1-3):299–320.

Chipman, H., George, E., and McCulloch, R. (2010). BART: Bayesian additive regression trees. *The Annals of Applied Statistics*, 4(1):266–298.

Chipman, H. and McCulloch, R. (2016). BayesTree: *Bayesian Additive Regression Trees.* R package version 0.3-1.4.

Chipman, H., Ranjan, P., and Wang, W. (2012). Sequential design for computer experiments with a flexible Bayesian additive model. *Canadian Journal of Statistics*, 40(4):663–678.

Chipman, H. A., George, E. I., and McCulloch, R. E. (1998). Bayesian CART model search. *Journal of the American Statistical Association*, 93(443):935–948.

Chung, M., Binois, M., Bardsley, J., Gramacy, R., Moquin, D., Smith, A., and Smith, A. (2019). Parameter and uncertainty estimation for dynamical systems using surrogate stochastic processes. *SIAM Journal on Scientific Computing*, 41(4):A2212–A2238. Preprint on arXiv:1802.00852.

Cioffi-Revilla, C. (2014). *Introduction to Computational Social Science.* Springer, New York, NY.

Cohn, D. (1994). Neural network exploration using optimal experiment design. In *Advances in Neural Information Processing Systems*, pages 679–686.

Conn, A., Scheinberg, K., and Vicente, L. (2009). *Introduction to Derivative-Free Optimization.* Society for Industrial and Applied Mathematics, Philadelphia, PA, USA.

Cox, D., Park, J., and Singer, C. (2001). A statistical method for tuning a computer code to a data base. *Computational Statistics & Data Analysis*, 37(1):77–92.

Craig, P., Goldstein, M., Seheult, A., and Smith, J. (1996). Bayes linear strategies for matching hydrocarbon reservoir history. *Bayesian Statistics*, 5:69–95.

Cressie, N. (1992). *Statistics for Spatial Data*. Wiley Online Library.

Cressie, N. and Johannesson, G. (2008). Fixed rank kriging for very large spatial data sets. *Journal of the Royal Statistical Society: Series B (Statistical Methodology)*, 70(1):209–226.

Currin, C., Mitchell, T., Morris, M., and Ylvisaker, D. (1991). Bayesian prediction of deterministic functions, with applications to the design and analysis of computer experiments. *Journal of the American Statistical Association*, 86(416):953–963.

Da Veiga, S., Wahl, F., and Gamboa, F. (2009). Local polynomial estimation for sensitivity analysis on models with correlated inputs. *Technometrics*, 51(4):452–463.

Damianou, A. and Lawrence, N. (2013). Deep Gaussian processes. In *Artificial Intelligence and Statistics*, pages 207–215.

Dancik, G. (2018). `mlegp`: *Maximum Likelihood Estimates of Gaussian Processes*. R package version 3.1.7.

Datta, A., Banerjee, S., Finley, A., and Gelfand, A. (2016). Hierarchical nearest-neighbor Gaussian process models for large geostatistical datasets. *Journal of the American Statistical Association*, 111(514):800–812.

de Micheaux, P. (2017). `CompQuadForm`: *Distribution Function of Quadratic Forms in Normal Variables*. R package version 1.4.3.

Denison, D., Mallick, B., and Smith, A. (1998). A Bayesian CART algorithm. *Biometrika*, 85(2):363–377.

Deville, Y., Ginsbourger, D., and Roustant, O. (2018). `kergp`: *Gaussian Process Laboratory*. R package version 0.4.0, with contributions by Nicholas Durrande.

DiDonato, A. and Morris Jr, A. (1986). Computation of the incomplete gamma function ratios and their inverse. *ACM Transactions on Mathematical Software*, 12(4):377–393.

Drake, R., Doss, F., McClarren, R., Adams, M., Amato, N., Bingham, D., Chou, C., DiStefano, C., Fidkowski, K., and Fryxell, B. (2011). Radiative effects in radiative shocks in shock tubes. *High Energy Density Physics*, 7(3):130–140.

Draper, N. (1963). Ridge analysis of response surfaces. *Technometrics*, 5(4):469–479.

Duchesne, P. and de Micheaux, P. (2010). Computing the distribution of quadratic forms: further comparisons between the Liu–Tang–Zhang approximation and exact methods. *Computational Statistics & Data Analysis*, 54(4):858–862.

Dunlop, M., Girolami, M., Stuart, A., and Teckentrup, A. (2018). How deep are deep Gaussian processes? *The Journal of Machine Learning Research*, 19(1):2100–2145.

Dupuy, D., Helbert, C., and Franco, J. (2015). `DiceDesign` and `DiceEval`: Two R packages for design and analysis of computer experiments. *Journal of Statistical Software*, 65(11):1–38.

Durrande, N., Ginsbourger, D., and Roustant, O. (2012). Additive covariance kernels for high-dimensional Gaussian process modeling. In *Annales de la Faculté des sciences de Toulouse: Mathématiques*, volume 21 (3), pages 481–499.

Eidsvik, J., Shaby, B., Reich, B., Wheeler, M., and Niemi, J. (2014). Estimation and prediction in spatial models with block composite likelihoods. *Journal of Computational and Graphical Statistics*, 23(2):295–315.

Emery, X. (2009). The kriging update equations and their application to the selection of neighboring data. *Computational Geosciences*, 13(3):269–280.

Erickson, C. B., Ankenman, B. E., and Sanchez, S. M. (2018). Comparison of Gaussian process modeling software. *European Journal of Operational Research*, 266(1):179–192.

Fadikar, A., Higdon, D., Chen, J., Lewis, B., Venkatramanan, S., and Marathe, M. (2018). Calibrating a stochastic, agent-based model using quantile-based emulation. *SIAM/ASA Journal on Uncertainty Quantification*, 6(4):1685–1706.

Finley, A. and Banerjee, S. (2019). spBayes: *Univariate and Multivariate Spatial-Temporal Modeling*. R package version 0.4-2.

Finley, A., Sang, H., Banerjee, S., and Gelfand, A. (2009). Improving the performance of predictive process modeling for large datasets. *Computational Statistics & Data Analysis*, 53(8):2873–2884.

Forrester, A., Sobester, A., and Keane, A. (2008). *Engineering Design via Surrogate Modelling: a Practical Guide*. John Wiley & Sons.

Franck, C. and Gramacy, R. (2020). Assessing Bayes factor surfaces using interactive visualization and computer surrogate modeling. *To appear in The American Statistician*. Preprint on arXiv:1809.05580.

Franco, J., Dupuy, D., Roustant, O., Kiener, P., Damblin, G., and Iooss, B. (2018). DiceDesign: *Designs of Computer Experiments*. R package version 1.8.

Francom, D. (2017). BASS: *Bayesian Adaptive Spline Surfaces*. R package version 0.2.2.

Franey, M., Ranjan, P., and Chipman, H. (2012). A short note on Gaussian process modeling for large datasets using graphics processing units. *Preprint on arXiv:1203.1269*.

Friedman, J. (1991). Multivariate adaptive regression splines. *The Annals of Statistics*, 19(1):1–67.

Furrer, R. (2018). spam: *SPArse Matrix*. R package version 2.2-0.

Furrer, R., Genton, M., and Nychka, D. (2006). Covariance tapering for interpolation of large spatial datasets. *Journal of Computational and Graphical Statistics*, 15(3):502–523.

Gardner, J., Pleiss, G., Weinberger, K., Bindel, D., and Wilson, A. (2018). GPyTorch: Blackbox matrix-matrix Gaussian process inference with GPU acceleration. In *Advances in Neural Information Processing Systems*, pages 7576–7586.

Gattiker, J., Myers, K., Williams, B., Higdon, D., Carzolio, M., and Hoegh, A. (2016). Gaussian process-based sensitivity analysis and Bayesian model calibration with gpmsa. *Handbook of Uncertainty Quantification*, pages 1–41.

Gauthier, B. and Pronzato, L. (2014). Spectral approximation of the IMSE criterion for optimal designs in kernel-based interpolation models. *SIAM/ASA Journal on Uncertainty Quantification*, 2(1):805–825.

Genz, A., Bretz, F., Miwa, T., Mi, X., and Hothorn, T. (2018). mvtnorm: *Multivariate Normal and t Distributions*. R package version 1.0-8.

Ginsbourger, D. and Le Riche, R. (2010). Towards Gaussian process-based optimization with finite time horizon. In *mODa 9–Advances in Model-Oriented Design and Analysis*, pages 89–96. Springer-Verlag.

Ginsbourger, D., Le Riche, R., and Carraro, L. (2010). Kriging is well-suited to parallelize optimization. In *Computational Intelligence in Expensive Optimization Problems*, pages 131–162. Springer-Verlag.

Gneiting, T. and Raftery, A. (2007). Strictly proper scoring rules, prediction, and estimation. *Journal of the American Statistical Association*, 102(477):359–378.

Goh, J., Bingham, D., Holloway, J., Grosskopf, M., Kuranz, C., and Rutter, E. (2013). Prediction and computer model calibration using outputs from multifidelity simulators. *Technometrics*, 55(4):501–512.

Goldberg, P., Williams, C., and Bishop, C. (1998). Regression with input-dependent noise: a Gaussian process treatment. In *Advances in Neural Information Processing Systems*, volume 10, pages 493–499, Cambridge, MA. MIT Press.

Gonzalez, J., Osborne, M., and Lawrence, N. (2016). GLASSES: Relieving the myopia of Bayesian optimisation. In *Proceedings of the 19th International Conference on Artificial Intelligence and Statistics*, pages 790–799.

Goovaerts, P. (1997). *Geostatistics for Natural Resources Evaluation*. Oxford University Press on Demand.

Gorodetsky, A. and Marzouk, Y. (2016). Mercer kernels and integrated variance experimental design: connections between Gaussian process regression and polynomial approximation. *SIAM/ASA Journal on Uncertainty Quantification*, 4(1):796–828.

Gramacy, R. (2005). *Bayesian treed Gaussian process models*. PhD thesis, Citeseer.

Gramacy, R. (2007). tgp: an R package for Bayesian nonstationary, semiparametric nonlinear regression and design by treed Gaussian process models. *Journal of Statistical Software*, 19(9):6.

Gramacy, R. (2014). plgp: *Particle Learning of Gaussian Processes*. R package version 1.1-7.

Gramacy, R. (2016). laGP: large-scale spatial modeling via local approximate Gaussian processes in R. *Journal of Statistical Software*, 72(1):1–46.

Gramacy, R. (2018a). monomvn: *Estimation for Multivariate Normal and Student-t Data with Monotone Missingness*. R package version 1.9-8.

Gramacy, R. (2018b). A shiny update to an old experiment game. *The American Statistician*, pages 1–6.

Gramacy, R. and Apley, D. (2015). Local Gaussian process approximation for large computer experiments. *Journal of Computational and Graphical Statistics*, 24(2):561–578.

Gramacy, R., Bingham, D., Holloway, J., Grosskopf, M., Kuranz, C., Rutter, E., Trantham, M., and Drake, R. (2015). Calibrating a large computer experiment simulating radiative shock hydrodynamics. *The Annals of Applied Statistics*, 9(3):1141–1168.

Gramacy, R., Gray, G., Le Digabel, S., Lee, H., Ranjan, P., Wells, G., and Wild, S. (2016). Modeling an augmented Lagrangian for blackbox constrained optimization. *Technometrics*, 58(1):1–11.

Gramacy, R. and Haaland, B. (2016). Speeding up neighborhood search in local Gaussian process prediction. *Technometrics*, 58(3):294–303.

Gramacy, R. and Lee, H. (2008a). Bayesian treed Gaussian process models with an application to computer modeling. *Journal of the American Statistical Association*, 103(483):1119–1130.

Gramacy, R. and Lee, H. (2008b). Gaussian processes and limiting linear models. *Computational Statistics & Data Analysis*, 53(1):123–136.

Gramacy, R. and Lee, H. (2009). Adaptive design and analysis of supercomputer experiments. *Technometrics*, 51(2):130–145.

Gramacy, R. and Lee, H. (2012). Cases for the nugget in modeling computer experiments. *Statistics and Computing*, 22(3):713–722.

Gramacy, R. and Lee, H. K. (2011). Optimization under unknown constraints. In *Bayesian Statistics*, volume 9. Oxford University Press.

Gramacy, R. and Lian, H. (2012). Gaussian process single-index models as emulators for computer experiments. *Technometrics*, 54(1):30–41.

Gramacy, R. and Ludkovski, M. (2015). Sequential design for optimal stopping problems. *SIAM Journal on Financial Mathematics*, 6(1):748–775.

Gramacy, R., Niemi, J., and Weiss, R. (2014). Massively parallel approximate Gaussian process regression. *SIAM/ASA Journal on Uncertainty Quantification*, 2(1):564–584.

Gramacy, R. and Pantaleo, E. (2010). Shrinkage regression for multivariate inference with missing data, and an application to portfolio balancing. *Bayesian Analysis*, 5(2):237–262.

Gramacy, R. and Polson, N. (2011). Particle learning of Gaussian process models for sequential design and optimization. *Journal of Computational and Graphical Statistics*, 20(1):102–118.

Gramacy, R., Samworth, R., and King, R. (2010). Importance tempering. *Statistics and Computing*, 20(1):1–7.

Gramacy, R. and Sun, F. (2018). laGP: *Local Approximate Gaussian Process Regression*. R package version 1.5-3.

Gramacy, R. and Taddy, M. (2010). Categorical inputs, sensitivity analysis, optimization and importance tempering with tgp version 2, an R package for treed Gaussian process models. *Journal of Statistical Software*, 33(6):1–48.

Gramacy, R. and Taddy, M. (2016). tgp: *Bayesian Treed Gaussian Process Models*. R package version 2.4-14.

Gramacy, R., Taddy, M., and Anagnostopoulos, C. (2017). dynaTree: *Dynamic Trees for Learning and Design*. R package version 1.2-10.

Gramacy, R., Taddy, M., and Wild, S. (2013). Variable selection and sensitivity analysis using dynamic trees, with an application to computer code performance tuning. *The Annals of Applied Statistics*, 7(1):51–80.

Greenwell, B., Boehmke, B., and Cunningham, C. (2019). gbm: *Generalized Boosted Regression Models*. R package version 2.1.5.

Gu, M. (2019). Jointly robust prior for Gaussian stochastic process in emulation, calibration and variable selection. *Bayesian Analysis*, 14(3):857–885.

Gu, M., Palomo, J., and Berger, J. (2018). RobustGaSP: *Robust Gaussian Stochastic Process Emulation*. R package version 0.5.6.

Gu, M. and Wang, L. (2018). Scaled Gaussian stochastic process for computer model calibration and prediction. *SIAM/ASA Journal on Uncertainty Quantification*, 6(4):1555–1583.

Guinness, J. and Katzfuss, M. (2019). GpGp: *Fast Gaussian Process Computation Using Vecchia's Approximation*. R package version 0.1.1.

Gul, E. (2016). *Designs for computer experiments and uncertainty quantification*. PhD thesis, Georgia Institute of Technology.

Haaland, B. and Qian, P. (2011). Accurate emulators for large-scale computer experiments. *The Annals of Statistics*, 39(6):2974–3002.

Hankin, R. (2013). BACCO: *Bayesian Analysis of Computer Code Output*. R package version 2.0-9.

Hankin, R. (2019). emulator: *Bayesian emulation of computer programs*. R package version 1.2-20.

Hastie, T., Tibshirani, R., and Friedman, J. (2009). *The elements of statistical learning: data mining, inference, and prediction*. Springer, New York, NY.

Heaton, M., Datta, A., Finley, A., Furrer, R., Guinness, J., Guhaniyogi, R., Gerber, F., Gramacy, R., Hammerling, D., Katzfuss, M., Lindgren, F., Nychka, D., Sun, F., and Zammit-Mangion, A. (2018). A case study competition among methods for analyzing large spatial data. *Journal of Agricultural, Biological and Environmental Statistics*, pages 1–28.

Herbei, R. and Berliner, L. M. (2014). Estimating ocean circulation: an MCMC approach with approximated likelihoods via the bernoulli factory. *Journal of the American Statistical Association*, 109(507):944–954.

Hernández-Lobato, J., Gelbart, M., Hoffman, M., Adams, R., and Ghahramani, Z. (2015). Predictive entropy search for Bayesian optimization with unknown constraints. In *Proceedings of the 32nd International Conference on Machine Learning*.

Higdon, D. (2002). Space and space-time modeling using process convolutions. In *Quantitative Methods for Current Environmental Issues*, pages 37–56. Springer, New York, NY.

Higdon, D., Gattiker, J., Williams, B., and Rightley, M. (2008). Computer model calibration using high-dimensional output. *Journal of the American Statistical Association*, 103(482):570–583.

Higdon, D., Kennedy, M., Cavendish, J., Cafeo, J., and Ryne, R. (2004). Combining field data and computer simulations for calibration and prediction. *SIAM Journal on Scientific Computing*, 26(2):448–466.

Hoff, P. (2009). *A First Course in Bayesian Statistical Methods*, volume 580. Springer, New York, NY.

Hong, L. and Nelson, B. (2006). Discrete optimization via simulation using COMPASS. *Operations Research*, 54(1):115–129.

Hu, R. and Ludkovski, M. (2017). Sequential design for ranking response surfaces. *SIAM/ASA Journal on Uncertainty Quantification*, 5(1):212–239.

Huan, X. and Marzouk, Y. (2016). Sequential Bayesian optimal experimental design via approximate dynamic programming. *Preprint on arXiv:1604.08320*.

Huang, D., Allen, T., Notz, W., and Zeng, N. (2006). Global optimization of stochastic black-box systems via sequential kriging meta-models. *Journal of Global Optimization*, 34(3):441–466.

Huang, J., Gramacy, R., Binois, M., and Librashi, M. (2020). On-site surrogates for large-scale calibration. *To appear in Applied Stochastic Models in Business and Industry*. Preprint on arXiv:1810.01903.

Jacquier, E., Polson, N., and Rossi, P. (2004). Bayesian analysis of stochastic volatility models with fat-tails and correlated errors. *Journal of Econometrics*, 122:185–212.

Jalali, H., Van Nieuwenhuyse, I., and Picheny, V. (2017). Comparison of kriging-based algorithms for simulation optimization with heterogeneous noise. *European Journal of Operational Research*, 261(1):279–301.

Johnson, L. (2008). Microcolony and biofilm formation as a survival strategy for bacteria. *Journal of Theoretical Biology*, 251:24–34.

Johnson, L., Gramacy, R., Cohen, J., Mordecai, E., Murdock, C., Rohr, J., Ryan, S., Stewart-Ibarra, A., and Weikel, D. (2018). Phenomenological forecasting of disease incidence using heteroskedastic Gaussian processes: a dengue case study. *The Annals of Applied Statistics*, 12(1):27–66.

Johnson, M., Moore, L., and Ylvisaker, D. (1990). Minimax and maximin distance designs. *Journal of Statistical Planning and Inference*, 26(2):131–148.

Jones, D., Schonlau, M., and Welch, W. (1998). Efficient global optimization of expensive black-box functions. *Journal of Global Optimization*, 13(4):455–492.

Joseph, V. (2006). Limit kriging. *Technometrics*, 48(4):458–466.

Joseph, V., Gul, E., and Ba, S. (2015). Maximum projection designs for computer experiments. *Biometrika*, 102(2):371–380.

Journel, A. and Huijbregts, C. (1978). *Mining Geostatistics*, volume 600. Academic Press.

Kamiński, B. (2015). A method for the updating of stochastic kriging metamodels. *European Journal of Operational Research*, 247(3):859–866.

Kannan, A. and Wild, S. (2012). Benefits of deeper analysis in simulation-based groundwater optimization problems. In *Proceedings of the XIX International Conference on Computational Methods in Water Resources (CMWR 2012)*, volume 4(5), page 10. Springer–Verlag.

Karatzoglou, A., Smola, A., and Hornik, K. (2018). kernlab: *Kernel-Based Machine Learning Lab*. R package version 0.9-27.

Kass, R., Carlin, B., Gelman, A., and Neal, R. (1998). Markov chain Monte Carlo in practice: a roundtable discussion. *The American Statistician*, 52(2):93–100.

Katzfuss, M. (2017). A multi-resolution approximation for massive spatial datasets. *Journal of the American Statistical Association*, 112(517):201–214.

Katzfuss, M. and Guinness, J. (2018). A general framework for Vecchia approximations of Gaussian processes. *Preprint on arXiv:1708.06302*.

Kaufman, C., Bingham, D., Habib, S., Heitmann, K., and Frieman, J. (2011). Efficient emulators of computer experiments using compactly supported correlation functions, with an application to cosmology. *The Annals of Applied Statistics*, 5(4):2470–2492.

Kennedy, M. and O'Hagan, A. (2001). Bayesian calibration of computer models. *Journal of the Royal Statistical Society: Series B (Statistical Methodology)*, 63(3):425–464.

Kersting, K., Plagemann, C., Pfaff, P., and Burgard, W. (2007). Most likely heteroscedastic Gaussian process regression. In *Proceedings of the International Conference on Machine Learning*, pages 393–400, New York, NY. ACM.

Kim, H., Mallick, B., and Holmes, C. (2005). Analyzing nonstationary spatial data using piecewise Gaussian processes. *Journal of the American Statistical Association*, 100(470):653–668.

Konomi, B., Karagiannis, G., Lai, K., and Lin, G. (2017). Bayesian treed calibration: an application to carbon capture with AX sorbent. *Journal of the American Statistical Association*, 112(517):37–53.

Lam, R., Willcox, K., and Wolpert, D. (2016). Bayesian optimization with a dinite budget: an approximate dynamic programming approach. In *Advances in Neural Information Processing Systems*, pages 883–891.

Lanckriet, G., Cristianini, N., Bartlett, P., Ghaoui, L., and Jordan, M. (2004). Learning the kernel matrix with semidefinite programming. *Journal of Machine Learning Research*, 5(Jan):27–72.

Larson, J., Menickelly, M., and Wild, S. (2019). Derivative-free optimization methods. *Preprint on arXiv:1904.11585*.

Law, A. (2015). *Simulation Modeling and Analysis*. McGraw-Hill, 5 edition.

Lazaro-Gredilla, M. and Titsias, M. (2011). Variational heteroscedastic Gaussian process regression. In *Proceedings of the International Conference on Machine Learning*, pages 841–848, New York, NY. ACM.

Le Gratiet, L. and Cannamela, C. (2015). Cokriging-based sequential design strategies using fast cross-validation techniques for multi-fidelity computer codes. *Technometrics*, 57(3):418–427.

Le Gratiet, L. and Garnier, J. (2014). Recursive co-kriging model for design of computer experiments with multiple levels of fidelity. *International Journal for Uncertainty Quantification*, 4(5).

Leatherman, E., Santner, T., and Dean, A. (2017). Computer experiment designs for accurate prediction. *Statistics and Computing*, pages 1–13.

Lee, H., Gramacy, R., Linkletter, C., and Gray, G. (2011). Optimization subject to hidden constraints via statistical emulation. *Pacific Journal of Optimization*, 7(3):467–478.

Lee, H. K. H. (2007). Chocolate chip cookies as a teaching aid. *The American Statistician*, 61(4):351–355.

Lee, M. and Owen, A. (2015). Single nugget kriging. Technical report, Stanford University. Preprint on arXiv:1507.05128.

Leisch, F., Hornik, K., and Ripley, B. (2017). mda: *Mixture and Flexible Discriminant Analysis*. R package version 0.4-10; S original by Trevor Hastie and Robert Tibshirani. Original R port by the authors.

Letham, B., Karrer, B., Ottoni, G., Bakshy, E., et al. (2019). Constrained bayesian optimization with noisy experiments. *Bayesian Analysis*, 14(2):495–519.

Lin, C. and Tang, B. (2015). Latin hypercubes and space-filling designs. *Handbook of Design and Analysis of Experiments*, pages 593–625.

Liu, F., Bayarri, M., and Berger, J. (2009). Modularization in Bayesian analysis, with emphasis on analysis of computer models. *Bayesian Analysis*, 4(1):119–150.

Liu, Y. (2014). *Recent advances in computer experiment modeling*. PhD thesis, Rutgers University.

Lyu, X., Binois, M., and Ludkovski, M. (2018). Evaluating Gaussian process metamodels and sequential designs for noisy level set estimation. *Preprint on arXiv:1807.06712*.

MacDonald, B., Chipman, H., and Ranjan, P. (2019). GPfit: *Gaussian Processes Modeling*. R package version 1.0-8.

MacKay, D. (1992). Information-based objective functions for active data selection. *Neural Computation*, 4(4):590–604.

Marrel, A., Iooss, B., Laurent, B., and Roustant, O. (2009). Calculations of Sobol indices for the Gaussian process metamodel. *Reliability Engineering & System Safety*, 94(3):742–751.

Matheron, G. (1963). Principles of geostatistics. *Economic Geology*, 58(8):1246–1266.

Matott, L., Leung, K., and Sim, J. (2011). Application of MATLAB and Python optimizers to two case studies involving groundwater flow and contaminant transport modeling. *Computers & Geosciences*, 37(11):1894–1899.

Matott, L., Rabideau, A., and Craig, J. (2006). Pump-and-treat optimization using analytic element method flow models. *Advances in Water Resources*, 29(5):760–775.

Mayer, A., Kelley, C., and Miller, C. (2002). Optimal design for problems involving flow and transport phenomena in saturated subsurface systems. *Advances in Water Resources*, 25(8-12):1233–1256.

McClarren, R., Ryu, D., Drake, R., Grosskopf, M., Bingham, D., Chou, C., Fryxell, B., Van der Holst, B., Holloway, J., and Kuranz, C. (2011). A physics informed emulator for laser-driven radiating shock simulations. *Reliability Engineering & System Safety*, 96(9):1194–1207.

McCulloch, R., Sparapani, R., Gramacy, R., Spanbauer, C., and Pratola, M. (2019). BART: *Bayesian Additive Regression Trees*. R package version 2.4.

McKeague, I. W., Nicholls, G., Speer, K., and Herbei, R. (2005). Statistical inversion of South Atlantic circulation in an abyssal neutral density layer. *Journal of Marine Research*, 63(4):683–704.

Mead, R. and Freeman, K. (1973). An experiment game. *Journal of the Royal Statistical Society. Series C (Applied Statistics)*, 22(1):1–6.

Mehta, P., Walker, A., Lawrence, E., Linares, R., Higdon, D., and Koller, J. (2014). Modeling satellite drag coefficients with response surfaces. *Advances in Space Research*, 54(8):1590–1607.

Milborrow, S. (2019). earth: *Multivariate Adaptive Regression Splines*. R package version 5.1.1; Derived from mda:mars by Trevor Hastie and Rob Tibshirani. Uses Alan Miller's Fortran utilities with Thomas Lumley's leaps wrapper.

Mitchell, T., Sacks, J., and Ylvisaker, D. (1994). Asymptotic Bayes criteria for nonparametric response surface design. *The Annals of Statistics*, 22(2):634–651.

Mitchell, T. and Scott, D. (1987). A computer program for the design of group testing experiments. *Communications in Statistics – Theory and methods*, 16(10):2943–2955.

Mockus, J., Tiesis, V., and Zilinskas, A. (1978). The application of Bayesian methods for seeking the extremum. *Towards Global Optimization*, 2(117-129):2.

Morgan, J. and Sonquist, J. (1963). Problems in the analysis of survey data, and a proposal. *Journal of the American Statistical Association*, 58(302):415–434.

Morris, M. (2010). *Design of Experiments: An Introduction Based on Linear Models*. Chapman and Hall/CRC.

Morris, M. and Mitchell, T. (1995). Exploratory designs for computational experiments. *Journal of Statistical Planning and Inference*, 43(3):381–402.

Morris, M., Mitchell, T., and Ylvisaker, D. (1993). Bayesian design and analysis of computer experiments: use of derivatives in surface prediction. *Technometrics*, 35(3):243–255.

Morris, R., Kottas, A., Taddy, M., Furfaro, R., and Ganapol, B. (2008). A statistical framework for the sensitivity analysis of radiative transfer models. *IEEE Transactions on Geoscience and Remote Sensing*, 46(12):4062–4074.

Myers, R., Montgomery, D., and Anderson-Cook, C. (2016). *Response Surface Methodology: Process and Product Optimization Using Designed Experiments*. John Wiley & Sons.

Neal, R. (1998). Regression and classification using Gaussian process priors. *Bayesian Statistics*, 6:475.

Nelder, J. (1966). Inverse polynomials, a useful group of multi-factor response functions. *Biometrics*, 22:128–141.

Nelder, J. and Mead, R. (1965). A simplex method for function minimization. *The Computer Journal*, 7(4):308–313.

Ng, S. and Yin, J. (2012). Bayesian kriging analysis and design for stochastic systems. *ACM Transations on Modeling and Computer Simulation (TOMACS)*, 22(3):article no. 17.

Nocedal, J. and Wright, S. (2006). *Numerical Optimization*. Springer Science & Business Media, New York, NY.

Nychka, D., Furrer, R., Paige, J., and Sain, S. (2019). `fields`*: Tools for Spatial Data*. R package version 9.7.

Oakley, J. and O'Hagan, A. (2004). Probabilistic sensitivity analysis of complex models: a Bayesian approach. *Journal of the Royal Statistical Society: Series B (Statistical Methodology)*, 66(3):751–769.

Opsomer, J., Ruppert, D., Wand, W., Holst, U., and Hossler, O. (1999). Kriging with nonparameteric variance function estimation. *Biometrics*, 55:704–710.

Owen, A. (2014). Sobol indices and Shapley value. *SIAM/ASA Journal on Uncertainty Quantification*, 2(1):245–251.

Paciorek, C. (2003). *Nonstationary Gaussian processes for regression and spatial modelling*. PhD thesis, Carnegie Mellon University.

Paciorek, C. and Schervish, M. (2006). Spatial modelling using a new class of nonstationary covariance functions. *Environmetrics: The Official Journal of the International Environmetrics Society*, 17(5):483–506.

Pamadi, B., Covell, P., Tartabini, P., and Murphy, K. (2004a). Aerodynamic characteristics and glide-back performance of Langley glide-back booster. In *22nd Applied Aerodynamics Conference and Exhibit*, page 5382.

Pamadi, B., Tartabini, P., and Starr, B. (2004b). Ascent, stage separation and glideback performance of a partially reusable small launch vehicle. In *42nd AIAA Aerospace Sciences Meeting and Exhibit*, page 876.

Park, C. and Apley, D. (2018). Patchwork kriging for large-scale Gaussian process regression. *The Journal of Machine Learning Research*, 19(1):269–311.

Park, T. and Casella, G. (2008). The Bayesian lasso. *Journal of the American Statistical Association*, 103(482):681–686.

Pav, S. (2017). `sadists`*: Some Additional Distributions*. R package version 0.2.3.

Peng, C. and Wu, C. (2014). On the choice of nugget in kriging modeling for deterministic computer experiments. *Journal of Computational and Graphical Statistics*, 23(1):151–168.

Picheny, V. and Ginsbourger, D. (2013). A nonstationary space-time Gaussian process model for partially converged simulations. *SIAM/ASA Journal on Uncertainty Quantification*, 1:57–78.

Picheny, V., Ginsbourger, D., and Krityakierne, T. (2016a). Comment: Some enhancements over the augmented Lagrangian approach. *Technometrics*, 58(1):17–21.

Picheny, V., Ginsbourger, D., Richet, Y., and Caplin, G. (2013). Quantile-based optimization of noisy computer experiments with tunable precision. *Technometrics*, 55(1):2–13.

Picheny, V., Ginsbourger, D., and Roustant, O. (2016b). DiceOptim: *Kriging-Based Optimization for Computer Experiments*. R package version 2.0, with contributions by M. Binois, C. Chevalier, S. Marmin and T. Wagner.

Picheny, V., Ginsbourger, D., Roustant, O., Haftka, R., and Kim, N. (2010). Adaptive designs of experiments for accurate approximation of a target region. *Journal of Mechanical Design*, 132(7):071008.

Picheny, V., Gramacy, R., Wild, S., and Le Digabel, S. (2016c). Bayesian optimization under mixed constraints with a slack-variable augmented Lagrangian. In *Advances in Neural Information Processing Systems*, pages 1435–1443.

Picheny, V., Wagner, T., and Ginsbourger, D. (2012). A benchmark of kriging-based infill criteria for noisy optimization. *Structural and Multidisciplinary Optimization*, pages 1–20.

Picone, J., Hedin, A., Drob, D. P., and Aikin, A. (2002). NRLMSISE-00 empirical model of the atmosphere: Statistical comparisons and scientific issues. *Journal of Geophysical Research: Space Physics*, 107(A12):SIA–15.

Plumlee, M. (2017). Bayesian calibration of inexact computer models. *Journal of the American Statistical Association*, 112(519):1274–1285.

Plumlee, M. and Tuo, R. (2014). Building accurate emulators for stochastic simulations via quantile kriging. *Technometrics*, 56(4):466–473.

Pratola, M., Chipman, H., George, E., and McCulloch, R. (2017a). Heteroscedastic BART using multiplicative regression trees. *Preprint on arXiv:1709.07542*.

Pratola, M., Harari, O., Bingham, D., and Flowers, G. (2017b). Design and analysis of experiments on nonconvex regions. *Technometrics*, pages 1–12.

Pratola, M. T. (2016). Efficient Metropolis–Hastings proposal mechanisms for Bayesian regression tree models. *Bayesian Analysis*, 11(3):885–911.

Qian, P. (2009). Nested Latin hypercube designs. *Biometrika*, 96(4):957–970.

Qian, P., Wu, H., and Wu, C. (2008). Gaussian process models for computer experiments with qualitative and quantitative factors. *Technometrics*, 50(3):383–396.

Quadrianto, N., Kersting, K., Reid, M., Caetano, T., and Buntine, W. (2009). Kernel conditional quantile estimation via reduction revisited. In *Proceedings of the 9th IEEE International Conference on Data Mining*, pages 938–943.

R Core Team (2019). *R: A Language and Environment for Statistical Computing*. R Foundation for Statistical Computing, Vienna, Austria.

Racine, J. and Nie, Z. (2018). crs: *Categorical Regression Splines*. R package version 0.15-31.

Ranjan, P., Bingham, D., and Michailidis, G. (2008). Sequential experiment design for contour estimation from complex computer codes. *Technometrics*, 50(4):527–541.

Rasmussen, C. and Williams, C. (2006). *Gaussian Processes for Machine Learning*. MIT Press, Cambridge, MA.

Raymer, D. (2012). *Aircraft Design: A Conceptual Approach.* American Institute of Aeronautics and Astronautics, Inc.

Richardson, S. and Green, P. (1997). On Bayesian analysis of mixtures with an unknown number of components (with discussion). *Journal of the Royal Statistical Society: Series B (Statistical Methodology)*, 59(4):731–792.

Ridgeway, G. (2007). Generalized boosted models: A guide to the gbm package. R package vignette: https://cran.r-project.org/web/packages/gbm/vignettes/gbm.pdf.

Ripley, B. (2015). spatial*: Functions for Kriging and Point Pattern Analysis.* R package version 7.3-11.

Rogers, S., Aftosmis, M., Pandya, S., Chaderjian, N., Tejnil, E., and Ahmad, J. (2003). Automated CFD parameter studies on distributed parallel computers. In *16th AIAA Computational Fluid Dynamics Conference*, page 4229.

Roustant, O., Ginsbourger, D., and Deville, Y. (2018). DiceKriging*: Kriging Methods for Computer Experiments.* R package version 1.5.6; Contributors: Clement Chevalier and Yann Richet.

Rullière, D., Durrande, N., Bachoc, F., and Chevalier, C. (2018). Nested kriging predictions for datasets with a large number of observations. *Statistics and Computing*, 28(4):849–867.

Rushdi, A., Swiler, L., Phipps, E., D'Elia, M., and Ebeida, M. (2017). VPS: Voronoi piecewise surrogate models for high-dimensional data fitting. *International Journal for Uncertainty Quantification*, 7(1).

Sacks, J., Welch, W., Mitchell, T., and Wynn, H. (1989). Design and analysis of computer experiments. *Statistical Science*, 4(4):409–423.

Saltelli, A. (2002). Making best use of model evaluations to compute sensitivity indices. *Computer Physics Communications*, 145(2):280–297.

Saltelli, A., Chan, K., and Scott, M. (2000). *Sensitivity Analysis.* John Wiley & Sons, New York, NY.

Saltelli, A. and Tarantola, S. (2002). On the relative importance of input factors in mathematical models: safety assessment for nuclear waste disposal. *Journal of the American Statistical Association*, 97(459):702–709.

Sampson, P. and Guttorp, P. (1992). Nonparametric estimation of nonstationary spatial covariance structure. *Journal of the American Statistical Association*, 87(417):108–119.

Sang, H. and Huang, J. (2012). A full scale approximation of covariance functions for large spatial data sets. *Journal of the Royal Statistical Society: Series B (Statistical Methodology)*, 74(1):111–132.

Santner, T., Williams, B., and Notz, W. (2003). *The Design and Analysis of Computer Experiments, First Edition.* Springer–Verlag.

Santner, T., Williams, B., and Notz, W. (2018). *The Design and Analysis of Computer Experiments, Second Edition.* Springer–Verlag, New York, NY.

Sasena, M. (2002). *Flexibility and efficiency enhancements for constrained global design optimization with kriging approximations.* PhD thesis, University of Michigan Ann Arbor.

Schmidt, A. and O'Hagan, A. (2003). Bayesian inference for non-stationary spatial covariance structure via spatial deformations. *Journal of the Royal Statistical Society: Series B (Statistical Methodology)*, 65(3):743–758.

Schmidt, S. and Launsby, R. (1989). *Understanding Industrial Designed Experiments*. Air Academy Press.

Schonlau, M. (1997). *Computer experiments and global optimization*. PhD thesis, University of Waterloo.

Schonlau, M., Welch, W., and Jones, D. (1998). Global versus local search in constrained optimization of computer models. *Lecture Notes-Monograph Series*, pages 11–25.

Seo, S., Wallat, M., Graepel, T., and Obermayer, K. (2000). Gaussian process regression: active data selection and test point rejection. In *Mustererkennung 2000*, pages 27–34. Springer–Verlag, New York, NY.

Sheather, S. (2009). *A Modern Approach to Regression with R*. Springer Science & Business Media, New York, NY.

Sherlock, C., Fearnhead, P., and Roberts, G. (2010). The random walk Metropolis: linking theory and practice through a case study. *Statistical Science*, 25(2):172–190.

Shewry, M. and Wynn, H. (1987). Maximum entropy sampling. *Journal of Applied Statistics*, 14(2):165–170.

Snelson, E. and Ghahramani, Z. (2006). Sparse Gaussian processes using pseudo-inputs. In *Advances in Neural Information Processing Systems*, pages 1257–1264.

Snoek, J., Larochelle, H., and Adams, R. (2012). Practical Bayesian optimization of machine learning algorithms. In *Advances in Neural Information Processing Systems*, pages 2951–2959.

Sobol, I. (1993). Sensitivity analysis for non-linear mathematical models. *Mathematical Modelling and Computational Experiment*, 1:407–414.

Sobol, I. (2001). Global sensitivity indices for nonlinear mathematical models and their Monte Carlo estimates. *Mathematics and Computers in Simulation*, 55(1-3):271–280.

Song, E., Nelson, B., and Staum, J. (2016). Shapley effects for global sensitivity analysis: theory and computation. *SIAM/ASA Journal on Uncertainty Quantification*, 4(1):1060–1083.

Srinivas, N., Krause, A., Kakade, S., and Seeger, M. (2009). Gaussian process optimization in the bandit setting: no regret and experimental design. *Preprint on arXiv:0912.3995*.

Stein, M. (2012). *Interpolation of Spatial Data: Some Theory for Kriging*. Springer Science & Business Media, New York, NY.

Stein, M., Chi, Z., and Welty, L. (2004). Approximating likelihoods for large spatial data sets. *Journal of the Royal Statistical Society: Series B (Statistical Methodology)*, 66(2):275–296.

Storlie, C. and Helton, J. (2008). Multiple predictor smoothing methods for sensitivity analysis: description of techniques. *Reliability Engineering & System Safety*, 93(1):28–54.

Storlie, C., Swiler, L., Helton, J., and Sallaberry, C. (2009). Implementation and evaluation of nonparametric regression procedures for sensitivity analysis of computationally demanding models. *Reliability Engineering & System Safety*, 94(11):1735–1763.

Stroud, J., Stein, M., and Lysen, S. (2017). Bayesian and maximum likelihood estimation for Gaussian processes on an incomplete lattice. *Journal of Computational and Graphical Statistics*, 26(1):108–120.

Sun, F. and Gramacy, R. (2019). maximin*: Space-Filling Design Under Maximin Distance*. R package version 1.0-2.

Sun, F., Gramacy, R., Haaland, B., Lawrence, E., and Walker, A. (2019a). Emulating satellite drag from large simulation experiments. *IAM/ASA Journal on Uncertainty Quantification*, 7(2):720–759. preprint arXiv:1712.00182.

Sun, F., Gramacy, R., Haaland, B., Lu, S., and Hwang, Y. (2019b). Synthesizing simulation and field data of solar irradiance. *Statistical Analysis and Data Mining*, 12(4):311–324. Preprint on arXiv:1806.05131.

Sung, C., Gramacy, R., and Haaland, B. (2018). Exploiting variance reduction potential in local Gaussian process search. *Statistica Sinica*, 28:577–600.

Surjanovic, S. and Bingham, D. (2013). Virtual library of simulation experiments: test functions and datasets. http://www.sfu.ca/~ssurjano.

Taddy, M., Gramacy, R., and Polson, N. (2011). Dynamic trees for learning and design. *Journal of the American Statistical Association*, 106(493):109–123.

Taddy, M., Lee, H., Gray, G., and Griffin, J. (2009). Bayesian guided pattern search for robust local optimization. *Technometrics*, 51(4):389–401.

Tan, M. (2013). Minimax designs for finite design regions. *Technometrics*, 55(3):346–358.

Ter Meer, J., Van Duijne, H., Nieuwenhuis, R., and Rijnaarts, H. (2007). Prevention and reduction of groundwater pollution at contaminated megasites: Integrated management strategy and its application on megasite cases. In *Groundwater Science and Policy*, pages 403–420. The Royal Society of Chemistry.

Thompson, W. (1933). On the likelihood that one unknown probability exceeds another in view of the evidence of two samples. *Biometrika*, 25(3/4):285–294.

Tuo, R. and Wu, C. (2015). Efficient calibration for imperfect computer models. *The Annals of Statistics*, 43(6):2331–2352.

Tuo, R. and Wu, C. (2016). A theoretical framework for calibration in computer models: parametrization, estimation and convergence properties. *SIAM/ASA Journal on Uncertainty Quantification*, 4(1):767–795.

Ubaru, S., Chen, J., and Saad, Y. (2017). Fast estimation of $\text{tr}(f(a))$ via stochastic Lanczos quadrature. *SIAM Journal on Matrix Analysis and Applications*, 38(4):1075–1099.

Vanhatalo, J., Riihimäki, J., Hartikainen, J., Jylänki, P., Tolvanen, V., and Vehtari, A. (2012). Bayesian modeling with Gaussian processes using the GPstuff toolbox. *Preprint on arXiv:1206.5754.*

Vapnik, V. (2013). *The Nature of Statistical Learning Theory*. Springer Science & Business Media, New York, NY.

Vecchia, A. (1988). Estimation and model identification for continuous spatial processes. *Journal of the Royal Statistical Society: Series B (Methodological)*, 50(2).297–312.

Ver Hoef, J. and Barry, R. (1998). Constructing and fitting models for cokriging and multivariate spatial prediction. *Journal of Statistical Planning and Inference*, 69:275–294.

Vernon, I., Goldstein, M., and Bower, R. (2010). Galaxy formation: a Bayesian uncertainty analysis. *Bayesian Analysis*, 5(4):619–669.

Wang, K., Pleiss, G., Gardner, J., Tyree, S., Weinberger, K., and Wilson, A. (2019). Exact Gaussian processes on a million data points. *Preprint on arXiv:1903.08114*.

Wang, W. and Chen, X. (2016). The effects of estimation of heteroscedasticity on stochastic kriging. In *Proceedings of the 2016 Winter Simulation Conference*, pages 326–337. IEEE Press.

Welch, W., Buck, R., Sacks, J., Wynn, H., Mitchell, T., and Morris, M. (1992). Screening, predicting, and computer experiments. *Technometrics*, 34(1):15–25.

Wendland, H. (2004). *Scattered Data Approximation*. Cambridge University Press, Cambridge, England.

Williamson, D., Goldstein, M., Allison, L., Blaker, A., Challenor, P., Jackson, L., and Yamazaki, K. (2013). History matching for exploring and reducing climate model parameter space using observations and a large perturbed physics ensemble. *Climate Dynamics*, 41(7-8):1703–1729.

Worley, B. (1987). Deterministic uncertainty analysis. Technical report, Oak Ridge National Lab., TN (USA).

Wu, C. and Hamada, M. (2011). *Experiments: Planning, Analysis, and Optimization*, volume 552. John Wiley & Sons.

Wu, J., Poloczek, M., Wilson, A., and Frazier, P. (2017). Bayesian optimization with gradients. In *Advances in Neural Information Processing Systems*, pages 5267–5278.

Wu, Y., Tjelmeland, H., and West, M. (2007). Bayesian CART: prior specification and posterior simulation. *Journal of Computational and Graphical Statistics*, 16(1):44–66.

Xie, J., Frazier, P., and Chick, S. (2012). Assemble to Order simulator. http://simopt.org/wiki/index.php?title=Assemble_to_Order&oldid=447.

Xie, Y. (2015). *Dynamic Documents with R and* knitr. Chapman and Hall/CRC, Boca Raton, Florida, 2nd edition. ISBN 978-1498716963.

Xie, Y. (2016). bookdown: *Authoring Books and Technical Documents with Rmarkdown*. Chapman and Hall/CRC.

Xie, Y. (2018a). bookdown: *Authoring Books and Technical Documents with Rmarkdown*. R package version 0.9.

Xie, Y. (2018b). knitr: *A General-Purpose Package for Dynamic Report Generation in R*. R package version 1.20.

Zhang, B., Cole, D., and Gramacy, R. (2020). Distance-distributed design for Gaussian process surrogates. *To appear in Technometrics*. Preprint on arXiv:1812.02794.

Zhang, Y., Tao, S., Chen, W., and Apley, D. (2018). A latent variable approach to Gaussian process modeling with qualitative and quantitative factors. *Preprint on arXiv:1806.07504*.

Zhao, Y., Amemiya, Y., and Hung, Y. (2018). Efficient Gaussian process modeling using experimental design-based subagging. *Statistica Sinica*, 28(3):1459–1479.

Zhou, Q., Qian, P., and Zhou, S. (2011). A simple approach to emulation for computer models with qualitative and quantitative factors. *Technometrics*, 53(3):266–273.

Index

F-test, 22, 361
t-test, 361
    pairwise, *see* training and testing
        exercise, pairwise *t*-test

ABM, *see* agent-based model
accept–reject, 131
acquisition function, 58, *237*, 261, 264, 284,
        315, 481
    believer, 493
active learning, 2, 188, 237, 254, 261, 416,
        see also design of experiments,
        sequential
    algorithm, 235, 264
    Cohn, *246*, 249, 255, 257, 259, 264, 277,
        285, 286, 408, 411, 414, 418, 421,
        428
    Fisher information, 251, 422, 438
    MacKay, *237*, 241, 249, 255, 258, 264,
        271, 277, 285, 408, 411, 414
    submodularity, 291
additive penalty method, *see* optimization,
        additive penalty method
advection–diffusion, 493
AEM, *see* analytic element method
agent-based model, 14, 15, 457, 493
AIC, *see* Akaike information criteria
AID, *see* automatic interaction detection
`aimprob`, 306, 307, 311, 315, 321
`aimprob2`, 324
Akaike information criteria, 22
`akima`, 33, 449
ALBO, *see* augmented Lagrangian Bayesian
        optimization
ALC, *see* active learning, Cohn
aleatoric uncertainty, 288
aliasing pattern, 123
ALM, *see* active learning, MacKay
`ALoptim`, 310, 321
`ALwrap`, 310
amplitude, *see* Gaussian process, scale
analytic element method, 53
anisotropy, 172

anistropic Gaussian, *see* kernel, separable
        Gaussian
APM, *see* optimization, additive penalty
        method
Apple OSX, 47, 54
    Accelerate Framework, 379, 499, 503
ARD, *see* automatic relevance determination
assemble to order, *see* example, ATO
ATLAS project, 502
ATO, *see* example, ATO
augmented Lagrangian, 308, 309
    algorithm, 308, 310, 314
    inner loop, 309, 312, 319
    method, 292, 308, 311
    outer loop, 309, 314, 320
    subproblem, 308, 309, 311
augmented Lagrangian Bayesian
        optimization, 314, 423
    algorithm, 319
    composite random variable, 315
    equality constraints, 316, 318, 320, 327
    expected improvement, *see* expected
        improvement, slack variable ALBO,
        316, 320, 321, 324, 327, 331
    EY, 315, 321, 331
    gestalt approach, 314
    implementation, 318
    mixed constraints, 316, 320, 327
    separate modeling, 315
augmenting design, 134, *see also* design of
        experiments, sequential
automatic interaction detection, 396
automatic relevance determination, 219
autotuning, 416

`BACCO`, 361
bagging, 417
bakeoff, *see* training and testing exercise
BART, *see* Bayesian additive regression trees
`BART`, 416
Basic Linear Algebra Subprograms, 379, 499,
        500, 502
`BASS`, 175, 220

Milton Keynes UK
Ingram Content Group UK Ltd.
UKHW052025141024
449569UK00016B/708